RATING OF ELECTRIC POWER CABLES

IEEE Press
445 Hoes Lane, P.O. Box 1331
Piscataway, NJ 08855-1331

BOOKS IN THE IEEE PRESS POWER ENGINEERING SERIES

ANALYSIS OF ELECTRIC MACHINERY
Paul C. Krause and Oleg Wasynczuk, *Purdue University*; Scott D. Sudhoff, *University of Missouri at Rolla*
1994 Hardcover 584 pp IEEE Order No. PC4556 ISBN 0-7803-1101-9

ANALYSIS OF FAULTED POWER SYSTEMS, Revised Printing
Paul M. Anderson, *Power Math Associates, Inc.*
1995 Hardcover 536 pp IEEE Order No. PC5616 ISBN 0-7803-1145-0

ELECTRIC POWER APPLICATIONS OF FUZZY SYSTEMS
Mohamed E. El-Hawary, *Technical University of Nova Scotia*
1997 Hardcover 500 pp IEEE Order No. PC5666 ISBN 0-7803-1197-3

ELECTRIC POWER SYSTEMS: Design and Analysis, Revised Printing
Mohamed E. El-Hawary, *Technical University of Nova Scotia*
1995 Hardcover 808 pp IEEE Order No. PC5606 ISBN 0-7803-1140-X

POWER SYSTEM CONTROL AND STABILITY, Revised Printing
Paul M. Anderson, *Power Math Associates, Inc.* and A. A. Fouad, *Iowa State University*
1993 Hardcover 480 pp IEEE Order No. PC3798 ISBN 0-7803-1029-2

POWER SYSTEM STABILITY, VOLUMES I, II, III: An IEEE Press Classic Reissue Set
Edward Wilson Kimbark
1995 Softcover 1008 pp IEEE Order No. PP5600 ISBN 0-7803-1135-3

PROTECTIVE RELAYING FOR POWER SYSTEMS, VOLUMES I & II
Stanley H. Horowitz, editor, *American Electric Power Service Corporation*
1992 Softcover 1184 pp IEEE Order No. PP3228 ISBN 0-7803-0426-8

SUBSYNCHRONOUS RESONANCE IN POWER SYSTEMS
Paul M. Anderson, *Power Math Associates, Inc.*; B. L. Agrawal, *Arizona Public Service Company*; and J. E. Van Ness, *Northwestern University*
1990 Hardcover 282 pp IEEE Order No. PP2477 ISBN 0-87942-258-0

RATING OF ELECTRIC POWER CABLES

Ampacity Computations for Transmission, Distribution, and Industrial Applications

George J. Anders
Ontario Hydro Technologies

IEEE Power Engineering Society, *Sponsor*

IEEE Press Power Engineering Series
Dr. Paul M. Anderson, *Series Editor*

McGRAW-HILL
New York San Francisco Washington, D.C. Auckland Bogotá
Caracas Lisbon London Madrid Mexico City Milan
Montreal New-Delhi San Juan Singapore
Sydney Tokyo Toronto

The Institute of Electrical
and Electronics Engineers, Inc.,
New York

McGraw-Hill books are available at special quantity discounts to use
as premiums and sales promotions, or for use in corporate training
programs. For more information, please write to the Director of Special
McGraw-Hill, 11 West 19th Street, New York, NY 10011. Or contact
your local bookstore.

To order or receive additional information on these or any other
McGraw-Hill titles in the United States, please call 2MCGRAW.
In other countries, contact your local McGraw-Hill representative.

This is a copublication of the IEEE Press in association with McGraw-Hill,
a division of the McGraw-Hill Companies, Inc.

Printed in the United States of America

10 9 8 7 6 5 4 3 2 1

IEEE ISBN 07803-1177-9

McGraw-Hill ISBN 0-07-001791-3

Library of Congress Cataloging-in-Publication Data

Anders, George J. (date)
Rating of electric power cables: ampacity computations for transmission, distribution, and industrial applications / George J. Anders.
p. cm. -- (IEEE Press power engineering series)
Includes index.
ISBN 0-7803-1177-9
1. Electric cables--Standards. 2. Electric cables--Mathematical models. 3. Electric currents--Measurements--Mathematics. I. Title. II. Series
TK3307.A69 1997

To Justyna and Adam

Contents

Preface

The subject matter of this book is the computation of current ratings (also called *current-carrying capacities* or *ampacities*) of electric power cables. Computations of cable ampacities are generally quite involved; therefore, cable engineers have traditionally used published ampacity tables or performed approximate calculations to determine the cable size and type required for a particular application or to assess the ratings of existing cables. This practice could lead to the installation of oversized cables and increased installation costs. The advent of inexpensive and powerful personal computers made the development of fast, user-friendly computer programs for cable ampacity calculations feasible. These give the engineer an opportunity to determine accurate cable ratings with an ease not available before.

The widespread use of personal computers shifted the burden of ampacity computations from a few experts to a wider group of cable engineers. Inevitably, without having the years of experience required to perform cable thermal analysis, engineers ask questions about the basis of the computations performed by the various cable rating programs. But even for the experienced cable experts, the background of many of the computations may not always be clear.

Currently, two major sources are used throughout the world as basic references about computations of steady-state ratings: (1) the classic paper by Neher and McGrath (1957), and (2) the IEC Standard 287 (1982, 1989, 1994, 1995). Equivalent sources for transient analysis are Neher (1964) and the IEC Standards 853-1 (1985) and 853-2 (1989).[1] The IEC standards, being the more recent documents, contain more up-to-date information. At the writing of this book, a new, revised edition of the IEC Standard 287 is being prepared. Since the method described in the Neher/McGrath (1957) paper is still widely used in the United States, the differences between this method and the one outlined in the IEC Publication 287 (1994–1995) are summarized in Appendix F.

[1] See the list of references at the end of Chapters 1 and 2 for full details about these publications.

All of these sources contain hundreds of formulas, many of them developed empirically. In the majority of cases, the equations are given without derivation or an explanation of their origin and of the assumptions which may restrict their applicability. In addition, many developments in cable rating computations which have taken place over the last few years, such as the application of numerical methods, are not covered by international standards.

These facts clearly indicate that there is a need for a reference book which would help cable engineers, researchers, and teachers to understand the theory behind the computation of cable ratings. The present book was written with this need in mind. It is hoped that by clearly describing different aspects of the theory and by providing numerical examples which illustrate the concepts, the book will promote a consistent approach to applying theory to the computations needed for standard and nonstandard cable installations.

The book is divided into three parts. In the first part, a general theory of heat transfer is briefly described and, based on these theoretical principles, the steady-state and transient rating equations are developed. In the second, computation of the parameters required in rating equations is discussed. Whenever the equations describing parameter calculations were developed empirically, they are simply reproduced in the text with the explanation of their origin; if they are developed from basic principles, the full development is presented. This is avoided only in a few cases where the theory is so complex that it would unnecessarily cloud the presentation. Complete references are provided for all cases.

The third part contains specialized applications and advanced computational procedures. In particular, cable installations in air requiring the solution of a set of heat transfer equations are discussed. Also in Part III, an introduction to numerical methods for cable rating computations is given, and an optimization problem is formulated for the selection of the most economic conductor cross section.

The book contains a large number of numerical examples which explain the various concepts discussed in the text. Each new concept is illustrated through examples based on practical cable constructions and installations. To facilitate the computational tasks, I have selected five model cables which will be used throughout the book. Three are transmission-class, high-voltage cables, and two are distribution cables. The model cables were selected to represent major constructions encountered in practice and are described in Appendix A.

Even though computer programs are now in common use for cable ampacity calculations, there are merits to performing some computations by hand, if only for the purpose of checking sophisticated computer software. To this end, I have assembled, at the end of the book, calculation sheets for steady-state ampacity computations. These sheets can be used as templates for rating power cables in the most common installations. There is considerable use of advanced mathematical derivations in the book. Therefore, in order to emphasize the equations which are later assembled in the computation sheets and which are important for the rating calculations, boxes are placed around some of the formulas in the main text.

All equations use SI (metric) units since, with the exception of the United States, this is common practice around the world (even in the United States, IEEE recommendations suggest the use of the metric system in engineering computations). Therefore, even for the cable system peculiar to North American installations (e.g., high-pressure oil-filled cables), I converted imperial units into the metric system. A conversion table from imperial to metric units is given at the end of the list of symbols.

While the book discusses a large number of subjects, there are still some topics which are not addressed. In particular, forced cooling of the underground cables, ampacity computations of dc cables, and rating of cryogenic cables are not considered. The first topic is well covered in the book *Thermal Design of Underground Systems* by Weedy (Wiley, 1988). The remaining subjects are too specialized to be included in a general reference book. The thermal analysis of cable joints is also not discussed, but the methods described herein could be used for such analysis. The joints have, in the majority of cases, better thermal characteristics than the other parts of the cable circuit, and thus do not require special attention from the ampacity point of view. Cable heating during the short circuits is not discussed in the book because it does not affect cable current-carrying capability in normal and emergency operating conditions.

Acknowledgments

A large part of the material covered in this book was derived from various International Electrotechnical Commission (IEC) standards and reports dealing with the thermal rating of power cables. These publications are being prepared by Working Group 10 of the Study Committee 20A (High Voltage Cables) of the IEC as an ongoing activity. The author is particularly indebted to Mr. Mark Coates from ERA Technology in Britain, the Convener of WG 10, who provided a substantial amount of background material used in Chapters 5, 8, and 9 of this book. Frequent discussions with the members of WG 10: Mr. J. Van Eerde from NKF Kabel in Holland, Dr. B. Harjes of Felten & Guilleaume in Germany, Dr. A. Orini of Pirelli Cavi in Italy, Mrs. S. Le Peurian of Electricite de France, Mrs. A. Van Geertruyden from Laborelec in Belgium, and Dr. R. Wlodarski from BBJ-SEP in Poland, as well as with the cable system designers in Ontario Hydro: Mr. D. J. Horrocks, Mr. J. Motlis, and Mr. M. Foty, contributed greatly to the development of many procedures described in the book. In addition, I could not have written this book without an involvement and close association with several individuals who contributed their ideas and took the time to read the manuscript. I am particularly indebted to Dr. G. L. Ford and Dr. J. Endrenyi from Ontario Hydro Technologies (OHT) and Mr. D. J. Horrocks from Ontario Hydro Transmission Design Department, who have reviewed the entire manuscript and provided several helpful comments. Discussions with Dr. S. Barrett and Dr. B. Gu of OHT helped to clarify several issues in the computation of joule losses and the analysis of cables on riser poles. All the drawings were prepared by Mrs. G. Gostkowski from the drafting unit of OHT.

I would also like to acknowledge the personal encouragement of Mr. M. Bourassa from the Standards Council of Canada (SCC) and Dr. G. L. Ford from OHT, as well as the financial assistance of SCC and OHT in supporting my participation in the activities of WG 10 of the IEC over many years. The Canadian Electrical Association (CEA) and Ontario Hydro have contributed very substantial funds to the development of the Cable Ampacity Program (CAP) used extensively in this book. Mr. Jacob Roiz from CEA was overseeing the development of this software, now used by over 200 users in 33 countries on five continents. A close association with Dr. T. Rodolakis from CYME Int. helped me to correct many problems in the program. Dr. Rodolakis also reviewed most of the manuscript.

Finally, but by no means last, I would like to thank my wife Justyna and my son Adam who supported wholeheartedly this difficult endeavor.

George J. Anders
Toronto, Canada

Symbols

The symbols used in this book and the quantities which they represent are given in the list below. Following the practice adopted in IEC Standard 287, all cable component diameters are given in millimeters. When the formulas shown in the text require these dimensions to be given in meters, an asterisk is added to the symbol. For example, D_e denotes the external diameter of the cable expressed in millimeters, whereas D_e^* denotes the same diameter expressed in meters.

Symbol	Definition	Unit
A_a	= cross-sectional area of the armor	mm^2
A	= cost parameter related to condutor cross section (Chapter 13)	$\$/m \cdot mm^2$
A_w	= area of the wall inside surface for unit length riser	m^2
A_s	= area of the jacket outside surface for unit length jacket	m^2
A_o	= area of the outside wall surface per unit length	m^2
A_{os}	= equivalent area of the wall per unit length perpendicular to sun rays	m^2
A_{sr}	= effective radiation area	m^2
B_0, B_1, B_2, B_3, B_4	= components of impedance due to the steel wires (Section 8.3.3)	Ω/m
B	= factor used in economic analysis of conductor size	
C	= electrical capacitance per core	F/m
CI	= installation cost of a cable	\$
CL	= cost of losses	\$
CT	= sum of the cost of installation and the cost of losses	\$
D	= demand charge per year	$\$/W \cdot year$
D'_a	= external diameter of armor	mm
D_b	= diameter over armor bedding	mm

D_d	= internal diameter of duct or pipe	mm
D_e	= external diameter of cable, or external diameter of pipe covering in pipe-type cable	mm
D_i	= diameter over insulation	mm
D_o	= external diameter of duct or pipe	mm
D_s	= external diameter of metal sheath	mm
D_T	= diameter over the reinforcing tape	mm
D_{oc}	= diameter of the imaginary coaxial cylinder which just touches the crests of a corrugated sheath	mm
D_{ot}	= diameter of the imaginary coaxial cylinder which would just touch the outside surface of the troughs of a corrugated sheath $= D_{it} + 2t_s$	mm
D_{ic}	= diameter of the imaginary cylinder which would just touch the inside surface of the crests of a corrugated sheath $= D_{oc} - 2t_s$	mm
D_{is}	= diameter over core screen	mm
D_{it}	= diameter of the imaginary cylinder which just touches the inside surface of the troughs of a corrugated sheath	mm
D_x	= fictitious diameter at which effect of loss factor commences	mm
E	= constant used in Chapter 9	
$-Ei(-x)$	exponential integral function	
F	= mutual heating coefficient for a group of cables	
F	= factor for eddy currrent losses for large segmental conductors (Chapter 8)	
F_1	= coefficient for belted cables defined in Chapter 9	
F_2	= coefficient for belted cables defined in Chapter 9	
F_c	= factor to take account of the length variations in a cross-bonded system	
$F(N)$	= auxiliary quantity in cost computations	
$F_i(N)$	$= i = 1, 2$, or 3, auxiliary quantities in cost computations	
$F_{J,IR}$	= thermal radiation shape factor between two long concentric cylinders	
G	= aspect ratio, $G = L/\delta$	
G	= geometric factor for belted cables	
G	= constant component of cost unaffected by size of cable (Chapter 13)	\$/m
$\bar{G}$	= geometric factor for SL and SA type cables	
Gr	= Grashoff Number based on δ	
H	= intensity of solar radiation	W/m^2
H	= magnetizing force (Section 8.3.1.4)	ampere turns/m
$\mathbf{H}$	= thermal conductivity matrix in finite-element analysis	
I	= current in one conductor (rms value)	A
I_0	= maximum load on a cable during the first year (Chapter 13)	A
I_1	= constant current applied to cable prior to emergency load	A
I_2	= emergency load current which may subsequently be applied for time t so that the conductor temperature rise above ambient at the end of the period of emergency load is θ_{max}	
I_c	= rms conductor current	A

I_s	= rms sheath current	A
I_a	= rms armor current	A
I_c	= charging current in a cable	A
I_{max}	= highest current of daily load cycle: used as denominator when determining loss-load ordinates	A
I_n	= the magnitude of the nth harmonic current	A
I_R	= sustained (100% load factor) rated current to attain, but not exceed, the permissible maximum conductor temperature	A
I_z	= current-carrying capacity for a maximum permitted temperature rise of $\theta - \theta_{amb}$ (Chapter 13)	A
K	= screening factor for the thermal resistance of screened cables	
K	= diameter ratio, $K = D_d/D_e$	
K	= boundary conditions and heat source vector in finite-element analysis	
K_A	= coefficient used in Section 9.6.8.1	
L	= depth of laying to cable axis or center of trefoil in steady-state rating	mm
L^*	= axial depth of burial of a cable in transient rating and in Section 10.6.1	m
L	= length of the cable route in economic computations	m
L	= length of the riser	m
L_G	= distance from the soil surface to the center of a duct bank	mm
L_T	= depth of tunnel centerline	m
LF	= daily load factor	
LF_u, LF_p	= ultimate and present load factors (Chapter 13)	
LLF	= load-loss factor	
M	= cyclic rating factor	
M, N	= coefficients defined in Section 8.4.10	
M_o, N_o	= coefficients used for calculating cable partial transient temperature rise (Section 5.2.1)	s, s^2
N	= number of cables in a group	
N	= economic life of a cable	years
N	= number of loaded cables in a duct bank (Section 9.6.5.1)	
N_0	= period during which the load on a cable grows	years
N_p	= number of phase conductors per circuit	
N_c	= number of circuits carrying the same value and type of load	
$\mathbf{N^e}$	= vector of shape coefficients in finite-element analysis	
Nu	= average Nusselt Number	
P	= cost of 1 watt hour of energy at the relevant voltage level	\$/Wh
Pr	= Prandtl Number, one of the air thermal properties	
P, Q	= coefficients defined in Section 8.3.5.3	Ω/m
Q	= total thermal capacitance of a cable	J/m · K
$\mathbf{Q}$	= thermal capacity matrix in finite-element analysis	
Q_a, Q_B	= elements of two-part thermal circuit (see Section 3.3.2)	J/m · K

Q_a	= thermal capacitance of armor	J/m · K
Q_c	= thermal capacitance of conductor	J/m · K
Q_d	= thermal capacitance of a duct	J/m · K
Q_f	= thermal capacitance of filling in SL type cables, or of filling between cores of a gas-pressure pipe-type cable	J/m · K
Q_i	= thermal capacitance of dielectric per conductor	J/m · K
Q_{i_1}	= thermal capacitance of first portion of insulation	J/m · K
Q_{i_2}	= thermal capacitance of second portion of insulation	J/m · K
Q_j	= thermal capacitance of a jacket	J/m · K
Q_o	= thermal capacitance of oil in pipe-type cable	J/m · K
Q_p	= thermal capacitance of a pipe	J/m · K
Q_s	= thermal capacitance of sheath and reinforcement	J/m · K
$Q_P(N)$, $Q_D(N)$	= coefficients taking into account the increase in load, the increase in cost of energy over the N years, and the discount rate	
R	= alternating current resistance of conductor at its maximum operating temperature	Ω/m
R_a	= resistance of armor	Ω/m
R_s	= resistance of sheath	Ω/m
R'	= dc resistance of conductor at maximum operating temperature	Ω/m
R_O	= dc resistance of conductor at 20°C	Ω/m
R_1	= ac resistance of conductor before application of emergency current	Ω/m
R_R	= ac resistance of conductor with sustained application of rated current I_R, i.e., at standard maximum permissible temperature	Ω/m
R_{max}	= ac resistance of conductor at end of period of emergency loading	Ω/m
R_m	= mean ac resistance of conductor during economic life	Ω/m
R_T	= resistance of reinforcing tape	Ω/m
Ra*	= modified Rayleigh Number	
Re	= Reynolds Number	
S	= cross-sectional area of conductor	mm^2
S_{ec}	= economic conductor cross section	mm^2
T	= total thermal resistance of a cable from conductor to outer surface	K · m/W
T	= operating time at maximum joule loss (Chapter 13)	h/year
T_A, T_B, T_C	elements of equivalent thermal circuit	K · m/W
T_1	= thermal resistance per core between conductor and sheath	K · m/W
T_2	= thermal resistance between sheet and armor	K · m/W
T_3	= thermal resistance of external serving	K · m/W
T_4	= thermal resistance of surrounding medium (ratio of cable surface temperature rise above ambient to the losses per unit length)	K · m/W
T_4^*	= external thermal resistance in free air, adjusted for solar radiation	K · m/W
T_4'	= thermal resistance between cable and duct (or pipe)	K · m/W
T_4''	= thermal resistance of the duct (or pipe)	K · m/W
T_4'''	= thermal resistance of the medium surrounding the duct (or pipe)	K · m/W
ΔT_4	= additional external thermal resistance caused by heating from other cables in a group	K · m/W
T_a, T_b	= apparent thermal resistances used to calculate cable partial transient temperature rise (Section 5.2.1)	K · m/W

T_f	= thermal resistance of filling between cores of SL type cable and of gas-pressure pipe-type cable	K · m/W
T_o	= thermal resistance of the oil in a pipe-type cable	K · m/W
THD	= total harmonic distortion (Section 7.3)	
U_O	= voltage between conductor and screen or sheath	V
U, V	= constants used in Section 9.6.4.1	
V_{air}	= local air velocity	m/s
W_A	= losses in armor per unit length	W/m
W_C	= losses in conductor per unit length	W/m
W_I	= total I^2R power loss of each cable	W/m
W_{int}	= amount of heat generated within the body	W/m^3
W_d	= dielectric losses per unit length per phase	W/m
W_k	= losses dissipated by cable k	W/m
W_s	= losses dissipated in sheath per unit length	W/m
W_{sol}	= solar radiation absorbed by the riser surface per unit length	W/m
$W_{(s+A)}$	= total losses in sheath and armor per unit length	W/m
W_{TOT}	= total power dissipated in the trough per unit length	W/m
W_t	= total power dissipated in the cable per unit length	W/m
$W_{cond,s-o}$	= heat conduction rate from the wall inner surface to its outside surface per unit length	W/m
$W_{conv,w}$	= natural convection heat transfer rate between the wall inside surface and the air per unit length	W/m
$W_{conv,s}$	= natural convection heat transfer rate between the jacket outside surface and the air per unit length	W/m
$W_{conv,o}$	= natural convection heat transfer rate between the wall outside surface and atmosphere air per unit length	W/m
$W_{rad,s-w}$	= thermal radiation heat transfer rate between the wall inner surface and the jacket outside surface per unit length	W/m
$W_{rad,o-sur}$	= thermal radiation heat transfer rate between wall outside surface and surrounding objects per unit length	W/m
X	= reactance of sheath (two-core cables and three-core cables in trefoil)	W/m
X_1	= reactance of sheath (cables in flat formation)	Ω/m
X_m	= mutual reactance between the sheath of one cable and the conductors of the other two when cables are in flat formation	Ω/m
Y	= coefficient used in Section 9.6.4.1	
$Y_o \ldots Y_{23}$	= scaled ordinate in (load)^2 graph used in calculation of cyclic rating factor	
a	= increase in load per year	
a	= length and width of square cable tunnel	m
a_0	= shortest minor length in a cross-bonded electrical section having unequal minor lengths	
a, b	= coefficients used for calculating cable partial transient temperature rise	1/s
b	= increase in cost of energy per year, not including the effect of inflation	

c = demand charge escalation factor
c = distance between the axes of conductors and the axis of the cable for three-core cables (= $0.55\ r_1 + 0.29\ t$ for sector-shaped conductors) mm
c = specific volumetric heat capacity of the material (Chapters 2 and 10) J/m^3
c_p = specific heat at constant pressure, J/kg · K
d = mass density, kg/m^3
d = mean diameter of sheath or screen mm
d = mean diameter of sheath and reinforcement mm
d_0 = dry density of soil kg/m^3
d_2 = mean diameter of reinforcement mm
d_a = mean diameter of armor mm
d_a = external diameter of belt insulation mm
d_c = external diameter of conductor mm
d_{cs} = external diameter of conductor shield mm
d_{cm} = minor diameter of an oval conductor mm
d_{cM} = major diameter of an oval conductor mm
d_c = external diameter of equivalent round solid conductor having the same central duct as a hollow conductor mm
d_c^e = conductor diameter of equivalent single-core cable having the same losses as the three-core cable (Section 3.3.1) mm
d_f = diameter of a steel wire mm
d_i = internal diameter of hollow conductor mm
d_M = major diameter of screen or sheath or an oval conductor mm
d_m = minor diameter of screen or sheath of an oval conductor mm
d_{pk} = distance from center of pth cable whose rating is being determined to an adjacent cable k mm
d'_{pk} = distance from center of pth cable, whose rating is being determined, to the image of an adjacent cable k (see Fig. 5.3) mm
d_{Tm} = mean diameter of the reinforcing tape mm
d_x = diameter of an equivalent circular conductor having the same cross-sectional area and degree of compactness as the shaped one mm
f = system frequency Hz
f_φ = factor by which thermal resistances are reduced for cables in trefoil
g = coefficient used in Section 9.6.8.1
g = acceleration of gravity m/s^2
g_s = coefficient used in Section 8.4.4
h = heat dissipation coefficient W/m^2 · K$^{5/4}$
h_1 = ratio I_1/I_R
h_{conv} = total convection coefficient W/K · m^2
h_o = natural or forced convection coefficient at wall outside surface W/K · m^2
h_r = radiation heat transfer coefficient W/K · m^2
h_s = natural convection coefficient at jacket outside surface W/K · m^2
h_w = natural convection coefficient at wall inside surface W/K · m^2
$\mathbf{h^e}$ = element conductivity matrix in finite-element analysis

i	= time	h
i	= discount rate	
k_0	= factor used in the calculation of hysteresis losses in armor or reinforcement (Section 8.3.6.3)	
k	= ratio of cable, pipe, or duct external-surface temperature rise above ambient to conductor temperature rise above ambient under steady conditions	
$\mathbf{k^e}$	= element heat source vector in finite-element analysis	
k_1	= value of k for a cable in a group	
k_{air}	= air thermal conductivity	$W/K \cdot m$
k_n	= lay-length factor of the wires in layer n of a stranded conductor	
k_p	= factor used in calculating x_p (proximity effect)	
k_s	= factor used in calculating x_s (skin effect)	
l	= length of a cable section (general symbol)	m
ℓ_a	= length of lay of a stranded wire	mm
ℓ_T	= length of lay of a tape	mm
ln	= natural logarithm (logarithm to base e)	
m	$= (\omega/R_s)10^{-7}$	
n	= number of conductors in a cable	
n_n	= number of wires in layer n of stranded conductor	
n_a	= number of armor wires	
n_t	= number of moisture barrier metallic tapes in pipe-type cable	
p	= factor for apportioning the thermal capacitance of a dielectric (long-duration transients—Section 3.3.1)	
p	= factor for computation of load loss factor (Chapter 13)	
p^*	= factor for apportioning the thermal capacitance of a dielectric (short-durations transients—Section 3.3.1)	
p_0	= part of the perimeter of the cable trough which is effective for heat dissipation (Section 10.6.2)	m
p'	= factor for apportioning the thermal capacitance of cable coverings	
p_d	= factor for apportioning thermal capacitance of dielectric when calculating transient caused by dielectric loss	
p_2, q_2	= coefficients used in Section 8.3.3.5	
q	= heat flux	W/m^2
$\mathbf{q^e}$	= element capacity matrix in finite-element analysis	
r	= ratio $\theta_{max}/\theta_R(\infty)$	
r_a	= mean radius of the armor	mm
r_c	= external radius of the conductor	mm
r_s	= mean radius of the sheath	mm
r_1	= circumscribing radius of two- or three-sector shaped conductors	mm
s	= axial separation of conductors	mm
s_1	= axial separation of two adjacent cables in a horizontal group of three, not touching	mm
t	= time from start of application of heating, a general symbol for time, usually in seconds ($t = 3600i$)	s

Symbol		Definition	Unit
t	=	insulation thickness between conductors	mm
t_1	=	insulation thickness between conductors and sheath	mm
t_2	=	thickness of the bedding	mm
t_3	=	thickness of the serving	mm
t_i	=	thickness of core insulation, including screening tapes plus half the thickness of any nonmetallic tapes over the laid up cores	mm
t_a	=	armor thickness	mm
t_J	=	thickness of the jacket	mm
t_s	=	thickness of the sheath	mm
t_t	=	thickness of moisture barrier metallic tape in pipe-type cables	mm
u	=	$(2L/D_e)$ in Section 9.6.1	
u	=	(L_G/r_b) in Section 9.6.5	
x_p	=	argument of a Bessel function used to calculate proximity effect	
x_s	=	argument of a Bessel function used to calculate skin effect	
x, y	=	sides of duct bank/backfill $(y > x)$ (Section 9.6.5.1)	mm
y_p	=	proximity effect factor	
y_s	=	skin effect factor	
α	=	temperature coefficient of electrical resistivity of conductor material	1/K
$\alpha(t), \beta(t)$	=	attainment factors for the conductor to cable surface and cable surface to ambient temperature rises, respectively	
α_{20}	=	temperature coefficient of electrical resistivity at 20°C per Kelvin	1/K
α_0	=	wall surface absorptivity to solar radiation	
β	=	coefficient of volumetric expansion	K^{-1}
β	=	reciprocal of temperature coefficient of electrical resistivity at 0°C	K
β	=	angle between axis of armor wires and axis of cable	
γ	=	angular time delay	
γ_n	=	ratio of the nth harmonic current to the fundamental (Section 7.3)	
$\nabla\theta$	=	temperature gradient vector in finite-element analysis	
Δ_1, Δ_2	=	coefficients used in Section 8.4.5	
$\Delta\tau$	=	time step in transient analysis	s
δ	=	soil thermal diffusivity	m^2/s
δ	=	derating factor for cable systems containing harmonics (Section 7.3)	
δ	=	air gap thickness between the cable and the wall	m
δ_0	=	equivalent thickness of armor or reinforcement	mm
δ_1	=	thickness of metallic screens in screened-type cables	mm
$\tan\delta$	=	loss factor of insulation	
ε	=	relative permittivity of insulation	
ε_w	=	emissivity of the wall inner surface	
ε_J	=	emissivity of the jacket outside surface	
ε_o	=	emissivity of the wall outside surface	
η	=	moisture content of soil in percent of dry weight	
θ	=	maximum operating temperature of conductor	°C
θ_i	=	conductor temperature at commencement of transient	°C

θ_m = mean operating temperature of a conductor during economic life °C
$\theta(t)$ = transient temperature rise of conductor above ambient, without correction for variation in conductor loss K
$\theta_c(t)$ = transient temperature rise of conductor above the outer surface cable K
$\theta_{cd}(t)$ = transient temperature rise of conductor above outer surface of the cable due to dielectric loss K
$\theta_d(t)$ = transient temperature rise of conductor above ambient due to dielectric loss K
$\theta_{de}(t)$ = transient temperature rise of cable surface above ambient due to dielectric loss K
$\theta_e(t)$ = transient temperature rise of outer surface of a cable (or hottest cable in a group of similarly loaded cables) above ambient temperature K
$\theta_o(t)$ = transient average temperature rise of oil in a pipe-type cable above the outer surface of the pipe K
$\theta_R(t)$ = conductor temperature rise above ambient at time t after application of current I_R, neglecting variation in conductor resistance K
$\theta_R(i)$ = $\theta(t)$ when the magnitude of the step function current is the sustained (100% load factor) rated current, and i is expressed in hours K
$\theta_R(\infty)$ = value of θ_R in the steady state, i.e., the standard maximum permissible temperature rise K
$\theta_a(t)$ = conductor transient temperature rise above ambient corrected for variation in conductor resistance with temperature K
$\theta(\infty)$ = conductor steady-state temperature rise above ambient K
θ_e = temperature of the external surface of cable or duct °C
θ_{gas} = air temperature in the gap (Chapter 10) °C
θ_{max} = maximum permissible temperature rise above ambient: 1) for a daily cyclic load, 2) at end of period of emergency loading K
θ_m = mean temperature of medium between a cable and duct or pipe °C
θ_o = average temperature at the wall outside surface (Chapter 10) °C
θ_s = average temperature of the jacket outside surface (Chapter 10) °C
θ_s = sheath temperature °C
θ_w = average temperature of the wall inner surface (Chapter 10) °C
$\Delta\theta$ = permissible temperature rise of conductor above the ambient temperature K
$\Delta\theta_d$ = factor to account for dielectric loss for calculating T_4 for cables in free air K
$\Delta\theta_{ds}$ = factor to account for both dielectric loss and direct solar radiation for calculating T_4^* for cables in free air K
$\Delta\theta_{duct}$ = difference between the mean temperature of air in a duct and ambient temperature K
$\Delta\theta_s$ = difference between the surface temperature of a cable in air and ambient temperature K
$\Delta\theta_{tr}$ = temperature rise of the air in a cable trough K
λ_1, λ_2 = ratio of the total losses in metallic sheaths and armor, respectively, to the total conductor losses (or losses in one sheath or armor to the losses in one conductor)

λ_1'	= ratio of the losses in one sheath caused by circulating currents in the sheath to the losses in one conductor	
λ_1''	= ratio of the losses in one sheath caused by eddy currents to the losses in one conductor	
λ_{lm}'	= loss factor for the middle cables	
λ_{ll}'	= loss factor for the outer cable with the greater losses	
λ_{12}'	= loss factor for the outer cable with the least losses	
μ'	= relative magnetic permeability of armor material	
μ	= loss-load factor of a load cycle	
μ_e	= longitudinal relative permeability	
μ_r	= relative permeability	
μ_t	= transverse relative permeability	
μ_0	= peremeability of nonmagnetic systems	
μ	= viscosity	kg/s · m
ν	= air kinematic viscosity	m^2/s
ρ_{20}	= conductor electrical resistivity at 20°C	$\Omega \cdot m$
ρ_e	= thermal resistivity of earth surrounding a duct bank/backfill	K · m/W
ρ_{el}	= electrical resistivity of metallic material	$\Omega \cdot m$
ρ_c	= thermal resistivity of concrete used for a duct bank	K · m/W
ρ_m	= thermal resistivity of metallic screens on multicore cables	K · m/W
ρ_m	= mean ac resistivity of conductor during economic life	$\Omega \cdot m$
ρ_s	= thermal resistivity of soil	K · m/W
ρ	= thermal resistivity of material	K · m/W
ρ_i	= dielectric thermal resistivity	K · m/W
σ	= absorption coefficient of solar radiation for the cable surface	
σ_w	= reflectivity of the wall inner surface	
σ_s	= reflectivity of the jacket outside surface	
ω	= angular frequency of system $(2\pi f)$	l/s

Table of North American Cable Sizes[1]

	Cross-Sectional Area				Cross-Sectional Area	
Conductor Size, AWG	kcmil[2]	Square inches	Square millimeters	Conductor Size, kcmil	Square inches	Square millimeters
8	16.51	0.01297	8.367	250	0.1964	126.7
6	26.24	0.02061	13.30	350	0.2749	177.3
4	41.74	0.03278	21.15	500	0.3927	253.4
2	66.36	0.05212	33.62	750	0.5891	380.0
1	83.69	0.06573	42.41	1000	0.7854	506.7
1/0	105.6	0.08291	53.49	1250	0.9818	633.4
2/0	133.1	0.1045	67.43	1500	1.178	760.0
3/0	167.8	0.1318	85.01	1750	1.374	886.7
4/0	211.6	0.1662	107.2	2000	1.571	1013
				2250	1.767	1140
				2500	1.964	1267
				3000	2.356	1520
				3500	2.749	1774
				4000	3.142	2028

[1] Other useful conversions are : 1 in = 0.0254 m, 1 ft = 12 in = 0.3048 m.

[2] Thousand circular mils (kcmil). A "mil" is a 1/1000 in, and circular mils represent the area of an equivalent solid rod having a diameter expressed in mils. To convert mils to millimeters, the mil value is multiplied by 0.0254.

PART I MODELING

1

Cable Constructions and Installations

1.1 INTRODUCTION

Underground cables are far more expensive to install and maintain than overhead lines. The greater cost of underground installations reflects the high cost of the equipment, labor, and time necessary to manufacture the cable, to excavate and backfill the trench, and to install the cable. Because of the extra expense, most underground installations are constructed in congested urban areas and as leads from generating plants and in substations. The large capital cost associated with cable installations also makes it necessary that particular care be applied in selecting the proper cable type and size to serve the load for the life of the installation.

Information about the maximum current-carrying capacity which a cable can tolerate throughout its life without risking deterioration or damage is extremely important in power cable engineering and operation. Ampacity values are required for every new cable installation, as well as for cable systems in operation. With some underground transmission cable circuits approaching the end of their design life, the development of a systematic method for determining the feasibility of extending cable life and/or increasing current ratings is of paramount importance.

The capacity of a transmission line is normally given in MVA (megavolt-amperes). MVA is made up of two components, namely, the MW (megawatt) component, which represents real power and is available to do work, and MVAR (megavolt-ampere-reactive), the reactive component, which is present in the system due to its inductive and capacitive elements, but cannot be utilized to produce work. Cable rating studies usually involve computation of the permissible current flowing in the conductor for a specified maximum operating temperature of the conductor. The current causes the cable to heat, and the limit of loading capacity is determined from the acceptable conductor temperature. Occasionally, the current value may be given, and the studies involve calculation of the temperature distribution within the cable and in the surrounding environment.

The cable must be able to carry large amounts of current without overheating, and must maintain an acceptable voltage profile. Cable heating presents one of the major problems associated with underground lines. While it is relatively easy to dissipate the heat generated by the flow of current through the conductor in overhead lines, the heat generated by losses in underground systems must pass through the electrical insulation system to the surrounding earth, and both the insulation system and the earth represent an obstacle to heat dissipation. Because the maximum temperature at which conductors can operate is limited by the conductor electrical insulation system and because these systems conduct heat poorly, the result is a much larger conductor than would be required for an overhead line of equal capacity.

The cable must carry the load currents without overheating, and also without producing excessive voltage drop. This voltage drop is known as IZ drop after the formula used to determine it, but in underground systems this is rarely a limiting factor.

In addition to the normal loads, a transmission system is customarily designed to carry overloads due to equipment and line outages or other abnormal system conditions for limited periods of time which may, nevertheless, be often 10 h or longer. Operation at higher than normal temperatures is permitted during these overload periods, and the response of the cable system to these overloads is evaluated in transient rating computations.

The rating of an electric power cable will depend on its construction and the method of installation. There is a great variety of cable constructions currently used around the world. Also, installation conditions vary widely. The Association of Edison Illuminating Companies (AEIC) in the United States issues specifications and guides for various cable constructions and installations. The current list of these specifications and guides is given in the References section at the end of this chapter.

To help the reader understand the terminology and the computational process adopted for cable ratings, in the following we will review briefly major cable components and installation methods. This review is not exhaustive, and its aim is to focus on those cable parameters and the installation requirements which affect thermal rating of the cable circuit. We will close this chapter with a historical note on standardization efforts of cable ampacity computations.

1.2 CABLE COMPONENTS

Every electric power cable is composed of at least two components: (1) an electrical conductor, and (2) the conductor insulation which prevents direct contact or unsafe proximity between conductor and other objects. The need to provide adequate electrical insulation which will also permit heat to be conducted and dissipated poses technical challenges at higher voltages. The problem of heat conduction is further aggravated by the fact that, in the majority of power cables, the primary electrical insulation has to be protected against mechanical, electromechanical, and chemical damage. This protection is provided by additional concentric layers over the insulation. The most common form is a *metallic sheath* which is often covered with a nonconducting material called an *oversheath* or *jacket.* Some cables do not have a metallic sheath, but only a nonconducting jacket. Some cables have *concentric neutral wires* instead of sheath. These wires serve mainly as a return path for a neutral current or for short-circuit current during system faults. Submarine and special-purpose cables usually have an additional metallic layer called *armor.* Armor may or may not be further protected by a special outer covering. Each of these components is briefly described below.

1.2.1 Conductors

Two materials are usually used in cable conductors: copper and aluminum. Since the price of aluminum is less than 50% of the price of the copper, savings are possible by utilizing aluminum. These savings are small since the cost of the conductor is only a fraction of the installed cost of the system. Some of the savings in metal cost are offset by the cost of additional insulation and accessories needed since the cross-sectional area of aluminum required is approximately 1.5 times that of copper to carry the same load, resulting in a cable with a larger diameter.

Conductor construction also plays an important role in rating computations. This topic is discussed in Chapter 7, and we will restrict ourselves here to showing only typical conductor constructions currently made by cable manufacturers (see Fig. 1.1).

An important parameter in rating computations is the conductor cross section. Conductors are designed to conform to a range of nominal areas in graduated steps. In practice, the following two ranges are used:

1. In North America and in some South American countries, cable cross section is based on American Wire Gauge (AWG) up to 4/0, that is, 107 mm^2, and larger sizes are in thousand circular mils (kcmil). A "mil" is a 1/1000 in, and circular mils represent the area of an equivalent solid rod having a diameter expressed in mils. To convert mils to millimeters, the mil value is multiplied by 0.0254.

2. In all other countries, the metric system is used to define conductor area (mm^2) and the dimensions specified in IEC Publication 228 (1978, 1982) are adopted.

In practice, the conductors are not manufactured to precise areas. Instead, manufacturers adjust their wire sizes and manufacturing processes to meet a specified maximum resistance rather than area. The IEC Standard 228 (1982) specifies a single maximum dc resistance for each size of conductor of a given material. American specifications adhere to a pattern of a "nominal" resistance, together with a tolerance to provide a maximum resistance for a single-core cable and a further tolerance for multicore cable. These tolerances also vary with conductor classification.

The nominal values of resistances used in North America as well as those specified in the IEC Standard 228 are quoted in Chapter 7.

In general, the larger the conductor cross section, the larger is the current-carrying capability of the cable. For conventional cables with natural cooling, we can state that doubling cable ampacity requires approximately a fourfold increase in conductor cross section. For example, increasing transmission capacity from 300 to 600 MVA in a 230 kV cable circuit requires a change of conductor cross section from 800 to 3200 mm^2 Cu or 4800 mm^2 Al. Since with the present technology the manufacture of a cable with a conductor cross section larger than 3000 mm^2 is practically impossible, special cooling facilities or several cables per phase are required when large transmission capacity is needed.

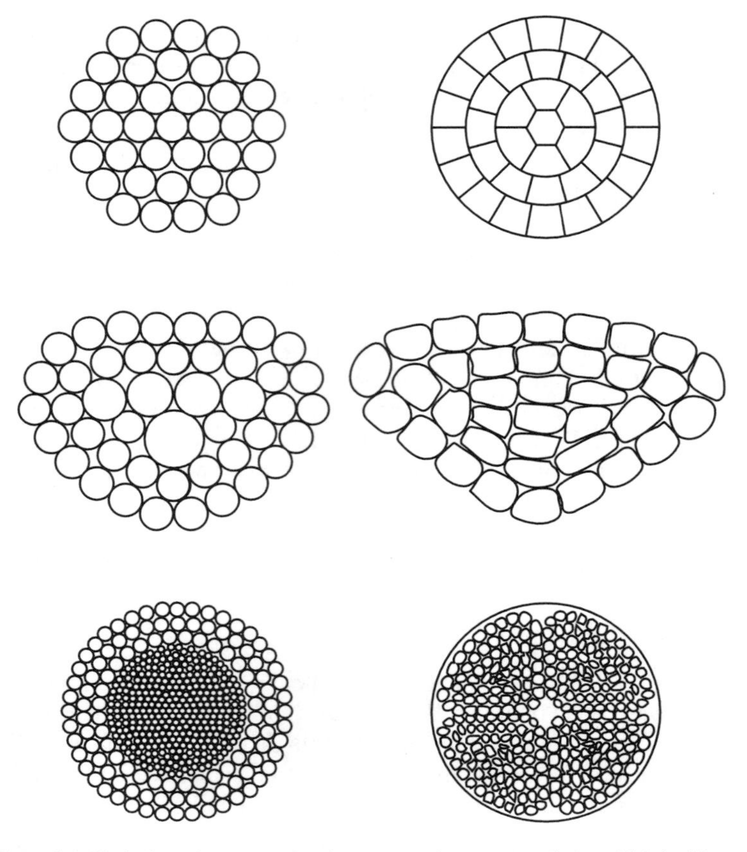

Figure 1.1 Typical conductor constructions contrasting compact designs (right) with conventional stranding (left) (Barnes, 1964).

1.2.2 Insulation

The purpose of the electrical insulation is to prevent the flow of electricity from the energized conductors to the ground or to an adjacent conductor. The insulation must be able to withstand the electrical stresses produced by the alternating voltage and any superimposed transient voltage stress on the conductor without dielectric failure and causing a short circuit.

There are many materials that have acceptable insulating properties. The most commonly used and the least expensive is the air surrounding the conductors on overhead lines. There are, however, only a few materials available that can be used in underground or enclosed applications to insulate conductors from ground and from each other. These are: oil-impregnated paper tapes, solid insulations such as polyethylenes, and ethylene–propylene rubber, and the more recently developed polypropylene (PPL) and compressed gas insulation (such as sulfur hexafluoride or SF_6) systems.

When paper and the solid dielectric insulations are subjected to alternating voltage, they act as large capacitors and charging currents flow in them. The energy required to effect the realignment of electrons each time the voltage direction changes (i.e., 50 or 60 times a second) produces heat and results in a loss of power which is called dielectric loss, and should be distinguished from conductive loss. The magnitude of the required charging currents is a function of the dielectric constant of the insulation, the cable length, the dimensions of the cable, and the operating voltage. Charging current produces also a resistive component of the losses in the insulation but, for ac applications, they are extremely small compared to the capacitive component.

Various cable types used in power systems are often distinguished on the basis of the type of insulation they have. For example, paper-insulated cables are classified as:

1. Mass-impregnated solid type where the insulation is saturated with impregnation fluid before shipment.
2. Low-pressure fluid-filled (LPFF) with a filling of thin oil at a pressure of about 1 atm (see Appendix A for an example of such a cable); the oil pressure is maintained by reservoirs feeding the cables along the route.
3. High-pressure fluid-filled (HPFF) cables, where the paper-insulated cores are installed in a steel pipe and the pipe is filled with insulating oil under a high pressure of 1.38 MN/m^2 (200 psi). The pressure is maintained by a special pumping plant consisting of an oil reservoir, controls, and a pump. The pump is activated automatically when the oil pressure drops below a specified value.

Both LPFF and HPFF cables are impregnated during their manufacturing. A newer insulating material used in high-voltage cables is paper–polypropylene–paper (PPL). It combines excellent insulating properties of paper with the low dielectric losses of polypropylene. Other cable types are those using solid insulation (extruded cables) or using gas, usually SF_6, as an insulating medium.

The insulation type has a strong effect on cable rating. From a thermal point of view, a good insulating material should have low thermal resistivity and should result in low dielectric losses. A cross-linked polyethylene (XLPE) is a good example of such material. It has primarily been used in low- and medium-voltage cables. However, recently, an increasing number of manufacturers are producing high- and extra-high-voltage cables with extruded insulation.

All modern electric power cables are constructed with semiconducting screens around the conductor and around the insulation. For thermal calculations, these screens are considered to be a part of the insulation.

1.2.3 Sheath/Concentric Neutral Wires

Metallic sheaths are essential for paper-insulated cables to exclude water from the insulation and to retain impregnating fluid in LPFF cables. For extruded insulations, there is not such an obvious need; however, many early extruded cables without metallic sheaths developed problems with integrity of the insulation. In addition, there are safety concerns related to unshielded cables. Now IEC 502 (1983) requires that such cables of rated voltages above 1 kV should have a metallic covering.

When a solid sheath is used in the cable construction, it is usually made of lead or aluminum. Lead sheath, especially for large cables, may require a reinforcing metallic tape. Aluminum sheaths, on the other hand, are lighter and, for additional flexibility, are often corrugated. In some special constructions, a corrugated copper sheath may be used. In addition to its insulation protection function, the sheath is used as the cable component to carry neutral and/or fault current to earth in the event of an earth fault on the system. Some cables may not have a solid sheath, but instead may be constructed with concentric neutral wires to carry the fault current. The wires are usually made of copper and in some instances of aluminum.

Because of safety considerations, metallic shields are always grounded in at least one place. The nature of earthing of metallic shields has a profound effect on cable rating computations. For three-phase systems composed of single core cables with metallic sheaths/concentric neutral wires, the bonding arrangement and the thermal resistivity of the trench fill are the most important factors influencing cable rating which can be controlled by the owner of the circuit. This topic is discussed briefly in Section 1.3.2 and in more detail in Chapter 8.

1.2.4 Armor

Protective armor is usually made of steel wires or tapes. Steel construction, when applied to single-core cables, may result in high magnetic hysteresis and circulating current losses which reduce cable rating. In order to reduce magnetic losses, for these types of cables, nonmagnetic materials such as aluminum or copper are preferably used.

The use of armor wires on cables with lead sheaths, installed in three-phase systems at close spacing, causes additional sheath losses because the presence of armor wires reduces sheath resistance (sheath and armor are connected in parallel), and the losses are largest when the sheath-circuit resistance is equal to its reactance. Without armor wires, the reactance of the sheath is always very much smaller than the resistance. To minimize this increase in losses, armor wires made of high-resistance material such as copper–silicon–manganese alloy are sometimes used. When the cables are spaced further apart, the reactance increases. In this case, low-resistance armor can be used (e.g., aluminum alloy) because the combined resistance is so much less than the reactance that the losses are reduced. Steel armor is mostly used in submarine cables, and the rating of these cables is discussed in detail in Section 8.3.

1.2.5 External Covering

Presently, most power cables are manufactured with external protective coverings. Usually, these are extruded over the sheath or armor. Polyethylene (PE) or polyvinyl chloride (PVC) are the materials used most often. On armored cables, compounded jute or fibrous materials are sometimes used as an *armor serving.* Armored cables usually have an additional nonconducting layer installed between the metallic sheath and the armor. This layer, called *armor bedding*, is usually made of the same material as the armor serving. The external covering provides additional restriction to the heat transfer from the conductor, and therefore reduces the cable rating. The thermal resistance of the cable external covering will depend on the material selected. Polyethylene has the best thermal conductivity from all of the materials used for this purpose. Computation of the thermal resistances of various nonconductive layers is discussed in Chapter 9.

Figure 1.2 shows an array of cross sections of various fluid-filled cables encountered in practice.

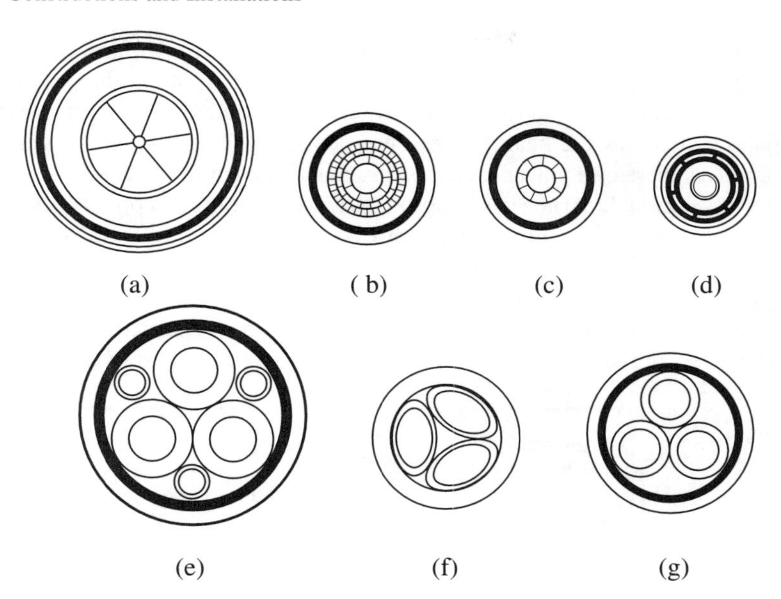

Figure 1.2 Cross sections of low-pressure fluid-filled cables. (a) – (d) Single core cables. (e) – (g) Three-core cables (from King and Halfter, 1983).

1.3 CABLE INSTALLATIONS

1.3.1 Laying Conditions

Insulated power cables are either installed underground or in air. It is important to note that the ambient temperature used in the computation of cable ratings refers to climatic conditions and the mode of cable installation—in the ground, in air, in an open space or in a building, or under water. Typical cable laying conditions are briefly reviewed below.

1.3.1.1 Underground Installations. The most common method for installing power cables underground is to lay them directly in the soil at a depth of about 1 m (3–4 ft). Typical installation configurations of a three-phase circuit composed of single-core cables laid directly in the soil are shown in Fig. 1.3. The arrangement in Fig. 1.3a is called

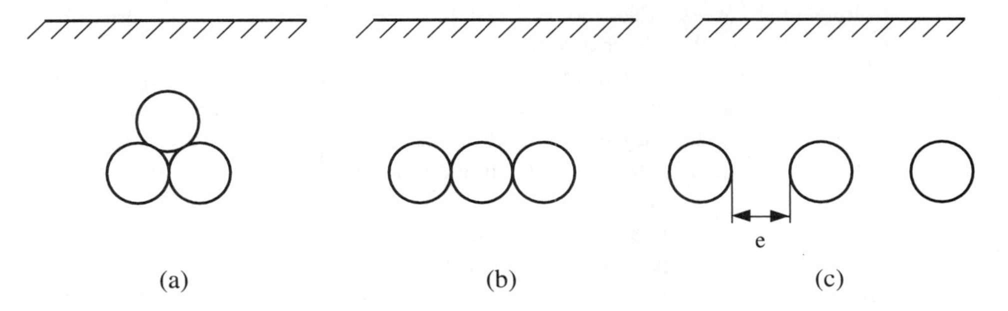

Figure 1.3 Typical installations of directly buried cables.

a *triangular* or *trefoil configuration*, and the remaining two arrangements are called *flat configurations*. In the arrangements shown in Fig. 1.3a and b, the cables are touching, while in the last configuration, the phases are separated by a distance e. Separation of the phases improves the heat dissipation process; however, in some cases, this arrangement produces increased power losses, as discussed later in this section.

Occasionally, when it is important to achieve the highest possible current ratings, cables installed underground are located in an envelope of a material characterized by better thermal heat conduction than the native soil. This additional material is called *thermal backfill.* A good backfill material can have a thermal conductivity two or more times greater than the native soil. Figure 1.4 shows a typical installation of a cable circuit in a thermal backfill. The effect of thermal backfill on cable ampacity is discussed in Chapter 9.

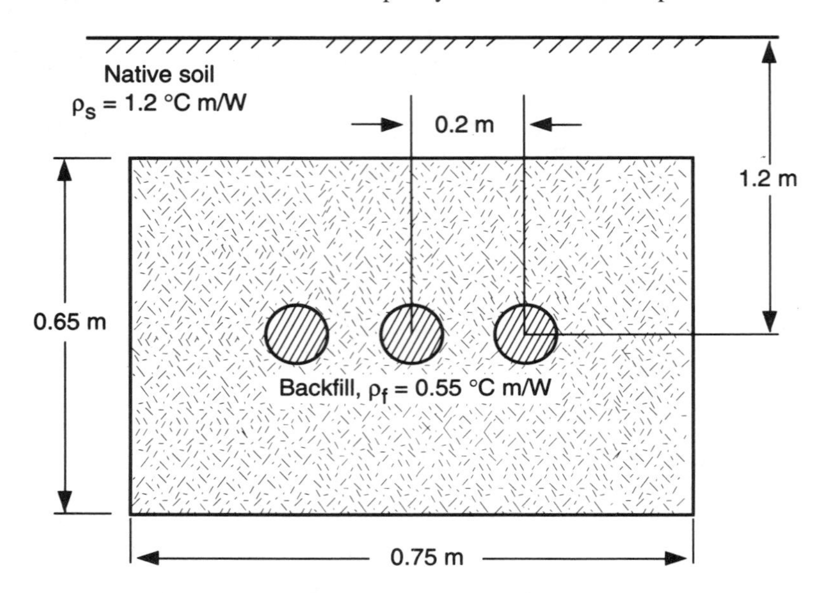

Figure 1.4 Cables installed in a thermal backfill.

In urban areas, there is often a need to install a large number of cable circuits in one trench. In such cases, a special concrete structure is built with uniformly spaced holes to house the cables. Each hole is usually lined with a plastic tube, most often made of polyethylene or PVC. Such a construction is called a *duct bank.* This construction permits installation, and subsequent removal when necessary, of a large number of circuits. For example, in Toronto, it is not uncommon to find 40 cables installed in one duct bank. Very often, several cables may be installed in one duct. Figure 1.5 shows a typical installation of two circuits in a duct bank. Several ducts are empty in this case, permitting installation of additional circuits in the future.

The principal method of transmitting power at higher voltages in the United States is through the use of pipe-type cable. This cable type makes use of three cables insulated by layers of oil-impregnated paper installed within a coated steel pipe which is then filled with an insulating oil or gas. The heat generated within the cable is conducted by the oil or gas and the steel pipe to the surrounding soil. The current rating of the pipe-type cable can

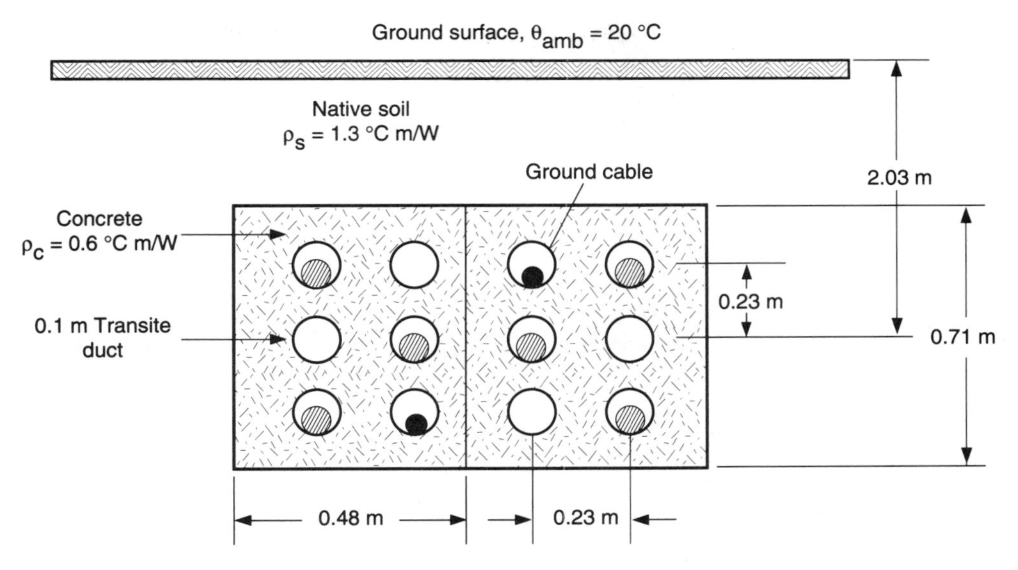

Figure 1.5 Two circuits in a duct bank.

be increased by forced cooling, using heat exchangers and pumps to circulate and cool the oil within the pipe. Pipe-type cables usually have a lower current-carrying capability than directly buried cables with the same conductor size and operating at the same voltage level because of the close proximity of the cores and the losses generated in the steel pipe. An advantage of HPFF cable is the longer lengths of insulated cores that can be placed on a reel, and therefore a fewer number of joints. There is also some advantage in using HPFF versus self-contained cables in congested city areas. For the HPFF cable, only a short stretch of roadway needs to be opened up at a time. After sections of a pipe are welded together, the cables are pulled through the pipe at a later date. Figure 1.6 shows a typical cross section of a cable circuit in a pipe.

1.3.1.2 Cables Installed in Air. Cables are often installed in air, outdoors and indoors, and it is usually necessary to provide a cleating system to support such cables. For example, power delivery systems frequently consist of a combination of overhead lines and underground cables. In most lower voltage installations, the underground cable system is connected to the overhead line through a short section of cable located in a *protective riser.* The cable is secured to the pole, and the riser is provided for mechanical protection. The current-carrying capacity of the composite system is limited by that segment of the system that operates at the maximum temperature. Very often, the riser–pole portion of the cable system will be the limiting segment.

Another type of cable installation in air can be found in an electric power generation plant and in distribution systems where a typical arrangement could consist of a 3 in deep, 24 in wide metal trough or raceway containing anywhere from to 20 to 400 randomly arranged cables ranging in size from #12 AWG to 750 kcmil. We refer to these installations as *cables in trays.* This array of cables is usually secured along the cable tray to prevent shifting should additional cables be pulled into the tray. In many cases, especially in nuclear

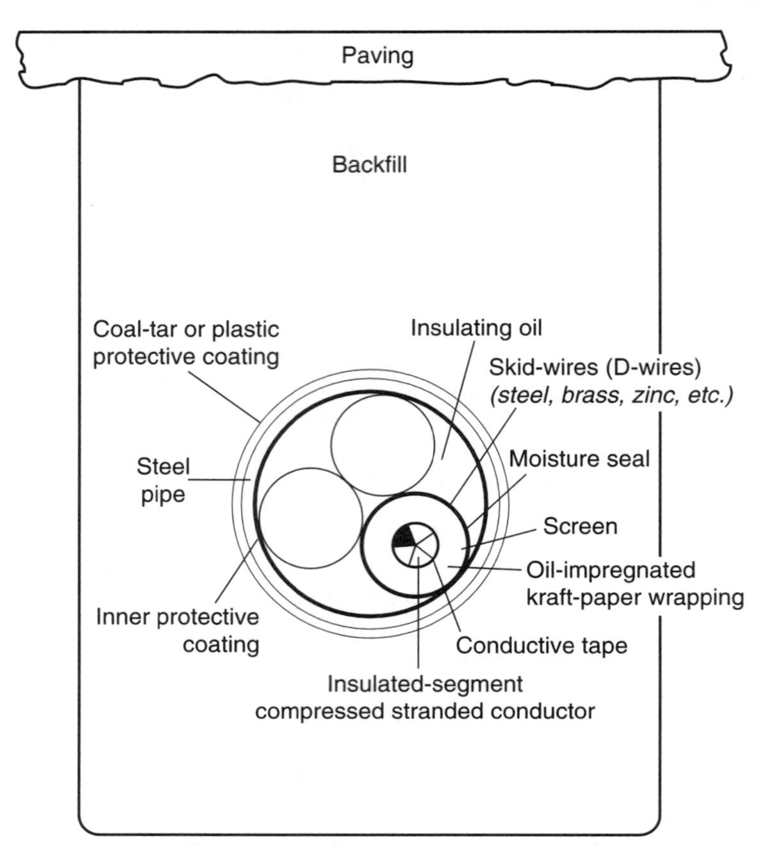

Figure 1.6 Cross section of pipe-type cable in trench (Weedy, 1988).

power plants, the trays are covered with fire protection wrap around the raceway. Because of the very strong mutual heating effects, the ampacity of cables in trays is usually lower than that of cables installed in free air.

Yet another type of installation of cables surrounded by air are cables installed in *tunnels and shafts.* Cables are sometimes installed in tunnels provided for other purposes. In generating stations, short tunnels are often used to convey a large number of cable circuits. Long tunnels are built or existing tunnels adapted solely for the purpose of carrying major EHV transmission circuits which for various reasons cannot be carried overhead. River crossings are obvious cases where tunnels would be used either for technical or environmental reasons. The costs of such installations are considerable, and it is desirable to optimize as far as possible the current-carrying capacity, groupings, and number of circuits to be installed to meet a given transmission capacity.

Rating computations of cables in air are generally more complex than for underground installations. Details of these computations are discussed in Chapters 9 and 10.

1.3.2 Bonding Arrangements

In a three-phase system, sheaths or concentric neutrals of power cables are always bonded and grounded at least at one end. If the sheath is in the form of a metallic tube, eddy currents will flow in it, producing additional heat which has to be dissipated from the cable surface. If the sheaths are bonded at two or more points, circulating currents will flow in them, producing additional losses and reducing cable ampacity. This current will increase with increasing phase separation. Circulating current losses are usually much greater than the losses induced by eddy currents. Therefore, from an ampacity point of view, single-point bonded installations or installations utilizing special bonding arrangements are preferred. These arrangements are referred to as *single-point bonding* or *cross bonding*.

In single-point bonded systems, the cable sheaths are bonded and directly grounded at one end, with the remote end sheaths grounded through a voltage-limiting device. The cable system must be designed to limit the sheath standing voltage to a locally permitted level, and these values determine the maximum length of the cable circuit.

In cross-bonded systems, the cable run is divided into groups, each consisting of three approximately equal sections. The cable sheaths are cross bonded so that induced voltages are canceled, and are bonded together at the end of each group of three sections. In addition, in some instances, the cables are physically transposed to enhance the cancellation of the induced sheath voltages. However, it is an expensive procedure, and is therefore applied mainly in installations with a conductor cross section above 500 mm^2. Figure 1.7 shows typical bonding and grounding arrangements of a transmission circuit composed of three single-phase cables.

The computation of eddy and circulating current losses and the effect of bonding arrangement are discussed in detail in Chapter 8. Various bonding arrangements are discussed in the ANSI/IEEE Standard 575 (1988).

1.3.3 Forced Cooling of Cable Circuits

As mentioned earlier, in order to increase the capacity of cable circuits, special cooling arrangements are sometimes made. The selection of the cooling method depends on the cable type, length and trench profile, and limiting temperatures, and is related to the need to provide adequate cooling of joints and terminations. The following are the most common cooling arrangements: (1) flow of fluid in the central duct in the conductor (LPFF cables), (2) forced cooling of high-pressure fluid-filled cables, (3) internal water cooling of a cable (cable installed in a pipe filled with water), and (4) external water cooling (water pipes installed externally to the cable). Figure 1.8 shows the permissible steady-state current rating for a 2500 mm^2 copper conductor, fluid-filled cable system. The maximum conductor temperature is 85°C, soil thermal resistivity is 1 K·m/W, and the cables are located 1 m below the ground surface.

Computation of cable ratings in the presence of forced cooling is quite involved and will not be discussed in this book. Some aspects of forced cooling are discussed by Weedy (1988) and in several publications (EPRI, 1984; Aabo *et al.*, 1995; Notaro and Webster, 1971; Flamand, 1966; Buckweitz and Pennell, 1976; CIGRE 1976, 1986).

1.4 HEAT SOURCES IN POWER CABLES

Depending on cable construction and installation conditions, there may be a number of sources generating heat inside the cable. The heat produced inside a cable is referred to as cable *heat loss*. In general, there are two types of losses generated in a cable:

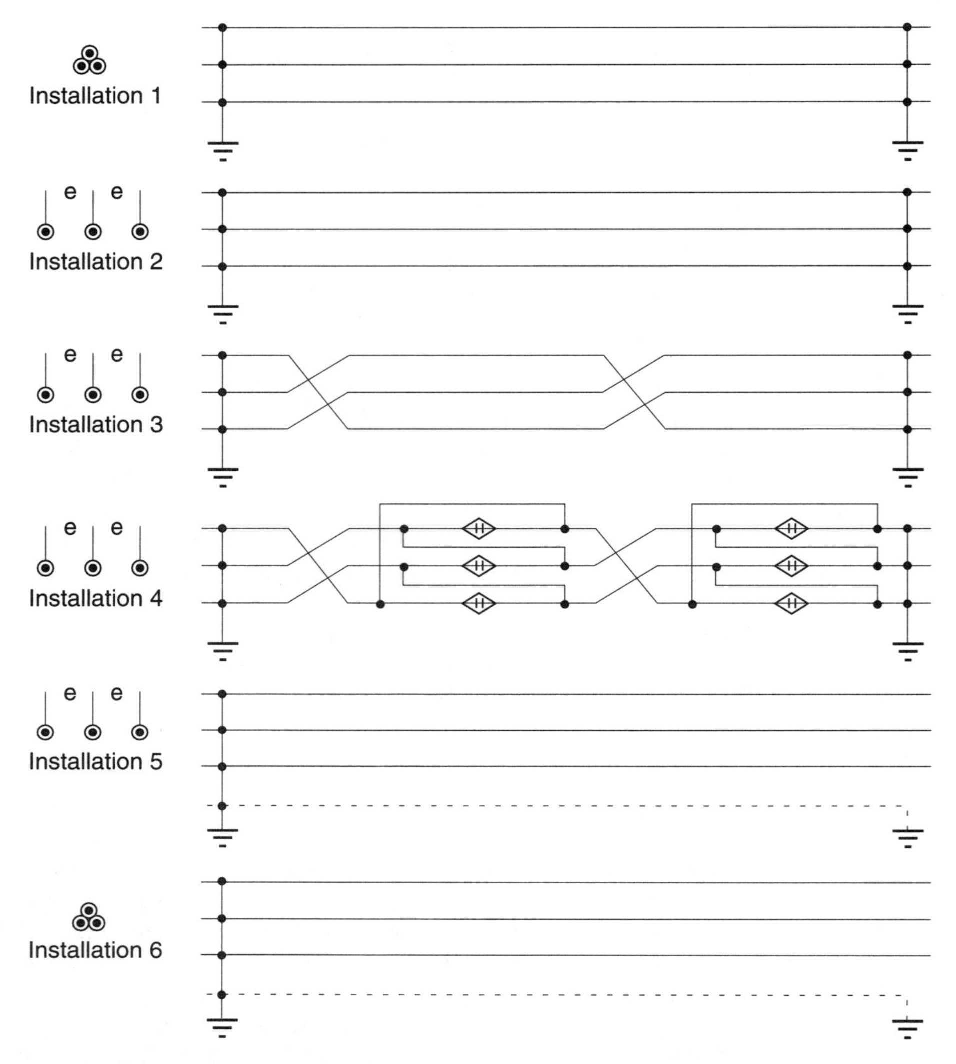

Figure 1.7 Installation of single-core cable circuits with bonding and grounding arrangements of cable sheaths.

1. Current-dependent losses
2. Voltage-dependent losses.

The losses belonging to both groups are briefly discussed below and treated in detail in later chapters of this book.

1.4.1 Current-dependent Losses

The current-dependent losses refer to the heat generated in metallic cable components, namely, the conductor, sheath or screens, armor, and pipe. Sheath, armor, and pipe losses are only generated, of course, when the cable is equipped with these components.

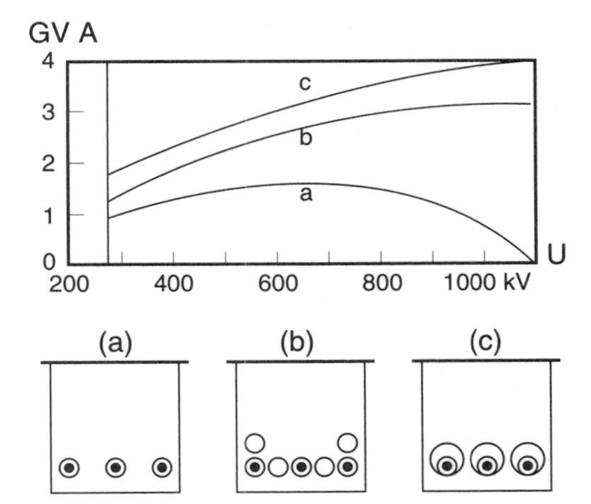

Figure 1.8 The effect of forced cooling on rating of a 2500 mm^2 LPFF cable. (a) Natural cooling in the ground. (b) Forced cooling with the water pipes located outside the cables. (c) Forced cooing with water in pipes containing the cables. The temperature of water intake and ambient soil is equal to 10°C (from Ball et al., 1977).

Conductor Losses. All cables will have losses generated by the current flowing in the conductor. These losses, denoted by W_c (W/m), are often referred to as joule, or I^2R losses from the equation from which they are computed. Conductor losses are a function of the load current, and in calculations of cyclic current rating, the losses are based on a load factor representing the load variation within a specified time period. If the load is constant at the maximum rating of the cable, the load factor is 100%.

Sheath Losses. Metallic parts of the cable other than the conductor may also be a source of joule losses caused by the currents induced in them. There are two parts of the cable which may generate these additional losses: (1) the sheath or screen (losses are denoted by W_s), and (2) the armor (the losses are denoted by W_a).

In general, there are two types of losses generated in sheaths and screens. The first type, called the *sheath eddy loss*, is caused by the induced eddy currents. It is known that, whenever an alternating flux penetrates a piece of conducting material, eddy currents will be produced therein. These currents circulate in the sheath. This loss occurs because no point of any sheath is equidistant from all three of the current-carrying conductors.

The second type of sheath loss, known as the *sheath circulating loss*, which occurs in single-core cable systems and in three-core SL cables (cables with lead sheath around each core), is caused by induced currents flowing along the metallic sheath and returning through the sheath of other cable phases or through earth. Sheath circulating loss only exists when the sheaths of two or three single-core cables are bonded together at two different positions, such as the ends of the cable route. In this case, the sheath eddy current is superimposed on the sheath circulating current, and therefore the actual current flowing in the sheath will not be uniformly distributed around the sheath circumference. However, in rating calculations, eddy losses are neglected when the circulating losses are present.

Armor and Pipe Losses. Some cable installations, for example, submarine cables, are constructed with protective armor. Since, in the majority of cases, this armor is multipoint grounded, circulating losses are produced therein. If the armor is made of a nonmagnetic material, the armor and sheath losses are considered together in rating computations. However, the majority of armored cables are constructed with magnetic steel wire or tape and, in this case, the hysteresis losses have to be considered. For pipe-type cables, hysteresis losses in the steel pipe are also included in rating computations.

1.4.2 Voltage-dependent Losses

Two different types of losses belong to this group: (1) dielectric losses, and (2) losses caused by the charging current. Both losses are always present when the cable is energized, and they are dependent on the cable electrical capacitance.

Dielectric losses produced in the cable insulation are the result of the energy storage capability of insulating materials subjected to alternating voltage. Cable acts as a large capacitor subject to charging currents. The work required to charge this capacitor is called dielectric loss (denoted by W_d).

The charging current produced by the cable capacitance generates ohmic losses in the cable that are present any time the cable is energized. Therefore, these losses operate at 100% load factor as do dielectric losses. The magnitude of these losses is determined by system reactive design and operation and is difficult to evaluate. The charging current must be supplied by the system, but the charging current losses are not strictly losses within the cable itself.

Dielectric losses can be neglected for distribution voltages, and the limiting voltage levels for each insulation type can be those given in IEC Publication 287 (1982) and discussed in Chapter 6 of this book. Charging current losses are usually neglected in rating computations.

1.5 HISTORICAL NOTE

Current rating techniques of electric power cables have as long a history as the cable itself. Methods presented in some of the first publications on this subject (Kennelly, 1893; Mie, 1905) are still used in today's standards. Over the last hundred years, many researchers and engineers have worked on various aspects of cable ratings. There are too many of them to even mention their names; therefore, only a few selected developments are highlighted here, with the emphasis on the work which has had the greatest influence on the standardization efforts.[1] The major contributors to the development of the cable rating techniques are mentioned throughout the book when the technical issues are discussed.

The first rating tables were issued in Britain in the early 1920s. As a matter of fact, the history of the calculation of current ratings in the United Kingdom can be traced through the reports of one company that is still active in the field of cable ratings. This company is ERA Technology Ltd., formerly the Electrical Research Association. The early involvement of ERA can best be seen by quoting the preface from a 1924 ERA Report, "Permissible Current Loading of British Standard Impregnated Paper Insulated Electric Cables." The preface reads as follows:

> *The work herein described was initiated in 1913 by the Institution of Electrical Engineers, continued under the auspices of the Electrical Research Committee, and finally transferred to the Electrical Research Association on its formation in 1921.*
>
> *There has been little change in the personnel of the working Committee since the commencement. The new research work herein described has been carried out by Mr S. W. Melsom of*

[1] The account of the history of standardization of cable rating computations in the United Kingdom is based on a private communiqué of Mr. Mark Coates, and that in the United States is based on the presentations by M. A. Martin (1989) and W. Z. Black (1989) during th IEEE/PES T&D Conference in New Orleans, LA.

the National Physical Laboratory, and Mr E. Fawssett, of the Newcastle-upon-Tyne Electric Supply Co., who are members of the directing Committee.

The committee structure, then in place for setting out standard methods of calculating, appears very similar to that currently operating in the IEC with its technical committees and working groups.

The first ERA report on the subject, published in 1923, set out a standard method of calculating current ratings and provided tabulated ratings. Work continued and calculation methods improved in many areas with papers being published by a number of workers; most notable is the work by Arnold on joule losses. The next major U.K. document was prepared by Whitehead and Hutchings and published in 1939. This report was limited to methods of calculation, with tabulated ratings being published in a separate document. In the next two decades, many reports were published which improved one aspect or another in the calculation of current ratings. These improvements were brought together by Goldenberg in a 1958 report numbered F/T 187. This was the last U.K. document to give a comprehensive guide to current rating calculations before the publication of IEC 287.

In the United States, the National Electric Light Association (NELA) adopted in 1931 a set of standard constants and reference conditions for the calculation of load capabilities of cables. This, in turn, resulted in the publication of the first set of current-carrying capacity tables for paper-insulated, lead-covered cables in underground ducts or in air. This publication was issued in March 1932 and was entitled, "Determination of Rating for Underground and Aerial Cables—NELA Publication #28."

In 1933, the tables were expanded and republished by the Edison Electric Institute (EEI), "*Current-Carrying Capacity of Impregnated Paper—Insulated Cable*" (EEI Publication A-14-1933). These tables lasted for ten years when, in 1943, after considerable improvements in cable insulation materials and growth in underground systems, the Insulated Power Cable Engineering Association (IPCEA) published three documents, P-29-226, P19-102, and P20-161, for impregnated paper, rubber, and varnished cambric insulated cables, respectively.

In 1951, W. A. Del Mar of Phelps Dodge Wire & Cable Company coined the word "ampacity" to replace "current-carrying capacity."

Concurrent with the development of power cable ampacity calculations over the years, the National Electrical Code (NEC) in the United States has published ampacity tables for building wires since the early 1900s. These ampacity tables have dealt primarily with cables rated 600 V and below. The technical basis for the development of the tables issued by NEC was developed by a committee chaired by S. J. Rosch, and dealt primarily with low-voltage, code-grade, rubber-insulated cables. The committee's final report was issued in June 1938. Unfortunately, the work completed by Rosch was very crude by today's engineering standards and, in fact, it was in error in several fundamental technical areas.

In the 1930s, the field of heat transfer was in its infancy, and Rosch did not appear to be aware of some of the heat transfer correlations that were first beginning to become available from early heat transfer studies conducted mostly in the United Kingdom. Lacking any heat transfer foundations and unaware of other similar work, he and his committee resorted to a series of experimental tests to determine the relationship between the current in a conductor and the conductor temperature. The committee's tests were limited to a relatively small number of all copper conductors insulated with code grade rubber and suspended horizontally in air. No buried cables were considered, and shield losses were not included. Rosch neglected the variation of the heat transfer coefficient with the cable surface temperature, which is an error that can be quite significant. He did, however, make two additional assumptions that result in a conservative value for the recommended ampacity: the air surrounding the cable is still and radiation from the surface to the environment is neglected.

Rosch recognized some of the shortcomings of his ampacity model, and in 1938 he published a paper that detailed a series of experimental test on cables placed in horizontal conduits. Temperatures of cable sizes between 12 AWG and 1 000 000 cmils were measured for a number of different current levels. Using these data, he was able to calculate a Q value that was to be incorporated in the NEMA tables, so that ampacity values predicted by the model were brought into line with the experimental data. In fact, the value of Q was a correction factor that pointed out that the correlation suggested by the ampacity model was actually in error; its use was necessary to correct this error. If correct heat transfer correlations had been used in the first place, the application of a correction factor would not have been necessary.

Ampacity work progressed steadily from the time of Rosch's final report until 1957 when Neher and McGrath published their paper which successfully summarized most of the important ampacity work by extending an earlier comprehensive work of Simmons. Their paper actually introduced no new advancements in the area of ampacity; it simply and effectively put all of the ampacity principles into a single, all-encompassing paper. Naturally, this is a tall order when we consider the complexity of the ampacity problem. Due to Neher and McGrath's successful summary paper, most engineers in North America refer to the calculation procedure used to determine ampacity values as the Neher–McGrath method. Actually, the technique that they described is based on a simple model of energy balance in the conductor, and on an analogy between the flow of electric current and the flow of heat. Both of these principles were well known long before 1957. Nonetheless, the Neher–McGrath paper is credited as the paper which forms the basis for modern ampacity standards.

The Neher–McGrath Model, as opposed to the Rosch Model, is based on a technically correct set of equations. It effectively accounts for a much greater diversity of cable designs and installation geometries. It considers heat generated in the shield material resulting from induced shield currents and it also accounts for dielectric heating which can become significant in high voltage cables. It uses more accurate and more technically correct heat transfer coefficients than the Rosch Model. It describes the equations that should be used for underground installations, as well as cables oriented vertically and horizontally in air. It describes a procedure that can be used to derate cables when the heat generated by one cable influences other cables. In other words, it permits ampacity calculations for multiple cable installations where mutual heating between cables can be significant. It also includes a technique for calculating the thermal resistance of a duct bank that may contain multiple cables.

Although the Neher–McGrath Model is technically correct, it does have some weaknesses, mainly in dealing with cables in air. Since the method is based on the analogy between the flow of heat and current, one must know the thermal resistances on the circuit before the ampacity value can be calculated. Unfortunately, convective heat transfer coefficients are temperature-dependent, and therefore the temperature at local points in the thermal circuit must be known before the problem is solved. Neher and McGrath use assumed temperature values to solve this dilemma. In some instances, these assigned temperatures can lead to unacceptable errors. Also, the Neher/McGrath Model uses experimental constants to aid in the computation of the thermal resistance of fluid layers in the thermal circuit. As Black (1989) pointed out, the accuracy of these constants has not been thoroughly verified. And finally, the model accounts for only a single value of the thermal resistance of the soil layer for buried installations. Changes in soil resistivity can occur when the soil adjacent to the cable–earth interface begins to dry. This change is known to be very influential in the cable ampacity because the soil thermal resistance is the largest single resistance in the composite circuit. Such changes in the resistance of the soil can lead to thermal runaway of the cable temperature resulting from a phenomenon referred to as the thermal instability of the soil.

After several dissenting views were resolved at a symposium devoted to the subject, the electric power industry adopted the Neher–McGrath methods for the calculation of power cable ampacity. In 1962, the American Institute of Electrical Engineers (AIEE) and IPCEA jointly published new ampacity tables based on the Neher/McGrath paper. These tables were published in two volumes, one for copper and one for aluminum cables (AIEE S-135-1 and S-135-2), providing ratings for impregnated paper, varnished cloth, rubber, thermoplastic polyethylene- and asbestos-insulated cables. Subsequently, these tables became known in industry circles as the *black books*. In 1967, the IPCEA published a revised version of ampacities for impregnated paper-insulated cables (IPCEA #48-426) because the Association of Edison Illuminating Companies (AEIC) revised upward their specification for conductor operating temperatures.

After the publication of Goldenberg's report in the United Kingdom and the Neher–McGrath paper in the United States, it was felt that sufficient methodology has been developed to warrant the issuing of an international standard. Such a standard was prepared by the International Electrotechnical Commission (IEC 287, 1969, 1982). The immediate predecessor of IEC 287 was a CIGRE report presented in Paris in 1964. The countries that participated in preparing the CIGRE document were the United Kingdom, the Netherlands, France, Germany, Italy, and the United States. The major contributions to the development of this standard were made by such well-known workers in this field as R. G. Parr and H. Goldenberg of the United Kingdom, A. Morello of Italy, M. McGrath of the United States, H. Brekelmann of Germany, and R. Wlodarski of Poland. It is the same countries, with the addition of Canada and Belgium, that are active in the continued development of IEC 287.

The CIGRE report was adopted by IEC in 1969, and after a number of amendments, a second edition was published in 1982. For the third edition, IEC 287 is being divided into a number of parts and sections, each of which covers a different aspect of the calculation of cable ratings. At the time of writing this book, the IEC working group responsible for current ratings is preparing tabulated current ratings for publication by IEC. This takes us back to 1923 when both methods of calculating current ratings and tabulated ratings were published in the United Kingdom.

Meanwhile, in the United States, after the publication of IPCEA tables in 1967, the work on any new ampacity tables remained dormant for approximately ten years. In 1972, due to a proliferation of shielded single-conductor cables and the absence of ratings for cables with circulating currents in the shields, a working group was formed within the

Insulated Conductors Committee (ICC). This group reviewed the industry needs and recommended that additional ampacity tables be published to supplement the "black books" (AIEE S-135-1, S-135-2).

During the late 1970s, the Insulated Conductors Committee requested and received a project authorization from IEEE to revise the "black books." This project was necessitated by the outdated parameters and cable constants and the major changes in heat transfer technology associated with modern cable systems. It was later decided to include in the revision the documents AIEE S135-1 and S135-2, IPCEA P-48-426, P-53-426, and P-54-440 (Cable Tray Ampacities).

To accomplish this task, another working group was organized in ICC, and in 1981 it published a set of new parameters for the new ampacity tables. The parameters were to be used to develop tables for extruded and laminar dielectric underground power cables of 600 V–500 kV. The tables, published in 1994 (IEEE, 1994), address new issues such as cable/earth interface temperature and limiting heat flux, cables in vented and nonvented risers, and cables in open and covered trays. Over 3000 ampacity tables for extruded dielectric power cables rated through 138 kV and laminar dielecric power cables rated through 500 kV are provided. The development of these tables has been spearheaded by M. A. Martin, Jr. of M & E Technology Inc.

The work on refining cable ampacity computations is being continued. It proceeds in two directions: (1) experimental studies are being performed to fine-tune some of the computational formulas and adjust the value of constants, and (2) numerical methods are being applied to overcome limitations inherent in the analytical approach. From the perusal of the book, the reader will be able to learn about some of the most important historical developments in power cable rating computations, and will also be able to follow the latest developments in the application of numerical methods in this field.

REFERENCES

Aabo, T., Lawson, W. G., and Pancholi, S. V. (1995), "Upgrading the ampacity of HPFF pipe-type cable circuits," *IEEE Trans. Power Delivery*, vol. 10, no. 1, pp. 3–8.

AEIC CS1-90 (1990), "Specifications for impregnated paper-insulated, metallic-sheathed cable, solid type."

AEIC CS2-90 (1990), "Specifications for impregnated paper and laminated paper polypropylene insulated cable, high pressure pipe type."

AEIC CS3-90 (1990), "Specifications for impregnated paper-insulated, metallic-sheathed cable, low-pressure gas-filled type."

AEIC CS4-93 (1993), "Specifications for impregnated-paper-insulated low and medium pressure self contained liquid filled cable."

AEIC CS5-94 (1994), "Specifications for cross-linked polyethylene insulated shielded power cables rated 5 through 46 kV."

AEIC CS6-87 (1987), "Specifications for ethylene propylene rubber insulated shielded power cables rated 5 through 69 kV."

AEIC CS7-93 (1993), "Specifications for cross-linked polyethylene insulated shielded power cables rated 69 through 138 kV."

AEIC CS31-84 (1984), "Specifications for electrically insulated low viscosity pipe filling liquids for high-pressure pipe-type cable."

AEIC G1-68 (1968), "Guide for application of AEIC maximum insulation temperatures at the conductor for impregnated-paper-insulated cables."

AEIC G2-72 (1972), "Guide for electrical tests of cable joints 138 kV and above."

AEIC G4-90 (1990), "Guide for installation of extruded dielectric insulated power cable systems rated 69 kV through 138 kV."

AEIC G5-90 (1990), "Underground extruded power cable pulling guide."

AEIC G7-90 (1990), "Guide for replacement and life extension of extruded dielectric 5-35 kV underground distribution cables."

ANSI/IEEE Standard 575 (1988), "Application of sheath-bonding methods for single conductor cables and the calculation of induced voltages and currents in cable sheaths."

Ball, E. H., Endacott, J. D., and Skipper, D. J., (1977), "U. K. Requirements and future prospects for forced-cooled cable systems," *Proc. IEE*, part C, no. 5.

Barnes, C. C. (1964), *Electric Cables*. London: Sir Isaac Pitman & Sons.

Black, W. Z. (1989), "The ampacity table dilemma: plotting a future course," presented at the 11th IEEE/PES Transmission and Distribution Conf. and Exposition, New Orleans, LA, Apr. 2–7, 1989.

Buckweitz, D. B., and Pennell, D. B. (Apr. 1976), "Forced cooling of UG lines," *Transmission and Distribution*, pp. 51–58.

CIGRE (1976), "The calculation of continous ratings of forced cooled cables," WG 08 of CIGRE SC 21, *Electra*, no. 66, pp. 59–84. Erratum in *Electra*, no. 76, p. 48, May 1981.

CIGRE (1986), "Forced cooled cables. Calculation of thermal transients and cyclic loads," WG 08 of CIGRE SC 21, *Electra*, no. 104, pp. 23–38.

Flamand, C. A. (1966), "Forced cooling of high-voltage feeders," *IEEE Trans. Power App. Syst.*, vol. PAS-85, no. 9, pp. 980–986.

IEC Standard 228 (1978), "Conductors of insulated cables," 2nd ed., 1st suppl., 1982.

IEC Standard 287 (1969, 1982), "Calculation of the continuous current rating of cables (100% load factor)," 1st ed. 1969, 2nd ed. 1982, 3rd ed. 1994–1995.

IEC Standard 502 (1983), "Extruded solid dielectric insulated power cables for rated voltages from 1 kV to 30 kV."

IEEE (1994) Standard 835, "IEEE Standard—Power cable ampacity tables."

EPRI (1984), "Designer handbook for forced-cooled high-pressure oil-filled pipe-type cable systems," EPRI EL-3624, Project 7801-5, Report July 1984.

Kennelly, A. E. (1893), "Minutes, ninth annual meeting, Association of Edison Illuminating Companies," New York, NY.

King, S. Y., and Halfter, N. A. (1983), *Underground Power Cables*. London: Longman.

Martin, Jr., M. A. (1989), "Historical perspective on power cable ampacity ratings," presented at the 11th IEEE/PES Transmission and Distribution Conf. and Exposition, New Orleans, LA, Apr. 2–7, 1989.

Mie, G. (1905), "Über die Warmeleitung in einem verseilten Kable," *Electrotechnische Zeitschrift*, p. 137.

Notaro, J., and Webster, D. J. (1971), "Thermal analysis of forced cooled cables," *IEEE Trans. Power Syst.*, vol. PAS-71, no. 3, pp. 1225–1231.

Weedy, B. M. (1988), *Thermal Design of Underground Systems*. Chichester, England: Wiley.

2

Modes of Heat Transfer and Energy Conservation Equations

2.1 INTRODUCTION

Ampacity computations of power cables require solution of the heat transfer equations which define a functional relationship between the conductor current and the temperature within the cable and in its surroundings. In the previous chapter, we discussed how the heat is generated within the cable. In this chapter, we will analyze how the heat is dissipated to the environment. We will also develop the basic heat transfer equations, and discuss how these equations are solved, thus laying the groundwork for cable rating calculations.

2.2 HEAT TRANSFER MECHANISM IN POWER CABLE SYSTEMS

The two most important tasks in cable ampacity calculations are the determination of the conductor temperature for a given current loading, or conversely, determination of the tolerable load current for a given conductor temperature. In order to perform these tasks, the heat generated within the cable and the rate of its dissipation away from the conductor for a given conductor material and given load must be calculated. The ability of the surrounding medium to dissipate heat plays a very important role in these determinations, and varies widely because of several factors such as soil composition, moisture content, ambient temperature, and wind conditions. The heat is transferred through the cable and its surroundings in several ways, and these are described in the following sections.

2.2.1 Conduction

For underground installations, the heat is transferred by conduction from the conductor and other metallic parts as well as from the insulation. It is possible to quantify heat transfer processes in terms of appropriate *rate equations*. These equations may be used to compute the amount of energy being transferred per unit time. For heat conduction, the rate equation is known as *Fourier's law*. For a wall having a temperature distribution $\theta(x)$, the rate equation is expressed as

$$q = -\frac{1}{\rho}\frac{d\theta}{dx} \tag{2.1}$$

The heat flux q (W/m^2) is the heat transfer rate in the x direction per unit area perpendicular to the direction of transfer, and is proportional to the temperature gradient $d\theta/dx$ in this direction. The proportionality constant ρ is a transport property known as thermal resistivity (K · m/W) and is a characteristic of the material. The minus sign is a consequence of the fact that heat is transferred in the direction of decreasing temperature (see Fig. 2.1).

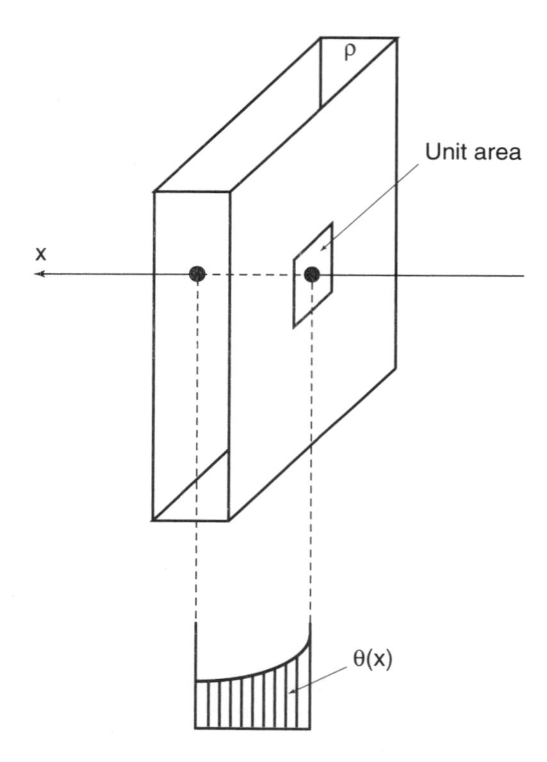

Figure 2.1 Illustration of Fourier's law.

2.2.2 Convection

For cables installed in air, convection and radiation are important heat transfer mechanisms from the surface of the cable to the surrounding air. Convection heat transfer may be classified according to the nature of the flow. We speak of *forced convection* when the flow is caused by external means, such as by wind, pump, or fan. In contrast, for *free* (or *natural*) *convection*, the flow is induced by buoyancy forces which arise from density differences caused by temperature variations in the air. In order to be somewhat conservative, in cable rating computations, we usually assume that only natural convection takes place at the outside surface of the cable (see Chapter 9). However, both convection modes will be considered in Chapter 10.

Regardless of the particular nature of the convection heat transfer process, the appropriate rate equation is of the form

$$q = h(\theta_s - \theta_{\text{amb}}) \tag{2.2}$$

where q, the convective heat flux (W/m^2), is proportional to the difference between the surface temperature and the ambient air temperature, θ_s and θ_{amb}, respectively. This expression is called *Newton's law of cooling* and the proportionality constant h (W/m^2 · K) is referred to as *the convection heat transfer coefficient*. Determination of the heat convection coefficient is perhaps the most important task in computation of ratings of cables in air. The value of this coefficient varies between 2 and 25 W/m^2 · K for free convection and between 25 and 250 W/m^2 · K for forced convection.

2.2.3 Radiation

Thermal radiation is energy emitted by cable or duct surface. The heat flux emitted by a cable surface is given by the *Stefan–Boltzmann law*:

$$q = \epsilon \sigma_B \theta_s^{*4} \tag{2.3}$$

where θ_s^* is the absolute temperature (K) of the surface,[1] σ_B is the *Stefan–Boltzmann constant* ($\sigma_B = 5.67 \cdot 10^{-8}$W/m^2 · K^4), and ϵ is a radiative property of the surface called the *emissivity*. This property, whose value is in the range $0 \le \epsilon \le 1$, indicates how efficiently the surface emits compared to an ideal radiator. Conversely, if radiation is incident upon a surface, a portion will be absorbed, and the rate at which energy is absorbed per unit surface area may be evaluated from the knowledge of surface radiative property known as *absorptivity* α; that is,

$$q_{\text{abs}} = \alpha q_{\text{inc}} \tag{2.4}$$

where $0 \le \alpha \le 1$. Equations (2.3) and (2.4) determine the rate at which radiant energy is emitted and absorbed, respectively, at a surface. Since the cable both emits and absorbs radiation, radiative heat exchange can be modeled as an interaction between two surfaces. Determination of the net rate at which radiation is exchanged between two surfaces is generally quite complicated. However, for cable rating computations, we may assume that a cable surface is small and the other surface is remote and much larger. Assuming this surface is one for which $\alpha = \epsilon$ (a *gray* surface), the net rate of radiation exchange between the cable and its surroundings, expressed per unit area of the cable surface, is

$$q = \epsilon \sigma_B \left(\theta_s^{*4} - \theta_{\text{amb}}^{*4}\right) \tag{2.5}$$

A variation of equation (2.5) will be used in the book for rating of cables on riser poles and in tunnels (see Chapter 10).

Throughout this book, we will use a notion of heat rate rather than heat flux. The heat transfer rate is obtained by multiplying heat flux by the area. Thus, the heat rate for

[1] Throughout this book, the temperature with an asterisk will denote absolute value in degrees Kelvin.

radiative heat transfer will be given by the following equation:[2]

$$W_{\text{rad}} = \epsilon \sigma_B A_{sr} \left(\theta_s^{*4} - \theta_{\text{amb}}^{*4}\right) \tag{2.6}$$

where A_{sr} (m^2) is the effective radiation area per meter length.

In power cable installed in air, the cable surface within the surroundings will simultaneously transfer heat by convection and radiation to the adjoining air. The total rate of heat transfer from the cable surface is the sum of the heat rates due to the two modes.[3] That is,

$$W = hA_s(\theta_s - \theta_{\text{amb}}) + \epsilon A_{sr}\sigma_B \left(\theta_s^{*4} - \theta_{\text{amb}}^{*4}\right) \tag{2.7}$$

where A_s (m^2) is the convective area per meter length.

For some special cable installations, the ambient temperature used for heat convection can be different from the one used for heat transfer by radiation. The appropriate temperatures to be used are described in Chapter 10.

2.2.4 Energy Balance Equations

In the analysis of heat transfer in a cable system, the *law of conservation of energy* plays an important role. We will formulate this law on a *rate basis*; that is, at any instant, there must be a balance between all energy rates, as measured in joules per second (W). The energy conservation law can be expressed by the following equation:

$$W_{\text{ent}} + W_{\text{int}} = W_{\text{out}} + \Delta W_{st} \tag{2.8}$$

W_{ent} is the rate of energy entering the cable. This energy may be generated by other cables located in the vicinity of the given cable or by solar radiation. W_{int} is the rate of heat generated internally in the cable by joule or dielectric losses, and ΔW_{st} is the rate of change of energy stored within the cable. The value of W_{out} corresponds to the rate at which energy is dissipated by conduction, convection, and radiation. For underground installations, the cable system will also include the surrounding soil.

In words, this relation says that the amount of energy inflow and generation act to increase the amount of energy stored within the cable, whereas outflow acts to decrease the stored energy. The inflow and outflow terms W_{ent} and W_{out} are surface phenomena, and these rates are proportional to the surface area. The thermal energy generation rate W_{int} is associated with the rate of conversion of electrical energy to thermal energy and is proportional to the volume. The energy storage is also a volumetric phenomenon, but it is simply associated with an increase ($\Delta W_{st} > 0$) or decrease ($\Delta W_{st} < 0$) in the energy of the cable. Under steady-state conditions, there is, of course, no change in energy storage ($\Delta W_{st} = 0$).

In analysis of cables installed in air, we will often apply the conservation of energy requirement at the surface of the cable. In this special case, we will only deal with the surface phenomena, and the conservation requirement becomes

$$W_{\text{ent}} - W_{\text{out}} = 0 \tag{2.9}$$

[2] Throughout the book, symbol W will be used for heat transfer rate.

[3] The heat conduction in air is often neglected in cable rating computations (see, however, Section 9.6.8).

Even though thermal energy generation will be occurring within the cable, the process will not affect the energy balance at the cable surface. Moreover, this conservation requirement holds for both steady-state and transient conditions.

We will use the fundamental equations described in this chapter to develop rating equations throughout the reminder of the book.

2.3 HEAT TRANSFER EQUATIONS

As we mentioned earlier, current flowing in the cable conductor generates heat which is dissipated through the insulation, metal sheath, and cable servings into the surrounding medium. The cable ampacity depends mainly upon the efficiency of this dissipation process and the limits imposed on the insulation temperature. To understand the nature of the heat dissipation process, we need to develop the relevant heat transfer equations. Energy balance equation (2.8) will be the focal point in our investigations.

2.3.1 Underground Directly Buried Cables

Let us consider an underground cable located in a homogeneous soil. In such cable, the heat is transferred by conduction through cable components and the soil. Since the length of the cable is much greater than its diameter, end effects can be disregarded, and the heat transfer problem can be formulated in two dimensions only as shown in Fig. 2.2.[4]

The terms in equation (2.8) are defined as follows. When heat passes through the body shown in Fig. 2.2, it meets certain thermal resistance in its path. Applying Fourier's law of

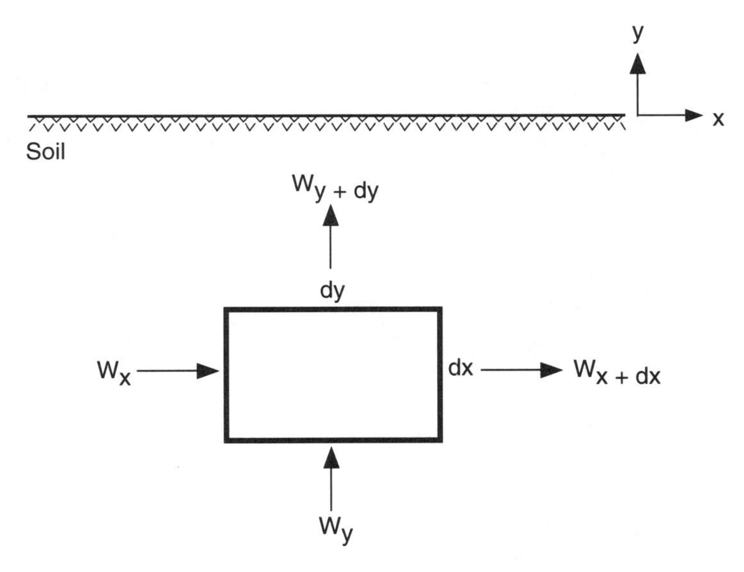

Figure 2.2 Illustration of a heat conduction problem.

[4] Because end effects are neglected, all thermal parameters will be expressed in this book on a per-unit length basis.

heat conduction [equation (2.1)], we have

$$W_x = -\frac{S}{\rho}\frac{\partial \theta}{\partial x} \tag{2.10}$$

where W_x = heat transfer rate through the area S in the x direction, W

ρ = thermal resistivity, K · m/W

S = surface area perpendicular to heat flow, m^2

$\frac{\partial \theta}{\partial x}$ = temperature gradient in x direction.

Consider the small element $dx\,dy$ in Fig. 2.2. If there are temperature gradients, conduction heat transfer will occur across each of the surfaces. The conduction heat rates perpendicular to each of the surfaces at the x and y coordinate locations are indicated by terms W_x and W_y, respectively. The conduction heat rates at the opposite surfaces can then be expressed as a Taylor series expansion where, neglecting higher order terms,

$$\begin{aligned} W_{x+dx} &= W_x + \frac{\partial W_x}{\partial x}dx \\ W_{y+dy} &= W_y + \frac{\partial W_y}{\partial y}dy \end{aligned} \tag{2.11}$$

Within the element $dx\;dy$, there may also be an energy source term associated with the rate of thermal energy generation. This term is represented as

$$W_g = W_{\text{int}}dx\;dy$$

where W_{int} is the rate at which energy is generated per unit volume of the body by resistive and capacitive currents ($W/m^3/m$). In addition, there may occur changes in the amount of internal energy stored by the material in the small body $dx\,dy$. These changes are related to the capacitive nature of cable insulation. On a rate basis, we express this energy storage term as

$$\Delta W_{st} = c\frac{\partial \theta}{\partial t}dx\;dy$$

where c is the volumetric thermal capacity of the material. Recognizing that the conduction rates constitute the energy inflow and outflow and there are no other energy transfer modes,[5] the energy balance equation (2.8) for this body can be written as

$$W_x + W_y + W_{\text{int}}dx\;dy - W_{x+dx} - W_{y+dy} = c\frac{\partial \theta}{\partial t}dx\;dy$$

Observing that, in this case, $S = dx\,dy$, we can rewrite the last equation by substituting from equations (2.10) and (2.11) to obtain:

$$\boxed{\frac{\partial}{\partial x}\left(\frac{1}{\rho}\frac{\partial \theta}{\partial x}\right) + \frac{\partial}{\partial y}\left(\frac{1}{\rho}\frac{\partial \theta}{\partial y}\right) + W_{\text{int}} = c\frac{\partial \theta}{\partial t}} \tag{2.12}$$

[5] Since the case of an underground cable is considered here, heat is transferred by conduction only.

For a cable buried in soil, equation (2.12) is solved with the boundary conditions usually specified at the soil surface. These boundary conditions can be expressed in two different forms. If the temperature is known along a portion of the boundary, then

$$\theta = \theta_B(s) \tag{2.13}$$

where θ_B is the boundary temperature that may be a function of the surface length s. If heat is gained or lost at the boundary due to convection $h(\theta - \theta_{\text{amb}})$ or a heat flux q, then

$$\frac{1}{\rho}\frac{\partial \theta}{\partial n} + q + h(\theta - \theta_{\text{amb}}) = 0 \tag{2.14}$$

where n is the direction of the normal to the boundary surface, h is a convection coefficient, and θ is an unknown boundary temperature.

Occasionally, it may be advantageous to express the heat transfer equation in cylindrical coordinates. In two dimensions, they become

$$\boxed{\frac{1}{r}\frac{\partial}{\partial r}\left(\frac{r}{\rho}\frac{\partial \theta}{\partial r}\right) + \frac{1}{r^2}\frac{\partial}{\partial \phi}\left(\frac{1}{\rho}\frac{\partial \theta}{\partial \phi}\right) + W_{\text{int}} = c\frac{\partial \theta}{\partial t}} \tag{2.15}$$

In cable rating computation, the temperature of the conductor is usually given, and the maximum current flowing in the conductor is sought. Thus, when the conductor heat loss is the only energy source in the cable, we have $W_{\text{int}} = I^2 R$, and equation (2.12) or (2.15) is used to solve for I with the specified boundary conditions.

The challenge in solving equations (2.12) and (2.15) analytically stems mostly from the difficulty of computing the temperature distribution in the soil surrounding the cable. An analytical solution can be obtained when a cable is represented as a line source placed in an infinite homogenous surrounding. Since this is not a practical assumption for cable installations, another assumption is often used, namely, that the earth surface is an isotherm. In practical cases, the depth of burial of the cables is on the order of ten times their external diameter, and for the usual temperature range reached by such cables, the assumption of an isothermal earth surface is a reasonable one. In cases where this hypothesis does not hold, namely, for large cable diameters and cables located close to the earth surface, a correction to the solution equation has to be used or numerical methods should be applied. Both are discussed later in this book.

EXAMPLE 2.1

In order to illustrate the solution process for the heat transfer equation, we will determine the temperature distribution within the insulation of a single core cable. Let as assume that the conductor temperature is θ_c and the temperature at the outer surface of the insulation is θ_i. The cable has a conductor radius r_c and a radius over insulation r_i. Steady-state conditions are assumed.

For steady-state conditions with no heat generation within the insulation, equation (2.15), with the assumption of radial symmetry, reduces to

$$\frac{1}{r}\frac{d}{dr}\left(\frac{r}{\rho}\frac{d\theta}{dr}\right) = 0 \tag{2.16}$$

Assuming the value of ρ to be constant, equation (2.16) may be integrated twice to obtain the general solution

$$\theta(r) = C_1 \ln r + C_2 \tag{2.17}$$

To obtain the constants of integration C_1 and C_2, we use the specified boundary conditions at the inner and outer surface of the insulation. Substituting these conditions into equation (2.17) and solving for C_1 and C_2, we obtain

$$\theta(r) = \frac{\theta_c - \theta_i}{\ln \frac{r_c}{r_i}} \ln \frac{r}{r_i} + \theta_i \tag{2.18}$$

We observe that the temperature distribution in the insulation associated with radial conduction through a cylinder is logarithmic.

An additional difficulty in finding the analytical solution to equation (2.12) arises in the case of cables being located in a nonuniform medium, for example, in a backfill or duct bank. In this case, the thermal resistivity is not constant. Here again, either correction factors or numerical techniques are applied, and these are discussed in detail in Chapters 9 and 11.

2.3.2 Cables in Air

For an insulated power cable installed in air, several modes of heat transfer have to be considered. Conduction is the main heat transfer mechanism inside the cable. Suppose that the heat generated inside the cable (due to joule, ferromagnetic, and dielectric losses) is W_t (W/m). Another source of heat energy can be provided by the sun if the cable surface is exposed to solar radiation. Energy outflow is caused by convection and net radiation from the cable surface. Therefore, the energy balance equation (2.9) at the surface of the cable can be written as

$$W_t + W_{\text{sol}} - W_{\text{conv}} - W_{\text{rad}} = 0 \tag{2.19}$$

where W_{sol} is the heat gain per unit length caused by solar heating, and W_{conv} and W_{rad} are the heat losses due to convection and radiation, respectively. Computation of the losses generated inside the cable is discussed in detail in Chapters 6–8. Substituting appropriate formulas for the remaining heat gains and losses [equation (2.7)], the following form of the heat balance equation is obtained:

$$W_t + \sigma D_e^* H - \pi D_e^* h \left(\theta_e^* - \theta_{\text{amb}}^*\right) - \pi D_e^* \epsilon \sigma_B \left(\theta_e^{*4} - \theta_{\text{amb}}^{*4}\right) = 0 \tag{2.20}$$

where
θ_e^* = cable surface temperature, K
σ = solar absorption coefficient
H = intensity of solar radiation, W/m^2
σ_B = Stefan–Boltzmann constant, equal to $5.67 \cdot 10^{-8}$ W/m$^2 \cdot$ K^4
ϵ = emissivity of the cable outer covering
D_e^* = cable external diameter,[6] m
θ_{amb}^* = ambient temperature, K

[6] We recall that the dimension symbols with an asterisk refer to the length in meters and without it to the length in millimeters.

This equation is usually solved iteratively. In steady-state rating computations, the effect of heat gain by solar radiation and heat loss caused by convection are taken into account by suitably modifying the value of the external thermal resistance of the cable. Computation of the global solar radiation H and the convection coefficient h can be quite involved (Morgan, 1991). This topic is discussed in Chapters 9 and 10 of the book. Recommended values are also discussed in the same chapters.

2.4 ANALYTICAL VERSUS NUMERICAL METHODS OF SOLVING HEAT TRANSFER EQUATIONS

Equations (2.12) and (2.20) can be solved either analytically, with some simplifying assumptions, or numerically. Analytical methods have the advantage of producing current rating equations in a closed formulation, whereas numerical methods require iterative approaches to find cable ampacity. However, numerical methods provide much greater flexibility in the analysis of complex cable systems and allow representation of more realistic boundary conditions. In practice, analytical methods have found much wider application than the numerical approaches. There are several reasons for this situation. Probably the most important one is historical: cable engineers have been using analytical solutions based on either Neher/McGrath (1957) formalism or IEC Publication 287 (1982) for a long time. Computations for a simple cable system could often be performed using pencil and paper or with the help of a hand-held calculator. Numerical approaches, on the other hand, require extensive manipulation of large matrices, and have only become popular with an advent of powerful computers. Both approaches will be described in this book; analytical methods are discussed in Parts I and II, whereas the numerical approaches are dealt with in Part III.

REFERENCES

IEC Standard 287 (1982), "Calculation of the continuous current rating of cables (100% load factor)," 2nd ed., 3rd amendment, 1993.

Neher, J. H., and McGrath, M. H. (Oct. 1957), "The calculation of the temperature rise and load capability of cable systems," *AIEE Trans.*, vol. 76, part 3, pp. 752–772.

3

Circuit Theory Network Analogs for Thermal Modeling

3.1 INTRODUCTION

Analytical solutions to the heat transfer equations are available only for simple cable constructions and simple laying conditions. In trying to solve the cable heat dissipation problem, electrical engineers noticed a fundamental similarity between the heat flow due to the temperature difference between the conductor and its surrounding medium and the flow of electrical current caused by a difference of potential (Pashkis and Baker, 1942). Using their familiarity with the lumped parameter method to solve differential equations representing current flow in a material subjected to potential difference, they adopted the same method to tackle the heat conduction problem. The method begins by dividing the physical object into a number of volumes, each of which is represented by a thermal resistance and a capacitance. The thermal resistance is defined as the material's ability to impede heat flow. Similarly, the thermal capacitance is defined as the material's ability to store heat. The thermal circuit is then modeled by an analogous electrical circuit in which voltages are equivalent to temperatures and currents to heat flows. If the thermal characteristics do not change with temperature, the equivalent circuit is linear and the superposition principle is applicable for solving any form of heat flow problem.

In a thermal circuit, charge corresponds to heat; thus, Ohm's law is analogous to Fourier's law. The thermal analogy uses the same formulation for thermal resistances and capacitances as in electrical networks for electrical resistances and capacitances. Note that there is no thermal analogy to inductance and in steady-state analysis; only resistance will appear in the network. The analogy between electrical and thermal networks is explored in the next section.

Since the lumped parameter representation of the thermal network offers a simple means for analyzing even complex cable constructions, it has been widely used in thermal analysis of cable systems. A full thermal network of a cable for transient analysis may consist of several loops. Before the advent of digital computers, the solution of the network equations was a formidable numerical task. Therefore, simplified cable representations were adopted and methods to reduce a multiloop network to a two-loop circuit were developed. A two-loop representation of a cable circuit turned out to be quite accurate for most practical applications and, consequently, was adopted in international standards. In this chapter, we will explain how the thermal circuit of a cable is constructed, and we will show how the required parameters are computed. We will also explain how full network equations are solved, and present several examples of the two-loop representation of major cable constructions.

3.2 ELECTRICAL AND THERMAL ANALOGY

3.2.1 Thermal Resistance

All nonconducting materials in the cable will impede heat flow away from the cables (the thermal resistance of the metallic parts in the cable, even though not equal to zero, is so small that it is usually neglected in rating computations). Thus, we can talk about material resistance to heat flow. In order to explain the concept of thermal resistance, let us consider a cylindrical nonconducting layer with constant thermal resistivity ρ_{th}. Cable insulation is a good example of such a layer. If the internal and external radii of this layer are r_1 and r_2, respectively, then the temperature distribution inside this layer is given by equation (2.18) as

$$\theta(r) = \frac{\theta_1 - \theta_2}{\ln \dfrac{r_1}{r_2}} \ln \frac{r}{r_2} + \theta_2 \tag{3.1}$$

where θ_1 and θ_2 are given temperatures at locations corresponding to r_1 and r_2, respectively. We may use this temperature distribution in Fourier's law, equation (2.10), to determine the conduction heat rate. Taking the derivative in equation (3.1), we obtain

$$\frac{d\theta(r)}{dr} = \frac{\theta_1 - \theta_2}{\ln \dfrac{r_1}{r_2}} \frac{1}{r}$$

The area of the cylinder per unit length is equal to $2\pi r$. Substituting this into equation (2.10) yields the following rate equation:

$$W = \frac{2\pi}{\rho_{th} \ln \dfrac{r_2}{r_1}} (\theta_1 - \theta_2) \tag{3.2}$$

From equation (3.2), we can observe an analogy between the diffusion of heat and electrical charge mentioned in the Introduction. Just as an electrical resistance is associated with the conduction of electricity, a thermal resistance may be associated with the conduction of heat. Defining resistance as the ratio of a driving potential to the corresponding transfer rate, it follows from equation (3.2) that the thermal resistance for conduction of a cylindrical layer per unit length is

$$\boxed{T = \frac{\rho_{th}}{2\pi} \ln \frac{r_2}{r_1}} \tag{3.3}$$

For a rectangular wall, we have

$$T = \rho_{th} \frac{l}{S} \tag{3.4}$$

where ρ_{th} = thermal resistivity of a material, K · m/W
S = cross section area of the body, m^2
l = thickness of the body, m

Similarly, for electrical conduction in the same system, Ohm's law provides an electrical resistance of the form

$$R = \frac{V_1 - V_2}{I} = \rho_{el} \frac{l}{S} \tag{3.5}$$

The analogy between equations (3.4) and (3.5) is obvious. We also can write that

$$\boxed{W = \frac{\Delta\theta}{T}} \tag{3.6}$$

which is the thermal equivalent of Ohm's law.

A thermal resistance may also be associated with heat transfer by convection at a surface. From Newton's law of cooling [equation (2.2)],

$$W = h_{conv} A_s (\theta_e - \theta_{amb}) \tag{3.7}$$

where A_s is the area of the outside surface of the cable for unit length, h_{conv} is the cable surface convection coefficient, and θ_e is the cable surface temperature.

The thermal resistance for convection is then

$$T_{conv} = \frac{\theta_e - \theta_{amb}}{W} = \frac{1}{h_{conv} A_s} \tag{3.8a}$$

Yet another resistance may be pertinent for a cable installed in air. In particular, radiation exchange between the cable surface and its surroundings may be important. It follows that a thermal resistance for radiation may be defined as

$$T_{rad} = \frac{\theta_e^* - \theta_{gas}^*}{W_{rad}} = \frac{1}{h_r A_{sr}} \tag{3.8b}$$

where A_{sr} is the area of the cable surface effective for heat radiation for unit length of the cable and θ_{gas}^* is the temperature of the air surrounding the cable which, when cable is installed in free air, is equal to the ambient temperature, θ^*_{amb}. h_r is the radiation heat transfer coefficient obtained from expression (2.6) for radiation heat transfer rate:

$$W_{\text{rad}} = \epsilon\sigma_B A_{sr}\left(\theta_e^{*4} - \theta_{\text{gas}}^{*4}\right)$$
$$= \epsilon\sigma_B A_{sr}\left(\theta_e^{*} - \theta_{\text{gas}}^{*}\right)\left(\theta_e^{*} + \theta_{\text{gas}}^{*}\right)\left(\theta_e^{*2} + \theta_{\text{gas}}^{*2}\right) = h_r A_{sr}\left(\theta_e^{*} - \theta_{\text{gas}}^{*}\right)$$

Hence,

$$h_r = \epsilon\sigma_B\left(\theta_e^{*} + \theta_{\text{gas}}^{*}\right)\left(\theta_e^{*2} + \theta_{\text{gas}}^{*2}\right) \tag{3.9}$$

The total heat transfer coefficient for a cable in air is given by

$$h_t = h_{\text{conv}} + h_r \tag{3.10}$$

EXAMPLE 3.1

Determine an expression for the external thermal resistance for a cable in air. The external diameter of the cable is D_e^* (m).[1]

The external thermal resistance for the cable is obtained from equation (3.6):

$$T_4 = \frac{\theta_e - \theta_{\text{amb}}}{W_t}$$

where W_t is the total heat loss per unit length, and is equal to

$$W_t = \pi D_e^* h_t\left(\theta_e - \theta_{\text{amb}}\right) \tag{3.10a}$$

Hence,

$$T_4 = \frac{1}{\pi D_e^* h_t} \tag{3.10b}$$

In the remainder of this book, we will use symbol h to denote the heat transfer coefficient. In the majority of cases, this transfer will be purely convective. In Chapter 9, we will show the relationship between coefficients h defined in IEC 287 and h_t as defined above.

Circuit representations provide a useful tool for both conceptualizing and quantifying heat transfer problems. The equivalent thermal circuit for the composite cylindrical wall is shown in Fig. 3.1. The heat transfer rate can be determined separately considering each element in the network. That is,

$$W = \frac{\theta_1 - \theta_2}{\dfrac{\rho_A \ln(r_2/r_1)}{2\pi}} = \frac{\theta_2 - \theta_3}{\dfrac{\rho_B \ln(r_3/r_2)}{2\pi}} = \frac{\theta_3 - \theta_4}{\dfrac{\rho_C \ln(r_4/r_3)}{2\pi}} = \frac{\theta_4 - \theta_{\text{amb}}}{\dfrac{1}{2\pi r_4 h}}$$

where all the radii are expressed in meters.

In terms of the *overall temperature difference* $\theta_1 - \theta_{\text{amb}}$ and the *total thermal resistance* T_{tot}, the heat transfer rate can also be expressed as

$$W = \frac{\theta_1 - \theta_{\text{amb}}}{T_{\text{tot}}}$$

[1] We will denote diameters expressed in meters with an asterisk.

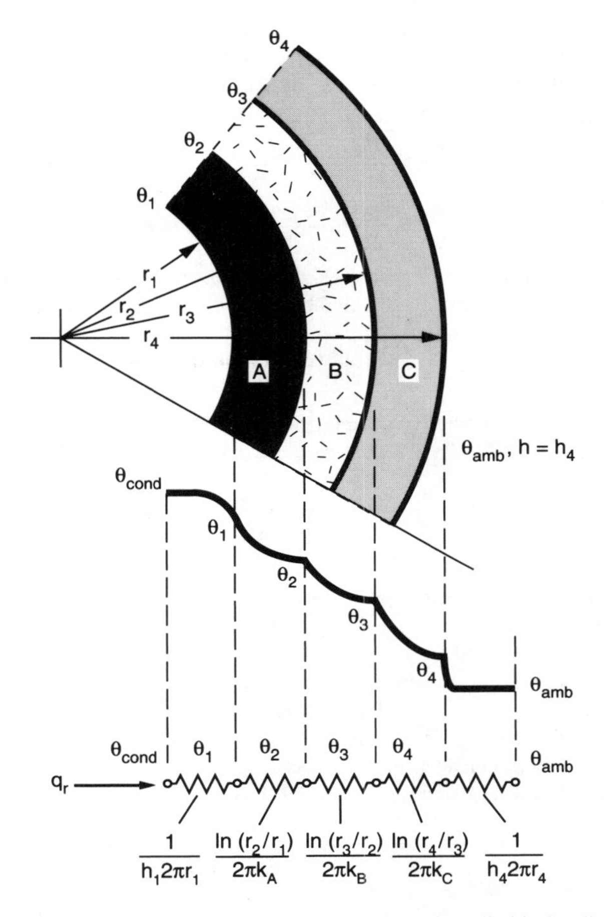

Figure 3.1 Temperature distribution for a composite cylindrical wall.

Because the conduction and convection resistances are in series and may be added up, we have

$$T_{\text{tot}} = \frac{\rho_A}{2\pi}\ln\frac{r_2}{r_1} + \frac{\rho_B}{2\pi}\ln\frac{r_3}{r_2} + \frac{\rho_C}{2\pi}\ln\frac{r_4}{r_3} + \frac{1}{2\pi r_4 h}$$

EXAMPLE 3.2

Consider cable model No. 1 (see Appendix A) installed in air. The possible existence of an optimum insulation thickness for a cable (from a thermal point of view) is suggested by the presence of the competing effects: on the one hand, the conduction resistance decreases with the reduction of insulation thickness, and on the other, a reduction in the overall cable diameter increases the external thermal resistance. Hence, an insulation thickness may exist that maximizes the heat transfer rate. The optimal insulation thickness must, of course, be sufficient to satisfy voltage insulation requirements.

Let the overall heat transfer coefficient be equal to 10 W/m$^2\cdot$ K. To facilitate the computations, we will make the following two assumptions: (1) the presence of concentric neutral wires will be neglected (they have negligible effect on cable's thermal resistance), and (2) the thermal resistance of the jacket remains constant as the diameters change. The last assumption implies that the ratio of the jacket thickness to the diameter under the jacket is constant, that is, $t_J/D_i = k$. We will also assume that the value of k is equal to the t_J/D_i ratio obtained for nominal conditions for this cable which are specified in Table A1. The insulation screen is treated as a part of the insulation.

Neglecting the thermal resistance of concentric wires, the total thermal resistance of the cable is equal to

$$T_{\text{tot}} = \frac{\rho_i}{2\pi}\ln\frac{D_i}{d_{cs}} + \frac{\rho_J}{2\pi}\ln\frac{D_i+t_J}{D_i} + \frac{1000}{\pi(D_i+t_J)h}$$

$$= \frac{\rho_i}{2\pi}\ln\frac{D_i}{d_{cs}} + \frac{\rho_J}{2\pi}\ln(1+k) + \frac{1000}{\pi D_i(1+k)h}$$

The optimum insulation thickness would be associated with the value of D_i that minimizes this resistance. This value can be obtained by solving the following equation:

$$\frac{DT_{\text{tot}}}{dD_i} = \frac{\rho_i}{2\pi D_i} - \frac{1000}{\pi h D_i^2(1+k)} = 0$$

For the cable model No. 1, the jacket has a thickness of 3.2 mm and the diameter under the jacket is 30.1 mm. The PVC jacket has a thermal resistivity of 3.5 K · m/W; hence,

$$D_i = \frac{2000}{\rho_i h(1+k)} = \frac{2000}{3.5\cdot 10\cdot\left(1+\dfrac{0.0032}{0.0301}\right)} = 51.7 \text{ mm.}$$

It remains to be verified that the thermal resistance is minimum at this point. This is achieved by computing the second derivative of T_{tot}. The second derivative is positive at $D_i = 51.7$ mm; hence, the task is accomplished.

We observe that this value is much greater than the actual diameter over the insulation for cable model No. 1. Therefore, in order to improve the heat dissipation process for this cable, we could increase the thickness of the insulation almost four times. This is, of course, not practical because of cost and weight considerations. Nevertheless, any increase of this thickness, up to the value of 51.7 mm, will result in an improvement of the heat dissipation process for this cable under assumed conditions.

3.2.2 Thermal Capacitance

Many cable rating problems are time-dependent. Consider, for example, two cable circuits operating in parallel under steady-state conditions. When one of the circuits is suddenly switched off, the second circuit will carry additional current. This sudden change of loading will cause slower changes in the temperature distribution within the cable and in the surrounding medium.

To determine the time dependence of the temperature distribution within the cable and its surroundings, we could begin by solving the appropriate form of the heat transfer equation, for example, equation (2.12). In the majority of practical cases, it is very difficult to obtain analytical solutions of this equation and, where possible, a simpler approach is preferred. One such approach may be used where temperature gradients within the cable components are small. It is termed the *lumped capacitance method.* Since this method offers the only hope for solving the heat equations analytically for a cable system, it is typically applied in practical cable rating computations. In order to satisfy the requirement that the temperature gradient within the body must be small, some components of the cable system, for example, the insulation and surrounding soil, must be subdivided into smaller entities. In Section 3.3, we discuss how this subdivision is performed for a cable's insulation and jacket. Treatment of the external cable environment is discussed in Chapter 5.

We will explain the concept of heat capacitance by considering heat transfer in the jacket of a cable installed in air. At time $t = 0$, the current is switched off and the cooling process begins. The essence of the lumped capacitance method is the assumption that the temperature of the solid is spatially uniform at any instant during the transient process. This assumption implies that the temperature gradients within the body are negligible.

From Fourier's law, heat conduction in the absence of a temperature gradient implies the existence of infinite thermal conductivity. Such a condition is clearly impossible. However, this condition can be closely approximated if the resistance to conduction within the jacket is small compared with the resistance to heat transfer between the cable and the surrounding air. Let us assume for a moment that this is, in fact, the case.

Applying equations (2.8) and (3.10a), the energy balance equation for the cable jacket with volume V now takes the form

$$-W_{\text{out}} = W_{st}$$

or

$$-h_t A\,(\theta - \theta_{\text{amb}}) = Vc\frac{d\theta}{dt}$$

where A = total area of the jacket exposed to heat transfer by convection and radiation, m^2

V = volume of the body, m^3

c = volumetric specific heat of the material, $\text{J/m}^3 \cdot {}^\circ\text{C}$

Assuming that the initial jacket temperature is $\theta(0) = \theta_0$, the solution to this equation is

$$\theta(t) = \theta_{\text{amb}} + (\theta_0 - \theta_{\text{amb}}) \exp\left[-\left(\frac{h_t A}{Vc}\right)t\right]$$

This result indicates that the difference between jacket and ambient temperatures must decay exponentially to zero as t approaches infinity. From the last equation, it is also evident that the quantity $(Vc/h_t A)$ may be interpreted as a *thermal time constant*. Utilizing equations (3.10b), this time constant can be written as

$$\tau = \left(\frac{1}{h_t A}\right)(Vc) = T \cdot Q \tag{3.11}$$

where Q is the *lumped thermal capacitance* of the jacket.

We can immediately observe that the foregoing behavior is analogous to the voltage decay that occurs when a capacitor is discharged through a resistor in an electrical RC circuit. This analogy suggests that RC electrical circuits may be used to determine the transient behavior of thermal systems. In fact, before the advent of digital computers, RC circuits were widely used to simulate the transient thermal behavior of cables. Even today, this is the preferred method for such analysis, and we will explore it further in Chapter 5.

Meanwhile, we observe that just as we found in the case of Ohm's law, an analogy between thermal and electrical networks exists, which is the following:

$$\begin{aligned} &\text{Electrical:} \quad \Delta V = \frac{Q}{C} \\ &\text{Thermal:} \quad \Delta\theta = \frac{W_{th}}{Q_{th}} \end{aligned} \tag{3.12}$$

where $C =$ electrical capacity, F
$Q_{th} =$ thermal capacity, J/°C
$Q =$ electrical charge stored in C, C
$W_{th} =$ heat stored in Q_{th}, J
$\Delta V =$ voltage rise across C due to Q, V
$\Delta\theta =$ temperature rise in Q_{th} due to W_{th}, °C

As mentioned above, an equivalent thermal network will contain only thermal resistances T and thermal capacitances Q_{th}. The thermal capacitance Q_{th} can be defined as the "ability to store the heat," and from equation (3.11) we have

$$\boxed{Q_{th} = Vc} \tag{3.13}$$

As an illustration, the formula for the thermal capacitance is established for a coaxial configuration with internal and external diameters D_1^* and D_2^* (m), respectively, which may represent, for example, a cylindrical insulation.

The thermal capacitance is calculated from equation (3.13):

$$Q_{th} = \frac{\pi}{4}\left(D_2^{*2} - D_1^{*2}\right)c \tag{3.14}$$

Thermal capacitances and resistances are used to construct a thermal ladder network to obtain the temperature distribution within the cable and its surroundings as a function of time. This topic is discussed in the next section.

3.3 CONSTRUCTION OF A LADDER NETWORK OF A CABLE

The electrical and thermal analogy discussed in Section 3.2 allows the solution of many thermal problems by applying mathematical tools well known to electrical engineers. An ability to construct a ladder network is particularly useful in transient computations. To build a ladder network, the cable is considered to extend as far as the inner surface of the soil for buried cables, and to free air for cables in air.

In constructing ladder networks dielectric losses, which are described in Chapter 6, require special attention. Although the dielectric losses are distributed throughout the insulation, it may be shown that for a single-conductor cable and also for multicore, shielded cables with round conductors, the correct temperature rise is obtained by considering for transients and steady state that all of the dielectric loss occurs at the middle of the thermal resistance between conductor and the sheath. For multicore belted cables, dielectric losses can generally be neglected, but if they are represented, the conductors are taken as the source of dielectric loss (Neher and McGrath, 1957).

Thermal capacitances of the metallic parts are placed as lumped quantities corresponding to their physical position in the cable. The thermal capacitances of materials with high thermal resistivity and possibly large temperature gradients across them (e.g., insulation and coverings) are allocated by the technique described in Section 3.3.1 below.

3.3.1 Representation of Capacitances of the Dielectric

In the early days of transient rating computations, the thermal capacity of the insulation was typically divided equally between the conductor and the sheath (Buller, 1951).

However, the thermal capacity of the insulation is not a linear function of the thickness of the dielectric. To improve the accuracy of the approximate solution using lumped constants, Van Wormer (1955) proposed a simple method for allocating the thermal capacity of the insulation between the conductor and the sheath so that the total heat stored in the insulation is represented. An assumption made in the derivation is that the temperature distribution in the insulation follows a steady-state logarithmic distribution for the period of the transient. We will present the derivation of the Van Wormer factor for a long-duration transient and simply state the result for the short-duration conditions. Whether the transient is long or short depends on the cable construction. For the purpose of transient rating computations, long-duration transients are those lasting longer than $\frac{1}{3}\Sigma T \cdot \Sigma Q$, where ΣT and ΣQ are the internal cable thermal resistance and capacitance, respectively. The methods for computing the values of T and Q are discussed in Chapter 9.

3.3.1.1 Van Wormer Coefficient for Long-duration Transients. The dielectric is represented by lumped thermal constants. The total thermal capacity of the dielectric (Q_i) is divided between the conductor and the sheath as shown in Fig. 3.2.

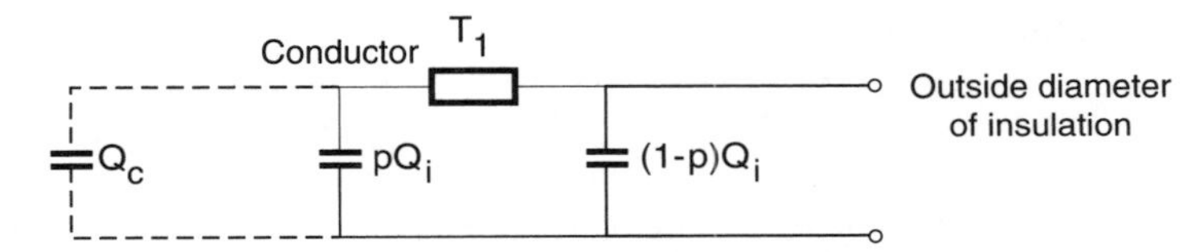

Figure 3.2 Representation of the dielectric for times greater than $\frac{1}{3}\Sigma T \cdot \Sigma Q$. T_1 = total thermal resistance of dielectric per conductor (or equivalent single-core conductor of a three-core cable; see below).

where Q_i = total thermal capacitance of dielectric per conductor (or equivalent single-core conductor of a three-core cable)

Q_c = thermal capacitance of conductor (or equivalent single-core conductor of a three-core cable).

(IEC Standard 853-2, 1989)

Note: When screening layers are present, metallic tapes are considered to be part of the conductor or sheath, while semiconducting layers (including metallized carbon paper tapes) are considered part of the insulation in thermal calculations.

The Van Wormer coefficient p is derived as follows.

Consider a part of the cable extending to the cable sheath as shown in Fig. 3.3. As indicated earlier, we will consider a cable 1 m long. Let q_i denote the capacitance of the insulation expressed per unit area, that is, $Q_i = A \cdot q_i$. If p represents the part of the insulation thermal capacity to be placed at conductor temperature θ_c and $(1 - p)$ the part to be placed at sheath temperature θ_s, the total heat stored in the insulation may be accounted for as follows:

$$pAq_i\theta_c + (1 - p)Aq_i\theta_s = q_i \int_{d_c/2}^{D_i/2} \theta_r 2\pi r \, dr \tag{3.15}$$

where D_i = external diameter of dielectric

d_c = external diameter of conductor.

From Fig. 3.3,

$$A = \frac{\pi}{4}\left(D_i^2 - d_c^2\right) \tag{3.16}$$

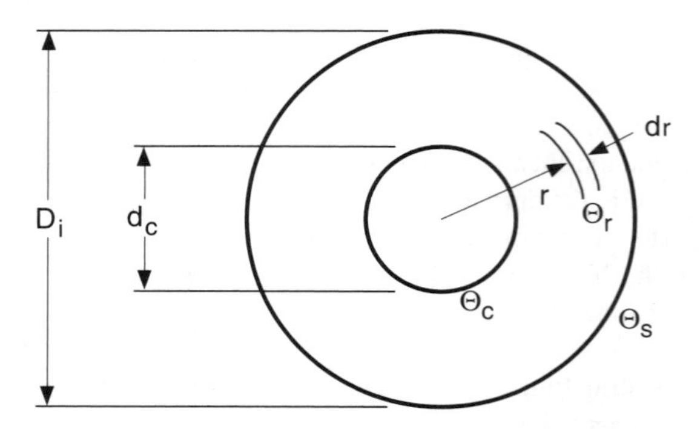

Figure 3.3 Cable conductor, insulation, and sheath.

In steady state, the temperature in the insulation at a distance r from the conductor is obtained from equation (3.2) as

$$\theta_c - \theta_r = W_c \frac{\rho_i}{2\pi} \ln \frac{r}{d_c} \tag{3.17}$$

and in particular,

$$\theta_c - \theta_s = W_c \frac{\rho_i}{2\pi} \ln \frac{D_i}{d_c} \tag{3.18}$$

where W_c = heat generated in the conductor
ρ_i = thermal resistivity of the insulation.

Combining equations (3.15)–(3.18) results in

$$pAq_i(\theta_c - \theta_s) + Aq_i\theta_s = 2\pi q_i \int_{d_c/2}^{D_i/2} \left(\theta_c - W_c \frac{\rho_i}{2\pi} \ln \frac{r}{d_c}\right) r\, dr$$

After performing integration, we obtain

$$pAq_i(\theta_c - \theta_s) + Aq_i\theta_s = 2\pi q_i \left[\frac{\theta_c\left(D_i^2 - d_c^2\right)}{8} - W_c \frac{\rho_i}{2\pi}\left(\frac{D_i^2}{8} \ln \frac{D_i}{d_c} - \frac{D_i^2 - d_c^2}{16}\right)\right]$$

Combining this equation with equations (3.15) and (3.17), the following solution for p is obtained:

$$p = \frac{1}{2 \ln\left(\frac{D_i}{d_c}\right)} - \frac{1}{\left(\frac{D_i}{d_c}\right)^2 - 1} \tag{3.19}$$

Equation (3.19) is also used to allocate the thermal capacitance of the outer covering in a similar manner to that used for the dielectric. In this case, the Van Wormer factor is given by

$$p' = \frac{1}{2 \ln\left(\frac{D_e}{D_s}\right)} - \frac{1}{\left(\frac{D_e}{D_s}\right)^2 - 1} \tag{3.20}$$

where D_e and D_s are the outer and inner diameters of the covering.

For long-duration transients and cyclic factor computations, the three-core cable is replaced by an equivalent single-core construction dissipating the same total conductor losses (Wollaston, 1949). The diameter d_c^* of the equivalent single-core conductor is obtained on the assumption that new cable will have the same thermal resistance of the insulation as the thermal resistance of a single core of the three-core cable; that is,

$$\frac{T_1}{3} = \frac{\rho_i}{2\pi} \ln \frac{D_i^*}{d_c^*}$$

where D_i^* is the same value of diameter over dielectric (under the sheath) as for the three-core cable, and T_1 is the thermal resistance of the three-conductor cable as given in Chapter 9; ρ_i is the thermal resistivity of the dielectric. Hence, we have

$$\boxed{d_c^* = D_i^* e^{-\frac{2\pi T_1}{3\rho_i}}} \tag{3.21}$$

Thermal capacitances are calculated on the following assumptions (see Example 3.8):

1. The actual conductors are considered to be completely inside the diameter of the equivalent single conductor, the remainder of the equivalent conductor being occupied by insulation.
2. The space between the equivalent conductor and the sheath is considered to be completely occupied by insulation (for fluid-filled cable, this space is filled partly by the total volume of oil in the ducts and the remainder is oil-impregnated paper).

Factor p is then calculated using the dimensions of the equivalent single-core cable, and is applied to the thermal capacitance of the insulation based on assumption (2) above.

Van Wormer has also suggested that when the insulation is thick, the thermal capacity of the insulation should be placed in part at the conductor, in part halfway through the insulation thermal resistance, and in part at the sheath.

3.3.1.2 Van Wormer Coefficient for Short-duration Transients. Short-duration transients last usually between 10 min and about 1 h. In general, for a given cable construction, the formula for the Van Wormer coefficient developed in this section applies when the duration of the transient is not greater than $\frac{1}{3}\Sigma T \cdot \Sigma Q$. The heating process for short-duration transients can be regarded as if the insulation were thick.

The method is the same as for long-duration transients except that the cable insulation is divided at diameter $d_x = \sqrt{D_i \cdot d_c}$, giving two portions having equal thermal resistances, as shown in Fig. 3.4.

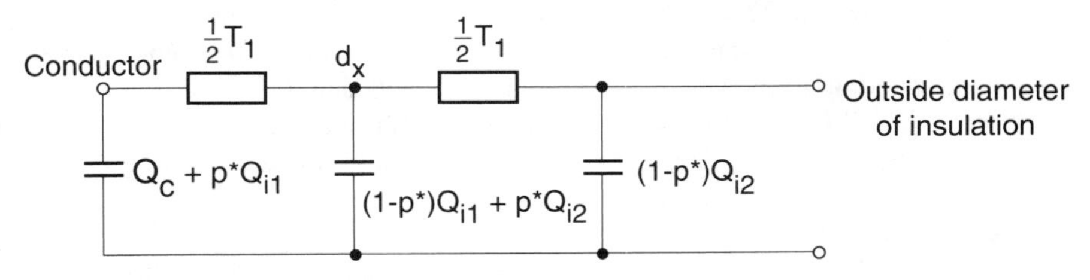

Figure 3.4 Representation of the dielectric for times less than or equal to $\frac{1}{3}\Sigma T \cdot \Sigma Q$. (IEC Standard 853-2, 1989)

The thermal capacitances Q_{i1} and Q_{i2} are defined in Chapter 9.

The Van Wormer coefficient is given by

$$p^* = \frac{1}{\ln\left(\frac{D_i}{d_c}\right)} - \frac{1}{\left(\frac{D_i}{d_c}\right) - 1} \tag{3.22}$$

EXAMPLE 3.3

Construct a ladder network for model cable No. 4 in Appendix A for a short-duration transient.

This network is shown in Fig. 3.5.

As shown in Fig. 3.5, the insulation thermal resistance is divided into two equal parts, the insulation capacitance into four parts, and the capacitance of the cable serving into two parts.

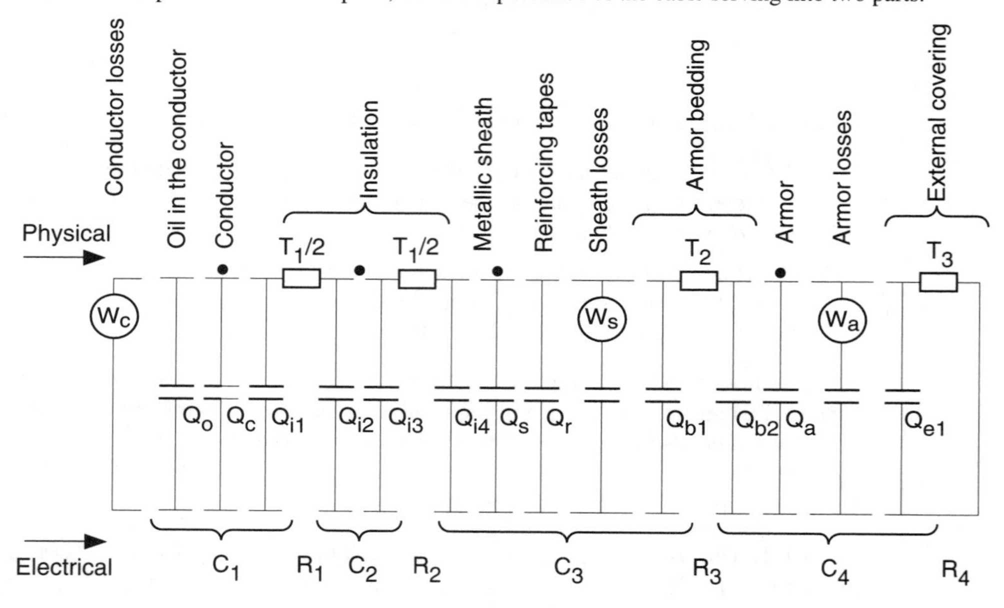

Figure 3.5 Thermal network for model cable No. 4 with electrical analogy.

A three-core cable is represented as an equivalent single-core cable as described above for durations of about $\frac{1}{2}\Sigma T \cdot \Sigma Q$ or longer (the quantities ΣT and ΣQ refer to the whole cable). However, for very short transients (i.e., for durations up to the value of the product $\Sigma T \cdot \Sigma Q$ where ΣT and ΣQ now refer to the single core), the mutual heating of the cores is neglected, and a three-core cable is treated as a single-core cable with the dimensions corresponding to the one core. For durations between these two limits, $\Sigma T \cdot \Sigma Q$ for one core and $\frac{1}{2}\Sigma T \cdot \Sigma Q$ for the whole cable, the transient is assumed to be given by a straight line interpolation in a diagram with axes of linear temperature rise and logarithmic times.

EXAMPLE 3.4

Consider cable No. 2 (Appendix A) located in a PVC duct in air shaded from solar radiation. Duct internal and external diameters are 100 and 102 mm, respectively. The external thermal resistance of this cable is composed of: (1) resistance of the air inside the duct (equal to 0.4167 K · m/W), (2) resistance of the duct (equal to 0.0216 K · m/W), and (3) resistance of the air outside the duct (equal to 0.259 K · m/W) (see Chapter 9). The external resistance per core of the air inside the duct and the duct itself is therefore

$$T_4' + T_4'' = 0.4167 + 0.0216 = 0.4383 \text{ K} \cdot \text{m/W}$$

The thermal capacitances on a per-core basis are: (1) conductor 1035 J/K · m, (2) insulation 915.6 J/K · m, (3) screen 4 J/K · m, (4) jacket 432.4 J/K · m, (5) air inside the duct 4.8 J/K · m, and (6) duct 179.8 J/K · m (see Chapter 9 for the derivation of these values). The total thermal capacitance of a single core of the cable is the sum of these values and is equal to 2571.6 J/K · m.

The time constant of a single core of cable No. 2 installed in a duct is therefore equal to 0.59 h. With the insulation and jacket thermal resistances equal to 0.307 and 0.078 K · m/W, respectively, the time constant of the entire three-core cable installed in the duct is equal to

$$\Sigma T \cdot \Sigma Q = [0.307 + 3 \cdot (0.078 + 0.4383)] \cdot [3 \cdot (1035 + 915.6 + 4 + 432.4 + 4.8 + 179.8)]/3600 = 4.0 \text{ h}$$

Hence, long-duration transients for this cable are those lasting about 2 h or more and short-duration transient last between 10 and 36 min. For durations between 36 min and 2 h, the transient is assumed to be given by straight line interpolation on linear temperature rise/logarithmic time axes.

3.3.1.3 Van Wormer Coefficient for Transients Due to Dielectric Loss. In the preceding sections, it has been assumed that the temperature rise of the conductor due to dielectric loss has reached its steady state, and that the total temperature at any time during the transient can be obtained simply by adding the constant temperature value due to the dielectric loss to the transient value due to the load current.

If a change in load current and system voltage occur at the same time, then an additional transient temperature rise due to the dielectric loss has to be calculated (Morello, 1958). For cables at voltages up to and including 275 kV, it is sufficient to assume that half of the dielectric loss is produced at the conductor and the other half at the insulation screen or sheath. The cable thermal circuit is derived by the method given above with the Van Wormer coefficient computed from equations (3.19) and (3.22) for long- and short-duration transients, respectively.

For paper-insulated cables operating at voltages higher than 275 kV, the dielectric loss is an important fraction of the total loss and the Van Wormer coefficient is calculated by (IEC 853-2, 1989)

$$p_d = \frac{\left[\left(\frac{D_i}{d_c}\right)^2 \ln\left(\frac{D_i}{d_c}\right)\right] - \left[\ln\left(\frac{D_i}{d_c}\right)\right]^2 - \frac{1}{2}\left[\left(\frac{D_i}{d_c}\right)^2 - 1\right]}{\left[\left(\frac{D_i}{d_c}\right)^2 - 1\right]\left[\ln\left(\frac{D_i}{d_c}\right)\right]^2} \tag{3.23}$$

In practical calculations for all voltage levels for which dielectric losses are important, half of the dielectric loss is added to the conductor loss and half to the sheath loss; therefore, the loss coefficients $(1 + \lambda_1)$ and $(1 + \lambda_1 + \lambda_2)$ used to evaluate thermal resistances and capacitances are set equal to 2 (see Example 5.4).

EXAMPLE 3.5

To illustrate formula 3.23, we will compute the Van Wormer coefficient for dielectric losses for cable No. 3 described in Appendix A. From Table A1, we have $D_j = 67.26$ mm and $d_c = 41.45$ mm. Hence,

$$p_d = \frac{\left[\left(\frac{67.26}{41.45}\right)^2 \ln\left(\frac{67.26}{41.45}\right)\right] - \left[\ln\left(\frac{67.26}{41.45}\right)\right]^2 - \frac{1}{2}\left[\left(\frac{67.26}{41.45}\right)^2 - 1\right]}{\left[\left(\frac{67.26}{41.45}\right)^2 - 1\right]\left[\ln\left(\frac{67.26}{41.45}\right)\right]^2} = 0.585$$

An example of the transient analysis with the voltage applied simultaneously with the current is given in Example 5.4.

3.3.2 Reduction of a Ladder Network to a Two-loop Circuit

CIGRE (1972, 1976) and later IEC (1985, 1989) introduced computational procedures for transient rating calculations employing a two-loop network with the intention of simplifying calculations and with the objective of standardizing the procedure for basic cable types. Even though with the advent and wide availability of fast desktop computers the advantage of simple computations is no longer so pronounced, there is some merit in performing some computations by hand, if only for the purpose of checking sophisticated computer programs. To perform hand computations for the transient response of a cable to a variable load, the cable ladder network has to be reduced to two sections. The procedure to perform this reduction is described below.

Consider a ladder network composed of v resistances and $(v+1)$ capacitances as shown in Fig. 3.6. If the last component of the network is a capacitance, the last capacitance Q_{v+1} is short circuited. An equivalent network, which represents the cable with sufficient accuracy, is derived with two sections $T_A Q_A$ and $T_B Q_B$ as shown in Fig. 3.7.

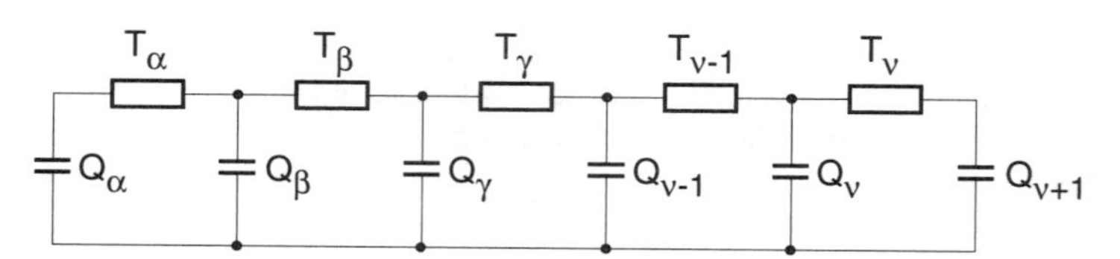

Figure 3.6 General ladder network representing a cable.

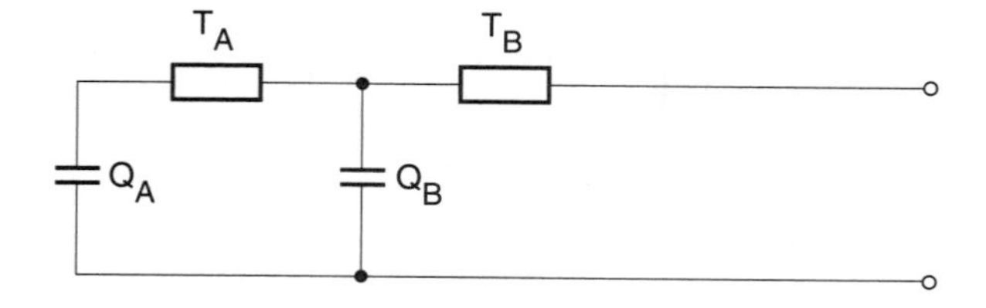

Figure 3.7 Two-loop equivalent network.

The first section of the derived network is made up of $T_A = T_\alpha$ and $Q_A = Q_\alpha$ without modification in order to maintain the correct response for relatively short durations.

The second section $T_B Q_B$ of the derived network is made up from the remaining sections of the original circuit by equating the thermal impedance of the second derived section to the total impedance of the multiple sections. The Laplace transform of the thermal impedance of the circuit $T_\beta Q_\beta$ to $T_v Q_v$ is

$$Z(s) = 1 \Bigg/ \overline{sQ_\beta + 1 \Big/ \overline{T_\beta + 1 \Big/ \overline{sQ_\gamma + 1 \Big/ \overline{T_\gamma + \cdots \Big/ \overline{+ \cdots 1 \Big/ \overline{sQ_v + \frac{1}{T_v}}}}}}} \tag{3.24}$$

and the corresponding operational equation of the simple equivalent network is

$$Z_B(s) = \frac{1}{sQ_B + \dfrac{1}{T_B}} \tag{3.25}$$

The total thermal resistance must be the same for each case; therefore,

$$T_B = T_\beta + T_\gamma + \cdots + T_v \tag{3.26}$$

Equating equations (3.24) and (3.25), an approximation for the equivalent capacitance is obtained by comparing the first degree terms in s and neglecting terms of higher degree:

$$\begin{aligned} Q_B = Q_\beta &+ \left(\frac{T_\gamma + T_\delta + \cdots + T_v}{T_\beta + T_\gamma + \cdots + T_v}\right)^2 Q_\gamma \\ &+ \left(\frac{T_\delta + T_\epsilon + \cdots + T_v}{T_\beta + T_\gamma + \cdots + T_v}\right)^2 Q_\delta + \cdots + \left(\frac{T_v}{T_\beta + T_\gamma + \cdots + T_v}\right)^2 Q_v \end{aligned} \tag{3.27}$$

Even though formulas (3.26)–(3.27) are straightforward, a great deal of care is required when the equivalent thermal resistances and capacitances are computed in the case when sheath, armor, and pipe losses are present (IEC, 1985). This is because the location of these losses inside the original network has to be carefully taken into account. The following section discusses several examples which illustrate this point.

3.3.3 Two-section Ladder Networks for Some Selected Cable Constructions

As indicated above, in order to make computations reasonably convenient and uniform throughout the industry, CIGRE (1972) proposed that two-section networks be used in transient rating computations. Earlier researchers (Buller, 1951; Van Wormer, 1955; Neher, 1964) used two-section networks for transient analysis as well since they afforded a relatively simple mathematical solution. In this section, we will derive several such networks for common cable constructions. It should be noted that, in a cable, all thermal resistances, capacitances, and losses are to be calculated as quantities per conductor, or per equivalent conductor in the case of multicore cables.

In a two-section network representing the cable, the first section always includes the thermal capacitance of the conductor and the inner portion of the dielectric along with the thermal resistance of the dielectric (or a part thereof for short-duration transients). The second section includes thermal resistances and capacitances of the remaining components of the cable.

EXAMPLE 3.6

3.3.3.1 Solid and Self-contained Fluid-filled Paper-insulated Cables, Extruded Cables, and Thermally Similar Constructions. We will construct a two-loop equivalent network for model cable No. 1 assuming: (1) a short-duration transient, and (2) a long-duration transient.

(1) *Short-duration transient*

From Table A1, we observe that, for this cable, short-duration transients are those lasting half an hour or less. The diagram of the full network for a short-duration transient is shown in Fig. 3.8.

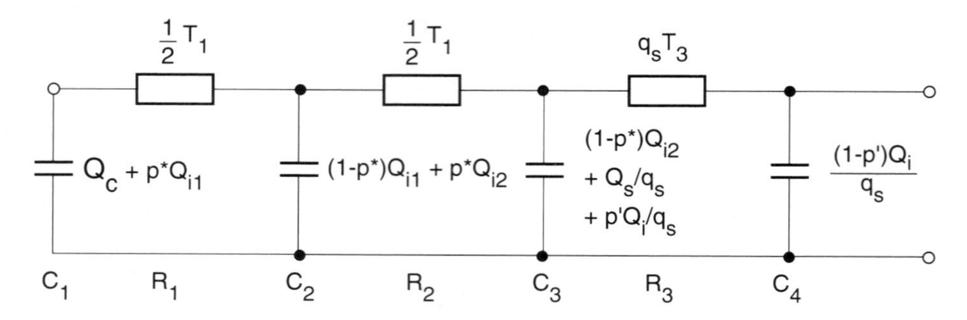

Figure 3.8 Network diagram for cable No. 1 for short-duration transients.

The method of dividing insulation and jacket capacitances into parts is discussed in Section 3.3.1. Before we apply the reduction procedure, we combine parallel capacitances into four equivalent capacitances. In the equivalent network, only conductor losses are represented. Therefore, to account for the presence of sheath losses, the thermal resistances beyond the sheath must be multiplied, and the thermal capacitances divided by the ratio of the losses in the conductor and the sheath to the conductor losses. By performing these multiplications and divisions, the time constants of the thermal circuits involved are not changed. Thus,

$$Q_1 = Q_c + p^* Q_{i1}, \quad Q_2 = (1 - p^*) Q_{i1} + p^* Q_{i2}$$

$$Q_3 = (1 - p^*) Q_{i2}, \quad Q_4 = \frac{Q_s - Q_{j1}}{1 + \lambda_1} \quad Q_5 = \frac{Q_{j2}}{1 + \lambda_1} \tag{3.28}$$

To compute numerical values, we will require expressions for Q_{i1} and Q_{i2}. These expressions are given in Chapter 9. The numerical values are as follows: $Q_{i1} = 763$ J/K · m and $Q_{i2} = 453.9$ J/K · m. With these values and with the additional numerical values in Table A1, $D_i = 30.1$ mm, $d_c = 20.5$ mm, $D_e = 35.8$ mm, and $D_s = 31.4$ mm, we have

$$p^* = \frac{1}{\ln\left(\frac{D_i}{d_c}\right)} - \frac{1}{\left(\frac{D_i}{d_c}\right) - 1} = \frac{1}{\ln\frac{30.1}{20.5}} - \frac{1}{\frac{30.1}{20.5} - 1} = 0.468$$

$$p' = \frac{1}{2\ln\left(\frac{D_e}{D_s}\right)} - \frac{1}{\left(\frac{D_e}{D_s}\right)^2 - 1} = \frac{1}{2\ln\left(\frac{35.8}{31.4}\right)} - \frac{1}{\left(\frac{35.8}{31.4}\right)^2 - 1} = 0.478$$

$$Q_1 = 1035 + 0.468 \cdot 763 = 1392.1 \text{ J/K} \cdot \text{m}$$

$$Q_2 = (1 - 0.468)763 + 0.468 \cdot 453.9 = 618.3 \text{ J/K} \cdot \text{m}$$

$$Q_3 = (1 - 0.468)453.9 = 241.5 \text{ J/K} \cdot \text{m}, \quad Q_4 = \frac{4 + 0.478 \cdot 394.8}{1.09} = 176.8 \text{ J/K} \cdot \text{m}$$

$$Q_5 = \frac{(1 - 0.478)394.8}{1.09} = 189.1 \text{ J/K} \cdot \text{m}$$

The final capacitance Q_5 is omitted in further analysis because the transient for the cable response is calculated on the assumption that the output terminals on the right-hand side are short circuited.

Since the first section of the network in Fig. 3.8 represents the conductor, and in rating computations the conductor temperature is of interest, the equivalent network will have the first section equal to the first section of the full network; that is,

$$T_A = \tfrac{1}{2}T_1 \quad \text{and} \quad Q_A = Q_1 \tag{3.29}$$

From equation (3.26), we have

$$T_B = \tfrac{1}{2}T_1 + (1 + \lambda_1)T_3 \tag{3.30}$$

Thermal capacitance of the second part is obtained by applying equation (3.27):

$$Q_B = Q_2 + \left[\frac{(1 + \lambda_1)T_3}{\frac{1}{2}T_1 + (1 + \lambda_1)T_3}\right]^2 (Q_3 + Q_4) \tag{3.31}$$

The sheath loss factor and thermal resistances for this cable are given in Table A1 as $\lambda_1 = 0.09$, $T_1 = 0.214$ K · m/W, and $T_3 = 0.104$ K · m/W. Substituting numerical values in equations (3.29)–(3.31), we obtain

$$T_A = 0.107 \text{ K} \cdot \text{m/W}, \quad Q_A = 1392.1 \text{ J/K} \cdot \text{m}$$

$$T_B = 0.107 + 1.09 \cdot 0.104 = 0.220 \text{ K} \cdot \text{m/W}$$

$$Q_B = 618.3 + \left(\frac{1.09 \cdot 0.104}{0.107 + 1.09 \cdot 0.104}\right)^2 (241.5 + 176.8) = 729.4 \text{ J/K} \cdot \text{m}$$

(2) *Long-duration transients*

Long-duration transient for this cable are those lasting longer than 0.5 h. The appropriate diagram is shown in Fig. 3.9.

In this case, we have

$$T_A = T_1 \qquad T_B = (1 + \lambda_1)T_3 \tag{3.32}$$

The insulation and jacket are split into two parts with the Van Wormer coefficients given by equations (3.19) and (3.20), respectively. Since the last part of the jacket capacitance is short circuited (see Section 3.3.2), Q_A and Q_B are simply obtained as the sums of relevant capacitances:

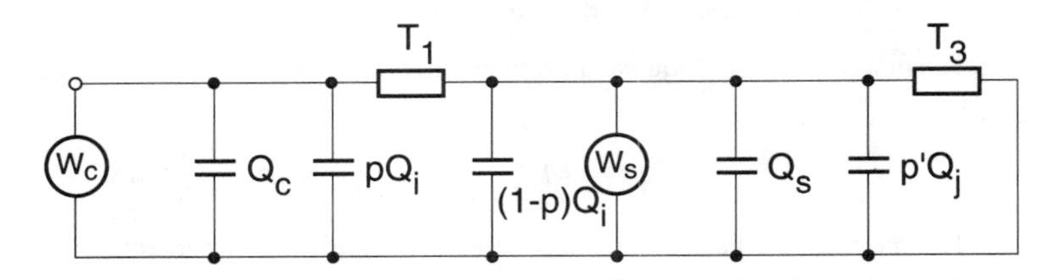

Figure 3.9 Network diagram for a long-duration transient for model cable No. 1.

$$Q_A = Q_c + pQ_i \qquad Q_B = (1-p)Q_i + \frac{Q_s + p'Q_j}{1+\lambda_1} \tag{3.33}$$

Substituting numerical values, we obtain

$$p = \frac{1}{2\ln\left(\frac{D_i}{d_c}\right)} - \frac{1}{\left(\frac{D_i}{d_c}\right)^2 - 1} = \frac{1}{2\ln\left(\frac{30.1}{20.5}\right)} - \frac{1}{\left(\frac{30.1}{20.5}\right)^2 - 1} = 0.437$$

$$T_A = 0.214 \text{ K} \cdot \text{m/W}, \quad Q_A = 1035 + 0.437 \cdot 915.6 = 1435.1 \text{ J/K} \cdot \text{m}$$

$$T_B = 1.09 \cdot 0.104 = 0.113 \text{ K} \cdot \text{m/W}$$

$$Q_B = (1 - 0.437)915.6 + \frac{4 + 0.478 \cdot 394.8}{1.09} = 692.3 \text{ J/K} \cdot \text{m}$$

EXAMPLE 3.7

3.3.3.2 High-pressure Fluid-filled Pipe-type (HPFF) Cables. Consider model cable No. 3 described in Appendix A. The parameters of this cable are: $D_i = 67.26$ mm, $d_c = 41.45$ mm, $D_e = 244.48$ mm, and $D_o = 219.08$ mm, $\lambda_1 = 0.010$, $\lambda_2 = 0.311$, $T_1 = 0.422$ K · m/W, $T_2 = 0.082$ K · mW, and $T_3 = 0.017$ K · m/W.

The thermal circuits are quite different for short- and long-duration transients.

(1) *Short-duration transients*

Short-duration transients for this cable are those lasting up to 6 h. The thermal circuit for short duration transients is shown in Fig. 3.10.

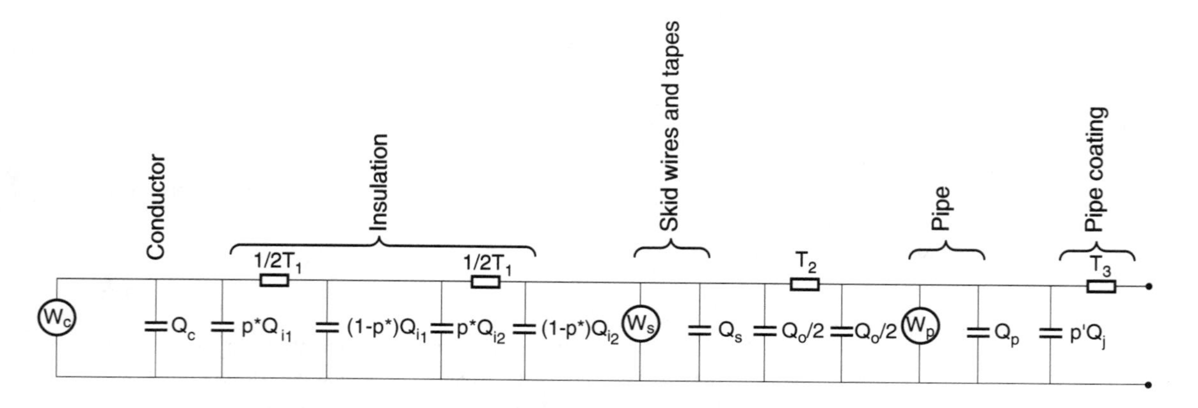

Figure 3.10 Ladder network for an HPLF cable for short-duration transients.

The first section of the equivalent network is composed of the conductor and half of the insulation; that is,

$$T_A = \tfrac{1}{6}T_1 \qquad Q_A = (Q_c + p^* Q_{i1})/3 \tag{3.34}$$

The thermal resistance of the second part is simply the summation of the remaining thermal resistances with appropriate multiplying factors to take account of losses in skid wires and screens as well as those in the pipe:

$$T_B = \tfrac{1}{6}T_1 + (1+\lambda_1)T_o + (1+\lambda_1+\lambda_2)T_3 \tag{3.35}$$

To compute Q_B, we apply equation (3.27):

$$Q_B = \frac{(1-p^*)Q_{i1} + p^* Q_{i2}}{3} + \left(\frac{T_B - \frac{1}{6}T_1}{T_B}\right)^2 \left[(1-p^*)Q_{i2}/3 + \frac{Q_s/3 + \frac{1}{2}Q_o}{1+\lambda_1}\right] + \left(\frac{(1+\lambda_1+\lambda_2)T_3}{T_B}\right)^2 \left(\frac{\frac{1}{2}Q_o}{1+\lambda_1} + \frac{Q_p + p'Q_j}{1+\lambda_1+\lambda_2}\right) \quad (3.36)$$

where Q_{i1} and Q_{i2} are defined in Chapter 9, T_B is given by (3.26), Q_s, Q_0, Q_p, and Q_j denote skid wire, oil, pipe, and pipe covering capacitances, respectively.

The numerical values of Q_{i1} and Q_{i2} are 1680.5 (J/K · m) and 2726.9 (J/K · m), respectively. Capacitance of the oil is equal to 38570 (J/K · m) (see Section 9.7.1). Therefore,

$$p^* = \frac{1}{\ln\left(\frac{D_i}{d_c}\right)} - \frac{1}{\left(\frac{D_i}{d_c}\right) - 1} = \frac{1}{\ln\frac{67.26}{41.45}} - \frac{1}{\frac{67.26}{41.45} - 1} = 0.460$$

$$p' = \frac{1}{2\ln\left(\frac{D_e}{D_o}\right)} - \frac{1}{\left(\frac{D_e}{D_o}\right)^2 - 1} = \frac{1}{2\ln\frac{244.48}{219.08}} - \frac{1}{\left(\frac{244.48}{219.08}\right)^2 - 1} = 0.482$$

$$T_A = \tfrac{1}{2}T_1/3 = \tfrac{1}{2}0.422/3 = 0.07 \text{ K} \cdot \text{m/W},$$

$$Q_A = (3484.5 + 0.46 \cdot 1680.5)/3 = 1419.2 \text{ J/K} \cdot \text{m}$$

$$T_B = 0.07 + 1.01 \cdot 0.082 + 1.321 \cdot 0.017 = 0.175 \text{ K} \cdot \text{m/W}$$

$$Q_B = (1-0.46)1680.5/3 + 0.46 \cdot 2726.9/3 + \left(\frac{0.175-0.07}{0.175}\right)^2 \left((1-0.46)2726.9/3 + \frac{1.2/3 + 0.5 \cdot 38\,570}{1.01}\right) + \left(\frac{1.321 \cdot 0.017}{0.175}\right)^2 \left(\frac{0.5 \cdot 38\,570}{1.01} + \frac{16\,126.4 + 0.482 \cdot 15\,720.9}{1.321}\right) = 8381.2 \text{ J/K} \cdot \text{m}.$$

(2) *Long-duration transient*

For a long-duration transient, oil thermal resistance is split in half as shown in Fig. 3.11.

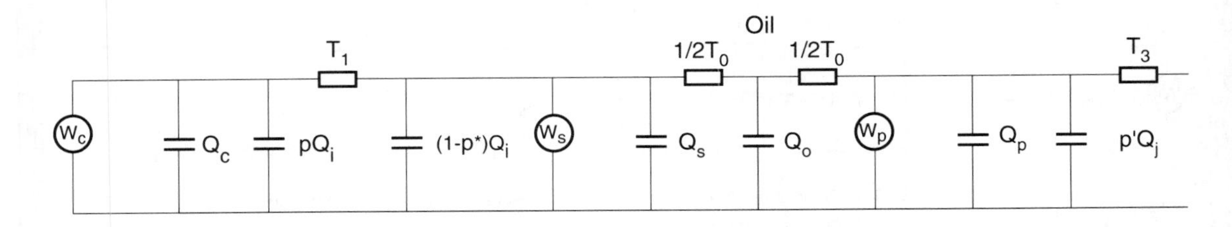

Figure 3.11 Ladder network for an HPLF cable for long-duration transients.

The parameters of an equivalent network are obtained from

$$T_A = T_1/3, \qquad Q_A = (Q_c + pQ_i)/3 \quad (3.37)$$

$$T_B = (1+\lambda_1)T_o + (1+\lambda_1+\lambda_2)T_3 \quad (3.38)$$

$$Q_B = (1-p)Q_i/3 + \frac{Q_s}{3(1+\lambda_1)} + \left[\frac{\frac{1}{2}(1+\lambda_1)T_o + (1+\lambda_1+\lambda_2)T_3}{T_B}\right]^2 \left(\frac{Q_o}{1+\lambda_1}\right)$$
$$+ \left[\frac{(1+\lambda_1+\lambda_2)T_3}{T_B}\right]^2 \frac{Q_p + p'Q_j}{1+\lambda_1+\lambda_2} \tag{3.39}$$

Using the numerical values obtained from Table A1, we have

$$p = \frac{1}{2\ln\left(\frac{D_i}{d_c}\right)} - \frac{1}{\left(\frac{D_i}{d_c}\right)^2 - 1} = \frac{1}{2\ln\frac{67.26}{41.45}} - \frac{1}{\left(\frac{67.26}{41.45}\right)^2 - 1} = 0.421$$

$T_A = 0.422/3 = 0.141$ K · m/W, $Q_A = (3484.5 + 0.421 \cdot 1458.4)/3 = 1366.2$ J/K · m

$T_B = 1.01 \cdot 0.082 + 1.321 \cdot 0.017 = 0.105$ K · m/W

$$Q_B = (1-0.421)1458.4/3 + \frac{1.2}{3 \cdot 1.01} + \left(\frac{0.5 \cdot 1.01 \cdot 0.082 + 1.321 \cdot 0.017}{0.105}\right)^2 \left(\frac{38\,570}{1.01}\right)$$
$$+ \left(\frac{1.321 \cdot 0.017}{0.105}\right)^2 \frac{16\,126.4 + 0.482 \cdot 15\,720.9}{1.321} = 15\,230.6 \text{ J/K} \cdot \text{m}$$

EXAMPLE 3.8

3.3.3.3 Cables in Ducts. Consider cable No. 2 (Appendix A) installed in a PVC duct as described in Example 3.4. The parameters of this cable are: $D_i = 30.1$ mm, $d_c = 20.5$ mm, $D_e = 72.9$ mm, and $D_s = 65.9$ mm, $\lambda_1 = 0.022$, $T_1 = 0.307$ K · m/W, $T_3 = 0.078$ K · m/W, and the thermal resistances of the air in the duct and the duct itself are equal to 0.4167 and 0.0216 K · m/W, respectively.

The following formulas, derived similarly as in previous examples, apply in this case:

(1) *Short-duration transients*

The diagram in Fig. 3.12 represents the ladder network for this case.

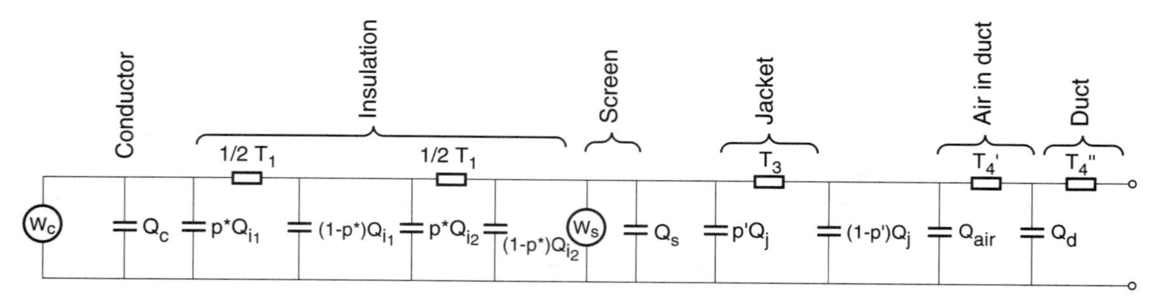

Figure 3.12 Ladder network for an equivalent single-core cable for cable No. 2 in duct for short-duration transient.

$$T_A = \tfrac{1}{2}T_1 \qquad Q_A = Q_c + p^* Q_{i1} \tag{3.40}$$

$$T_B = \tfrac{1}{2}T_1 + (1+\lambda_1)\left(T_3 + T_4' + T_4''\right) \tag{3.41}$$

To compute Q_B, we apply equation (3.27):

$$Q_B = (1-p^*)Q_{i1} + p^* Q_{i2} + \left[\frac{T_B - \frac{1}{2}T_1}{T_B}\right]^2 \left[(1-p^*)Q_{i2} + \frac{Q_s + p' Q_j}{1+\lambda_1}\right]$$
$$+ \left[\frac{(1+\lambda_1)(T_4' + T_4'')}{T_B}\right]^2 \left[\frac{(1-p')Q_j}{1+\lambda_1}\right] + \left[\frac{(1+\lambda_1)T_4''}{T_B}\right]^2 \frac{Q_d}{1+\lambda_1} \tag{3.42}$$

where T_B is given by equation (3.41), T_4' and T_4'' are the thermal resistances of the air space in duct and the duct itself, respectively, and Q_d is the capacitance of the duct.

To obtain numerical values, we recall that for short-duration transients, the mutual heating of the cores is neglected, and equations (3.40)–(3.42) apply to a single core of the cable. The time limit for the short-duration transient is equal to $\Sigma T \cdot \Sigma Q$ where ΣT and ΣQ refer to one core only. Short-duration transients for cable No. 2 located in duct in air last less then 36 min. The long-duration transients are those longer than 2 h (see Example 3.4). For transients having durations between these two values, a network diagram for short-duration transients must be solved. The numerical values required for this case are as follows:

$$p^* = \frac{1}{\ln\left(\frac{D_i}{d_c}\right)} - \frac{1}{\left(\frac{D_i}{d_c}\right) - 1} = \frac{1}{\ln\frac{30.1}{20.5}} - \frac{1}{\frac{30.1}{20.5} - 1} = 0.468$$

The Van Wormer coefficient for the cable jacket is equal to

$$p' = \frac{1}{2\ln\left(\frac{D_e}{D_s}\right)} - \frac{1}{\left(\frac{D_e}{D_s}\right)^2 - 1} = \frac{1}{2\ln\frac{72.9}{65.9}} - \frac{1}{\left(\frac{72.9}{65.9}\right)^2 - 1} = 0.483$$

For the duct dimensions as specified above, the values of T_4' and T_4'' are equal to 0.4167 and 0.0216 K · m/W, respectively. The thermal capacitances of the first and second part of the insulation are as follows (see Chapter 9 for the derivation of these values):

$$Q_{i1} = 371.0 \text{ J/K} \cdot \text{m} \qquad Q_{i2} = 544.7 \text{ J/K} \cdot \text{m}$$

With these values and the remaining parameters specified in Table A1, we obtain

$$T_A = 0.307/6 = 0.051 \text{ K} \cdot \text{m/W} \quad Q_A = (1035 + 0.468 \cdot 371)/3 = 402.9 \text{ J/K} \cdot \text{m}$$

$$T_B = \tfrac{1}{2} \cdot 0.307/3 + 1.022(0.078 + 0.4167 + 0.0216) = 0.580 \text{ K} \cdot \text{m/W}$$

$$Q_B = (1-0.468)371/3 + 0.468 \cdot 544.7/3 + \left(\frac{0.580 - 0.312/6}{0.580}\right)^2$$
$$+ \left[(1-0.468)544.7/3 + \frac{4 + 0.483 \cdot 1297.3}{1.022}\right]$$
$$+ \left[\frac{1.022(0.4167 + 0.0216)}{0.580}\right]^2 \frac{(1-0.483)1297.3}{1.022}$$
$$+ \left(\frac{1.022 \cdot 0.0216}{0.580}\right)^2 \frac{539.4}{1.022} = 1134.4 \text{ J/K} \cdot \text{m}$$

(2) *Long-duration transients*

Long-duration transients for this cable last 2 h or more. The diagram in Fig. 3.13 represents the network diagram for this case.

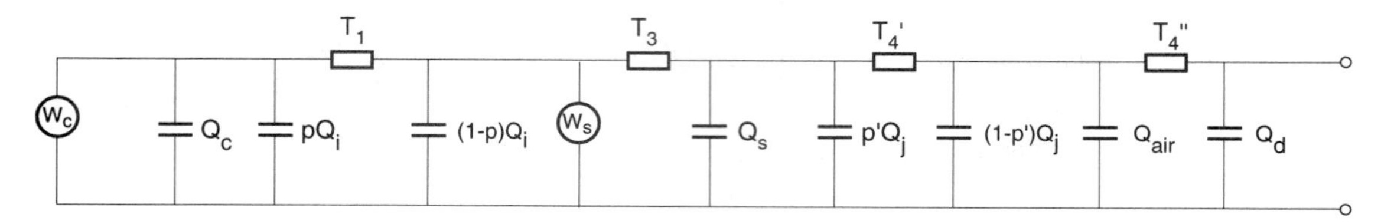

Figure 3.13 Ladder network for an equivalent single-core cable for cable No. 2 in duct for long-duration transient.

$$T_A = T_1 \qquad Q_A = Q_c + pQ_i \tag{3.43}$$

$$T_B = (1+\lambda_1)(T_3 + T_4' + T_4'') \tag{3.44}$$

$$Q_B = (1-p)Q_i + \frac{Q_s + p'Q_j}{1+\lambda_1} + \left[\frac{(1+\lambda_1)(T_4' + T_4'')}{T_B}\right]^2 \left(\frac{(1-p')Q_j}{1+\lambda_1}\right) + \left[\frac{(1+\lambda_1)T_4''}{T_B}\right]^2 \frac{Q_d}{1+\lambda_1} \tag{3.45}$$

To obtain numerical values, we will represent the three-conductor cable as an equivalent single-core cable. Thermal resistance of a single cable is equal to one third of the value given in Table A1. The equivalent conductor diameter is obtained from equation (3.21) as

$$d_c^* == D_i^* e^{-\frac{2\pi T_1}{3\rho_i}} = 65.9e^{-\frac{2\pi 0.307}{3\cdot 3.5}} = 54.8 \text{ mm}$$

The capacitance of the equivalent conductor is equal to the sum of the capacitances of the three conductors plus the capacitance of that portion of the insulation which is enclosed within the perimeter of the equivalent conductor.

The insulation area enclosed within the equivalent conductor is equal to

$$S_{\text{ins}} = \pi \cdot (54.8^2 - 3 \cdot 20.5^2)/4 = 1368.4 \text{ mm}^2$$

Hence, from equation (3.13) and the material properties as given in Table 9.1, the capacitance of the equivalent conductor is given by

$$Q_c = 3 \cdot 1035 + 2.4 \cdot 1368.4 = 6389.2 \text{ J/K} \cdot \text{m}$$

The equivalent cable has a diameter over insulation equal to 65.9 mm. The capacitance of the insulation is obtained from equation (3.14) as

$$Q_i = \frac{\pi}{4}(65.9^2 - 54.8^2)2.4 = 2525.4 \text{ J/K} \cdot \text{m}$$

The Van Wormer coefficient is thus equal to

$$p = \frac{1}{2\ln\left(\frac{D_i}{d_c}\right)} - \frac{1}{\left(\frac{D_i}{d_c}\right)^2 - 1} = \frac{1}{2\ln\frac{65.9}{54.8}} - \frac{1}{\left(\frac{65.9}{54.8}\right)^2 - 1} = 0.469$$

The equivalent network parameters are now obtained from equations (3.43)–(3.45):

$$T_A = 0.307/3 = 0.102 \text{ K} \cdot \text{m/W} \quad Q_A = (6368 + 0.469 \cdot 2525.4) = 7552.4 \text{ J/K} \cdot \text{m}$$

$$T_B = 1.022(0.078 + 0.4167 + 0.0216) = 0.5277 \text{ K} \cdot \text{m/W}$$

$$Q_B = (1 - 0.469)2525.4 + \frac{0.483 \cdot 1297.3}{1.022} + \left[\frac{1.022(0.4167 + 0.0216)}{0.5277}\right]^2 \frac{(1 - 0.483)1297.3}{1.022}$$

$$+ \left(\frac{1.022 \cdot 0.0216}{0.5277}\right)^2 \frac{539.4}{1.022} = 2427.9 \text{ J/K} \cdot \text{m}$$

The evaluation of the losses, thermal resistances and capacitances shown in the ladder network is discussed in detail in Chapters 6–9. Once all the parameters for the ladder network are determined, the network equations are solved as with electrical RC networks. Solution of the network equations yields the temperatures at various nodes of the network corresponding to points in the cable. This task is fairly straightforward for steady-state computations and somewhat more laborious for transient analysis.

In closing this chapter, we note that the lumped parameter network is not the only model that can be applied to solve heat conduction equations. In some publications, fully distributed models using Fourier integrals have been used. In the fully distributed model, the temperature is a function of time and the radial distance from the cable conductor. The lumped parameter model allows the computation of the temperature of each cable region as a function of time only. Since this book is focused on the most common methods used by cable engineers, we will consider the lumped parameter model only. Readers interested in an example where Fourier integrals are used for transient cable ratings are referred to Bernath and Olfe (1986).

REFERENCES

Bernath, A., and Olfe, D. B., (July 1986), "Cyclic temperature calculations and measurements for underground power cables," *IEEE Trans. Power Delivery*, vol. PWRD-1, pp. 13–21.

Bernath, A., Olfe, D. B., and Martin, F. (July 1986), "Short term transient temperature calculations and measurements for underground power cables," *IEEE Trans. Power Delivery*, vol. PWRD-1, pp. 22–27.

Buller, F. H. (1951), "Thermal transients on buried cables," *Trans. Amer. Inst. Elect. Eng.*, vol. 70, pp. 45–52.

CIGRE (Oct. 1972), "Current ratings of cables for cyclic and emergency loads. Part 1. Cyclic ratings (load factor less than 100%) and response to a step function," *Electra*, no. 24, pp. 63–96.

CIGRE (Jan. 1976), "Current ratings of cables for cyclic and emergency loads. Part 2. Emergency ratings and short duration response to a step function," *Electra*, no. 44, pp. 3–16.

IEC Standard (1985), "Calculation of the cyclic and emergency current ratings of cables. Part 1: Cyclic rating factor for cables up to and including 18/30 (36) kV," Publication 853-1.

IEC Standard (1989), "Calculation of the cyclic and emergency current ratings of cables. Part 2: Cyclic rating factor of cables greater than 18/30 (36) kV and emergency ratings for cables of all voltages," Publication 853-2.

Neher, J. H., and McGrath, M. H. (1957), "Calculation of the temperature rise and load capability of cable systems," *AIEE Trans.*, vol. 76, part 3, pp. 755–772.

Neher, J. H. (1964), "The transient temperature rise of buried power cable systems," *IEEE Trans.*, vol. PAS-83, pp. 102–111.

Pashkis, V., and Baker, H. D. (1942), "A method for determining the unsteady-state heat transfer by means of an electrical analogy," *ASME Trans.*, vol. 104, pp. 105–110.

Van Wormer, F. C. (1955), "An improved approximate technique for calculating cable temperature transients," *Trans. Amer. Inst. Elect. Eng*, vol. 74, part 3, pp. 277–280.

Wollaston, F. O. (1949), "Transient temperature phenomena of 3 conductor cables," *AIEE Trans.*, vol. 68, part 2, pp. 1248–1297.

4

Rating Equations—Steady-State Conditions

The current-carrying capability of a cable system will depend on several parameters. The most important of these are (1) the number of cables and the different cable types in the installation under study, (2) the cable construction and materials used for the different cable types, (3) the medium in which the cables are installed, (4) cable locations with respect to each other and with respect to the earth surface, and (5) the cable bonding arrangement. For some cable constructions, the operating voltage may also be of significant importance. All of the above issues are taken into account; some of them explicitly, the others implicitly, in the rating equations presented in this chapter. The lumped parameter network representation of the cable system discussed in Chapter 3 is used for the development of steady-state and transient rating equations. These equations are developed for a single cable either with one core or with multiple cores. However, they can be applied to multicable installations, for both equally and unequally loaded cables, by suitably selecting the value of the external thermal resistance as shown in Section 9.6.

The development of cable rating equations is quite different for steady-state and transient conditions. Therefore, we will discuss these issues in two separate chapters, Chapter 4 and Chapter 5.

4.1 BURIED CABLES WITH NO MOISTURE MIGRATION IN THE SOIL

Steady-state rating computations involve solving the equation for the ladder network shown in Fig. 3.1 with the thermal capacitances neglected. The resulting diagram which also includes the external thermal resistance is shown in Fig. 4.1.

The unknown quantity is either the conductor current I or its operating temperature θ_c. In the first case, the maximum operating conductor temperature is given, and in the second

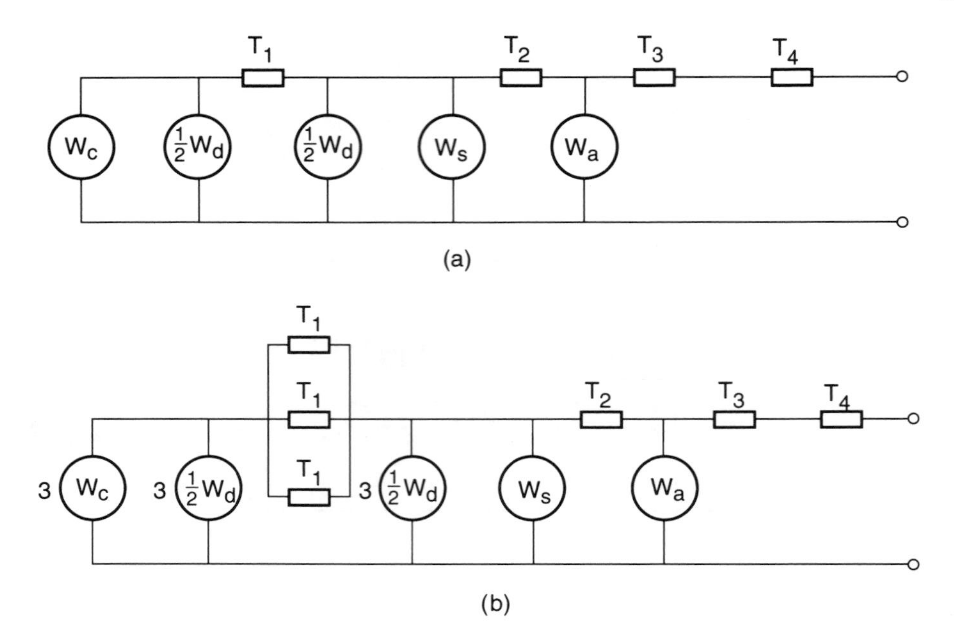

Figure 4.1 The ladder diagram for steady-state rating computations. (a) Single-core cable. (b) Three-core cable.

case, the conductor current is specified. Since losses occur at several positions in the cable system (for this lumped parameter network), the heat flow in the thermal circuit shown in Fig. 4.1 will increase in steps. Thus, the total joule loss W_I in a cable can be expressed as

$$W_I = W_c + W_s + W_a = W_c(1 + \lambda_1 + \lambda_2) \tag{4.1}$$

where W_c, W_s, and W_a are conductor, sheath, and armor losses, respectively. The quantity λ_1 is called the sheath loss factor, and is equal to the ratio of the total losses in the metallic sheath to the total conductor losses. Similarly, λ_2 is called the armor loss factor, and is equal to the ratio of the total losses in the metallic armor to the total conductor losses. Incidentally, it is convenient to express all heat flows caused by the joule losses in the cable in terms of the loss per meter of the conductor.

Referring now to the diagram in Fig. 4.1, and remembering the analogy between the electrical and thermal circuits, we can write the following expression for $\Delta\theta$, the conductor temperature rise above the ambient temperature:

$$\begin{aligned}\Delta\theta = (W_c + \tfrac{1}{2}W_d)T_1 &+ [W_c(1 + \lambda_1) + W_d]nT_2 \\ &+ [W_c(1 + \lambda_1 + \lambda_2) + W_d]n(T_3 + T_4)\end{aligned} \tag{4.2}$$

where W_c, λ_1, and λ_2 are defined above, and n is the number of load-carrying conductors in the cable (conductors of equal size and carrying the same load). W_d represents dielectric losses of which the evaluation is discussed in Chapter 6. The ambient temperature is the temperature of the surrounding medium under normal conditions at the location where the cables are installed, or are to be installed, including any local sources of heat, but not the increase of temperature in the immediate neighborhood of the cable due to the heat arising therefrom. T_1, T_2, T_3, and T_4 are the thermal resistances where T_1 is the thermal resistance per unit length between one conductor and the sheath, T_2 is the thermal resistance per unit length of the bedding between sheath and armor, T_3 is the thermal resistance per unit length of the external serving of the cable, and T_4 is the thermal resistance per unit length between the cable surface and the surrounding medium.

The permissible current rating is obtained from equation (4.2). Remembering that $W_c = I^2R$, we have

$$I = \left[\frac{\Delta\theta - W_d[0.5T_1 + n(T_2 + T_3 + T_4)]}{RT_1 + nR(1+\lambda_1)T_2 + nR(1+\lambda_1+\lambda_2)(T_3+T_4)}\right]^{0.5} \tag{4.3}$$

where R is the ac resistance per unit length of the conductor at maximum operating temperature.

EXAMPLE 4.1

Determine the steady-state rating of a cable system composed of three single-core cables in flat formation using cable model No.1 in Appendix A. The parameters of this system are specified in Table A1 as $\lambda_1 = 0.09$, $T_1 = 0.214$ K·m/W, $T_3 = 0.104$ K·m/W, and $T_4 = 1.933$ K·m/W. Since for this cable dielectric losses are negligible, applying equation (4.3) results in

$$I = \left[\frac{75}{7.81\cdot 10^{-5}(0.214 + 1\cdot(1+0.09)\cdot 0 + 1\cdot(1+0.09)(0.104+1.933))}\right]^{0.5} = 629 \text{ A}$$

Equation (4.2) is often written in a simpler form which clearly distinguishes between internal and external heat transfers in the cable. Denoting

$$T = \frac{T_1}{n} + (1+\lambda_1)T_2 + (1+\lambda_1+\lambda_2)T_3$$
$$T_d = \frac{T_1}{2n} + T_2 + T_3 \tag{4.4}$$

equation (4.2) becomes

$$\Delta\theta = n(W_cT + W_tT_4 + W_dT_d) \tag{4.5}$$

where W_t are the total losses generated in the cable defined by

$$W_t = W_I + W_d = W_c(1+\lambda_1+\lambda_2) + W_d \tag{4.6}$$

and T computed from (4.4) is an equivalent cable thermal resistance. This is an internal thermal resistance of the cable which depends only on the cable construction. The external thermal resistance, on the other hand, will depend on the properties of the surrounding medium as well as on the overall cable diameter, as explained in Chapter 9.

The last term in (4.5) is the temperature rise caused by dielectric losses. Denoting it by $\Delta\theta_d$,

$$\Delta\theta_d = nW_dT_d \tag{4.7}$$

4.2 BURIED CABLES WITH MOISTURE MIGRATION TAKEN INTO ACCOUNT

4.2.1 Introduction

The current-carrying capacity of buried power cables depends to a large extent on the thermal conductivity of the surrounding medium. In fact, results reported by El-Kady (1985) indicate that the sensitivity of cable temperature to variations in thermal conductivities of the surrounding medium is at least an order of magnitude greater than sensitivity to variations in other parameters such as ambient temperature, heat convection coefficient, or cable current.

Soil thermal conductivity is not a constant, but is highly dependent on its moisture content (Mochlinski, 1976). Under unfavorable conditions, the heat flux from the cable entering the soil may cause significant migration of moisture away from the cable. A dried-out zone may develop around the cable, in which the thermal conductivity is reduced by a factor of three or more over the conductivity of the bulk. This, in turn, may cause an abrupt rise in temperature of the cable sheath which may lead to damage to the cable insulation.

Studies of this problem have made it apparent that a need exists for a well-formulated procedure for calculating cable ampacities taking into account heat and moisture migration in the soil. The problem of thermal runaway has been studied by Groeneveld *et al.* (1984), Black *et al.* (1982), and Arman *et al.* (1964). In the first two cases, an analytical solution to the moisture migration problem was proposed and the governing equations were solved as single-dimensional approximations using the finite-difference method.

However, in typical engineering practice adopted by power utilities, the current ratings of cables are established on the basis of an assumed thermal conductivity of the surrounding medium. One of the reasons for this is that strict mathematical explanations and physical models describing moisture migration phenomena are very complicated, and adequate evaluations of the quantities concerned have not yet been made. In order to give some guidance on the effect of moisture migration on cable ratings, CIGRE (1986) has proposed a simple two-zone model for the soil surrounding loaded power cables, resulting in a minor modification of equation (4.3). Subsequently, this model has been adopted by the IEC as an international standard (IEC, 1993).

The concept on which the method proposed by CIGRE relies can be summarized as follows. Moist soil is assumed to have a uniform thermal resistivity; but if the heat dissipated from a cable and its surface temperature are raised above certain critical limits, the soil will dry out, resulting in a zone which is assumed to have a uniform thermal resistivity higher than the original one. The critical conditions, that is, the conditions for the onset of drying, are dependent on the type of soil, its original moisture content, and temperature.

Given the appropriate conditions, it is assumed that, when the surface of a cable exceeds the critical temperature rise above ambient, a dry zone will form around it. The outer boundary of the zone is on the isotherm related to that particular temperature rise (see Fig. 4.2). An additional assumption states that the development of such a dry zone does not change the shape of the isothermal pattern from what it was when all the soil was moist, only that the numerical values of some isotherms change. Within the dry zone, the soil has a uniformly high value of thermal resistivity, corresponding to its value when the soil is "oven dried" at not more than 105°C. Outside the dry zone, the soil has uniform thermal resistivity corresponding to the site moisture content. The essential advantages of these assumptions are that the resistivity is uniform over each zone, and that the values are both convenient and sufficiently accurate for practical purposes.

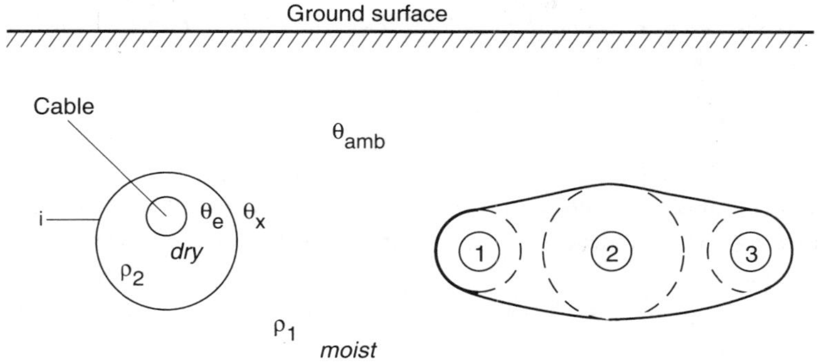

Figure 4.2 Illustration of the concept of dry zone within an isothermal circular boundary.

The method presented below assumes that the entire region surrounding a cable or cables has uniform thermal characteristics prior to drying out, the only nonuniformity being that caused by drying. As a consequence, the method should not be applied without further consideration to installations where special backfills, having properties different from the site soil, are used.

4.2.2 Development of Rating Equation

Let θ_e be the cable surface temperature corresponding to the moist soil thermal resistivity ρ_1. Then, without moisture migration, we obtain the following relations by applying equation (3.6) and remembering that soil thermal resistance is directly proportional to the value of resistivity:

$$nW_t = \frac{\theta_e - \theta_{\text{amb}}}{T_4} = \frac{(\theta_e - \theta_x) + (\theta_x - \theta_{\text{amb}})}{T_4} \tag{4.8}$$

and

$$nW_t = \frac{\theta_e - \theta_x}{C\rho_1} \tag{4.9}$$

where C is a constant and n is the number of cores in the cable. θ_e, θ_{amb}, and θ_x are cable surface temperature, ambient temperature, and the temperature of an isotherm at distance x, respectively. The total losses are given by equation (4.6)

If we now assume that the region between the cable and the θ_x isotherm dries out so that its resistivity becomes ρ_2, and that the power losses W_t remain unchanged, we have

$$nW_t = \frac{\theta'_e - \theta_x}{C\rho_2} \tag{4.10}$$

where θ'_e is the cable surface temperature when moisture migration has taken place.

Combining equations (4.9) and (4.10), we obtain the following form of equation (4.8):

$$nW_t = \frac{\rho_1/\rho_2(\theta'_e - \theta_x) + (\theta_x - \theta_{\text{amb}})}{T_4} \tag{4.11}$$

We can now define $\Delta\theta_x = \theta_x - \theta_{amb}$ as the *critical temperature rise* of the boundary between the moist and dry zones above ambient temperature. From equation (4.5), the cable surface temperature is equal to

$$\theta_e' = \theta_c - n(W_c T + W_d T_d) \tag{4.12}$$

Substituting (4.12) into (4.11) and recalling that $W_c = I^2 R$ and $W_t = W_c(1+\lambda_1+\lambda_2)+W_d$, we obtain the following equation for the current rating with moisture migration taken into account :

$$I = \left[\frac{\Delta\theta - W_d[0.5T_1 + n(T_2 + T_3 + \nu T_4)] + (\nu - 1)\Delta\theta_x}{RT_1 + nR(1+\lambda_1)T_2 + nR(1+\lambda_1+\lambda_2)(T_3 + \nu T_4)}\right]^{0.5} \tag{4.13}$$

where $\nu = \rho_2/\rho_1$ and the remaining variables are defined above. Note that T_4 is the external thermal resistance of the cable when it is laid in soil having a uniform resistivity ρ_1.

We can observe that equation (4.3) has been modified by the addition of the term $(\nu - 1)\Delta\theta_x$ in the numerator, and the substitution of νT_4 for T_4 in both the numerator and the denominator.

EXAMPLE 4.2

We will illustrate the effect of moisture migration on the rating of model cable No.1 located in standard conditions described in Appendix A. All the numerical values required in equation (4.13) are given in Table A1 as $\lambda_1 = 0.09$, $T_1 = 0.214$ K·m/W, $T_3 = 0.104$ K·m/W, and $T_4 = 1.933$ K·m/W. In addition, we assume the values $\rho_2 = 2.5$ K·m/W for the dry soil thermal resistivity and $\Delta\theta_x = 35°$C for the critical temperature rise. Then, applying equation (4.13),

$$I = \left[\frac{75 + (2.5-1)\cdot 3.5}{7.81\cdot 10^{-5}(0.214 + 1\cdot(1+0.09)\cdot 0 + 1\cdot(1+0.09)(0.104 + 2.5\cdot 1.933)}\right]^{0.5} = 541 \text{ A}$$

Thus, when the moisture migration is taken into account a 14% reduction in cable ampacity is required in comparison with the result obtained in Example 4.1.

The question of determination of $\Delta\theta_x$, the critical temperature rise above ambient, has been addressed by Donnazi *et al.* (1979). In practice, values between 35 and 50°C are used in most countries for the critical isotherm temperature value θ_x. The following section discusses this subject in more detail.

4.2.3 Determination of the Critical Temperature Rise $\Delta\theta_x$

There is not a great deal of direct information on the practical behavior of moisture in soil under the influence of a varying temperature distribution. Determination of the temperature, and hence the position of the critical isotherm, is a complicated matter, but examples of theoretical and experimental derivations are given in the literature (Donnazi *et al.*, 1979; Brakelmann, 1984; Groeneveld *et al.*, 1984; Black *et al.*, 1982). The following method is an adaptation of a work published by Donnazi *et al.* (1979), (CIGRE, 1992). It is the only practical method known at present which provides values of critical temperature and resistivity (apart from empirical values adopted in some countries), and which is relatively simple and backed up by experimental evidence.

The method relies on two experimentally determined quantities: (1) a critical moisture content expressed as a critical degree of saturation, and (2) a migration parameter. The critical degree of saturation can be determined by the use of a migration test or, for most sandy materials, from thermal resistivity/moisture content measurements. The migration parameter is obtained from a migration experiment. In the description of the parameters and their use, it is assumed that the soil surrounding the cable is homogeneous except for the moisture distribution derived from the two zones.

The critical temperature rise of a soil is then related to the critical degree of saturation s_{cr} (expressed as a fraction) and a parameter $\eta(\mathrm{K}^{-1})$, as follows:

$$\Delta\theta_x = \frac{1}{\eta}\left[(1-s_{cr})^2 \ln\frac{1-s_{cr}}{1-s_a} - \frac{1}{2}(s_a^2 - s_{cr}^2) - (s_a - s_{cr})(1-2s_{cr})\right] \tag{4.14}$$

where s_a is the degree of saturation of the soil controlled by the ambient moisture at the site.

An experimental method of deriving the values of s_{cr} and η is described by Donnazi *et al.* (1979) and in CIGRE (1992). Table 4.1 (CIGRE, 1992) gives the values of the parameters s_{cr} and η and the critical temperature rise $\Delta\theta_x$ as a function of s_{cr} for some selected soils[1]:

TABLE 4.1 Values of Critical Temperature Rise (K)

	Type of Soil			
	Sand/Clay	Sandy	Crushed Rock	Selected Sands
Porosity	0.47	0.42	0.24	0.25
s_{cr}	0.5	0.3	0.18	0.25
η, K^{-1}	$2\cdot10^{-4}$	$1\cdot10^{-4}$	$0.6\cdot10^{-4}$	$0.6\cdot10^{-4}$
s_a		Critical Temperature Rise		
0.25			2.5	
0.3			13	1
0.35		0.6	40	8
0.4		5	90	30
0.45		19		74
0.5		49		
0.55	0.4			
0.6	4			
0.65	15			
0.7	39			
0.75	85			

In closing this section, we would like to point out that the crucial assumption used in the above developments, that the critical temperature rise is independent of the heat flux at the surface of the cable, may not be valid when the soil becomes thermally unstable. In fact, developments presented by Hartley and Black (1982) suggest that the heat flux at the

[1] The soil descriptions given in Table 4.1 are not very precise, but they give a good indication of the desired thermal properties of soils. Crushed rock has the best thermal properties since its critical temperatures are high at low moisture contents. Selected sands follow closely.

surface of the cable is an important factor in determining the time required for a soil to become unstable. Therefore, equation (4.10) should be used with caution, and only in a completely static case where drying out occurs and soil reaches an equilibrium condition.

4.3 CABLES IN AIR

When cables are installed in free air, the same ladder network is used as discussed in Section 4.1. However, the external thermal resistance now accounts for the radiative and convective heat loss. For cables exposed to solar radiation, there is an additional temperature rise caused by the heat absorbed by cable external covering. The heat gain by solar absorption is equal to $\sigma D_e H$, with the meaning of the variables defined below. In this case, the external thermal resistance is different than for shaded cables in air, and the current rating is computed from the following modification of equation (4.3):

$$I = \left[\frac{\Delta\theta - W_d[0.5T_1 + n(T_2 + T_3 + T_4^*)] - \sigma D_e^* H T_4^*}{RT_1 + nR(1 + \lambda_1)T_2 + nR(1 + \lambda_1 + \lambda_2)(T_3 + T_4^*)} \right]^{0.5} \tag{4.15}$$

where σ = absorption coefficient of solar radiation for the cable surface; the values of this coefficient are given in Table 9.2

H = intensity of solar radiation; if this value is unknown locally, it could be assumed to be equal to 1000 W/m^2

T_4^* = external thermal resistance of the cable in free air, adjusted to take account of solar radiation (see Section 9.6.8).

We will illustrate the application of equation (4.13) in Section 9.6.8.1.

Computation of the values of several quantities appearing in equations (4.3) and (4.15) is discussed in detail in Chapters 6 through 9. Computation of dielectric losses is discussed in Chapter 6. Chapter 7 discusses computation of joule losses in the conductor, giving formulas for the computation of the conductor resistance R. Chapter 8 describes evaluation of sheath and armor loss factors, and Chapter 9 deals with the computation of thermal resistances.

REFERENCES

Arman, A. N., Cherry, D. M., Gosland, L., and Hollingsworth, P. M. (1964), "Influence of soil moisture migration on power rating of cables in H.V. transmission," *Proc. IEE*, vol. 111, pp. 1000–1016.

Black, W. Z. *et al.* (Sept. 1982), "Thermal stability of soils adjacent to underground transmission power cables," Final Report of EPRI Research Project 7883.

Brakelmann, H. (1984), "Physical principles and calculation methods of moisture and heat transfer in cable trenches," *ETZ Report 19.* Berlin: VDE Verlag.

CIGRE (1986), "Current ratings of cables buried in partially dried out soil. Part 1: Simplified method that can be used with minimal soil information: (100% load factor)," *Electra*, no. 104, pp. 11–22.

CIGRE (Dec. 1992), "Determination of a value of critical temperature rise for a cable backfill material," *Electra*, vol. 145, pp. 15–29.

Donnazi, F., Occhini, E., and Seppi, A. (1979), "Soil thermal and hydrological characteristics in designing underground cables," *Proc. IEE*, vol. 126, no. 6.

El-Kady, M. A. (Aug. 1985), "Calculation of the sensitivity of power cable ampacity to variations of design and environmental parameters," *IEEE Trans. Power App. Syst.*, vol. PAS-103, no. 8, pp. 2043–2050.

Groeneveld, G. J., Snijders, A. L., Koopmans, G., and Vermeer, J. (Mar. 1984), "Improved method to calculate the critical conditions for drying out sandy soils around power cables," *Proc. IEE*, vol. 131, part C, no. 2, pp. 42–53.

Hartley, J. G., and Black, W. Z. (May 1981), "Transient simultaneous heat and mass transfer in moist unsaturated soils," *ASME Trans.*, vol. 103, pp. 376–382.

IEC (1993), "Amendment 3 to IEC publication 287: Calculation of the continuous current rating of cables (100% load factor)."

Mochlinski, K. (1976), "Assessment of the influence of soil thermal resistivity on the ratings of distribution cables," *Proc. IEE*, vol. 123, no. 1, pp. 60–72.

5

Rating Equations—Transient Conditions

5.1 INTRODUCTION

J. H. Neher began his famous 1964 paper (Neher, 1964) by stating: "*The calculation of the transient temperature rise of buried cable systems, that is, the determination of the curve of conductor temperature rise versus time after the application of a constant conductor current, has intrigued the more mathematically minded cable engineers for many years.*" Thirty years have passed since this paper was presented in Toronto, and the fascination with the subject by mathematically minded cable engineers has not abated. On the contrary, judging by the number of recent publications dealing with this topic, one might conclude that either the number of mathematically inclined cable engineers must have risen dramatically in the past 30 years, or that the subject has increased significantly in prominence after steady-state computations were more or less standardized or, possibly both.

In transient rating computations, the following three questions are of particular importance:

1. Given current operating conditions (i.e., conductor temperature), what will be the final component temperature if loading is increased by a given amount and if the new loading is maintained for a specified period of time?
2. What is the maximum current a cable can carry for a given period of time if the temperature of the conductor cannot exceed a specified limit?
3. Given current operating conditions, how long can a new higher loading be applied without exceeding a specified temperature limit?

The first question requires computation of the temperature of the component of interest after a specified time period (including the steady-state conditions). This can be achieved by the methods described in Sections 5.2 and 5.3.

The second question relates to the classical problem of transient ratings which also includes cyclic conditions. The initial operating temperatures, which represent specified preloading conditions, are given. These conditions can be obtained either from steady-state computations, or from on-line monitoring. The time period of interest can vary between 10 min and infinity. In the latter case, one obtains the steady-state cyclic rating of the cables. The load curves, possibly different for each cable circuit, could represent the same curves which existed before the onset of transient conditions, but scaled appropriately to reflect changes in the loading. Alternatively, special curves may apply for the duration of the transient (e.g., a step function could be applied to some cables).

The third question reflects operating concerns when an emergency condition exists in the system and the system operator needs to know how long this condition can be tolerated.

To answer the first question, computations of component temperature variation with time are required. These computations are described by the mathematical model presented in this chapter. Special attention will be paid to: (1) the selection of an appropriate time step, (2) proper handling of the mutual heating of cables, and (3) the necessity to account for the variation of electrical resistances of metallic components with temperature. The treatment of the last two items is discussed in Sections 5.2 and 5.5. The selection of the time step depends on the procedure adopted for the discretization of the load curve. A brief discussion of this topic is presented in Section 11.2.3.

In general, computation of transient ratings (question 2 above) requires an iterative procedure. The main iteration loop involves the adjustment of cable loadings. At each step of iteration, the loading of each cable is selected and the temperature of the desired component is calculated (see Fig. 5.1). This step requires the same procedure as that of question 1.

To answer the third question, the main iteration loop must allow positive or negative increments in time. Once the time horizon is set for a particular iteration, the block computing the temperatures is called. If the temperature of the component of interest is within the specified tolerance from the desired temperature, the process is stopped; otherwise, a new time horizon is set and the algorithm is repeated.

Based on the above discussion, the procedure to evaluate temperatures is the main computational block in all three cases being considered. This block requires a fairly complex programming procedure to take into account self and mutual heating, and to make suitable adjustments in the loss calculations to reflect changes in the conductor resistance with temperature.

Transient rating of power cables requires the solution of the equations for the network in Fig. 3.5. The unknown quantity in this case is the variation of the conductor temperature rise with time,[1] $\theta(t)$. Unlike in the steady-state analysis, this temperature is not a simple function of the conductor current $I(t)$. Therefore, the process for determining the maximum value of $I(t)$ so that the maximum operating conductor temperature is not exceeded requires an iterative procedure. An exception is the simple case of identical cables carrying equal current located in a uniform medium. Approximations have been proposed for this case, and explicit rating equations developed. We will discuss this case later in this chapter.

[1] Unless otherwise stated, in this chapter, we will follow the notation in (IEC, 1989), and we will use the symbol θ to denote temperature rise and not $\Delta\theta$ as in Chapter 4 and (IEC, 1982).

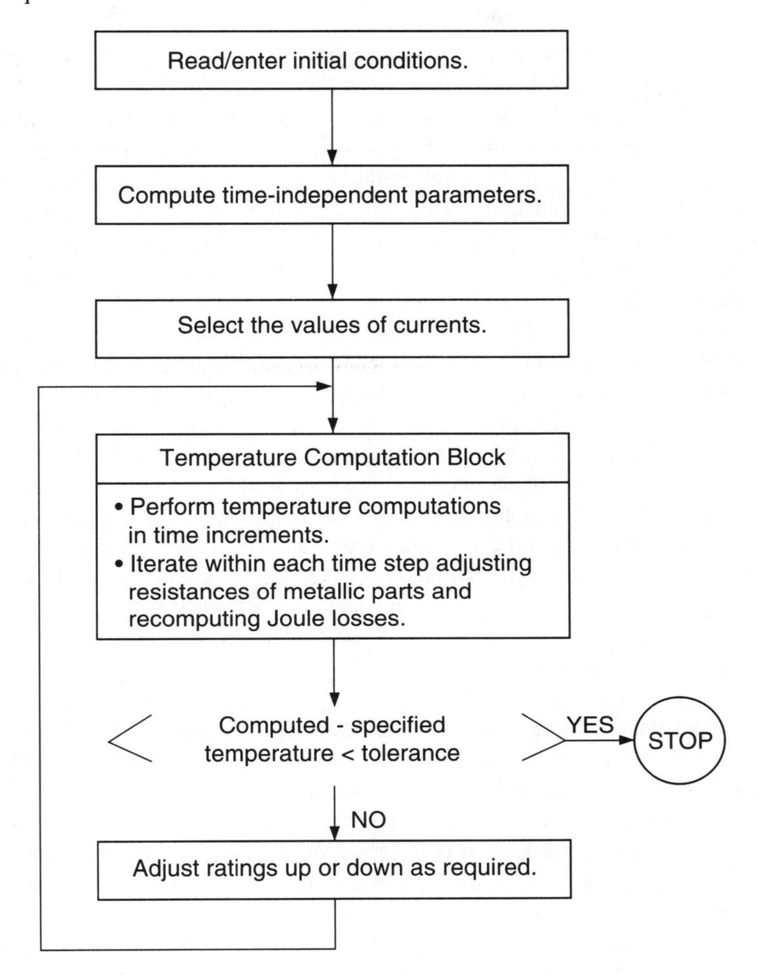

Figure 5.1 Flowchart for rating computations in transient analysis.

5.2 RESPONSE TO A STEP FUNCTION

5.2.1 Preliminaries

Whether we consider the simple cable systems mentioned above, or a more general case of several cable circuits in a backfill or duct bank, the starting point of the analysis is the solution of the equations for the network in Fig. 3.5. Our aim is to develop a procedure to evaluate temperature changes with time for the various cable components. As observed by Neher (1964), the transient temperature rise under variable loading may be obtained by dividing the loading curve at the conductor into a sufficient number of time intervals, during any one of which the loading may be assumed to be constant. Therefore, the response of a cable to a step change in current loading will be considered first.

This response depends on the combination of thermal capacitances and resistances formed by the constituent parts of the cable itself and its surroundings. The relative importance of the various parts depends on the duration of the transient being considered. For example, for a cable laid directly in the ground, the thermal capacitances of the cable, and the way in which they are taken into account, are important for short-duration transients, but can be neglected when the response for long times is required. The contribution of the surrounding soil is, on the other hand, negligible for short times, but has to be taken into account for long transients. This follows from the fact that the time constant of the cable itself is much shorter than the time constant of the surrounding soil.

The thermal network considered in this work is a derivation of the lumped parameter ladder network introduced early in the history of transient rating computations (Buller, 1951; Van Wormer, 1955; Neher, 1964; CIGRE, 1972; IEC 1985, 1989). For computational purposes, Baudoux (1962) and then Neher (1964) proposed to represent a cable in just two loops. Baudoux provided procedures for combining several loops, to obtain a two-section network which was latter adopted by CIGRE WG 02 and published in Electra (CIGRE, 1972). This topic is addressed in Section 3.3.2. However, transformation of a multiloop network into a two-loop equivalent not only requires substantial manual work before the actual transient computations can be performed, but also inhibits the computation of temperatures at parts of the cable other than the conductor. A procedure is given below for analytical solution of the entire network. Generally, the network will be somewhat different for short- and long-duration transients, and usually, the limiting duration to distinguish these two cases can be taken to be 1 h. Short transients are assumed to last at least 10 min. A more detailed time division between short and long transients can be found in Section 3.3.

The temperature rise of a cable component (e.g., conductor, sheath, jacket, etc.) can be represented by the sum of two components: the temperature rise inside and outside the cable. The method of combining these two components, introduced by Morello (Morello, 1958; CIGRE, 1972; IEC, 1985, 1989), makes allowance for the heat which accumulates in the first part of the thermal circuit and which results in a corresponding reduction in the heat entering the second part during the transient. The reduction factor, known as the attainment factor $\alpha(t)$, of the first part of the thermal circuit is computed as a ratio of the temperature rise across the first part at time t during the transient to the temperature rise across the same part in the steady state. Then, the temperature transient of the second part of the thermal circuit is composed of its response to a step function of heat input multiplied by a reduction coefficient (variable in time) equal to the attainment factor of the first part. Evaluation of these temperatures is discussed below. The validity of Morello's hypothesis was demonstrated in experiments carried out by Wlodarski and Cabiac (1966).

5.2.2 Temperature Rise in the Internal Parts of the Cable

The internal parts of the cable encompass the complete cable including its outermost serving or anti-corrosion protection. If the cable is located in a duct or pipe, the duct and pipe (including pipe protective covering) are also included. For cables in air, the cable extends as far as the free air.

Analysis of linear networks, such as the one in Fig. 3.5, involves the determination of the expression for the response function caused by the application of a forcing function. In our case, the forcing function is the conductor heat loss, and the response sought is the temperature rise above the cable surface at node i. This is accomplished by utilizing a mathematical quantity called the *transfer function* of the network. It turns out that this transfer function is the Fourier transform of the unit-impulse response of the network. The

Laplace transform of the network's transfer function is given by a ratio

$$H(s) = \frac{P(s)}{Q(s)} \tag{5.1}$$

$P(s)$ and $Q(s)$ are polynomials, their forms depending on the number of loops in the network. Node i can be the conductor or any other layer of the cable. In terms of time, the response of this network is expressed as (Van Valkenburg, 1964)[2]

$$\theta_i(t) = W_c \sum_{j=1}^{n} T_{ij} \left(1 - e^{P_j t}\right) \tag{5.2}$$

where $\theta_i(t)$ = temperature rise at node i at time t, °C

W_c = conductor losses including skin and proximity effects, W/m

T_{ij} = coefficient, °C · m/W

P_j = time constant, s^{-1}

t = time from the beginning of the step, s

n = number of loops in the network

i = node index

j = index from 1 to n

The coefficients T_{ij} and the time constants P_j are obtained from the poles and zeros of the equivalent network transfer function given by equation (5.1). Poles and zeros of the function $H(s)$ are obtained by solving equations $Q(s) = 0$ and $P(s) = 0$, respectively. From the circuit theory, the coefficients T_{ij} are given by

$$T_{ij} = -\frac{a_{(n-i)i}}{b_n} \frac{\prod_{k=1}^{n-i} (Z_{ki} - P_j)}{P_j \prod_{\substack{k=1 \\ k \neq j}}^{n} (P_k - P_j)} \tag{5.3}$$

where $a_{(n-i)i}$ = coefficient of the numerator equation of the transfer function

b_n = first coefficient of the denominator equation of the transfer function

Z_{ki} = zeros of the transfer function

P_j = poles of the transfer function

An algorithm for the computation of the coefficients of the transfer function equation is given in Appendix B.

Before desktop computers became widely available, several researchers presented simplified versions of equations (5.2) and (5.3). An interesting review of the work conducted in this area by Baudoux (1962), Van Wormer (1955), Morello (1958), and Wlodarski (1963)

[2] Unless otherwise stated, in the remainder of this chapter, all the temperature rises are caused by the joule losses in the cable.

can be found in Wlodarski and Cabiac (1966). They also present the results of experiments carried out to verify various models. The results of their analysis were later adopted by CIGRE and IEC and form the basis of today's standards.

EXAMPLE 5.1

A simple expression of equation (5.2) is obtained for the case of $n = 2$. Construction of such a network is discussed in Section 3.3.2 and the network is shown in Fig. 5.2.

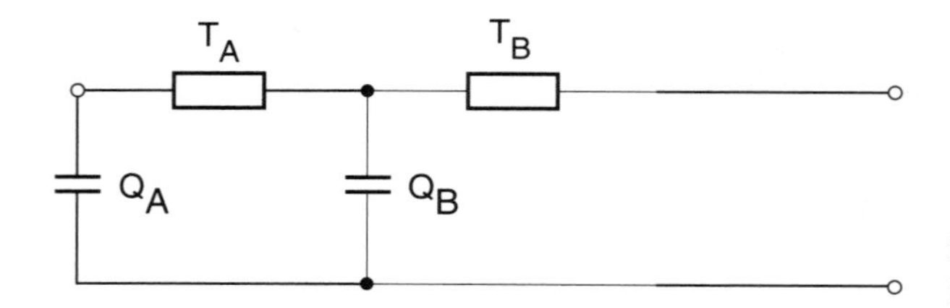

Figure 5.2 An equivalent thermal network composed of two loops.

In this simple case, the time-dependent solution for the conductor temperature can easily be obtained directly. However, to illustrate the procedure outlined above, we will compute this temperature from equations (5.1)–(5.3).

The transfer function for this network is given by

$$H(s) = \frac{(T_A + T_B) + sT_AT_BQ_B}{1 + s\,(T_AQ_A + T_BQ_B + T_BQ_A) + s^2T_AQ_AT_BQ_B} \tag{5.4}$$

Since we are interested in obtaining conductor temperature, $i = 1$ and $j = 1, 2$. To simplify the notation, we will use the following substitutions:

$$T_a = T_{11}, \quad T_b = T_{12}, \quad M_0 = 0.5\,(T_AQ_A + T_BQ_B + T_BQ_A), \quad N_0 = T_AQ_AT_BQ_B \tag{5.5}$$

The zeros and poles of the transfer function (5.4) are easily obtained as

$$Z_{11} = \frac{T_A + T_B}{T_AT_BQ_B}, \quad P_1 = -a, \quad P_2 = -b$$

where

$$a = \frac{M_0 + \sqrt{M_0^2 - N_0}}{N_0}, \qquad b = \frac{M_0 - \sqrt{M_0^2 - N_0}}{N_0} \tag{5.6}$$

From equation (5.4),

$$a_{(2-1)1} = T_AT_BQ_B, \quad b_2 = T_AT_BQ_AQ_B.$$

Thus,

$$\frac{a_{11}}{b_2} = \frac{1}{Q_A}.$$

From equation (5.3), we have

$$T_a = -\frac{1}{Q_A}\,\frac{-\dfrac{T_A + T_B}{T_AT_BQ_B} + a}{-a(-b + a)} = \frac{1}{a - b}\left(\frac{1}{Q_A} - \frac{T_A + T_B}{aT_AT_BQ_AQ_B}\right)$$

but

$$ab = \frac{1}{T_AT_BQ_AQ_B}$$

Hence,

$$T_a = \frac{1}{a-b}\left[\frac{1}{Q_A} - b\left(T_A + T_B\right)\right] \quad \text{and} \quad T_b = T_A + T_B - T_a \tag{5.7}$$

Finally, the conductor temperature as a function of time is obtained from equation (5.2):

$$\theta_c(t) = W_c\left[T_a\left(1 - e^{-at}\right) + T_b\left(1 - e^{-bt}\right)\right] \tag{5.8}$$

where W_c is the power loss per unit length in a conductor based on the maximum conductor temperature attained. The power loss is assumed to be constant during the step of the transient. Further,

$$\alpha(t) = \frac{\theta_c(t)}{W_c\left(T_A + T_B\right)} \tag{5.9}$$

Because the solution of network equations for a two-loop network is quite simple, IEC publications 853-1 (1985) and 853-2 (1989) recommend that this form be used in transient analysis. Examples of conversion of multiloop networks to their two-section equivalents are given in Section 3.3.3. The two-loop computational procedure was published at a time when access to fast computers was very limited (Wlodarski, 1966; CIGRE, 1972). Today, this limitation is no longer a problem, and a full network representation is recommended in transient analysis computations. This recommendation is particularly applicable in the case when temperatures of cable components, other than the conductor, are of interest.

5.2.3 Second Part of the Thermal Circuit—Influence of the Soil

The transient temperature rise $\theta_e(t)$ of the outer surface of the cable can be evaluated exactly in the case when the cable is represented by a line source located in a homogeneous, infinite medium with uniform initial temperature. Under these assumptions, equation (2.15) can be written as

$$\frac{\partial^2\theta}{\partial r^2} + \frac{1}{r}\frac{\partial\theta}{\partial r} + \rho_s W_t = \frac{1}{\delta}\frac{\partial\theta}{\partial t}$$

where ρ_s = soil thermal resistivity, K · m/W

$\delta = 1/\rho_s c$ = soil thermal diffusivity (see Section 5.4), m²/s.

Integrating this equation, we obtain

$$\theta_e(t) = -W_t\frac{\rho}{4\pi}\int_t^{\infty}\frac{1}{u}e^{-\frac{r^2}{4\delta u}}\,du$$

and making the change of variables,

$$\theta_e(t) = W_t\frac{\rho}{4\pi}\left[-Ei\left(-\frac{r^2}{4\delta t}\right)\right]$$

where $-Ei(-x) = \int_x^{\infty}\frac{e^{-v}}{v}\,dv$ is called the exponential integral. The value of the exponential integral can be developed in the series

$$-Ei(-x) = -0.577 - \ln x + x - \frac{x^2}{2\cdot 2!} + \frac{x^3}{3\cdot 3!}\cdots \tag{5.9a}$$

When $x < 0.1$,

$$-Ei(-x) = -0.577 - \ln x + x$$

to within 1% accuracy. For large x,

$$Ei(-x) = -\frac{e^{-x}}{x}\left(1 - \frac{1}{x} + \frac{2!}{x^2} - \frac{3!}{x^3} + \cdots\right)$$

The National Bureau of Standards published in 1940 *Tables of Exponential Integrals, Vol. 1*, in which values of $-Ei(-x)$ can be found. IEC has also published nomograms from which $-Ei(-x)$ can be obtained (IEC 853-2, 1989).

The mathematical solution obtained so far is valid under the assumption that the cable is treated as a line source with the internal thermal resistivities equal to that of the surrounding infinite soil. This result is valid for very short times and very deep cable locations only. However, for practical applications, we have to use another hypothesis, namely, the hypothesis of Kennelly, which assumes that the earth surface must be an isotherm. Under this hypothesis, the temperature rise at any point M in the soil is, at any time, the sum of the temperature rises caused by the heat source W_t and by its fictitious image placed symmetrically with the earth surface as the axis of symmetry and emitting heat $-W_t$ (see Fig. 5.3).

In this case, we have

$$\theta_M(t) = W_t\frac{\rho_s}{4\pi}\left[-Ei\left(-\frac{r^2}{4\delta t}\right) + Ei\left(-\frac{r'^2}{4\delta t}\right)\right]$$

Placing the point M at the surface of the cable and assuming identical material in the inside and outside of the cable, we obtain

$$\boxed{\theta_e(t) = W_t\frac{\rho_s}{4\pi}\left[-Ei\left(-\frac{D_e^{*2}}{16\delta t}\right) + Ei\left(-\frac{L^{*2}}{\delta t}\right)\right]} \tag{5.10}$$

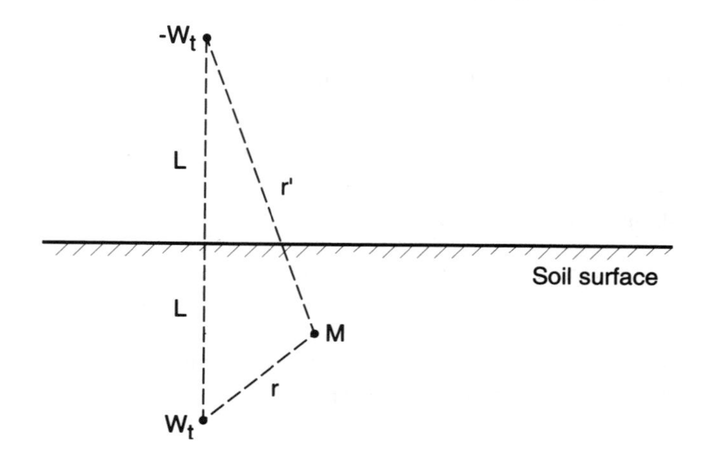

Figure 5.3 Illustration of Kennelly's hypothesis.

where D_e^* = external surface diameter of cable, m
L^* = axial depth of burial of the cable, m

Under steady-state conditions, $t \to \infty$ and x approaches zero. In this case, the terms after the logarithm in (5.9a) can be neglected. Summing now the exponential integrals in equation (5.10), we obtain

$$-\ln \frac{D_e^{*2}}{16\delta t} + \ln \frac{L^{*2}}{\delta t} = 2 \ln \frac{4L^*}{D_e^*}$$

Hence, in this case, equation (5.10) becomes

$$\theta_e(\infty) = W_t \frac{\rho_s}{2\pi} \ln \frac{4L^*}{D_e^*}$$

Several researchers proposed modifications of equation (5.10) to remove the assumption that the cable is a line source (Whitehead and Hutchings, 1938; Goldenberg, 1964). However, in experiments carried out to test various formulas, Wlodarski and Cabiac (1966) have shown that the results obtained with equation (5.10) are very close to the measured values for a wide range of times, and therefore this formula has been adopted by CIGRE and IEC as the standard equation.

5.2.4 Second Part of the Thermal Circuit—Cables in Air

For cables in air, it is unnecessary to calculate a separate response for the cable environment. The complete transient $\theta(t)$ is obtained from equation (5.1), but the external thermal resistance T_4, computed as described in Chapter 9, is included in the cable network.

EXAMPLE 5.2

For a cable in Example 5.1, obtain the transient conductor temperature if the cable is located in air.

In this case, the transient response of the internal part of the cable is given by equation (5.8). It is sufficient to modify the value of T_B to obtain the required answer. Thus, according to the discussion related to Example 5.1, we have

$$T_B = \frac{1}{2} T_1 + (1 + \lambda_1)(T_3 + T_4)$$

5.2.5 Representation of Dielectric Losses

The temperature rise caused by dielectric losses is of significance only for high-voltage cables because, as explained in Chapter 6, they are strongly voltage dependent. If the transient analysis is performed with a starting conductor temperature different from the ambient, the dielectric losses are assumed to be included in the temperature rise obtained at the onset of the transient conditions. If the application of load current and system voltage occur at the same time, then an additional transient temperature rise due to the dielectric loss has to be calculated. Assuming that the dielectric power factor is constant and equal to an appropriate value in the temperature range of interest, dielectric losses can be taken into account in transient conductor temperature rise computations as follows.

For cables at voltages up to and including 275 kV, it is sufficient to assume that half of the dielectric loss is produced at the conductor and the other half at the insulation screen or sheath. The cable thermal circuit is derived by the method given in Chapter 3 with the Van Wormer coefficient computed from equations (3.19) and (3.22) for long- and short-duration transients, respectively.

For cables operating at voltages higher than 275 kV, the dielectric loss is an important fraction of the total loss (i.e., for paper-insulated cables) and the appropriate Van Wormer coefficient is given by equation (3.23). Also, in this case, it is sufficient to assume that half the dielectric loss is produced at the conductor and the other half at the sheath.

Hence,

$$\theta_d(t) = \begin{cases} W_d\left[\frac{1}{2}T_1 + n\left(T_2 + T_3 + T_4\right)\right] & \text{if cable energized before } t = 0 \\ \theta_{cd}(t) + \alpha_d(t)\theta_{ed}(t) & \text{if cable energized at } t = 0 \end{cases} \tag{5.11}$$

where the subscript d refers to the circuit in which only dielectric losses are represented. This subject is illustrated in Example 5.4 below.

5.2.6 Groups of Equally or Unequally Loaded Cables

In a typical installation, several power cables are laid in a trench. The mutual heating effect reduces the current-carrying capacity of the cables, and this effect must be taken into account in rating computations. A major limitation of the methods presented by Neher (1964), Morello (1958), CIGRE (1972, 1976), and IEC (1985, 1989) is that they apply only in the case of identical, equally loaded, directly buried cables. However, one of the most common cases of interest is that associated with the loss of one circuit and a corresponding increase in loading in the remaining circuits. A method to account for the presence of other cables is discussed below (Anders and El-Kady, 1992; Affolter, 1987).

When there is more than one cable, for example, in a three-phase circuit composed of single-core cables, each will have its own thermal diagram and equivalent network from which the transient temperature rises caused by its own losses are evaluated. If the duration of the transient is sufficiently long for mutual heating (thermal coupling) from other cables to be significant, this must be taken into account when calculating the true temperatures used to update cable losses at each time step.

The temperature for each cable is obtained at each point in time by adding to its own temperature the temperature rise caused by other nearby cables. To achieve better accuracy in calculations with multiple time steps, the effect of other cables should be added at each time step so that their effect can be included with that caused by the temperature rise of the cable itself. Thus, the temperature rise in the cable of interest "p" due to one other adjacent cable "k" can be computed from

$$\boxed{\theta_{pk}(t) = W_{Ik}\frac{\rho_s}{4\pi}\left[-Ei\left(-\frac{d_{pk}^{*2}}{4\delta t}\right) + Ei\left(-\frac{d_{pk}^{*'2}}{4\delta t}\right)\right]} \tag{5.12}$$

in which W_{Ik} are the total joule losses in cable k, and d_{pk}^* and $d_{pk}^{*'}$ (m) denote the distance from the center of cable p to the center of cable k and its image, respectively, as shown in Fig. 5.4.

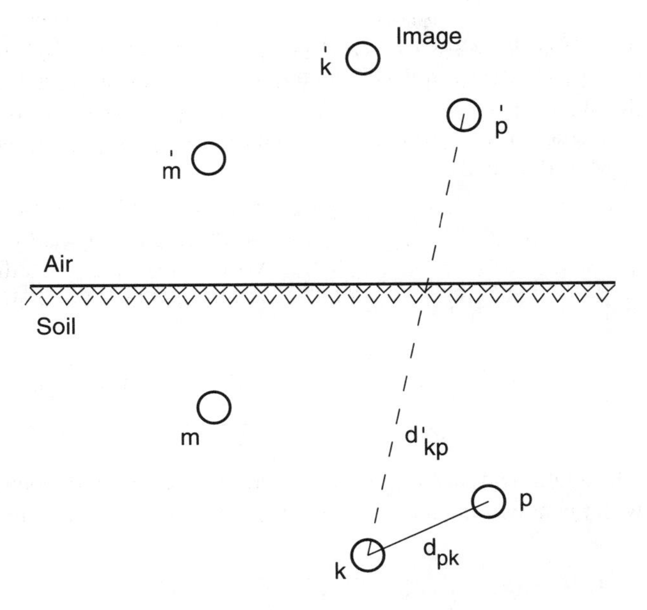

Figure 5.4 Example of cable configuration and image cables.

The temperature rise in the cable of interest caused by the dielectric losses of cable "k" is obtained from

$$\theta_{pdk}(t) = \begin{cases} W_{dk}\rho_s \ln \dfrac{d^{*\prime 2}_{pk}}{d^{*2}_{pk}} & \text{if cable energized before } t = 0 \\ W_{dk}\dfrac{\rho_s}{4\pi}\left[-Ei\left(-\dfrac{d^{*2}_{pk}}{4\delta t}\right) + Ei\left(-\dfrac{d^{*\prime 2}_{pk}}{4\delta t}\right)\right] & \text{if cable energized at } t = 0 \end{cases} \tag{5.13}$$

5.2.7 Total Temperature Rise

The total transient temperature rise of a cable at any time is the sum of the rise due to its own losses, given by its own network, and the rises due to mutual heating given by the networks of other cables and image sources, as appropriate. Thus, the final temperature rise at any layer of the cable of interest (that is, at any node of the equivalent network) at time t after the beginning of the load step is obtained from

$$\theta_{ptot}(t) = \theta_i(t) + \alpha(t)\theta_e(t) + \theta_d(t) + \alpha(t)\sum_{k=1}^{N-1}\left[\theta_{pk}(t) + \theta_{pdk}(t)\right] \tag{5.14}$$

where $\alpha(t)$ is the attainment factor for the transient temperature rise between the conductor and outside surface of the cable and N is the number of cables. The temperature rise $\theta_{pdk}(t)$ in equation (5.14) is multiplied by $\alpha(t)$ only if cable k is energized at time $t = 0$.

The internal temperature rise $\theta_d(t)$ is given by equation (5.11). In equation (5.14), θ_{ptot} is defined for any layer of the cable, and in the above formulation, only $\theta_i(t)$ is different for each layer.

The attainment factor varies in time, and a reasonable approach for obtaining $\alpha(t)$ is to use (Morello, 1958)

$$\alpha(t) = \frac{\text{tempertaure across cable at time } t}{\text{steady-state temperature rise across the cable}} \tag{5.15}$$

The conductor attainment factor is computed from this definition using equations (5.2) and (4.5):

$$\alpha(t) = \frac{\theta_c(t)}{\theta_c(\infty)} = \frac{W_c \sum_{j=1}^{n} T_{cj}\left(1 - e^{P_j t}\right)}{W_c T} = \frac{\sum_{j=1}^{n} T_{cj}\left(1 - e^{P_j t}\right)}{T} \tag{5.16}$$

The attainment factor associated with dielectric losses is obtained from a similar equation with the network parameters reflecting the presence of dielectric loss only.

5.3 TRANSIENT TEMPERATURE RISE UNDER VARIABLE LOADING

In order to perform computations for variable loading, a daily load curve is divided into a series of steps of constant magnitude. For different successive steps, the computations are done repeatedly, and the final result is obtained using the principle of superposition. The diagram in Fig. 5.5 illustrates how the temperature rise caused by an application of a single step of current lasting 1 h is computed.

From the diagram in Fig. 5.5, the temperature rise above ambient at time τ can be represented as

$$\theta(\tau) - \theta(\tau - 1) \tag{5.17}$$

5.4 DIFFUSIVITY OF SOIL

In the majority of cases, the soil diffusivity is not known. In such a case, a diffusivity value of $0.5 \cdot 10^{-6}$ m^2/s can be used. This value is based on a soil thermal resistivity of 1.0 K · m/W and a moisture content of about 7% of dry weight. When the density, moisture content, and thermal resistivity are known, the diffusivity of the soil can be computed from an empirical equation (IEC, 1985):

$$\delta = \frac{10^{-3}}{\rho_s d_0 (0.82 + 0.042\eta)} \tag{5.18}$$

where d_0 = dry density of soil, kg/m^3

η = moisture content, % of dry weight.

If only the thermal resistivity is known, then the diffusivity can be obtained from another empirical formula proposed by Neher (1964):

$$\delta = \frac{4.68}{\rho_s^{0.8}} \cdot 10^{-7} \tag{5.19}$$

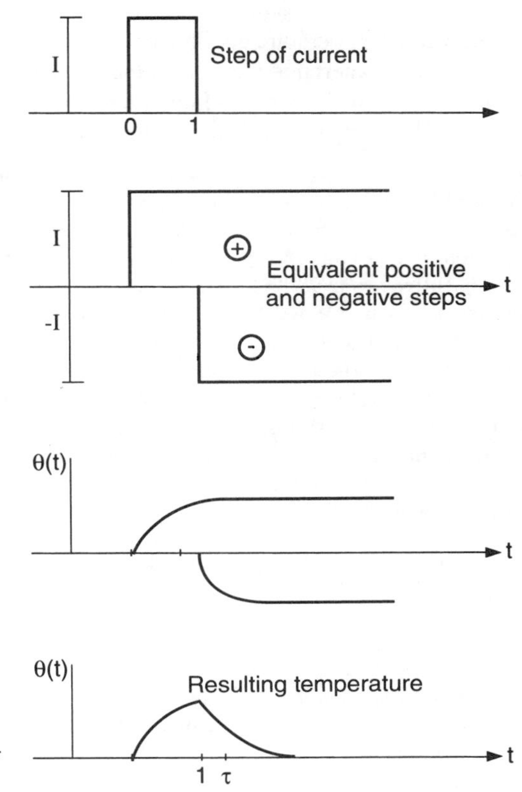

Figure 5.5 Temperature rise due to one step of current.

5.5 CONDUCTOR RESISTANCE VARIATIONS DURING TRANSIENTS

Since the conductor electrical resistance, as well as the resistance of other metallic parts of the cable, changes with temperature, the effect of these changes should be taken into account when computing conductor and sheath losses. Goldenberg (1967, 1971) developed a technique for obtaining arbitrarily close upper and lower bounds for the temperature rise of the conductor, taking into account the changes of the resistance of metallic parts with temperature. His upper bound formula has been adopted by CIGRE (1972, 1976) and IEC (1989). The derivation of this formula is very complex; therefore, it is simply quoted below:

$$\theta_a(t) = \frac{\theta(t)}{1 + a[\theta(\infty) - \theta(t)]} \tag{5.20}$$

where $\theta(t)$ = conductor transient temperature rise above ambient without correction for variation in conductor loss, based on the conductor resistance at the end of the transient

$\theta(\infty)$ = conductor steady-state temperature rise above ambient

a = temperature coefficient of electrical resistivity of the conductor material at the start of the transient. $a = 1/[\beta + \theta(0)]$ with β being a reciprocal of temperature coefficient at 0°C and $\theta(0)$ the conductor temperature at the start of the transient.

EXAMPLE 5.3

In this example, we present calculations for a circuit of three cables (cable model No. 1) in flat formation, not touching. The parameters of this cable are given in Appendix A as $\lambda_1 = 0.09$, $T_1 = 0.214$ K · m/W, $T_3 = 0.104$ K · m/W, and $T_4 = 1.933$ K · m/W. Laying conditions are as follows: cables are located 1 m below the ground in a flat configuration. Uniform soil properties are assumed throughout. Spacing between cables is equal to one cable diameter (spacing between centers equal to two cable diameters). Ambient soil temperature is 15°C. The thermal resistivity of the soil is equal to 1.0 K · m/W. The cables are solidly bonded and not transposed. We will compute the transient temperature response of the center cable for the first 24 h after the application of a step function of rated current. The temperature rise due to the dielectric loss is negligible for this cable.

Since we will perform computations without the help of a computer, we will use a two-section equivalent network.

(1) The thermal circuit of the cable

The thermal circuit for this cable has been derived in Example 3.6. The following values were obtained there:

$$T_A = 0.214 \text{ K} \cdot \text{m/W}, \quad Q_A = 1435.1 \text{ J/K} \cdot \text{m}$$

$$T_B = 0.113 \text{ K} \cdot \text{m/W}, \quad Q_B = 692.3 \text{ J/K} \cdot \text{m}$$

(2) Calculation of the response of the cable circuit

The response of the cable circuit to a step function of rated current is computed from equations given in Example 5.1:

$$M_0 = 0.5\,(T_A Q_A + T_B Q_B + T_B Q_A), \quad N_0 = T_A Q_A T_B Q_B$$

$$a = \frac{M_0 + \sqrt{M_0^2 - N_0}}{N_0}, \quad b = \frac{M_0 - \sqrt{M_0^2 - N_0}}{N_0}$$

$$T_a = \frac{1}{a-b}\left[\frac{1}{Q_A} - b\,(T_A + T_B)\right] \quad \text{and} \quad T_b = T_A + T_B - T_a$$

Substituting numerical values, we obtain

$$M_0 = \tfrac{1}{2}(0.214 \cdot 1435.1 + 0.113 \cdot 692.3 + 0.113 \cdot 1435.1) = 273.8 \text{ s}$$

$$N_0 = 0.214 \cdot 1435.1 \cdot 0.113 \cdot 692.3 = 24025.3 \text{ s}^2$$

$$a = \frac{273.8 + \sqrt{273.8^2 - 24025.3}}{24025.3} = 0.0208 \text{ s}^{-1}$$

$$b = \frac{273.8 - \sqrt{273.8^2 - 24025.3}}{24035.3} = 0.0020 \text{ s}^{-1}$$

$$T_a = \frac{1}{0.0208 - 0.0020}\left[\frac{1}{1435.1} - 0.0020(0.214 + 0.113)\right] = 0.0023 \text{ K} \cdot \text{m/W}$$

$$T_b = 0.214 + 0.113 - 0.0023 = 0.3247 \text{ K} \cdot \text{m/W}$$

The transient temperature rise for each step is determined from equation (5.8); it is calculated for 1 h (3600 s) below; the results for other times are tabulated in Table 5.2.

$$\theta_c(1) = W_c\left[T_a\left(1 - e^{-at}\right) + T_b\left(1 - e^{-bt}\right)\right]$$
$$= 30.82\left[0.0023\left(1 - e^{-0.0208\cdot 1\cdot 3600}\right) + 0.3247\left(1 - e^{-0.002\cdot 1\cdot 3600}\right)\right] = 10.1 \text{ K}$$

The conductor to cable surface attainment factor is determined for each time step from equation (5.9); it is calculated for 1 h below; results for the other times are shown in Table 5.2.

$$\alpha(1) = \frac{\theta_c(t)}{W_c\,(T_A + T_B)} = \frac{10.1}{30.82(0.214 + 0.113)} = 0.999$$

We can observe that because of the small value of the cable time constant, the conductor temperature rise above the cable surface temperature, reaches its steady state value within 1 h.

(3) Calculation of the response of cable environment

The response of the cable environment is given by equation (5.12). A sample calculation for 1 h is presented below. The other values are given in Table 5.1.

$$d^*_{pk} = 0.072 \text{ m}$$
$$d^{*'}_{pk} = \sqrt{0.072^2 + 2^2} = 2.001 \text{ m}$$
$$\delta = 0.5 \cdot 10^{-6} \text{ m}^2/\text{s}$$

On a term-by-term basis for 1 h (3600 s), we have

$$x = \frac{D_e^{*2}}{16t\delta} = \frac{0.0358^2}{16 \cdot 3600 \cdot 0.5 \cdot 10^{-6}} = 0.0445, \quad -Ei(-x) = 2.579$$

$$x = \frac{L^{*2}}{t\delta} = \frac{1^2}{3600 \cdot 0.5 \cdot 10^{-6}} = 555.6, \quad -Ei(-x) = 0$$

$$x = \frac{d_{pk}^{*2}}{4t\delta} = \frac{0.072^2}{4 \cdot 3600 \cdot 0.5 \cdot 10^{-6}} = 0.72, \quad -Ei(-x) = 0.365$$

$$x = \frac{\left(d^{*'}_{pk}\right)^2}{4t\delta} = \frac{2.001^2}{4 \cdot 3600 \cdot 0.5 \cdot 10^{-6}} = 556.1, \quad -Ei(-x) = 0$$

$$\theta_e(1) = \frac{1 \cdot 33.38}{4\pi}[(2.579 - 0) + 2(0.365 - 0)] = 8.8 \text{ K}$$

The second terms in equations (5.10) and (5.12) represent the effect of the image sources, and are usually negligible at normal depths of laying and for durations of less than 24 h.

TABLE 5.1 Components for Cable Environment Partial Transient

x	Item	Time (h) 1	2	3	4	5	6	12	24
$\frac{D_e^{*2}}{16t\delta}$	x	0.045	0.022	0.015	0.011	0.009	0.007	0.004	0.002
	$-E(-x)$	2.579	3.25	3.648	3.932	4.153	4.334	5.024	5.715
$\frac{d_{pk}^{*2}}{4t\delta}$	x	0.720	0.36	0.24	0.18	0.144	0.12	0.06	0.03
	$-E(-x)$	0.365	0.775	1.076	1.31	1.50	1.66	2.295	2.959

(4) Complete temperature rise

The complete transient is calculated from equation (5.14). A sample calculation for 1 h is shown below. The values for this and other times are summarized in column 5 of Table 5.2.

$$\theta(1) = 10.1 + 0.999 \cdot 8.8 = 18.9 \text{ K}$$

Correcting for variation of conductor losses using equation (5.20), we obtain

$$\theta_a(1) = \frac{\theta(t)}{1 + a[\theta(\infty) - \theta(t)]} = \frac{18.9}{1 + \frac{1}{234.5 + 10}(75 - 18.9)} = 15.4 \text{ K}$$

The corrected values of conductor temperature rise are summarized in column 6 of Table 5.2, along with the actual conductor temperature in column 7.

TABLE 5.2 Summary of Temperature Transient Components

Time (h)	$\theta_c(t)$ (K)	$\alpha(t)$	$\theta_e(t)$ (K)	$\theta(t)$ (K)	$\theta_a(t)$ (K)	Actual Conductor Temperature (°C)
(1)	(2)	(3)	(4)	(5)	(6)	(7)
1	10.1	0.999	8.8	18.9	15.4	30.4
2	10.1	1	12.8	22.9	18.9	33.9
3	10.1	1	15.4	25.5	21.3	36.3
4	10.1	1	17.4	27.5	23.1	38.1
5	10.1	1	19.0	29.1	24.6	39.6
6	10.1	1	20.3	30.4	25.8	40.8
12	10.1	1	25.5	35.6	30.8	45.8
24	10.1	1	30.9	41.0	36.1	51.1

EXAMPLE 5.4

In this example, we present calculations for a system of two cables (cable model No. 2 and cable model No. 3 described in Appendix A) buried underground. Laying conditions are shown in Fig. 5.6. We will compute the transient temperature response of both cables for the first 6 h after the application of a step function of current. The losses in each conductor of the pipe-type cable are assumed to be equal to 9 W/m and in each conductor of the three-core cable 22 W/m. Total joule losses in the pipe-type cable and the three-core cable are 40 and 72 W/m, respectively. The dielectric loss in the pipe-type cable is 4.83 W/m per cable. System frequency is 60 Hz. We will assume that the voltage is applied simultaneously with the current.

Since the response of the system in 1 h steps for the first 6 h is required, long-duration transient conditions can be assumed for the three-core cable and short-duration transient conditions for the pipe-type cable (see Examples 3.8 and 3.7). The temperature rise due to the dielectric loss is negligible for the three-core cable, but has to be evaluated for the pipe-type cable. Therefore, we will start by constructing an equivalent network for the pipe-type cable for the computation of the temperature rises caused by dielectric loss.

(1) Thermal network for dielectric loss of pipe-type cable

Parameters of this cable are given in Table A1 as $\lambda_1 = 0.010$, $\lambda_2 = 0.311$, $T_1 = 0.422$ K · m/W, $T_2 = T_o = 0.082$ K · m/W, and $T_3 = 0.017$ K · m/W. The Van Wormer coefficient for dielectric loss

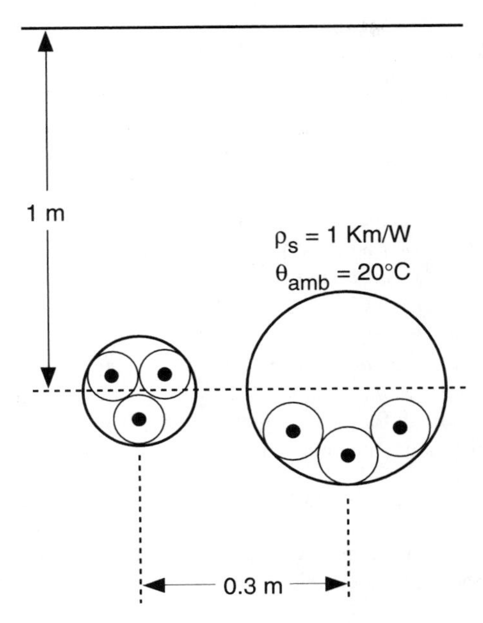

Figure 5.6 Laying conditions for two cables in Example 5.4.

for this cable has been computed in Example 3.5 and is equal to 0.585. Therefore, we have

$$T_A = \tfrac{1}{6}T_1, \quad Q_A = (Q_c + p_d Q_{i1})/3$$
$$T_B = \tfrac{1}{6}T_1 + (1+\lambda_1)T_o + (1+\lambda_1+\lambda_2)T_3$$

$$Q_B = \frac{(1-p_d)Q_{i1} + p_d Q_{i2}}{3} + \left(\frac{T_B - \frac{1}{6}T_1}{T_B}\right)^2 \left[(1-p_d)Q_{i2}/3 + \frac{Q_s/3 + \frac{1}{2}Q_o}{1+\lambda_1}\right]$$
$$+ \left(\frac{(1+\lambda_1+\lambda_2)T_3}{T_B}\right)^2 \left(\frac{\frac{1}{2}Q_o}{1+\lambda_1} + \frac{Q_p + p'Q_j}{1+\lambda_1+\lambda_2}\right) \tag{5.21}$$

Substituting numerical values, we obtain

$$T_A = \tfrac{1}{2}0.422/3 = 0.07 \text{ K}\cdot\text{m/W}, \quad Q_A = (3484.5 + 0.585 \cdot 1680.5)/3 = 1489.2 \text{ J/K}\cdot\text{m}$$
$$T_B = 0.07 + 2\cdot(0.082 + 0.017) = 0.268 \text{ K}\cdot\text{m/W}$$
$$Q_B = (1-0.585)1680.5/3 + 0.585\cdot 2726.9/3$$
$$+ \left(\frac{0.268 - 0.07}{0.268}\right)^2 \left((1-0.585)2726.9/3 + \frac{1.2/3 + 38570/2}{2}\right)$$
$$+ \left(\frac{2\cdot 0.017}{0.268}\right)^2 \frac{16126.4 + 0.482\cdot 15720.9 + 38570/2}{2} = 6734.2 \text{ J/K}\cdot\text{m}$$

The remaining cable parameters are obtained from equations (5.5)–(5.7) and are equal to

$$a = 0.012 \text{ s}^{-1}, \quad b = 0.00045 \text{ s}^{-1}, \quad T_a = 0.046 \text{ K}\cdot\text{m/W}, \quad T_b = 0.292 \text{ K}\cdot\text{m/W}$$

The transient temperature rise caused by dielectric loss for each step is determined from equation (5.8). Half of the dielectric losses are located at the conductor and half at the sheath. The calculation for 1 h (3600 s) is given below; the values for this and the other times are summarized in Table 5.4.

$$\theta_{cd}(1) = W_d\left[T_a\left(1 - e^{-at}\right) + T_b\left(1 - e^{-bt}\right)\right]$$
$$= \frac{3}{2}\cdot 4.83\left[0.046\left(1 - e^{-0.012\cdot 1\cdot 3600}\right) + 0.292\left(1 - e^{-0.00045\cdot 1\cdot 3600}\right)\right] = 2.0\text{ K}$$

The conductor to cable surface attainment factor is determined at each time step from equation (5.9); the calculation for 1 h is given below. The values for this and for the other times are shown in Table 5.4.

$$\alpha_d(1) = \frac{2.0}{7.245(0.07 + 0.268)} = 0.817$$

(2) Thermal circuit for the temperature rise caused by joule losses, and calculation of the response of the cable circuit

The thermal circuit for the pipe-type cable is obtained from the data in Table A.1 and Example 3.7. The following values are obtained:

$$a = 0.0119\text{ s}^{-1},\quad b = 0.00058\text{ s}^{-1},\quad T_a = 0.0498\text{ K}\cdot\text{m/W},\quad T_b = 0.1952\text{ K}\cdot\text{m/W}$$

The transient temperature rise for each step is determined again from equation (5.8); the calculations for 1 h (3600 s) is given below; the values for this and the other times are assembled in Table 5.5.

$$\theta_c(1) = 27\left[0.0498\left(1 - e^{-0.0119\cdot 1\cdot 3600}\right) + 0.1952\left(1 - e^{-0.00058\cdot 1\cdot 3600}\right)\right] = 6.0\text{ K}$$

The conductor to cable surface attainment factor is determined at each time step from equation (5.9); the value for 1 h is computed below; otherwise, see Table 5.4.

$$\alpha(1) = \frac{\theta_c(t)}{W_c\left(T_A + T_B\right)} = \frac{6.0}{27(0.07 + 0.175)} = 0.907$$

For the three-core cable, new values of T_A, T_B, Q_A, and Q_B have to be computed because the system frequency for this example is 60 Hz and the values in Table A.1 and Example 3.8 were obtained for 50 Hz. The only difference between these two cases is the value of the sheath loss factor. This factor is computed by a method discussed in Section 8.3. For a system frequency of 60 Hz, it is equal to 0.103. For illustration purposes, we will assume a single set of capacitance values. Proceeding now similarly as in Example 3.8 and remembering that the cable is directly buried, we obtain the following network parameters:

$$T_A = 0.104\text{ K}\cdot\text{m/W},\quad Q_A = 743.1\text{ J/K}\cdot\text{m}$$
$$T_B = 0.086\text{ K}\cdot\text{m/W},\quad Q_B = 1018.7\text{ K}\cdot\text{m}$$

and

$$a = 0.0286\text{ s}^{-1},\quad b = 0.0052\text{ s}^{-1},\quad T_a = 0.0156\text{ K}\cdot\text{m/W},\quad T_b = 0.1744\text{ K}\cdot\text{m/W}$$

The transient temperature rise for each step is determined again from equation (5.8); the sample calculation for 1 h (3600 s) is given below; the values for other times are assembled in Table 5.5.

$$\theta_c(1) = 66\left[0.0156\left(1 - e^{-0.0286\cdot 1\cdot 3600}\right) + 0.1744\left(1 - e^{-0.0052\cdot 1\cdot 3600}\right)\right] = 12.5\text{ K}$$

The conductor to cable surface attainment factor for $t = 1$ h is equal to

$$a(1) = \frac{12.5}{66(0.104 + 0.086)} = 1.0$$

The remaining values are given in Table 5.5.

(3) Calculation of the response of the environment for the pipe-type cable

The response of the cable environment is given by equation (5.10). A sample calculation is given for 1 h. The other values are assembled in Table 5.4.

$$x = \frac{D_e^{*2}}{16t\delta} = \frac{0.2445^2}{16 \cdot 3600 \cdot 0.5 \cdot 10^{-6}} = 2.075, \quad -Ei(-x) = 0.044$$

$$x = \frac{L^{*2}}{t\delta} = \frac{1^2}{3600 \cdot 0.5 \cdot 10^{-6}} = 555.6, \quad -Ei(-x) = 0$$

$$\theta_e(1) = W_t \frac{\rho_s}{4\pi}\left[-Ei\left(-\frac{D_e^{*2}}{16\delta t}\right) + Ei\left(-\frac{L^{*2}}{\delta t}\right)\right] = 40\frac{1}{4\pi} \cdot 0.044 = 0.1 \text{ K}$$

The external temperature rise due to dielectric loss is equal to

$$\theta_{de}(1) = \frac{1 \cdot 14.49}{4\pi} \cdot 0.044 = 0.05 \text{ K}$$

(4) Calculation of the response of the environment for the three-core cable

The response of the cable environment is given again by equation (5.10). A sample calculation is given for 1 h. The other values are assembled in Table 5.5.

$$x = \frac{D_e^{*2}}{16t\delta} = \frac{0.0729^2}{16 \cdot 3600 \cdot 0.5 \cdot 10^{-6}} = 0.1845, \quad -Ei(-x) = 1.289$$

$$x = \frac{L^{*2}}{t\delta} = \frac{1^2}{3600 \cdot 0.5 \cdot 10^{-6}} = 555.6, \quad -Ei(-x) = 0$$

$$\theta_e(1) = \frac{1 \cdot 72}{4\pi} \cdot 1.289 = 7.4 \text{ K}$$

(5) Calculation of the mutual heating effect

Since we have only two cables located at the same depth, the exponential integrals will be the same for the mutual heating effect for both cables.

From Fig. 5.6, we obtain

$$d_{pk}^* = 0.3 \text{ m}, \quad d_{pk}^{*'} = \sqrt{0.3^2 + 2^2} = 2.02 \text{ m}$$

The values of the exponential integral for $t = 1$ h are computed below, and for other values are shown in Table 5.3.

$$x = \frac{d_{pk}^{*2}}{4t\delta} = \frac{0.3^2}{4 \cdot 3600 \cdot 0.5 \cdot 10^{-6}} = 12.5, \quad -Ei(-x) = 0$$

$$x = \frac{\left(d_{pk}^{*'}\right)^2}{4t\delta} = \frac{2.02^2}{4 \cdot 3600 \cdot 0.5 \cdot 10^{-6}} = 566.7, \quad -Ei(-x) = 0$$

The temperature rise in the pipe-type cable due to joule losses in the three-core cable is obtained from equation (5.12). As can be seen from Table 5.3, the mutual heating effect will take place from $t = 5$ h on. Computations for 6 h are shown below, with the remaining values summarized in Table 5.4.

Heating of the pipe-type cable by the three-core cable

$$\theta_{pk}(6) = 72 \cdot \frac{1}{4\pi} \cdot 0.044 = 0.3 \text{ K}$$

TABLE 5.3 Components for Cable Environment Partial Transient Due to Mutual Heating

x	Item	Time 1	2	3	4	5	6
$\frac{d_{pk}^{*2}}{4t\delta}$	x	12.5	6.25	4.17	3.12	2.5	2.08
	$-E(-x)$	0	0	0	0	0.025	0.044

TABLE 5.4 Summary of Temperature Transient Components as a Function of Time for the Pipe-Type Cable

Time (h)	$\theta_c(t)$ (K)	$\alpha(t)$	$\theta_e(t)$ (K)	$\theta_d(t)^*$ (K)	$\theta_{pk}(t)$ (K)	$\theta(t)$ (K)	$\theta_a(t)$ (K)	Actual Cond. Temp. (°C)
(1)	(2)	(3)	(4)	(5)	(6)	(7)	(8)	(9)
1	6.0	0.907	0.1	2.1	0	8.2	7.0	27.0
2	6.5	0.988	0.7	2.5	0	9.7	8.4	28.4
3	6.6	0.998	1.2	2.7	0	10.5	9.1	29.1
4	6.6	1.0	1.7	2.8	0	11.1	9.6	29.6
5	6.6	1.0	2.2	2.9	0.1	11.8	10.3	30.3
6	6.6	1.0	2.6	3.0	0.3	12.5	10.9	30.9

*This column contains total temperature rise due to dielectric as computed from equation (5.11).

Heating of the three-core cable by the pipe-type cable

For the computation of the mutual heating effect, the total losses in the pipe-type cables are entered in equation (5.12). Thus, we have

$$\theta_{pk}(6) = 54.49 \cdot \frac{1}{4\pi} \cdot 0.044 = 0.2 \text{ K}$$

$$\theta_{pdk}(6) = 14.49 \cdot \frac{1}{4\pi} \cdot 0.044 = 0.05 \text{ K}$$

(6) Complete temperature rise

The complete transient is calculated by using equation (5.14). A sample calculation for 1 h is shown below; the values for other times are summarized in columns 7 of Tables 5.4 and 5.5 for the pipe-type and the three-core cable, respectively.

Pipe-type cable

$$\theta(1) = 6.0 + 0.907 \cdot 0.1 + 2.1 + 0.817 \cdot 0.05 + 0.901 \cdot 0 = 8.2 \text{ K}$$

Correcting for the variation of conductor losses using equation (5.20), we obtain

$$\theta_a(1) = \frac{8.2}{1 + \frac{1}{234.5 + 20}(50 - 8.2)} = 7.0 \text{ K}$$

Three-core cable

$$\theta(1) = 12.5 + 1.0 \cdot 7.4 + 1.0 \cdot (0 + 0) = 19.9 \text{ K}$$

TABLE 5.5 Summary of Temperature Transient Components as a Function of Time for the Three-Core Cable

Time (h)	$\theta_c(t)$ (K)	$\alpha(t)$	$\theta_e(t)$ (K)	$\theta_{pdk}(t)^*$ (K)	$\theta_{pk}(t)$ (K)	$\theta(t)$ (K)	$\theta_a(t)$ (K)	Actual Cond. Temp. (°C)
(1)	(2)	(3)	(4)	(5)	(6)	(7)	(8)	(9)
1	12.5	1.0	7.4	0	0	19.9	16.6	36.6
2	12.5	1.0	10.9	0	0	23.4	19.8	39.8
3	12.5	1.0	13.0	0	0	25.6	21.8	41.8
4	12.5	1.0	14.6	0	0	27.1	23.2	43.2
5	12.5	1.0	15.8	0.03	0.1	28.4	24.4	44.4
6	12.5	1.0	16.8	0.05	0.3	29.8	25.7	45.7

* This column contains total temperature rise due to dielectric as computed from equation (5.13).

Correcting for the variation of conductor losses using equation (5.20), we obtain

$$\theta_a(1) = \frac{19.9}{1 + \dfrac{1}{234.5 + 20}(70 - 19.9)} = 16.6 \text{ K}$$

The corrected values of conductor temperature rise are summarized in columns 8 of Table 5.4 and 5.5, along with the actual conductor temperature in columns 9.

Goldenberg's approximate formulas pertain to the application of a step function, and would have to be repeated for each new time step considered. However, taking advantage of fast desktop computers which are now in wide use, one can perform the computations iteratively, adjusting the resistances at each computational step. This is performed in the following way. At each time step, the temperature at the end of the step is computed based on the losses at the beginning of the step. New resistance values are then obtained and the computations for this time step are repeated. The process is carried out until convergence is obtained. As pointed out by Thomann *et al.* (1991), seldom are more than two iterations required to achieve the desired accuracy.

5.6 CALCULATION OF CYCLIC RATINGS

5.6.1 Introduction

The complexity of cyclic rating computations varies depending on the shape of the load curve and the amount of detail known for the load cycle. If only the load-loss factor or a daily load factor is known, a method proposed by Neher and McGrath (1957) can be used. This method involves modification of the cable external thermal resistance as discussed in Section 9.6.7. This modified value is then used in equation (4.3). Another possibility is to make the assumption that the load curve has a flat top lasting for at least 6 h and use the method described in Example 5.6 in this section.

If a more detailed analysis is required, the algorithm described in Sections 5.2 and 5.3 can be used to handle different cable types with different loading patterns (Anders *et al.*,

1990). For the case of a single cable or a group of identical cables, the simplified approach described below gives satisfactory results.

5.6.2 Cyclic Rating Computations

The general algorithm described in Sections 5.2 and 5.3 is applicable to a group of equally or unequally loaded cables with each circuit carrying a different cyclic load. Since the algorithm requires intensive computations, simplified approaches have evolved which are quite suitable for hand calculations. The first simplified approach was proposed by McGrath (1964) in response to a complex method involving application of the Fourier integral introduced by Neher (1964). This simplified approach uses the steady-state rating equation (4.3), but requires modification of the external thermal resistance of the cable. McGrath's approach continues to be the basis for the majority of cyclic loading computations performed in North America. Since it only requires modification of the value of T_4, we introduce it in Section 9.6.7 where thermal resistances are discussed.

Another simplified approach was introduced by Goldenberg (1957) and later adopted by the IEC (1985, 1989). This approach, applicable to a single cable or a cable system composed of identical, equally loaded cables located in a uniform medium, is presented for both cases below.

The cyclic rating of a single three-core cable or a group of equally loaded identical cables located in a uniform soil requires computation of a cyclic rating factor M by which the permissible steady-state rated current (100% load factor) may be multiplied to obtain the permissible peak value of current during a daily (24 h) cycle such that the conductor temperature attains, but does not exceed, the standard permissible maximum temperature during the cycle. A factor derived in this way uses the steady-state temperature, which is usually the permitted maximum temperature, as its reference. The cyclic rating factor depends only on the shape of the daily cycle, and is independent of the actual magnitudes of the current.

Goldenberg (1956) derived an expression for M taking into account the form of the whole cycle. Later, he showed that it is sufficient to consider only the loss-load factor of the cycle and the detail of the load current for the 6 h prior to the maximum conductor temperature (Goldenberg, 1957). Earlier values can be represented with sufficient accuracy by using an average. The loss-load factor μ provides this average (computation of this factor is illustrated in Example 5.5 below). Location of the time of maximum temperature is done by inspection, bearing in mind that although it usually occurs at the end of the period of maximum current, this may not always be the case. The loss-load factor μ of the daily current cycle is determined by decomposing the cycle into hourly rectangular pulses. The temperature response to the complete cycle of losses of the cable and the soil is found by adding together the responses to each hourly rectangular pulse, taking into account the time period between each pulse and the time to maximum temperature.

EXAMPLE 5.5

Determine a load-loss factor μ for a daily load curve shown in Fig. 5.7 (IEC, 1989).

The daily load cycle is given as a fraction of the maximum current in Table 5.6 (IEC, 1989) (the last column will be used in Example 5.6 later in this section).

The load-loss factor is then equal to

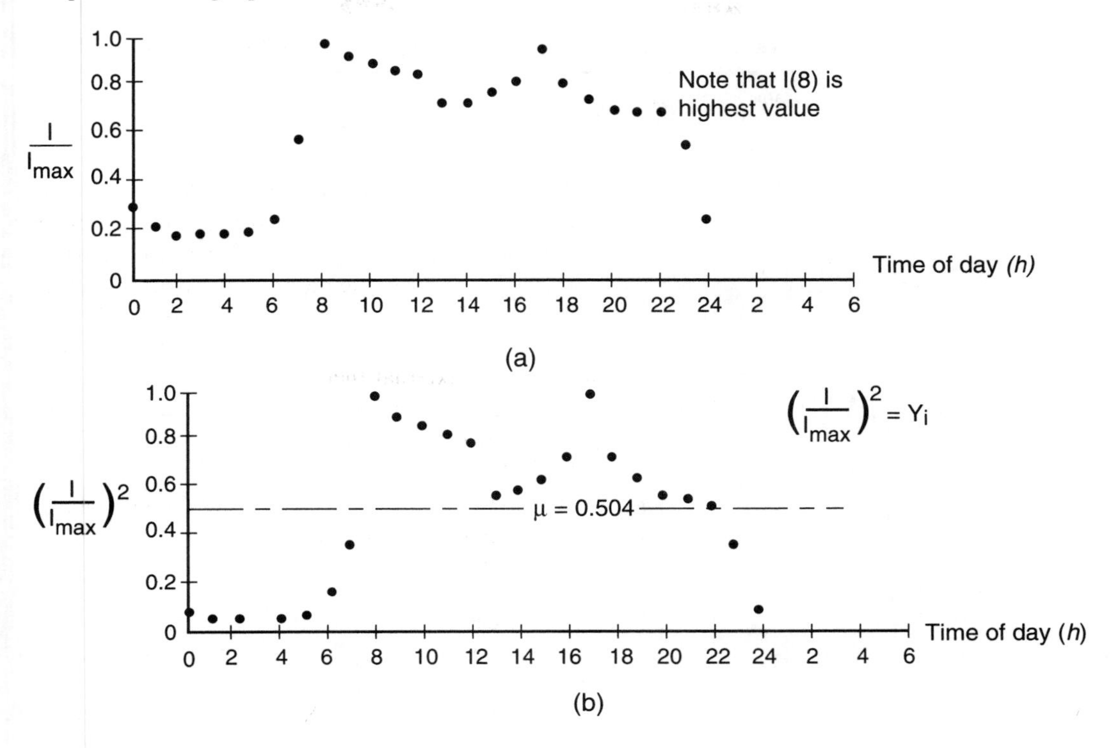

Figure 5.7 (a) Cyclic load divided by highest load; (b) load-loss graph. (IEC Standard 853-2, 1989)

$$\mu = \frac{1}{24} \frac{\sum_{i=0}^{23} I_i^2}{I_{\max}^2} = \frac{1}{24} \sum_{i=0}^{23} Y_i \tag{5.22}$$

For the numerical values of the load curve in Table 5.6, the load-loss factor is equal to

$$\mu = \frac{0.091 + 0.061 + 0.052 + \cdots + 0.360}{24} = 0.504$$

5.6.2.1 Single cable. The developments presented in this section apply to a single-conductor single cable located in a uniform, semi-infinite medium. The three-core cable is replaced in cyclic computations by an equivalent single-core construction dissipating the same total conductor losses as described in Section 3.3.1, with the thermal capacitances of an equivalent cable calculated with the assumptions specified there.

To obtain an expression for the conductor temperature rise at an arbitrary time τ, we will make an assumption that the heat dissipated in the cable is directly proportional to the square of the applied current. This simplifying assumption gives quite accurate results, as confirmed by the author by performing finite-element studies. The required temperature is then obtained as an algebraic sum of the hourly temperature rises as given by expression (5.17).

TABLE 5.6 Details of the Load Cycle in Fig. 5.7

Time (h)	Load (p.u.)	Value of Y_i	Order of Y_i
0	0.302	0.091	Y_{17}
1	0.247	0.061	Y_{16}
2	0.227	0.052	Y_{15}
3	0.232	0.054	Y_{14}
4	0.235	0.056	Y_{13}
5	0.246	0.061	Y_{12}
6	0.290	0.084	Y_{11}
7	0.600	0.360	Y_{10}
8	1.000	1.000	Y_9
9	0.950	0.902	Y_8
10	0.940	0.884	Y_7
11	0.910	0.828	Y_6
12	0.892	0.796	Y_5
13	0.770	0.593	Y_4
14	0.772	0.596	Y_3
15	0.800	0.640	Y_2
16	0.853	0.728	Y_1
17	0.996	0.992	Y_0
18	0.853	0.728	Y_{23}
19	0.790	0.624	Y_{22}
20	0.740	0.548	Y_{21}
21	0.740	0.548	Y_{20}
22	0.722	0.521	Y_{19}
23	0.600	0.360	Y_{18}

Let the conductor temperature rise at time i after the application of a step function of losses corresponding to the rated current I_R be $\theta_R(i)$. The temperature rise corresponding to the cyclic current with a maximum value I will be denoted by $\theta(i)$. The maximum conductor temperature rise at time τ after an application of a rectangular current pulse of 1 h duration is given by expression (5.17). Since the temperature rise is assumed to be proportional to the square of the current, the temperature rise at time $t = 0$ is equal to

$$\theta(0) = \frac{I^2}{I_R^2}\mu\theta_R(\infty) \tag{5.23}$$

Equation (5.23) assumes that a uniform current $\sqrt{\mu}\ I$ has been applied for a long time before $t = 0$. At the end of time $t = \tau$, the temperature rise caused by this average current is equal to

$$\frac{I^2}{I_R^2}\mu\left[\theta_R(\infty) - \theta_R(\tau)\right] \tag{5.24}$$

If at $t = 0$, a current pulse with magnitude I_0 is applied, the maximum conductor temperature rise at time $t = 1$ is obtained from (5.17) and (5.24) as

$$\theta_{\max} = \frac{I^2}{I_R^2}\left\{\mu\left[\theta_R(\infty) - \theta_R(1)\right] + Y_0\left[\theta_R(1) - \theta_R(0)\right]\right\}$$

where Y_0 is the ratio I_0^2/I^2 and $\theta_R(0) = 0$.

Similarly, when a cyclic load is applied to the conductor, at the end of τ hours, the maximum conductor temperature rise is equal to

$$\theta_{\max} = \frac{I^2}{I_R^2}\left\{\mu\left[\theta_R(\infty) - \theta_R(\tau)\right] + \sum_{i=0}^{\tau-1} Y_i\left[\theta_R(i+1) - \theta_R(i)\right]\right\}$$

$$= \frac{I^2\theta_R(\infty)}{I_R^2}\left\{\mu\left[1 - \frac{\theta_R(\tau)}{\theta_R(\infty)}\right] + \sum_{i=0}^{\tau-1} Y_i\left[\frac{\theta_R(i+1)}{\theta_R(\infty)} - \frac{\theta_R(i)}{\theta_R(\infty)}\right]\right\} \tag{5.25}$$

where each of the Ys is expressed as a fraction of their maximum value, and Y_i is a measure of the equivalent square current between i and $(i+1)$ hours prior to the expected time of maximum conductor temperature (see Example 5.5).

The ratio $\theta_R(t)/\theta_R(\infty)$ is computed as follows. Consider a step current I_R applied at time $t = 0$. Conductor temperature rise above ambient is obtained from (5.14):

$$\theta_R(t) = \theta_c(t) + \alpha(t)\cdot\theta_e(t) \tag{5.26}$$

From the definitions of the attainment factors $\alpha(t)$ and $\beta(t)$ of the cable conductor and the cable outer surface, respectively, we have

$$\theta_c(t) = \alpha(t)\theta_c(\infty) \quad \text{and} \quad \theta_e(t) = \beta(t)\theta_e(\infty) \tag{5.27}$$

The steady-state temperature rises, on the other hand, are obtained from (4.5) as follows:

$$\theta_R(\infty) = W_c T + W_I T_4, \quad \theta_c(\infty) = W_c T, \quad \theta_e(\infty) = W_I T_4 \tag{5.28}$$

where W_I is the total joule loss in the cable.

Combining now (5.26)–(5.28), the required ratio is obtained as

$$\frac{\theta_R(t)}{\theta_R(\infty)} = [1 - k + k\beta(t)]\alpha(t) \quad \text{for } i \geq 1$$
$$\theta_R(0) = 0 \tag{5.29}$$

where

$$k = \frac{\theta_e(\infty)}{\theta_R(\infty)} = \frac{W_I T_4}{W_c T + W_I T_4} \tag{5.30}$$

Whitehead and Hutchings (1939) computed the attainment factor for the cable surface temperature assuming that a cable can be represented by a thin cylindrical constant source of heat in a semi-infinite volume of soil. The source was located along the circumference of the cable. The thermal properties of the cable were assumed to be those of the surrounding soil. Goldenberg (1967) has shown that a better approximation of the value of $\beta(t)$ can be obtained by applying the exponential-integral formula. Thus, from equations (5.10) and from (3.6), we obtain

$$\beta(t) = \frac{\theta_e(t)}{\theta_e(\infty)} = \frac{\dfrac{\rho_s W_I}{4\pi}\left[-Ei\left(-\dfrac{D_e^{*2}}{16t\delta}\right) + Ei\left(-\dfrac{L^{*2}}{t\delta}\right)\right]}{W_I T_4} \tag{5.31}$$

As will be shown in Chapter 9, the external thermal resistance of a single cable buried in a uniform soil is equal to $(\rho_s/2\pi)\ln(4L^*/D_e^*)$, where L^* is the depth of cable burial and D_e^* is its external diameter, both in meters. Substituting this into the last equation, we obtain

$$\beta(t) = \frac{-Ei\left(-\frac{D_e^{*2}}{16t\delta}\right) + Ei\left(-\frac{L^{*2}}{t\delta}\right)}{2\ln\left(\frac{4L^*}{D_e^*}\right)} \tag{5.32}$$

For standard cyclic-rating-factor calculations, the maximum conductor temperature rise reached during cyclic load is assumed to be equal to the maximum conductor temperature rise reached for steady-state (100% load factor) current. The cyclic rating factor M is defined so that $I = MI_R$. Therefore, from equation (5.25), we obtain

$$\boxed{M = \frac{1}{\sqrt{\mu\left[1 - \frac{\theta_R(\tau)}{\theta_R(\infty)}\right] + \sum_{i=0}^{\tau-1} Y_i\left[\frac{\theta_R(i+1)}{\theta_R(\infty)} - \frac{\theta_R(i)}{\theta_R(\infty)}\right]}}} \tag{5.33}$$

with the temperature ratios given by (5.29) and τ is usually taken to equal 6 h.

Calculations are simplified considerably when the conductor attainment factor can be assumed to be equal to one. In this case, equation (5.33) takes the form

$$M = \frac{1}{\sqrt{(1-k)Y_0 + k\{B + \mu[1 - \beta(\tau)]\}}} \tag{5.34}$$

where

$$B = \sum_{i=0}^{\tau-1} Y_i \Phi_i, \qquad \Phi_m = \beta(m+1) - \beta(m) \tag{5.35}$$

IEC (1989) identifies the following cases when the internal cable capacitances can be neglected. If the period from the initiation of the thermal transient is longer than

1. 12 h for all cables,
2. the product $\Sigma T \cdot \Sigma Q$ when dealing with fluid-pressure pipe-type cables and all types of self-contained cables where the product $\Sigma T \cdot \Sigma Q \le 2$ h, and
3. the product $2 \cdot \Sigma T \cdot \Sigma Q$ when dealing with gas-pressure pipe-type cables and all types of self-contained cables where the product $\Sigma T \cdot \Sigma Q > 2$ h

where ΣT and ΣQ are the total internal thermal resistance (simple sum of all resistances) and capacitance (simple sum of all capacitances), respectively, of the cable.

Table 5.7, based on design values commonly used at present for the determination of cable dimensions, shows when cases 2 and 3 apply (IEC, 1989).

5.6.2.2 Groups of Identical, Equally Loaded Cables. In this section, we consider groups of N cables with equal losses where the cables or ducts do not touch. In this case,

TABLE 5.7 Cases When the Attainment Factor Can Be Assumed to Be Equal to 1

Type of Cable	Case 2	Case 3
Fluid-filled cables	1. All voltages < 220 kV 2. 220 kV: sections ≤150 mm^2	1. 220 kV: sections > 150 mm^2 2. All voltages > 220 kV
Pipe-type, fluid-pressure cables	1. All voltages < 220 kV 2. 220 kV: sections ≤ 800 mm^2	1. 220 kV: sections > 800 mm^2 2. All voltages > 220 kV
Pipe-type, gas-pressure cables		1. ≤ 220 kV 2. Sections ≤ 1000 mm^2
Cables with extruded insulation	1. All voltages < 60 kV 2. 60 kV: sections ≤ 150 mm^2	1. 60 kV: sections > 150 mm^2 2. All voltages > 60 kV

$\theta_R(i)$ is the conductor temperature rise of the hottest cable in the group. The external thermal resistance of the hottest cable in equations (5.30) and (5.31) will now include the effect of the other $(N-1)$ cables and will be denoted by $T_4 + \Delta T_4$. Applying equation (5.12), we obtain the following new form of equation (5.31):

$$\beta_1(t) = \frac{\rho_s}{4\pi} \frac{-Ei\left(-\frac{D_e^{*2}}{16t\delta}\right) + Ei\left(-\frac{L^{*2}}{t\delta}\right) + \sum_{\substack{k=1 \\ k\neq p}}^{N}\left[-Ei\left(-\frac{d_{pk}^2}{4t\delta}\right) + Ei\left(-\frac{d_{pk}^{'2}}{4t\delta}\right)\right]}{(T_4 + \Delta T_4)} \tag{5.36}$$

The value of ΔT_4 is obtained in Section 9.6.2.1 as

$$\Delta T_4 = \frac{\rho_s \ln F}{2\pi}$$

where

$$F = \frac{d'_{p1} \cdot d'_{p2} \cdot \cdots \cdot d'_{pk} \cdot \cdots \cdot d'_{pN}}{d_{p1} \cdot d_{p2} \cdot \cdots \cdot d_{pk} \cdot d_{pN}} \tag{5.37a}$$

with factor d'_{pp}/d_{pp} excluded, leaving $(N-1)$ factors in (5.37a). The distances d'_{pp} and d_{pp} are defined in Fig. 5.4.

Introducing the notation

$$d_f = \frac{4L^*}{F^{1/(N-1)}}, \tag{5.37b}$$

equation (5.36) can be approximated by

$$\beta_1(t) = \frac{-Ei\left(-\frac{D_e^{*2}}{16\theta\delta}\right) + Ei\left(-\frac{L^{*2}}{t\delta}\right) + (N-1)\left[-Ei\left(-\frac{d_f^2}{16t\delta}\right) + Ei\left(-\frac{L^{*2}}{t\delta}\right)\right]}{2\ln\frac{4L^*F}{D_e^*}}. \tag{5.38}$$

Also, equation (5.30) becomes

$$k_1 = \frac{W_I(T_4 + \Delta T_4)}{W_c T + W_I(T_4 + \Delta T_4)} \tag{5.39}$$

The cyclic rating factor is given by equation (5.33) with

$$\frac{\theta_R(t)}{\theta_R(\infty)} = [1 - k_1 + k_1\beta_1(t)]\alpha(t) \tag{5.40}$$

EXAMPLE 5.6

Determine the cyclic rating factor of the cable system analyzed in Example 5.3. First, we have to identify the time during the 24 h period at which we expect that the conductor temperature rise will reach its maximum value. From analysis of the values in Table 5.6, we select $i = 18$ as the hour at which this maximum occurs. Note that this is not the time of maximum load, and our selection is based on engineering judgment.[3] The six preceding hours are underlined in Table 5.6 and the values of Y reordered accordingly. The new order is shown in the last column of Table 5.6.

Much of the work for the determination of the cyclic rating factor for the cable system under consideration has already been performed in Example 5.3. In particular, the conductor attainment factors $\alpha(t)$ and the exponential integral values have been computed for all hours. The load-loss factor has been evaluated in Example 5.5. The next step is to evaluate the cable surface attainment factors $\beta_1(t)$. A sample computation is performed below for $t = 1$ h with the remaining values summarized in column 3 of Table 5.8.

$$D_e^* = 0.0358 \text{ m}, \quad L^* = 1.0 \text{ m}, \quad d_{pk} = 0.072 \text{ m}$$

$$d'_{pk} = \sqrt{0.072^2 + 2.0^2} = 2.001 \text{ m}, \quad \delta = 0.5 \cdot 10^{-6} \text{ m}^2/\text{s}, \quad N = 3$$

The auxiliary variables defined in equations (5.37a) and (5.37b) are equal to

$$F = \frac{d'_{p1} \cdot d'_{p2}}{d_{p1} \cdot d_{p2}} = \frac{2.001 \cdot 2.001}{0.072 \cdot 0.072} = 772.4, \quad d_f = \frac{4L^*}{F^{1/(N-1)}} = \frac{4 \cdot 10}{\sqrt{772.4}} = 0.144$$

From equation (5.38) and Tables 5.1 and 5.2, we obtain

$$\beta_1(1) = \frac{-Ei\left(-\frac{D_e^{*2}}{16t\delta}\right) + Ei\left(-\frac{L^{*2}}{t\delta}\right) + (N-1)\left[-Ei\left(-\frac{d_f^2}{16t\delta}\right) + Ei\left(-\frac{L^{*2}}{t\delta}\right)\right]}{2 \ln \frac{4L^*F}{D_e^*}}$$

$$= \frac{(2.579 - 0) + 2 \cdot (0.365 - 0)}{2 \ln \frac{4 \cdot 1.0 \cdot 772.4}{0.0358}} = 0.146$$

Next, from equation (5.39), we compute factor k_1 with the value of $T_4 + \Delta T_4$ given above (the external thermal resistance given in Table A1 already includes the mutual heating effect) and equivalent thermal resistance T_A and T_B from Example 5.3. We have

$$k_1 = \frac{W_I(T_4 + \Delta T_4)}{W_c T + W_I(T_4 + \Delta T_4)} = \frac{33.38 \cdot 1.933}{30.82(0.214 + 0.113) + 33.38 \cdot 1.933} = 0.865$$

The ratios $\theta_R(i)/\theta_R(\infty)$ are computed from equation (5.40). As an example, we will calculate this ratio for $i = 1$. The remaining values, corresponding to the underlined Y values in Table 5.6, are shown in column 5 in Table 5.8.

[3] If no other information is available, the time of the highest loading should be selected.

TABLE 5.8 Evaluation of Cyclic Rating Factor

Time (h)	Y_i	$\alpha(t)$	$\beta(t)$	$\theta_R(i)/\theta_R(\infty)$
(1)	(2)	(3)	(4)	(5)
0	0.992	0.999	0.146	0
1	0.728	1	0.211	0.261
2	0.640	1	0.255	0.318
3	0.596	1	0.288	0.356
4	0.593	1	0.315	0.384
5	0.796	1	0.337	0.407
6	—	1	0.356	0.426

$$\frac{\theta_R(1)}{\theta_R(\infty)} = [1 - k_1 + k_1\beta_1(t)]\,\alpha(t) = (1 - 0.865 + 0865 \cdot 0.146) \cdot 1.0 = 0.261$$

The cyclic rating factor is now computed from equation (5.33) as follows:

$$M = \frac{1}{\sqrt{\mu\left[1 - \frac{\theta_R(\tau)}{\theta_R(\infty)}\right] + \sum_{i=0}^{\tau-1} Y_i\left[\frac{\theta_R(i+1)}{\theta_R(\infty)} - \frac{\theta_R(i)}{\theta_R(\infty)}\right]}}$$

$$= [0.504(1 - 0.426) + 0.992(0.261) + 0.728(0.318 - 0.261) + 0.640(0.356 - 0.318) + 0.596(0.384 - 0.356) + 0.593(0.407 - 0.384) + 0.796(0.426 - 0.407)]^{1/2} = 1.23$$

Thus, the permissible peak value of the cyclic load current is

$$1.23 \cdot 629 = 774 \text{ A}$$

EXAMPLE 5.7

Determine a cyclic rating factor for a single three-core cable with the load curve having the following characteristics: (1) a sustained maximum current lasts for a minimum of 6 h, and (2) there are no restrictions on the shape of the reminder of the cycle, except that the maximum conductor temperature rise occurs at the end of the duration of the sustained high current.

Since in this case $Y_0 = Y_1 = \cdots = Y_6 = 1$, equation (5.33) simplifies to

$$M \frac{1}{\sqrt{\mu\left[1 - \frac{\theta_R(6)}{\theta_R(\infty)}\right] + \frac{\theta_R(6)}{\theta_R(\infty)}}} \tag{5.41}$$

Substituting now equation (5.29), we obtain

$$\boxed{M \frac{1}{\sqrt{\mu + (1-\mu)[1 - k + k\beta(6)]\alpha(6)}}} \tag{5.42}$$

EXAMPLE 5.8

Assume that a cyclic rating factor M has been derived for the cable system located in a soil with thermal resistivity ρ_s. We will determine a new cyclic factor for the same cable system with the soil thermal resistivity equal to ρ_s'. We will then use the resulting formula to determine the cyclic rating of the system in Example 5.6 with a soil thermal resistivity of 0.85 K · m/W.

The factors k and k_1 [see equations (5.30) and (5.39)] are dependent on the depth of laying and soil resistivity. We will first rewrite equation (5.30) as follows:

$$k = \frac{W_I T_4}{W_c T + W_I T_4} = \frac{W_I T_4}{W_I \dfrac{T}{1 + \lambda_1 + \lambda_2} + W_I T_4} = \frac{W_I T_4}{W_I T_c + W_I T_4} = \frac{T_4}{T_c + T_4} \tag{5.43}$$

where T_c is the cable internal thermal resistance as if all the joule losses were generated at the conductor. Now, if the laying conditions "a," for which a value of k is known, change to conditions "b," then

$$k(a) = \frac{T_4^a}{T_c + T_4^a} \tag{5.44}$$

and

$$k(b) = \frac{T_4^b}{T_c + T_4^b} \tag{5.45}$$

where T_4^a and T_4^b are the external thermal resistances for conditions "a" and "b," respectively.

Computing T_c from (5.44) and substituting this in (5.45), we obtain

$$k(b) = \frac{1}{1 + \dfrac{T_4^a}{T_4^b}\left[\dfrac{1 - k(a)}{k(a)}\right]} \tag{5.46}$$

The new value of k_1 is obtained in the same way.

In our example, the depth is not changed. Therefore,

$$\frac{T_4^a}{T_4^b} = \frac{\rho_s}{\rho_s'} = \frac{1}{0.85} = 1.176$$

and

$$k_1' = \frac{1}{1 + 1.176\left[\dfrac{1 - 0.865}{0.965}\right]} = 0.845$$

The new ratio $\theta_R(1)/\theta_R(\infty) = 1 - 0.845 + 0.845 \cdot 0.211 = 0.333$. The remaining ratios are as follows:

i	1	2	3	4	5	6
$\dfrac{\theta_R(i)}{\theta_R(\infty)}$	0.333	0.370	0.398	0.421	0.440	0.456

The revised cyclic loading factor is thus equal to

$$M = [0.504(1 - 0.456) + 0.992(0.333) + 0.728(0.370 - 0.333) + 0.640(0.398 - 0.370) + 0.596(0.421 - 0.398) + 0.593(0.440 - 0.421) + 0.796(0.456 - 0.440)]^{-1/2} = 1.21$$

The steady-state ampacity of this cable system is now larger than before. The new steady-state rating is equal to 674 A. Therefore, the new cyclic rating is equal to $1.21 \cdot 674 = 815$ A.

5.6.3 Cyclic Rating with Partial Drying of the Surrounding Soil

5.6.3.1 Assumptions. The method to be presented in this section, based on CIGRE (1992), applies a two-zone soil model and is an extension of the techniques described previously for calculating the cyclic rating factor for a cable in uniform soil. We should stress again that there are several assumptions associated with an application of a two-zone model. These assumptions were reviewed in Section 4.2. Additional assumptions, related to the application of the method for cyclic factor computations, are listed below.

- Once the soil around a cable has dried out, it will not rewet while there is any heat flux flowing in a direction away from the cable.
- The position of the dry zone boundary will be where the peak soil temperature during a load cycle just attains the critical value.
- Cyclic load has persisted for a duration long enough so that the temperature variations in the soil have attained their final values.

Based on the results of numerous calculations on different types of cables, Parr (1987, 1988) has observed that the cyclic rating factor is essentially independent of the value of the critical temperature or the size of the dry zone. For a given conductor temperature, the relationship between the sizes of the dry zones for both cyclic and sustained load varies in such a way that, although the steady-state rating changes substantially with the critical temperature, the cyclic rating factor multiplying that steady-state rating is largely unaffected.

Further, calculations for a wide range of soil and load characteristics indicate that a factor derived for cyclic conditions where the critical temperature of the soil is just equal to the peak cyclic temperature of the cable surface (that is, a dry zone has just not developed) will be on the safe side by not more than 5% from the cases where there is a substantial dry zone (Parr, 1987, 1988; CIGRE, 1992), and in most cases the error is not greater than 2%. Such an error is well within the accuracy with which soil and load characteristics are usually known.

This feature makes it possible to considerably simplify the computations of cyclic rating factors when soil dry out is expected. In the developments presented below, it is first assumed that the soil critical temperature is equal to the peak cyclic value of the cable surface temperature; that is, a dry zone is at the point of occurring. At this point, the soil surrounding the cable has uniform properties appropriate to its wet, *in situ*, state. The cyclic rating factor for these conditions is derived by equation (5.33). The factor is then adjusted so that it applies to the steady-state rating for the same assumed value of critical temperature when there is a dried out zone.

In general, the size of a dry zone where the boundary just achieves a certain critical temperature rise with cyclic loading is smaller than the zone which will form for the same critical temperature rise with steady-state loading. We also observe that the size of dry zone, and hence the cable external thermal resistance, changes with the type of loading. The last observation has an important implication for the computational procedure adopted in this section. As indicated above, the procedure is to make an adjustment to the cyclic rating factor computed for uniform soil when the external thermal resistance is the same for cyclic and the steady-state conditions. Therefore, for cyclic rating computations, an adjustment is made by simply using the ratio of the appropriate external thermal resistances. This is shown below.

5.6.3.2 Development of the Cyclic Rating Factor. The computation of a cyclic rating factor is performed as follows. The thermal response of the cable and its environment, including the effect of its internal thermal capacitance, is obtained by using the appropriate formulas in Section 5.2. According to the convention adopted at the beginning of this chapter, the temperature rises referred to here are those due to the joule losses only, and the equations quoted take this into account at the appropriate stages. The factor M is adjusted for the presence of a dry zone under steady-state loading conditions; Parr (1987) developed the following procedure to find the new cyclic factor M_1 for two-zone soils. Let I be the maximum current permitted by applying the factor from equation (5.33) to the steady-state rating based on uniform soil around the cable. Then

$$I = MI_R \tag{5.47}$$

where I_R is the rating based on an external thermal resistance with no dry zone. A load equal to I_R can be carried only if the cable surface temperature under steady-state loading does not exceed the critical temperature for the soil immediately around the cable. In this case, factor M is applicable without correction.

If such is not the case, a dry zone is presumed to form, and the cable external thermal resistance will be increased. Let I_R' denote the steady-state rating current when there is a dry zone. This current is obtained from equation (4.13). In order to maintain the same peak value of current I, we should have

$$I = MI_R = M_1 I_R' \tag{5.48}$$

From equation (5.48), we obtain

$$M_1 = \frac{I_R}{I_R'} M \tag{5.49}$$

Using the special notation adopted in this chapter, the rated current I_R is obtained from (4.3):

$$I_R = \left[\frac{\theta(\infty) - W_d\,[0.5T_1 + n(T_2 + T_3 + T_4)]}{R_c(T_c + T_4)} \right]^{0.5} \tag{5.50}$$

where

$$R_c = nR(1 + \lambda_1 + \lambda_2)$$

$$T_c = \frac{\frac{T_1}{n} + (1 + \lambda_1)T_2 + (1 + \lambda_1 + \lambda_2)T_3}{1 + \lambda_1 + \lambda_2}$$

and $\theta(\infty)$ represents permitted conductor temperature rise (due to all losses). T_c represents the internal thermal resistance of the cable computed under the assumption that all the joule losses are produced at the conductor. The external thermal resistance T_4 is that of the wet soil.

The rated current with moisture migration taken into account is given by equation (4.13), rewritten below with the special notation adopted in this chapter:

$$I_R' = \left[\frac{\theta(\infty) - W_d\,[0.5T_1 + n(T_2 + T_3 + \nu \cdot T_4)] + (\nu - 1)\theta_x}{R_c(T_c + \nu \cdot T_4)} \right]^{0.5} \tag{5.51}$$

where

$$\nu = \frac{\rho_2}{\rho_1} \tag{5.52}$$

and ρ_1 and ρ_2 are the thermal resistivities of the wet and dry soil, respectively.

We also have

$$\frac{T_c + \nu T_4}{T_c + T_4} = \frac{T_c + T_4 + (\nu - 1)T_4}{T_c + T_4} = 1 + k(\nu - 1) \tag{5.53}$$

because, from equation (5.43),

$$k = \frac{T_4}{T_c + T_4}$$

Dividing (5.50) by (5.51) with the ratio of thermal resistances given by (5.53), we obtain the required expression for M_1. In developing this ratio, we are taking into account the fact mentioned above, that a factor derived for cyclic conditions where the critical temperature of the soil is just equal to the peak cyclic temperature of the cable surface (that is, a dry zone has just not developed) will be on the safe side by a small margin. This means that we make the following substitution:

$$\theta_x = \theta_e(\tau) + nW_d T_4$$

Hence, M_1 can be expressed as

$$\boxed{M_1 = M\sqrt{\frac{1 + k(\nu - 1)}{1 + \dfrac{\theta_e(\tau)}{\theta_R(\infty)}(\nu - 1)}}} \tag{5.54}$$

where $\theta_e(\tau)$ is the peak cable surface temperature rise occurring at time $t = \tau$. Also, in accordance with the notation adopted in this chapter, all θs represent temperature rises above ambient due to joule losses.

Equation (5.54) is applicable only when the total cable surface temperature rise $[\theta_e(\tau) + nW_d T_4]$ is greater than the critical temperature rise of the soil; otherwise, there will be no drying and the cyclic rating factor is M without any correction. Thus, in order to determine which rating factor to use, the value of $[\theta_e(\tau) + nW_d T_4]$ has to be determined first. This is accomplished as follows.

From the definition of the cyclic rating factor

$$\theta_{\max} = \theta_R(\infty) \tag{5.55}$$

or

$$\theta_c(\tau) + \theta_e(\tau) = \theta_R(\infty)$$

On the other hand, assuming that the conductor-above-cable-surface temperature rise is proportional to the square of the current, we have from equation (5.25)

$$\begin{aligned}\theta_c(\tau) &= \frac{I^2\theta_c(\infty)}{I_R^2}\left\{\mu\left[1 - \frac{\theta_c(\tau)}{\theta_c(\infty)}\right] + \sum_{i=0}^{\tau-1} Y_i\left[\frac{\theta_c(i+1)}{\theta_c(\infty)} - \frac{\theta_c(i)}{\theta_c(\infty)}\right]\right\}\\ &= M^2\theta_c(\infty)\{\mu[1 - \alpha(\tau)] + A'\} = M^2\theta_R(\infty)(1 - k)\{\mu[1 - \alpha(\tau)] + A'\}\end{aligned} \tag{5.56}$$

because

$$M^2 I^2 = I_R^2 \quad \text{and} \quad \theta_c(\infty) = (1-k)\theta_R(\infty) \tag{5.57}$$

where

$$A' = \sum_{i=0}^{\tau-1} Y_i \left[\frac{\theta_c(i+1)}{\theta_c(\infty)} - \frac{\theta_c(i)}{\theta_c(\infty)} \right] = \sum_{i=0}^{\tau-1} Y_i \left[\alpha(i+1) - \alpha(i)\right] \tag{5.58}$$

Substituting (5.56) into (5.55), we obtain

$$\theta_e(\tau) = \theta_R(\infty) \left\{1 - M^2(1-k)\{A' + \mu[1-\alpha(\tau)]\}\right\} \tag{5.59}$$

EXAMPLE 5.9

Determine the cyclic rating factor for the cable system examined in Example 5.6, taking moisture migration into account. Assume a dry soil thermal resistivity value equal to 2.5 K · m/W and a critical soil temperature rise of 35 K.

First we compute from equation (5.58) the value of A' with the temperature ratios and attainment factors given in Table 5.8:

$$\begin{aligned} A' &= \sum_{i=0}^{\tau-1} Y_i \left[\frac{\theta_c(i+1)}{\theta_c(\infty)} - \frac{\theta_c(i)}{\theta_c(\infty)} \right] = \sum_{i=0}^{\tau-1} Y_i \left[\alpha(i+1) - \alpha(i)\right] \\ &= 0.992(1.0) + 0.728(1.0 - 1.0) + 0.640(1.0 - 1.0) \\ &\quad + 0.596(1.0 - 1.0) + 0.593(1.0 - 1.0) + 0.796(1.0 - 1.0) = 0.992 \end{aligned}$$

The conductor temperature rise due to joule losses is equal to $90 - 15 = 75$ K because the dielectric losses are negligible. From Example 5.6, $M = 1.23$, $k = 0.856$, and $\alpha(6) = 1$. The loss factor was computed in Example 5.5 and $\mu = 0.504$. Therefore, the peak temperature rise of the cable surface is equal to

$$\begin{aligned} \theta_e(6) = \theta_e(\tau) &= \theta_R(\infty) \left\{1 - M^2(1-k)\{A' + \mu[1-\alpha(\tau)]\}\right\} \\ &= 75 \cdot \{1 - 1.23^2(1 - 0.865)[0.992 + 0.504(1 - 1.0)]\} = 59.8 \text{ K} \end{aligned}$$

This is greater than the critical value of 35 K; hence, drying can be expected. The corrected rating factor for this condition is, from equation (5.54), equal to

$$M_1 = M \sqrt{\frac{1 + k(\nu - 1)}{1 + \dfrac{\theta_e(\tau)}{\theta_R(\infty)}(\nu - 1)}} = 1.23 \sqrt{\frac{1 + 0.865(2.5 - 1)}{1 + \dfrac{59.8}{75} \cdot (2.5 - 1)}} = 1.26$$

The steady-state rating with drying out when the critical temperature rise is 35 K was obtained in Example 4.2 and is equal to 541 A; hence, the peak current is $1.26 \cdot 541 = 682$ A.

5.7 CALCULATION OF EMERGENCY RATINGS

During emergency conditions (e.g., loss of a companion circuit in a two-circuit transmission line), power cables may be required to carry substantially higher currents than permitted by the steady-state rating. Emergency conditions usually last only a few hours, and the conductor temperature is often permitted to reach a higher value than that allowed in a steady-state operation. In this section, we will develop formulas for calculating the short time rating of a single circuit based on knowledge of the conductor temperature transient as derived in Section 5.2.

5.7.1 Thermally Isolated Circuits

Consider an isolated buried circuit carrying a constant current I_1 applied for a sufficiently long time so that steady-state conditions are effectively reached. If a cyclic load with a peak value of current equal to I amperes has been applied for a long time, then $I_1 = \sqrt{\mu} I$ where μ is the load-loss factor of the cyclic load. Subsequently, from a time defined by $t = 0$, an emergency load current I_2 (greater than I_1) is applied. If I_2 is applied for any given time t, the question is how large may I_2 be so that conductor temperature does not exceed a specified value, taking into account the variation of the electrical resistivity of the conductor with temperature. The effect of dielectric loss is first neglected, but is taken into account at the end of this discussion.

In the following developments, we will assume that the heat generation per unit volume of the conductor for the time t of emergency loading is constant and equal to its value W_{max} at the end of the period of emergency loading, at which time the conductor temperature rise above ambient is θ_{max}. Goldenberg (1971) has shown that with this assumption, a safe value of emergency current can be obtained. We can write

$$\frac{\theta_{max} - \theta_R(0)}{\theta_R(t)} = \frac{W_{max} - W_0}{W_R} \tag{5.60}$$

where $W_0 = I_1^2 R_1$ and $W_R = I_R^2 R_R$ are the heat generation per unit volume of conductor at time $t = 0$ and during the steady-state, respectively. I_R is the steady-state rated current and the conductor ac resistance corresponding to this current is denoted in this section by R_R. Substituting now the definition of the conductor losses, equation (5.60) can be rewritten as

$$\frac{r - h_1^2 R_1/R_R}{\theta_R(t)/\theta_R(\infty)} = \frac{\dfrac{I_2^2}{I_R^2} R_{max}}{R_R} - \frac{h_1^2 R_1}{R_R} \tag{5.61}$$

where $$r = \frac{\theta_{max}}{\theta_R(\infty)}$$

$$h_1 = \frac{I_1}{I_R}$$

$\theta_R(\infty)$ = steady-state temperature rise corresponding to current I_R

θ_{max} = maximum permissible temperature rise above ambient at the end of the emergency period

$\theta_R(t)$ = conductor temperature rise above ambient at time t after application of current I_R neglecting the variation of conductor resistance with temperature from $t = 0$

$\theta_R(0)$ = steady-state conductor temperature rise above ambient following application of current I_1

The emergency rating current is obtained by solving equation (5.61) for I_2:

$$I_2 = I_R \left\{ \frac{h_1^2 R_1}{R_{\max}} + \frac{\frac{R_R}{R_{\max}}\left[r - h_1^2 \frac{R_1}{R_R}\right]}{\frac{\theta_R(t)}{\theta_R(\infty)}} \right\}^{\frac{1}{2}} \tag{5.62}$$

The ratio $\theta_R(t)/\theta_R(\infty)$ is obtained by computing both temperatures rises separately. The steady-state conductor temperature rise above ambient is simply

$$\theta_R(\infty) = \theta_c - \theta_{\text{amb}} - \theta_d \tag{5.63}$$

with θ_{amb} and θ_d denoting the ambient temperature and the temperature rise caused by the dielectric loss, respectively. θ_c is the maximum steady-state conductor temperature expressed in °C. The value of $\theta_R(t)$ is obtained by applying equation (5.26) while neglecting dielectric losses.

The effect of heating due to dielectric loss is taken into account by calculating from equation (5.11) the steady-state conductor temperature rise θ_d due to the dielectric loss and subtracting this value from $\theta_{\max}$, $\theta_R(t)$, and $\theta_R(\infty)$. Any change in dielectric loss with temperature is neglected. The calculation of I_2 then proceeds as above, using the modified values of $\theta_{\max}$, $\theta_R(t)$, and $\theta_R(\infty)$. The values of R_2, $R_{\max}$, R_1, and R_2 are not altered.

Goldenberg (1971) has shown that the emergency rating current obtained from equation (5.62) is on the safe side with an error not exceeding 2% for an emergency duration of 3 h or more, 3% for an emergency duration of 2 h, and 1–5% for an emergency duration of 1 h.

EXAMPLE 5.10

Assume that the cable system examined in Example 5.3 has been operating at 550 A continuously, which gives a steady-state temperature of 70°C (the ampacity of this circuit at 90°C is given in Appendix A as $I_2 = 629$ A). Determine the emergency current level that can be carried for 6 h without exceeding the maximum operating temperature of 105°C.

From the statement of the problem, we have

$$r = \frac{\theta_{\max}}{\theta_R(\infty)} = \frac{105 - 15}{70 - 15} = 1.64, \quad t = 6\text{ h } (21\,600\text{ s}), \quad h_1 = \frac{I_1}{I_R} = \frac{550}{629} = 0.874$$

$$R_1 = 7.338 \cdot 10^{-5}\Omega/\text{m (ac resistance at 70°C)}$$

$$R_{\max} = 8.151 \cdot 10^{-5}\Omega/\text{m (ac resistance at 105°C)}$$

$$R_R = 7.810 \cdot 10^{-5}\Omega/\text{m (ac resistance at 90°C)}$$

Temperature rise due to dielectric loss = 0 K. Thus, from equation (5.63)

$$\theta_R(\infty) = \theta_c - \theta_{\text{amb}} - \theta_d = 90 - 15 = 75\text{ K}$$

$\theta_R = 30.4$ (see Table 5.2, column 5) for application of a step function of rated current. The emergency current is obtained from equation (5.62):

$$I_2 = I_R \left\{ \frac{h_1^2 R_1}{R_{\max}} + \frac{\frac{R_R}{R_{\max}}\left[r - h_1^2 \frac{R_1}{R_R}\right]}{\frac{\theta_R(t)}{\theta_R(\infty)}} \right\}^{\frac{1}{2}}$$

$$= 629 \left\{ \frac{0.874^2 \cdot 7.338}{8.151} + \frac{\frac{7.810}{8.151}\left[1.29 - 0.874^2 \frac{7.338}{7.810}\right]}{\frac{30.4}{75}} \right\}^{\frac{1}{2}} = 899 \text{ A}$$

5.7.2 Groups of Circuits

Where groups of circuits which are not thermally independent carry loads behaving in the same way, the methods for calculating the emergency ratings given in Section 5.7.1 can be applied.

When the loads do not behave in the same way, for example, the first circuit has an increased load while the second goes off load, the matter requires careful consideration. In most cases of buried cables, the time lag associated with the mutual heating between circuits is so great that any worthwhile reduction in temperature of the first circuit due to the second going off load does not occur for many hours.

Figure 5.8 (CIGRE, 1978) shows the time taken for a 1 K reduction in temperature of the first circuit of a double-circuit 1935 mm^2, 400 kV installation when the current in the second circuit is reduced to zero. Thus, any increase in load for the first circuit during this period will depend entirely on what increase in conductor temperature can be permitted above the value the first circuit had attained before the change in load.

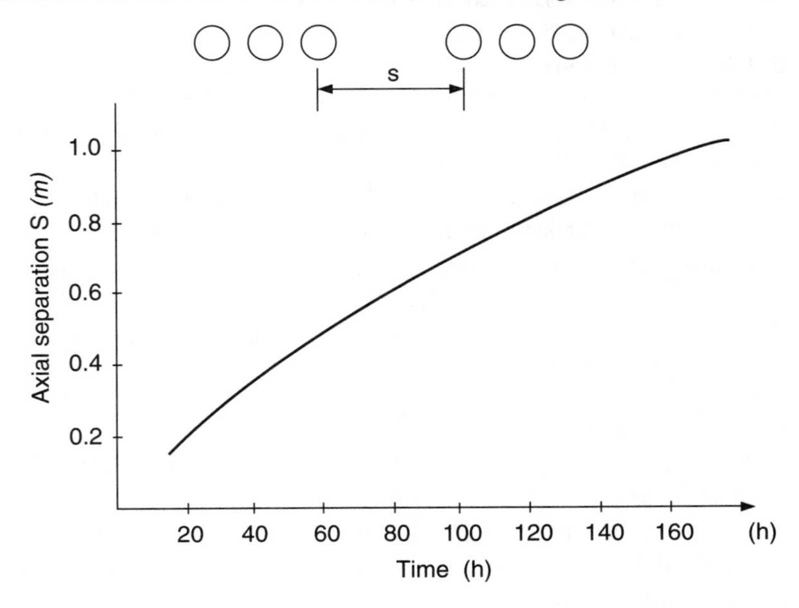

Figure 5.8 Time for 1 K reduction in temperature at the axis of the central cable of a first circuit due to current reduced to zero in a second circuit (IEC, 1989).

REFERENCES

Affolter, J. F. (1987), "An improved methodology for transient analysis of underground power cables using electrical network analogy," M.E. Thesis, McMaster University, Hamilton, Canada.

Anders, G. J., Moshref, A., and Roiz, J. (July 1990), "Advanced computer programs for power cable ampacity calculations," *IEEE Comput. Appl. in Power*, vol. 3, no. 3, pp. 42–45.

Anders, G. J., and El-Kady, M. A. (Oct. 1992), "Transient ratings of buried power cables—Part 1: Historical perspective and mathematical model," *IEEE Trans. Power Delivery*, vol. 7, no. 4, pp. 1724–1734.

Baudoux, A., Verbisselet, J., and Fremineur, A. (1962), "Etude des regimes thermiques des machines electriques par des modeles analogiques, application aux cables et aux transformateurs," Rapport CIGRE, No. 126.

Buller, F. H. (1951), "Thermal transients on buried cables," *Trans. Amer. Inst. Elect. Eng.*, vol. 70, pp. 45–52.

CIGRE (Oct. 1972), "Current ratings of cables for cyclic and emergency loads. Part 1. Cyclic ratings (load factor less than 100%) and response to a step function," *Electra*, No. 24, pp. 63–96.

CIGRE (Jan. 1976), "Current ratings of cables for cyclic and emergency loads. Part 2. Emergency ratings and short duration response to a step function," *Electra*, no. 44, pp. 3–16.

CIGRE (Dec. 1992), "Methods for calculating cyclic ratings for buried cables with partial drying of the surrounding soil," *Electra*, no. 145, pp. 33–67.

Goldenberg, H. (1957), "The calculation of cyclic rating factors for cables laid direct or in ducts," *Proc. IEE*, vol. 104, part C, pp. 154–166.

Goldenberg, H. (1958), "The calculation of cyclic rating factors and emergency loading for one or more cables laid direct or in ducts," *Proc. IEE*, vol. 105, part C, pp. 46–54.

Goldenberg, H. (1964), "Cyclic loading and transient temperatures of power cables," Working Paper of CIGRE WG02 of Study Committee 21.

Goldenberg, H. (1967), "Thermal transients in linear systems with heat generation linearly temperature-dependent: application to buried cables," *Proc. IEE*, vol. 114, pp. 375–377.

Goldenberg, H. (1971), "Emergency loading of buried cable with temperature-dependent conductor resistance," *Proc. IEE*, vol. 118, pp. 1807–1810.

IEC Standard (1985), "Calculation of the cyclic and emergency current ratings of cables. Part 1: Cyclic rating factor for cables up to and including 18/30 (36) kV," Publication 853-1.

IEC Standard (1989), "Calculation of the cyclic and emergency current ratings of cables. Part 2: Cyclic rating factor of cables greater than 18/30 (36) kV and emergency ratings for cables of all voltages," Publication 853-2.

King, S. Y., and Halfter, N. A. (1982), *Underground Power Cables*. New York: Longman.

McGrath, M. H. (1964), Discussion contribution to Neher, 1964, *IEEE Trans. Power App. Syst.*, vol. 83, p. 113.

Morello, A. (1958), "Variazioni transitorie die temperatura nei cavi per energia," *Elettrotecnica*, vol. 45, pp. 213–222.

Neher, J. H. (1964), "The transient temperature rise of buried power cable systems," *IEEE Trans. Power App. Syst.*, vol. PAS-83, pp. 102–111.

Parr, R. G. (1987), "Cyclic ratings for cables. A simple adaptation for partly dried soil," ERA Report No. 87-0265.

Parr, R. G. (1988), "Cyclic ratings for cables. A simple adaptation for partly dried soil. Supplement to Report No. 87-0265," ERA Report No. 88-0127.

Thomann, G. C., Aabo, T., Ghafurian, R., McKernan, T., and Bascom, E. C. (1991), "A Fourier transform technique for calculating cable and pipe temperatures for periodic and transient conditions," *IEEE Trans. Power Delivery*, vol. 6, no. 4, pp. 1345–1351.

Van Valkenburg, M. E. (1964), *Network Analysis*. Englewood Cliffs, NJ: Prentice-Hall.

Van Wormer, F. C. (1955), "An improved approximate technique for calculating cable temperature transients," *Trans. Amer. Inst. Elect. Eng.*, vol. 74, part 3, pp. 277–280.

Whitehead, S., and Hutchings, E. E. (1938), "Current ratings of cables for transmission and distribution," *J. IEE*, vol. 38, pp. 517–557.

Wlodarski, R. (1963), "Echauffement de cables souterrains de transport d'énergie," Edition de l'École Polytechnique de Varsovie.

Wlodarski, R., and Cabiac, M. (1966), *Études et Éxperiences Récentes Concernant la Détermination de l'Échauffement Transitoire des Càbles Enterrés*. Warsaw: Panstwowe Wydawnictwo Naukowe.

PART II EVALUATION OF PARAMETERS

6

Dielectric Losses

When paper and solid dielectric insulations are subjected to alternating voltage, they act as large capacitors and charging currents flow in them. The work required to effect the realignment of electrons each time the voltage direction changes (i.e., 50 or 60 times a second) produces heat and results in a loss of real power which is called dielectric loss, and which should be distinguished from reactive loss. For a unit length of a cable, the magnitude of the required charging current is a function of the dielectric constant of the insulation, the dimensions of the cable, and the operating voltage. For some cable constructions, notably for high-voltage, paper-insulated cables, this loss can have a significant effect on the cable rating. In this chapter, we will develop formulas for calculating the dielectric losses and examine the effect of cable construction on the value of these losses. Before doing so, however, we will review a few basic concepts related to the behavior of dielectrics subjected to ac voltages.

Cable insulation is a material whose dielectric response is a result of its capacitive nature (ability to store charge) and its conductive nature (ability to pass charge). The material can be represented by a resistor and capacitor in parallel (Fig. 6.1).

When a voltage U_0 is applied to this circuit, the current I will form an angle φ with the voltage as shown in Fig. 6.1. This current is composed of two components: capacitive (charging) current I_c and the resistive (leakage) current I_r. Since, in the case of good insulating materials, the magnitude of the leakage current vector is much smaller than that of the capacitance current vector, the loss angle δ is very small. The charging and leakage currents are equal to

$$I_c = j\omega C U_0 \quad \text{and} \quad I_r = \frac{U_0}{R_i} \tag{6.1}$$

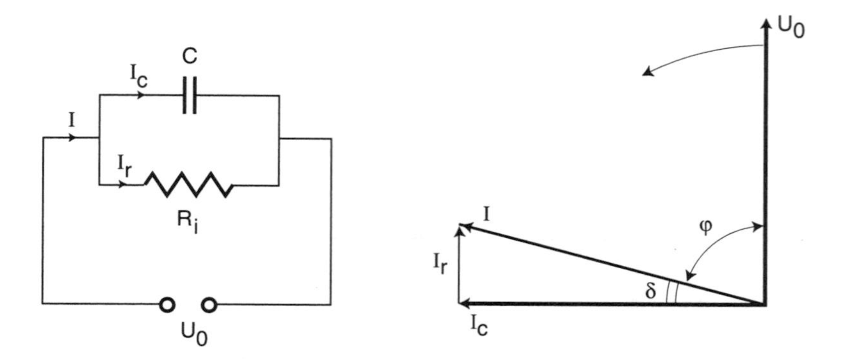

Figure 6.1 Representation of a cable insulation.

where C is the capacitance of the insulation and $\omega = 2\pi f$, f being the system frequency and $j = \sqrt{-1}$. To compute C, we observe that the effect of dielectric material has been traditionally described by introducing the concept of relative permittivity, denoted by ε and defined as

$$\varepsilon = \frac{C}{C_0}$$

where C_0 is the capacitance of identical size and construction capacitor with vacuum as the dielectric. The quantity ε is often referred to as the static or low-frequency value of the permittivity or the *dielectric constant*. Then,

$$\boxed{C = \varepsilon \cdot C_0 = \frac{\varepsilon}{18 \ln\left(\dfrac{D_i}{d_c}\right)} \cdot 10^{-9}} \tag{6.2}$$

where D_i is the external diameter of the insulation excluding screen and d_c is the diameter of the conductor, including screen. The same formula can be used for oval conductors if the geometric mean of the appropriate major and minor diameters is substituted for D_i and d_c.

Another measure of a dielectric is the dissipation factor denoted by $\tan\delta$ and often referred to as the *loss factor* of the insulation at power frequency. From Fig. 6.1, we can see that

$$\tan\delta = \frac{|I_r|}{|I_c|} = \frac{U_0}{R_i C \omega U_0} = \frac{1}{R_i C \omega} \tag{6.3}$$

Evidently, the smaller the value of $\tan\delta$, the more the dielectric material approaches the condition of a perfect insulator. For a given set of electric field and system frequency, the loss factor will undergo change with temperature. This relationship is shown in Fig. 6.2 (Westinghouse, 1957).

In practice, ε and $\tan\delta$ are assumed constant in computations of cable ratings. Their values are given in Table 6.1 (IEC 287, 1989; IEC 287-1-1, 1994).

The dielectric loss per unit length in each phase is then obtained from equation (6.4) as

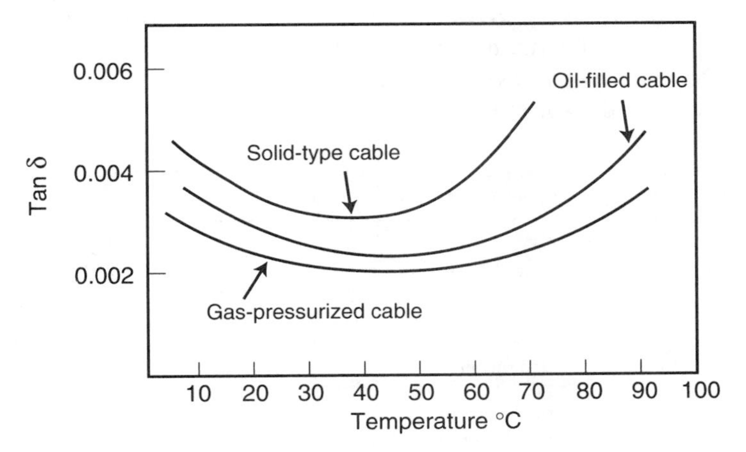

Figure 6.2 Variation of tan δ with temperature.

TABLE 6.1 Values of ε and tan δ for the Insulation of Power Cables at Power Frequency

Type of Cable	ε	tan δ
Cables insulated with impregnated paper	4	0.01
Solid type, fully impregnated, preimpregnated or mass-impregnated nondraining		
Fluid-filled low-pressure		
up to $U_0 = 36$ kV	3.6	0.0035
up to $U_0 = 87$ kV	3.6	0.0033
up to $U_0 = 160$ kV	3.5	0.0030
up to $U_0 = 220$ kV	3.5	0.0028
Fluid-pressure, pipe-type	3.7	0.0045
Internal gas-pressure	3.4	0.0045
External gas-pressure	3.6	0.0040
Cables with other kinds of insulation		
Butyl rubber	4	0.050
EPR—up to 18/30 kV	3	0.020
EPR—above 18/30 kV	3	0.005
PVC	8	0.1
PE (HD and LD)	2.3	0.001
XLPE up to and including 18/30 (36) kV—unfilled	2.5	0.004
XLPE above 18/30 (36) kV—unfilled	2.5	0.001
XLPE above 18/30 (36) kV—filled	3	0.005
Paper–polypropylene–paper (PPL)*	2.8	0.001

*The dielectric constant and the loss factor of PPL insulation have not been standardized yet.

$$W_d = \frac{U_0^2}{R_i} = \omega C U_0^2 \tan\delta \tag{6.4}$$

Dielectric loss is voltage dependent, and thus only becomes important at higher voltage levels. Table 6.2 (IEC 287-1-1, 1994) gives, for the insulation materials in common use, the value of U_0 at which the dielectric loss should be taken into account if three-core screened or single-core cables are used. It is not necessary to calculate the dielectric loss for unscreened multicore or dc cables.

TABLE 6.2 Phase-to-Ground Voltage above Which Dielectic Losses Must Be Calculated

Type of Cable	U_0 (kV)
Cables insulated with solid paper	38
Cables insulated with fluid/gas	63.5
Butyl rubber	18
EPR	63.5
PVC	6
PE (HD and LD)	127
XLPE unfilled	127
XLPE filled	63.5
PPLP	63.5

EXAMPLE 6.1

We will determine dielectric losses of the model cable model No. 3 (pipe-type cable) and compare them with the dielectric losses of model cable model No. 5 (400 kV PPL cable) assuming that both cables operate at 60 Hz. The parameters of the pipe-type cable are (see Appendix A) $D_i = 67.1$ mm, $d_c = 41.45$ mm, and $\varepsilon = 3.5$. The parameters of the cable model No. 5 are $D_i = 94.6$ mm, $d_c = 58.6$ mm, and $\varepsilon = 2.8$.

The capacitances of cables 3 and 5 are

$$C_{cab3} = \frac{\varepsilon}{18\ln\left(\dfrac{D_i}{d_c}\right)} \cdot 10^{-9} = \frac{3.5}{18\ln\left(\dfrac{67.1}{41.45}\right)} \cdot 10^{-9} = 0.4036 \cdot 10^{-9} \quad \text{F/m}$$

$$C_{cab5} = \frac{\varepsilon}{18\ln\left(\dfrac{D_i}{d_c}\right)} \cdot 10^{-9} = \frac{2.8}{18\ln\left(\dfrac{94.6}{58.6}\right)} \cdot 10^{-9} = 0.3248 \cdot 10^{-9} \quad \text{F/m}$$

Dielectric losses are obtained from equation (6.4) with the loss factor from Table 6.1:

$$W_{dcab3} = \frac{U_0^2}{R_i} = \omega C U_0^2 \tan\delta = 2\pi 60 \cdot 0.4267 \cdot 10^{-9} \cdot \left(\frac{138\,000}{\sqrt{3}}\right)^2 \cdot 0.005 = 4.83 \quad \text{W/m}$$

$$W_{dcab5} = \frac{U_0^2}{R_i} = \omega C U_0^2 \tan\delta = 2\pi 60 \cdot 0.3248 \cdot 10^{-9} \cdot \left(\frac{400\,000}{\sqrt{3}}\right)^2 \cdot 0.0001 = 6.53 \quad \text{W/m}$$

We can observe that in spite of a voltage level about three times higher, the dielectric losses in the PPL cable are almost the same as the corresponding losses in the pipe-type cable with standard paper insulation for these examples.

REFERENCES

IEC 287 (1982), "Calculation of the continuous current rating of cables (100%) load factor," IEC Publication 287.

IEC 287-1-1 (1994), "Electric cables—Calculation of the current rating—Part 1: Current rating equations (100% load factor) and calculation of losses. Section 1: General," IEC Publication 287.

Westinghouse (1957), *Underground Systems Handbook.*

Conductor cross-section $= \frac{\pi}{4} \times 67.1^2 = 3{,}536 \text{ mm}^2$

At current density of 2 A/mm^2 the loss in 1 m length is

$$(2 \times 10^6)^2 \times \underbrace{1.7 \times 10^{-8}}_{\rho_{cu}\ 20°C} \times 3.536 \times 10^{-3} \times 1$$

$$= 240 \text{ W}$$

So the dielectric loss is not so high but maybe 2 A/mm^2 is rather high. At 1 A/mm^2 the Joule loss would be 60 W and the dielectric loss would be 8% of this.

7

Joule Losses in the Conductor

7.1 INTRODUCTION

The majority of losses in the power system are the result of a natural physical characteristic of electrical conductors, referred to as resistance. The resistance causes electrical energy to be converted to heat energy whenever current flows. Although this conversion process can be harnessed as is done in electric stoves, clothes dryers, electric heaters, and so on, such conversions in the conductors of a power system are largely wasted or "lost" since the heat is usually dissipated into the atmosphere or to the ground. These losses, often referred to as joule losses and denoted by W_c (W/m), are computed from the equation

$$W_c = I^2 R$$

where I is the conductor current and R is its ac resistance at the operating temperature of the conductor.

In power cable installations, the heat generated by these losses has to be dissipated through the surrounding soil or through air. Assuming that the current to be transmitted through a cable is given, joule losses can be reduced by reducing the ac resistance of the conductor. Alternating current (ac) resistance is always higher than the direct current (dc) resistance mainly because of the presence of skin and proximity effects which are described in Section 7.2.2. Therefore, the ac resistance should be kept as low as possible by a careful design of the construction of the conductor. In this chapter, we will review how the conductor resistance is computed for rating calculations.

7.2 RESISTANCE OF CABLE CONDUCTOR

The relation between the dc resistance per unit length at 20°C, R_{20}, and the cross section S of a solid conductor is expressed by the well-known equation

$$R_{20} = \frac{\rho_{20}}{S} \tag{7.1}$$

where ρ_{20} is the electrical resistivity of the conductor material at 20°C. The conductivity of copper used as a cable conductor material was standardized in 1913 by the IEC. Table 7.1 gives the values of ρ_{20} for conductors and sheaths listed in the IEC Standard 287 (1982). For stranded conductors, we will assume that S is an effective area equal to the area of an equivalent solid conductor.

With changes of temperature, copper and aluminum change their dimensions and resistivity. The latter can affect the thermal design of cables. For the practical temperature range between −40 and 125°C, a linear relationship holds approximately between the resistivity and temperature; hence,

$$R' = R_{20}[1 + \alpha_{20})\theta - 20)] \tag{7.2}$$

where α_{20} is the temperature coefficient of resistance at 20°C and θ is the actual conductor temperature (°C). Besides temperature, the temperature coefficient of resistance varies with annealing and purity. The values given in Table 7.1 (IEC 287, 1982) are recommended in rating computations.

Equations (7.1) and (7.2) give reasonably accurate values for the dc resistance of a solid conductor. However, the vast majority of conductors in electric power cables are of stranded construction, and the resistance of stranded conductors is not so accurately known. It is commonly assumed that the current is confined to the individual strands, and does not transfer from strand to strand in the direction parallel to the axis of the conductor. To take account of an increased resistance of stranded conductors, the resistance computed from equation (7.1) is in practice multiplied by an empirical factor of 1.02.

We will review the computation of stranded conductor resistances in the following sections.

TABLE 7.1 Thermal Resistivities and Temperature Coefficients of Metals Used in Cable Construction

Material	Resistivity $(\rho_{20}) \cdot 10^{-8}$ $\Omega \cdot$m at 20°C	Temperature Coefficient $(\alpha_{20}) \cdot 10^{-3}$ per K at 20°C
Conductors		
Copper	1.7241	3.93
Aluminum	2.8264	4.03
Sheaths and armor		
Lead or lead alloy	21.4	4.0
Steel	13.8	4.5
Bronze	3.5	3.0
Stainless steel	70	Negligible
Aluminum	2.84	4.03

7.2.1 DC Resistance of Stranded Conductors

Because of the additional length due to stranding, the dc resistance of stranded conductors is larger than that of solid conductors of equal area. The resistance at 20°C, R_{20}^{n} of the wires of diameter d_w in layer n is given by

$$\frac{1}{R_{20}^{n}} = \frac{\pi d_w^2 n_n}{4\rho_{20} k_n} \tag{7.3}$$

where n_n is the number of wires in layer n and k_n is the lay-length factor of the wires in layer n, given by

$$k_n = \sqrt{1 + \left(\frac{\pi d_n}{\ell_n}\right)^2} \tag{7.3a}$$

where ℓ_n is the lay length of layer n and d_n is the mean diameter of layer n. The quantity $\ell_n/\pi d_n$ is known as the lay ratio of layer n. The total resistance of the conductor with N layers is

$$\frac{1}{R_{20}} = \sum_{n=1}^{N} \frac{1}{R_{20}^{n}} \tag{7.4}$$

EXAMPLE 7.1

We will compute the dc resistance of the parallel combination of the skid wire and tape for the model cable No. 3. The cable shield consists of a mylar tape intercalated with a 7/8(0.003) in bronze tape—1 in lay, and a single 0.1 (0.2) in D-shaped bronze skid wire—1.5 in lay.[1] The diameter over the tape is equal to 2.648 in. The resistances of the tape and skid wire at an operating temperature of 60°C are obtained as follows.

The mean diameter of the tape equals[2]

$$d_{Tm}^{*} = 0.06726 - 1 \cdot 0.762 \cdot 10^{-4} = 0.0672 \text{ m}$$

The cross-sectional area of the tape is obtained from

$$A_T = 0.762 \cdot 10^{-4} \cdot 0.0222 \cdot 1 = 0.169 \cdot 10^{-5} \text{ m}^2$$

Tape resistance at 60°C is obtained by combining equations (7.1)–(7.3a), and is equal to

$$R_T = \frac{\rho_{20}\sqrt{1 + \left(\frac{\pi d_n}{\ell_n}\right)^2}}{A_T}[1 + \alpha_{20}(\theta - 20)]$$

$$= \frac{0.35 \cdot 10^{-7} \cdot \sqrt{\left(\frac{\pi \cdot 0.0672}{0.0254}\right)^2 + 1}}{0.169 \cdot 10^{-5}}[1 + 0.003(60 - 20)] = 0.194\ \Omega/\text{m}$$

[1] 1 in = 0.0254 m.

[2] We recall that throughout the book, the distance symbols with an asterisk represent values in meters.

Similarly, for the D-shaped bronze skid wire, we obtain

$$d^*_{wm} = 0.06759 - 1 \cdot 0.005/2 = 0.0651 \text{ m}$$

$$\text{area} = \frac{\pi \cdot 0.005^2 \cdot 1/4}{2} = 0.101 \cdot 10^{-4} \text{ m}^2$$

$$R_w = \frac{0.35 \cdot 10^{-7} \cdot \sqrt{\left(\frac{\pi \cdot 0.0651}{0.0381}\right)^2 + 1}}{0.101 \cdot 10^{-4}} [1 + 0.003(60 - 20)] = 0.0211 \text{ } \Omega/\text{m}$$

The combined resistance of skid wire and the reinforcing tape is

$$R_{wT} = \frac{0.194 \cdot 0.0211}{0.194 + 0.0211} = 0.0190 \text{ } \Omega/\text{m}$$

7.2.2 AC Resistance of Conductors: Skin and Proximity Effects

The resistance of a conductor when carrying an alternating current is higher than that of the conductor when carrying a direct current. The principal reasons for the increase are: skin effect, proximity effect, hysteresis, and eddy current losses in nearby ferromagnetic materials, and induced losses in short-circuited nonferromagnetic materials nearby. The degree of complexity of the calculations that can economically be justified varies considerably. Except in very high-voltage cables consisting of large segmental conductors, it is common to consider only skin effect, proximity effect, and in some cases, an approximation of the effect of metallic sheath and/or conduit. Figure 7.1 illustrates the flow of eddy-type skin and proximity currents which are superimposed onto the load current, resulting in an uneven current density in the conductor.

The ac resistance R with the skin and proximity effects is given by

$$R = R' k_{sk} k_{pr} \approx R'(k_{sk} + k_{pr} - 1)$$

with R' defined in equation (7.2).

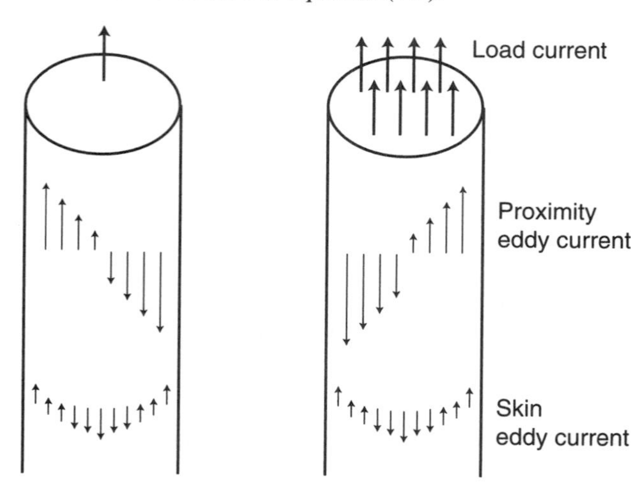

Figure 7.1 Illustration of the skin effect in electrical conductors.

Denoting $y_s = k_{sk} - 1$ and $y_p = k_{pr} - 1$ (the notation used in IEC 287, 1982), we have

$$R = R'(1 + y_s + y_p) \tag{7.5}$$

The factors k_{sk}, y_s, k_{pr}, and y_p are called *skin and proximity effect factors*, respectively. The ratio R/R' is usually close to unity at power frequency, and may be a few percent above unity for large diameter conductors. The derivation of these factors is very complex and involves the use of Bessel functions. Therefore, in the following sections, only the applicable final results are presented with the references to the full mathematical treatment provided as appropriate.

7.2.2.1 Skin Effect. Since not all the magnetic flux due to filaments of current near the center of a homogeneous conductor cuts the whole conductor, the inductance per unit area will decrease towards the surface; hence, the current per unit area will increase towards the surface (see Fig. 7.1).

The skin effect phenomenon was investigated by such people as Maxwell, Heaviside, Rayleigh, and Russell. The following approximate formulas for the skin effect factor are due to Arnold (1946) and Goldenberg (1961). Let

$$m = \sqrt{2\pi \mu_r \mu_0 f / \rho} \tag{7.6}$$

where ρ_{el} is the electrical resistivity at the operating temperature, μ_r is the relative permeability of the conductor ($\mu_r = 1$ for copper and aluminum conductors), μ_0 is the permeability of the free space ($\mu_0 = 4\pi \cdot 10^{-7}$), and f is the frequency. m is $\sqrt{2}$ times the reciprocal of the skin (penetration) depth. Let us consider a circular conductor with the diameter d_c. We will use the following notation:

$$x^2 = \left(\frac{md_c}{2}\right)^2 = \frac{8\pi f}{R'} \cdot 10^{-7} \tag{7.7}$$

To take into account the stranding and treatment of a conductor, a factor $k_s \leq 1$ is introduced in equation (7.7) to yield

$$x_s^2 = x^2 k_s = \frac{8\pi f}{R'} \cdot 10^{-7} k_s \tag{7.8}$$

The values of k_s are tabulated in Table 7.2. The skin effect factor is obtained as follows. For $0 < x_s \leq 2.8$

$$\boxed{y_s = \frac{x_s^4}{192 + 0.8x_s^4}} \tag{7.9}$$

For $2.8 < x_s \leq 3.8$

$$y_s = -0.136 - 0.0177x_s + 0.0563x_s^2 \tag{7.10}$$

For $3.8 < x_s$

$$y_s = \frac{x_s}{2\sqrt{2}} - \frac{11}{15} \tag{7.11}$$

In the absence of alternative formulas, IEC 287 (1982) recommends that the same expressions should be used for sector and oval-shaped conductors. Since for the majority of practical cases $x_s \leq 2.8$, equation (7.9) is used in the standards.

For a tubular conductor with inner and outer diameters d_i and d_c, respectively, the IEC standard (IEC 287, 1982) gives the following approximate value of the constant k_s to be used in conjunction with equation (7.8):

$$k_s = \frac{d'_c - d_i}{d'_c + d_i}\left(\frac{d'_c + 2d_i}{d'_c + d_i}\right)^2 \tag{7.12}$$

where d'_c is the diameter of an equivalent solid conductor having the same central duct.

Equations for the skin effect factor for a tubular conductor were developed by Dwight (1922). The rigorous solution of the problem of skin effect involves a Bessel equation for the determination of current distribution. Dwight's equations were rearranged by Lewis and Tuttle (1959) to permit convenient programming. To avoid the laborious evaluation of the Bessel functions, Arnold (1936) proposed the following remarkably accurate approximations. Let

$$\beta = 1 - \frac{d_i}{d_c} \quad \text{and} \quad z = 0.25m^2 k_s (d_c - d_i)^2 \tag{7.13}$$

Then

$$y_s = a(z)\left[1 - \frac{\beta}{2} - \beta^2 b(z)\right] \tag{7.14}$$

where, for $0 < z \leq 5$,

$$a(z) = \frac{7z^2}{315 + 3z^2} \quad \text{and} \quad b(z) = \frac{56}{211 + z^2}$$

For $5 < z \leq 30$, the values of $a(z)$ and $b(z)$ are calculated by the following polynomials (useful for computer programming):

$$\begin{aligned} a(z) &= 0.19701 - 0.1546295z + 0.073796z^2 - 9.02854 \cdot 10^{-3}z^3 + 6.27032 \cdot 10^{-4}z^4 \\ &\quad - 2.69028 \cdot 10^{-5}z^5 + 7.0647 \cdot 10^{-7}z^6 - 1.04301 \cdot 10^{-8}z^7 + 6.62315 \cdot 10^{-11}z^8 \\ b(z) &= 0.5356 - 0.21030734z + 6.495563 \cdot 10^{-2}z^2 - 1.089373 \cdot 10^{-2}z^3 \\ &\quad + 1.03728739 \cdot 10^{-3}z^4 - 5.8238557 \cdot 10^{-5}z^5 + 1.91099645 \cdot 10^{-6}z^6 \\ &\quad - 3.38936767 \cdot 10^{-8}z^7 + 2.509622 \cdot 10^{-1}z^8 \end{aligned} \tag{7.15}$$

For $z > 30$

$$a(z) = \sqrt{z/2 - 1} \quad \text{and} \quad b(z) = \frac{2}{4\sqrt{2z - 5}}$$

Dwight (1922) gave curves for the skin effect ratio k_{sk}. These were later recomputed and extended by Lewis and Tuttle (1959) and are shown in Fig. 7.2.

7.2.2.2 Skin Effect of Large Segmental Conductors. Special conductor constructions have been applied for underground cable systems with large conductor cross sections.

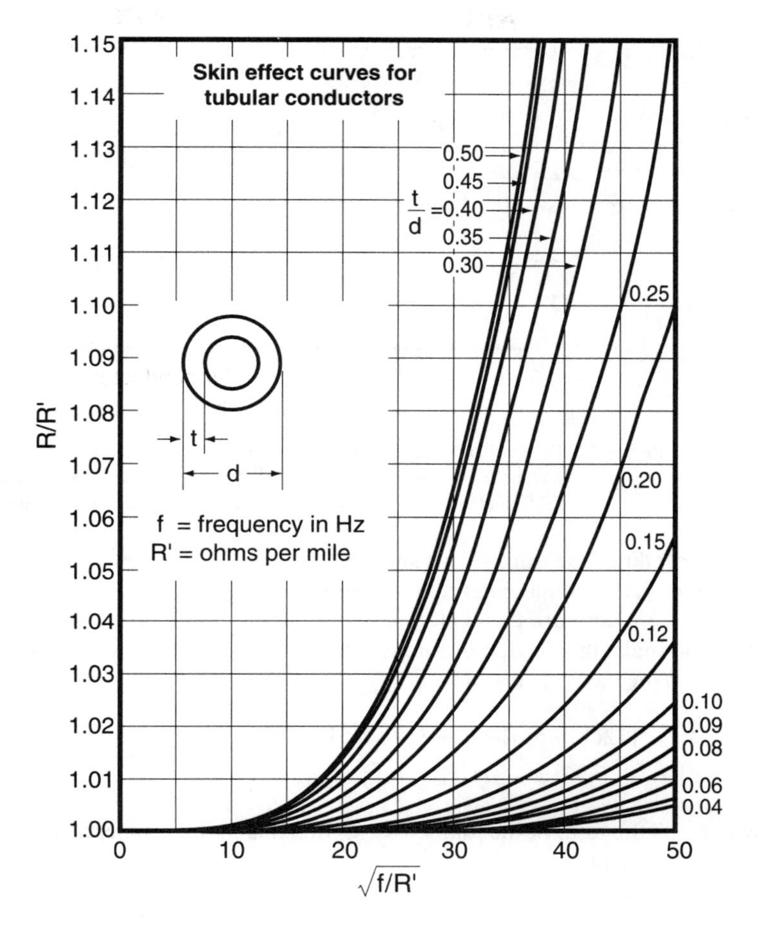

Figure 7.2 Skin effect resistance ratio for circular cylindrical conductors. R' is in ohms per mile (Lewis and Tuttle, 1959).

Conventional conductors are not suitable for large cross sections as their current-carrying capacity would be seriously reduced by skin and proximity effects. A segmental conductor design (British Patent 412 017), which is commonly referred to as the "*Milliken*" or "*type M*" conductor, has lightly insulated segments which reduce the magnitude of these effects. Figure 7.3 shows the reduction of ac resistance achieved with the Milliken conductor in comparison with conventional conductor construction (Ball and Maschio, 1968).

Recent work conducted in Germany and France resulted in the following empirical expression for constant k_s for Milliken conductors (IEC WG10, 1991):

$$k_s = -0.4 + 0.145 \ln(S) \tag{7.16}$$

where S (mm^2) is the nominal cross-sectional area of the conductor. This expression applies to conductors up to 2000 mm^2 having uninsulated wires with four, five, or six segments, irrespective of the direction of lay of the wires. Since no experimental data are available for larger conductors, equation (7.16) can be used for conductors with the cross-sectional

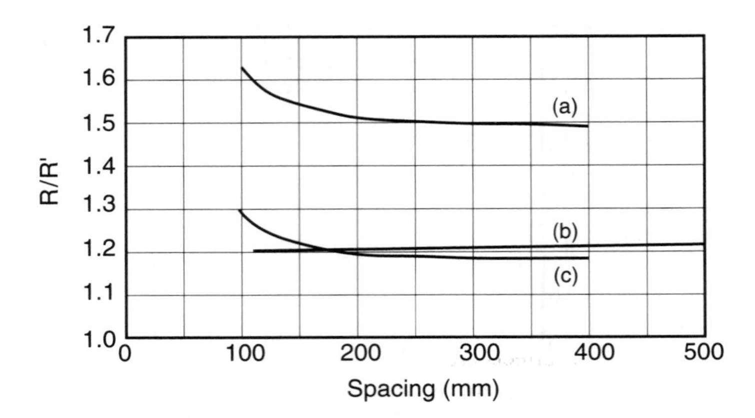

Figure 7.3 Comparison of ac to dc resistance ratio for segmented and nonsegmented conductors. (a) Annular conductor. (b) Measured values for segmental conductor. (c) Calculated values for segmental conductor. (Bell and Maschio, 1968)

area exceeding 2000 mm^2. This subject continues to be reviewed and developed; therefore, equation (7.16) should be used with caution.[3]

Large aluminum conductors are constructed with peripheral strands around the segments to make the conductor round. The constant k_s is then computed from the following formula (IEC WG10, 1991):

$$k_s = \{12(1-b)\left([\alpha(1-b)-0.5]^2 + [\alpha(1-b)-0.5](\psi-\alpha)(1-b) + 0.33(\psi-\alpha)^2(1-b)^2\right) + b(3-6b+4b^2)\}^{0.5} \tag{7.17}$$

where b is the ratio of the total cross-sectional area of peripheral strands to the total cross-sectional area of the conductor,

$$\alpha = \frac{1}{(1+\sin\pi/n)^2}, \qquad \psi = \frac{2\pi/n + 2/3}{2(1+\pi/n)} \tag{7.18}$$

and n is the number of segments. Equation (7.17) is applicable to aluminum conductors up to 1600 mm^2. If the total cross-sectional area of the peripheral strands exceeds 30% of the total area of the conductor, then k_s may be regarded as unity.

EXAMPLE 7.2

Compute the skin effect factor for a 1200 mm^2 aluminum conductor with two layers of peripheral strands. The ratio of the total cross-sectional area of peripheral strands to the total cross-sectional area of the conductor equals 27%.

From equation (7.18),

$$\alpha = \frac{1}{(1+\sin\pi/2)^2} = 0.25, \quad \psi = \frac{2\pi/2 + 2/3}{2(1+\pi/2)} = 0.7407$$

and from equation (7.17),

[3] Preliminary results from the experiments carried out in the United Kingdom indicate that the value of k_s computed from equation (7.16) may be too large.

$$k_s = \{12(1-0.27)([0.25(1-.27)-0.5]^2 + [0.25(1-0.27)-0.5](0.7407-0.25)(1-0.27) + 0.33(0.7407-0.25)^2(1-0.27)^2) + 0.27(3-6\cdot 0.27 + 4\cdot 0.27^2)\}^{0.5} = 0.84$$

7.2.2.3 Proximity Effect. When two conductors carrying alternating currents are parallel and close to each other, the current densities on the sides facing each other are decreased, and those on the remote sides are increased because of the difference in magnetic flux densities. This results in an increase in the conductor ac resistance and is called the *proximity effect.* The skin and proximity effects are seldom separable in cable work, and the combined effects are not directly cumulative. The proximity effect in three-conductor cables ordinarily is to slightly reduce the effect of skin effect alone. However, for computational convenience, these effects are considered separately.

To make the foregoing equations applicable to stranded conductors, the empirical transverse conductance factor k_p has been introduced (Neher and McGrath, 1957). This coefficient plays a role in proximity effect computations similar to that which the coefficient k_s plays in skin effect calculations. In analogy to equation (7.7), we have

$$x_p^2 = x^2 k_p = \frac{8\pi f}{R'} \cdot 10^{-7} k_p \tag{7.19}$$

In the majority of practical applications, $x_p \leq 2.8$. In this case, the following approximate formulas are given in IEC 287 (1982):

For two-core cables and two single-core cables,

$$\boxed{y_p = 2.9ay} \tag{7.20}$$

and for three core-cables and for three single-core cables,

$$\boxed{y_p = ay^2\left(0.312y^2 + \frac{1.18}{a+0.27}\right)} \tag{7.21}$$

where

$$a = \frac{x_p^4}{192 + 0.8x_p^4}, \qquad y = \frac{d_c}{s}$$

and s is the spacing between conductor centers.

Just as with the skin effect factor, the exact expressions for the proximity factor require the solution of Bessel equations. Arnold (1941) proposed the following approximations for various ranges of x_p:

For two-core cables and two circular single-core cables,

$$y_p = \frac{y^2 G(x_p)}{1 - y^2 A(x_p) - y^4 B(x_p)} \tag{7.22}$$

For three-core cables and for circular three single-core cables,

$$y_p = \frac{3y^2 G(x_p)}{2 - \frac{5}{12} y^2 H(x_p)} \tag{7.23}$$

For three stranded segmental conductors carrying three-phase current,

$$y_p = \frac{2.5y^2 G(x_p)}{2 - \frac{5}{12}y^2 H(x_p)} \tag{7.24}$$

where, for $0 < x_p \leq 2.8$,

$$\begin{aligned} A(x_p) &= \frac{0.042 + 0.012x_p^4}{1 + 0.0236x_p^4}, \qquad B(x_p) = 0 \\ G(x_p) &= \frac{11x_p^4}{704 + 20x_p^4} \quad \text{and} \quad H(x_p) = \frac{1}{3} \cdot \frac{1 + 0.0283x_p^4}{1 + 0.0042x_p^4} \end{aligned} \tag{7.25}$$

For $2.8 < x_p \leq 3.8$,

$$\begin{aligned} A(x_p) &= -0.223 + 0.237x_p - 0.0154x_p^2, \qquad B(x_p) = 0 \\ G(x_p) &= -1.04 + 0.72x_p - 0.08x_p^2 \\ \text{and} \quad H(x_p) &= 0.095 + 0.119x_p + 0.0384x_p^2 \end{aligned} \tag{7.26}$$

For $3.8 < x_p$,

$$\begin{aligned} A(x_p) &= 0.75 - 1.128x_p^{-1}, \qquad B(x_p) = 0.094 - 0.376x_p^{-1} \\ G(x_p) &= \frac{x_p}{4\sqrt{2}} - \frac{1}{8} \quad \text{and} \quad H(x_p) = \frac{2x_p - 4.69}{x_p - 1.16} \end{aligned} \tag{7.27}$$

The values of the constant k_p are shown in Table 7.2.

7.2.2.4 Skin and Proximity Effects in Pipe-type Cables. For pipe-type cables, the skin and proximity effects calculated by the above formulas are increased by a factor of 1.5 (Silver and Seman, 1982). For these cables,

$$\boxed{R = R'[1 + 1.5(y_s + y_p)]} \tag{7.28}$$

This is an empirical relation obtained for cables operating at voltages of up to 345 kV. Some cable designers use a value of 1.7 for this factor as recommended by Neher and McGrath (1957) and the IEC 287 (1982). The latter value can be used if we prefer a conservative design.

7.2.3 Summary of ac Resistance Computations

In rating computations, an engineer usually specifies the required conductor cross section and its construction. Since computation of the dc resistance of stranded conductors from this information is not very straightforward, a common practice is to use the tabulated values for R_{20} in equation (7.2). Tables 7.3 and 7.4 give the appropriate values as proposed in IEC 228 (1982) and IPCEA (1978), respectively. When the conductor size is outside the range covered by these references, the value of R_{20} may be chosen by agreement between the manufacturer and purchaser. Otherwise, it is computed from either equation (7.1) or (7.3) for solid and stranded conductors, respectively.

The procedure for computation of the ac resistance of cable conductor can be summarized as follows:

TABLE 7.2 Skin and Proximity Effects

Type of Conductor	Whether Dried and Impregnated or Not	k_s	k_p
Copper			
Round, stranded	Yes	1	0.8
Round, stranded	No	1	1
Round, compact	Yes	1	0.8
Round, compact	No	1	1
Round, segmental		0.435	0.37
Hollow, helical stranded	Yes	Eq.(712)	0.8
Sector-shaped	Yes	1	0.8
Sector-shaped	No	1	1
Aluminum			*
Round, stranded	Either	1	
Round, four segment	Either	0.28	
Round, five segment	Either	0.19	
Round, six segment	Either	0.12	
Segmental with periperal strands	Either	Eq.(7.17)	

*Since there are no accepted experimental results dealing specifically with aluminum stranded conductors, IEC 287 recommends that the values of k_p given in Table 7.2 for copper conductors also be applied to aluminum stranded conductor of similar design to copper conductors.

1. Read the dc resistance at 20°C from Tables 7.3/7.4 or use equations (7.1) or (7.4).
2. For given conductor operating temperature θ (°C), compute the dc resistance R' from equation (7.2).
3. Compute the skin effect factor $y_s (y_s = k_{sk} - 1)$ from equations (7.7)–(7.11) with the constant k_s obtained from Table 7.2. For segmental hollow core conductors, equation (7.12) should be used to obtain k_s or, alternatively, equations (7.16)–(7.18) can be applied. For more accurate calculations of k_{sk}, equations (7.13)–(7.14) and/or Fig. 7.2 can be used. Note that IEC 287 lists equations (7.9) and (7.12) only.
4. Compute the proximity effect factor $y_p (y_p = k_{pr} - 1)$ from equation (7.20) for two-core or two single-core cables or from equation (7.21) for three-core or three single-core cables. Constant k_p is read from Table 7.2. For more precise computations, equations (7.22)–(7.27) may be used.
5. Compute the ac resistance from equation (7.5) or (7.28).

EXAMPLE 7.3

We will compute the ac resistance (at 90°C) of model cable No. 1 using the IEC 287 method (assume that the cable is not dried or impregnated).

The dc resistance at 20°C of a 300 mm^2 conductor is read from Table 7.3 and the dc resistance at 90°C is computed from equation (7.2): $R_{20} = 6.01 \cdot 10^{-5}$ Ω/m and $R' = 6.01 \cdot 10^{-5}(1+0.00393 \cdot 70) = 7.663 \cdot 10^{-5}$ Ω/m. With the factor $k_s = 1$, the value of x_s is obtained from equation (7.8):

$$x_s = \sqrt{\frac{8\pi f}{R'} \cdot 10^{-7} k_s} = \sqrt{\frac{8\pi 50 \cdot 10^{-7} \cdot 1}{7.663 \cdot 10^{-5}}} = 1.281$$

TABLE 7.3 Maximum Resistance of Stranded Conductors at 20°C

Nominal Cross-Sectional Area (mm^2)	Copper Conductor: Plain Wires $\Omega/m \cdot 10^{-3}$	Copper Conductor: Metal-Coated Wires $\Omega/m \cdot 10^{-3}$	Aluminum Conductor: Plain, Metal-Coated, or Metal-Clad Wires $\Omega/m \cdot 10^{-3}$
0.5	36.0	36.7	
0.75	24.5	24.8	
1	18.1	18.2	
1.5	15.1	15.2	
2.5	7.41	7.56	
4	4.61	4.70	7.41
6	3.08	3.11	4.61
10	1.83	1.84	3.08
16	1.15	1.16	1.91
25	0.727	0.734	1.20
35	0.524	0.529	0.868
50	0.387	0.391	0.641
70	0.268	0.270	0.443
95	0.193	0.195	0.320
120	0.153	0.154	0.253
150	0.124	0.126	0.206
185	0.0991	0.100	0.164
240	0.0754	0.0762	0.125
300	0.0601	0.0607	0.100
400	0.0470	0.0475	0.0778
500	0.0366	0.0369	0.0605
630	0.0283	0.0286	0.0469
800	0.0221	0.0224	0.0367
1000	0.0176	0.0177	0.0291
1200	0.0151	0.0151	0.0247
1400	0.0129	0.0129	0.0212
1600	0.0113	0.0113	0.0186
1800	0.0101	0.0101	0.0165
2000	0.0090	0.0090	0.0149

*IEC 228 (1978) includes similar tables for solid and flexible conductors.

The skin effect factor is obtained from equation (7.9) as

$$y_s = \frac{x_s^4}{192 + 0.8x_s^4} = \frac{1.281^4}{192 + 0.8 \cdot 1.281^4} = 0.0138$$

The proximity effect factor is obtained from equation (7.21). Since $a = 0.0138$ ($a = y_s$ in this case) and $y = 20.5/71.6 = 0.286$, we have

$$y_p = ay^2\left(0.312y^2 + \frac{1.18}{a + 0.27}\right) = 0.0138 \cdot 0.286^2\left(0.312 \cdot 0.286^2 + \frac{1.18}{0.0138 + 0.27}\right) = 0.0047$$

Thus, the ac resistance is equal to

$$R = R'(1 + y_s + y_p) = 7.663 \cdot 10^{-5}(1 + 0.0138 + 0.0047) = 7.81 \cdot 10^{-5}\ \Omega/\text{m}$$

TABLE 7.4 Maximum Resistance of Stranded Conductors at 20°C

Conductor Size		Copper Conductor Ω /m· 10^{-3}			Aluminum Conductor Ω /m· 10^{-3}		
kcmil	mm^2	Hollow Core	No Core	Any Type	Hollow Core	No Core	Any Type
162.8	85			0.2065			0.3377
211.6	107.2			0.1637			0.2689
250	127			0.1386			0.2277
300	152			0.1158			0.1898
350	177			0.0990			0.1624
400	203			0.0868			0.1422
450	228			0.0772			0.1264
500	253			0.0695			0.1139
550	279			0.0630			0.1036
600	304			0.0579			0.0949
650	329			0.0534			0.0875
700	355			0.0495			0.0814
750	380	0.0460	0.0463		0.0766	0.0759	
800	405	0.0437	0.0434		0.0717	0.0711	
900	456	0.0389	0.0386		0.0640	0.0634	
1000	507	0.0351	0.0347		0.0576	0.0569	
1250	633	0.0280	0.0277		0.0460	0.0475	
1500	760	0.0234	0.0231		0.0383	0.0380	
1750	887	0.0200	0.0198		0.0328	0.0325	
2000	1013	0.0175	0.0172		0.0288	0.0285	
2250	1140			0.0156			0.0255
2500	1267			0.0140			0.0230
2750	1393			0.0126			0.0209
3000	1520			0.0117			0.0191
3500	1773			0.0101			0.0166
4000	2027			0.0088			0.0145

EXAMPLE 7.4

Compute the ac resistance at 90°C of a six-segment, copper Milliken conductor with a cross section of 2500 mm^2, external diameter of 69.1 mm, and hollow core diameter of 25 mm. Three single-conductor cables are in flat formation. Assume that the two neighboring cables are spaced 27 cm apart. The dc resistance at 20°C is equal to $R_{20} = 6.896 \cdot 10^{-6}$ Ω/m.

The dc resistance at 90°C is given by

$$R' = 6.896 \cdot 10^{-6}[1 + 0.00393(90 - 20)] = 8.794 \cdot 10^{-6}\ \Omega/\text{m}$$

We will compute the ac resistance using two approaches. In one case, we will apply the formulas recommended by the IEC 287 (Approach 1). In the second case, we will use more precise formulas presented in this chapter (Approach 2).

Approach 1. The constant k_s is obtained from equation (7.12). The equivalent diameter d_c' is obtained by solving the following equation:

$$\frac{25^2 \cdot \pi}{4} + 2500 = \frac{d_c'^2 \cdot \pi}{4}$$

which yields $d_c' = 71.7$ mm. Thus,

$$k_s = \frac{d_c' - d_i}{d_c' + d_i}\left(\frac{d_c' + 2d_i}{d_c' + d_i}\right) = \frac{61.7 - 25}{61.7 - 25}\left(\frac{61.7 + 2 \cdot 25}{61.7 + 25}\right)^2 = 0.7026$$

The skin effect factor is computed from equations (7.8) and (7.9):

$$x_s = \sqrt{\frac{8\pi f}{R'} \cdot 10^{-7} k_s} = x\sqrt{k_s} = \sqrt{\frac{8\pi 50 \cdot 10^{-7}}{8.794 \cdot 10^{-6}}} \cdot \sqrt{0.7026} = 3.78 \cdot 0.8382 = 3.168$$

$$y_s = \frac{x_s^4}{192 + 0.8x_s^4} = \frac{3.168^4}{192 + 0.8 \cdot 3.168^4} = 0.370$$

The proximity effect is computed from equation (7.21) with $k_p = 0.8$ obtained from Table 7.2. When the proximity factor is computed from equation (7.21), we first need the values of x_p, y, and a. These are equal to

$$x_p = x\sqrt{k_p} = 3.78\sqrt{0.8} = 3.381, \quad a = \frac{x_p^4}{192 + 0.8x_p^4} = \frac{3.381^4}{192 + 0.8 \cdot 3.381^4} = 0.4407$$

$$y = \frac{d_c}{s} = \frac{69.1}{270} = 0.2559$$

The proximity factor is now obtained as

$$y_p = ay^2\left(0.312y^2 + \frac{1.18}{a + 0.27}\right) = 0.4407 \cdot 0.2559^2\left(0.312 \cdot 0.2559^2 + \frac{1.18}{0.4407 + 0.27}\right) = 0.0485$$

The ac resistance in this case is obtained from equation (7.5) and is equal to

$$R = 8.794 \cdot 10^{-6}(1 + 0.370 + 0.0485) = 1.247 \cdot 10^{-5}\Omega/\text{m}$$

Approach 2. The constant k_s is obtained from equation (7.16):

$$k_s = -0.4 + 0.145\ln(S) = -0.4 + 0.145\ln(2500) = 0.7345$$

The skin effect factor y_s is computed from equations (7.13) and (7.14):

$$\beta = 1 - \frac{d_i}{d_c} = 1 - \frac{25}{69.1} = 0.6382$$

$$m^2 = \frac{2\pi\mu_r\mu_0 f}{\rho} = \frac{8\pi^2 50 \cdot 10^{-7}}{1.7241 \cdot 10^{-8}([1 + 0.00393(90 - 20)]} = 17\,958$$

$$z = 0.25m^2 k_s(d_c - d_i)^2 = 0.25 \cdot 17\,958 \cdot 0.7345(0.0691 - 0.025)^2 \cdot 10^{-6} = 6.413$$

Since $5 < z \leq 30$, the values of $a(z)$ and $b(z)$ are computed from equation (7.15). Thus, we have

$$a(z) = 0.19701 - 0.156295z + 0.073796z^2 - 9.02854 \cdot 10^{-3}z^3 + 6.27032 \cdot 10^{-4}z^4$$
$$- 2.69028 \cdot 10^{-5}z^5 + 7.0647 \cdot 10^{-7}z^6 - 1.04301 \cdot 10^{-8}z^7 + 6.62315 \cdot 10^{-11}z^8$$

$$b(z) = 0.5356 - 0.21030734z + 6.495563 \cdot 10^{-2}z^2 - 1.089373 \cdot 10^{-2}z^3 + 1.03728739 \cdot 10^{-3}z^4$$
$$- 5.8238557 \cdot 10^{-5}z^5 + 1.91099645 \cdot 10^{-6}z^6 - 3.38936767 \cdot 10^{-8}z^7 + 2.509622 \cdot 10^{-10}z^8$$

Substituting the value of z, we obtain

$$a(z) = 0.6619, \qquad b(z) = 0.2264$$

$$y_s = a(z)\left[1 - \frac{\beta}{2} - \beta^2 b(z)\right] = 0.6619(1 - 0.6382/2 - 0.6382^2 \cdot 0.2264) = 0.3896$$

We will compute the proximity factor using equation (7.24). From equation (7.19), we have

$$x_p = \sqrt{\frac{8\pi f}{R'} \cdot 10^{-7} k_p} = \sqrt{\frac{8\pi 50 \cdot 10^{-7} \cdot 0.8}{8.794 \cdot 10^{-6}}} = 3.381$$

Since $2.8 < x_p \leq 3.8$, from equation (7.26), we have

$$G(x_p) = -1.04 + 0.72x_p - 0.08x_p^2 = -1.04 + 0.72 \cdot 3.381 - 0.08 \cdot 3.381^2 = 0.4798$$
$$H(x_p) = 0.095 + 0.119x_p + 0.0384x_p^2 = 0.095 + 0.119 \cdot 3.381 + 0.0384 \cdot 3.381^2 = 0.9363$$

The value of y is simply $y = d_c/s = 69.1/270 = 0.2559$. Thus, from equation (7.24), we have

$$y_p = \frac{2.5y^2 G(x_p)}{2 - \frac{5}{12}y^2 H(x_p)} = \frac{2.5 \cdot 0.2559^2 \cdot 0.4798}{2 - \frac{5}{12} \cdot 0.2559^2 \cdot 0.9363} = 0.0398$$

The ac resistance in this case is equal to

$$R = 8.794 \cdot 10^{-6}(1 + 0.3896 + 0.0398) = 1.257 \cdot 10^{-5}\ \Omega/\text{m}$$

Thus, in this example, both approaches give similar results.

7.3 THE EFFECT OF HARMONICS

Because the resistance ratios, and hence the system losses, increase dramatically at higher frequencies, consideration of the effect of harmonics on cable losses is clearly justified. Three recent publications address this subject (Hiranandani, 1992; Meliopoulos and Martin, 1992; Palmer *et al.*, 1993).

Since the harmonic currents may appear not only in the conductor, but also in the screens and pipes, we will write the total joule losses in the cable at a given frequency as

$$W_I = I^2 R_I$$

where R_I is the apparent ac resistance of a conductor per unit length, taking into account both the skin and proximity effects and losses in metal screens, armor, and pipe. The value of R_I can be obtained from

$$R_I = R'\left(1 + y_s + y_p\right)(1 + \lambda_1 + \lambda_2) \tag{7.29}$$

where y_s and y_p are the skin and proximity effect factors and λ_1 and λ_2 are the loss factors defined in Chapter 8.

Due to the orthogonality of sinusoids having different frequencies, the effect of harmonics on losses may be calculated at each frequency and summed, that is,

$$W_I = \sum_{n=1}^{\infty} I_n^2 (R_I)_n = R' \sum_{n=1}^{\infty} I_n^2 \left(\frac{R_I}{R'}\right)_n \tag{7.30}$$

where n is the index of the harmonic. If γ_n is taken as the ratio of the nth harmonic current to the fundamental, equation (7.30) becomes

$$W_I = R' I^2 \sum_{n=1}^{\infty} \gamma_n^2 \left(\frac{R_I}{R'} \right)_n \tag{7.31}$$

so that the effective ac/dc resistance ratio, with the ac resistance including the effect of harmonics, can be given as

$$\left(\frac{R_I}{R'} \right)_{\text{dist}} = \sum_{n=1}^{\infty} \gamma_n^2 \left(\frac{R_I}{R'} \right)_n \tag{7.32}$$

This effective resistance ratio is a function of the magnitude of each harmonic flow in the conductor of the cable. Thus, the losses are dependent not only on the total harmonic distortion (THD), but also on the magnitude of each harmonic current. Because of this dependence, IEEE Standard 519 (1993) recommends limitations on both the total harmonic distortion and the distortion caused by any single harmonic. The total harmonic distortion is calculated as

$$\text{THD} = \frac{\sqrt{\sum_{n=2}^{\infty} I_n^2}}{I_1} \tag{7.33}$$

and the limits dictated by IEEE Standard 519 are shown in Tables 7.5–7.7.

Since harmonic currents produce additional joule losses in the cable, the ampacity of a cable will be lower than in the case when only the fundamental frequency is considered. The derating factor is obtained from

$$(I_{\text{derated}})^2 R' \left(\frac{R_I}{R'} \right)_{\text{dist}} = (I_{\text{rated}})^2 R' \left(\frac{R_I}{R'} \right)_{\text{fund}} \tag{7.34}$$

where

$$I_{\text{derated}} = (1 - \delta) I_{\text{rated}}$$

and the distorted resistance ratio is calculated by equation (7.32). This cable derating factor δ assumes that when currents are measured on a system containing harmonics, only the fundamental is measured, that is, the derating is applied against the fundamental current. The current derating factor δ further assumes that an acceptable current rating has been determined for a system free of harmonics, and a harmonics profile was later detected on that system. From equation (7.34), the factor is calculated as

$$\boxed{\delta = 1 - \sqrt{\frac{\left(\frac{R_I}{R'} \right)_{\text{fund}}}{\left(\frac{R_I}{R'} \right)_{\text{dist}}}}} \tag{7.35}$$

TABLE 7.5 Current Distortion Limits for General Distribution Systems (120–69 000 V)

	Maximum Harmonic Current Distortion in Percent of I_L Individual Harmonic Order (Odd Harmonics)					
I_{sc}/I_L	$n<11$	$11\leq n<17$	$17\leq n<23$	$23\leq n<35$	$35\leq n$	*THD*
<20*	4.0	2.0	1.5	0.6	0.3	5.0
20<50	7.0	3.5	2.5	1.0	0.5	8.0
50<100	10.0	4.5	4.0	1.5	0.7	12.0
100<1000	12.0	5.5	5.0	2.0	1.0	15.0
>1000	15.0	7.0	6.0	2.5	1.4	20.0

Even Harmonics are limited to 25% of the odd harmonic limits above.

Current distortions that result in a dc offset, e.g., half-wave converters, are not allowed.

*All power generation equipment is limited to these values of current distortion, regardless of actual I_{sc}/I_L.

I_{sc} = maximum short-circuit current at point of common coupling.

I_L = maximum demand current (fundamental frequency component) at point of common coupling.

TABLE 7.6 Current Distortion Limits for General Subtransmission Systems (60 001 V–161 000 V)

	Maximum Harmonic Current Distortion in Percent of I_L Individual Harmonic Order (Odd Harmonics)					
I_{sc}/I_L	$n<11$	$11\leq n<17$	$17\leq n<23$	$23\leq n<35$	$35\leq n$	*THD*
<20*	2.0	1.0	0.75	0.3	0.15	2.5
20<50	3.5	1.75	1.25	0.5	0.25	4.0
50<100	5.0	2.25	2.0	0.75	0.35	6.0
100<1000	6.0	2.75	2.5	1.0	0.5	7.5
>1000	7.5	3.5	3.0	1.25	0.7	10.0

Even Harmonics are limited to 25% of the odd harmonic limits above.

Current distortions that result in a dc offset, e.g., half-wave converters, are not allowed.

*All power generation equipment is limited to these values of current distortion, regardless of actual I_{sc}/I_L.

I_{sc} = maximum short-circuit current at point of common coupling.

I_L = maximum demand current (fundamental frequency component) at point of common coupling.

EXAMPLE 7.5

Consider cable model No. 1 with laying conditions specified in Appendix A. Let us assume that the system load has a third harmonic which constitutes 20% of the fundamental current. We will compute the derating factor for this cable system.

We start by computing the conductor resistance and loss factors for the third harmonic. From equation (7.7), we have

TABLE 7.7 Current Distortion Limits for General Transmission Systems (> 161 kV), Dispersed Generation and Cogeneration

	Individual Harmonic Order (Odd Harmonics)					
I_{sc}/I_L	$n<11$	$11\leq n<17$	$17\leq n<23$	$23\leq n<35$	$35\leq n$	*THD*
<50*	2.0	1.0	0.75	0.3	0.15	2.5
>50	3.0	1.5	1.15	0.45	0.22	3.75

Even Harmonics are limited to 25% of the odd harmonic limits above.

Current distortions that result in a dc offset, e.g., half-wave converters, are not allowed.

I_{sc} = maximum short-circuit current at point of common coupling.

I_L = maximum demand current (fundamental frequency component) at point of common coupling.

$$\left(x_s^2\right)_3 = \frac{8\pi f}{R'}\cdot 10^{-7} = \frac{8\pi\cdot 150}{0.0601\cdot 10^{-3}(1+0.00393\cdot 70)}\cdot 1\cdot 10^{-7} = \frac{8\pi\cdot 150}{0.07663\cdot 10^{-3}}\cdot 1\cdot 10^{-7} = 4.92$$

Since $(x_s)_3 < 2.8$, the skin effect factor is obtained from equation (7.9) as

$$(y_s)_3 = \frac{x_s^4}{192+0.8x_s^4} = \frac{4.92^2}{192+0.8\cdot 4.92^2} = 0.11$$

To compute the proximity effect factor, we observe that, since the cable is assumed to be neither dried nor impregnated, by equation (7.19), $\left(x_p^2\right)_3 = \left(x_s^2\right)_3 = 4.92$. Thus, from equation (7.25), we obtain

$$G(x_p) = \frac{11x_p^4}{704+20x_p^4} = \frac{11\cdot 4.92^2}{704+20\cdot 4.92^2} = 0.224$$

$$H(x_p) = \frac{1}{3}\cdot\frac{1+0.0283x_p^4}{1+0.0042x_p^4} = \frac{1}{3}\cdot\frac{1+0.0283\cdot 4.92^2}{1+0.0042\cdot 4.92^2} = 0.510$$

The proximity effect factor is obtained from equation (7.23):

$$y = \frac{d_c}{s} = \frac{20.5}{71.6} = 0.286$$

$$\left(y_p\right)_3 = \frac{2.5y^2G(x_p)}{2-\frac{5}{12}y^2H(x_p)} = \frac{2.5\cdot 0.286^2\cdot 0.224}{2-\frac{5}{12}\cdot 0.286^2\cdot 0.510} = 0.023$$

Hence, the ac resistance of the conductor for the third harmonic is equal to

$$(R)_3 = 0.07663\cdot 10^{-3}(1+0.011+0.023) = 0.0792\cdot 10^{-3}\ \Omega/\text{m}$$

When the conductor reaches an operating temperature of 90°C, the temperature of the concentric neutral wires computed using internal thermal resistances is equal to 83°C. Thus, the resistance of the screen is equal to

$$R_s = 0.759\cdot 10^{-3}(1+0.00393\cdot 63) = 0.947\cdot 10^{-3}\ \Omega/\text{m}$$

The circulating current loss factors are computed in Example 8.2 and are equal to

$$\left(\lambda'_{11}\right)_3 = 1.48,\qquad \left(\lambda'_{12}\right)_3 = 1.82,\qquad \left(\lambda'_{1m}\right)_3 = 0.195$$

The ac resistance of all three cables for the third harmonic is thus

$$(R_I)_3 = 0.0792 \cdot 10^{-3} \cdot (3 + 1.48 + 1.82 + 0.195) = 0.514 \cdot 10^{-3} \ \Omega/\mathrm{m}$$

The dc resistance was computed in Example 7.3 as $0.0766 \cdot 10^{-3}$ Ω/m. The skin and proximity factors were also computed in the same example, and are equal to 0.0139 and 0.0047, respectively. The concentric wires loss factor for this cable is equal to 0.09 (see Section 8.3.5.3). Hence, the effective distorted ac/dc resistance ratio, computed from equation (7.32), is

$$\left(\frac{R_I}{R'}\right)_{\text{dist}} = \sum_{n=1}^{\infty} \gamma_n^2 \left(\frac{R_I}{R'}\right)_n = 1^2 \cdot (1 + 0.0139 + 0.0047)(1 + 0.09) + 0.2^2 \cdot \frac{0.514 \cdot 10^{-3}}{3 \cdot 0.0766 \cdot 10^{-3}} = 1.2$$

Finally, the ampacity reduction factor is obtained from equation (7.35) as[4]

$$\delta = 1 - \sqrt{\frac{\left(\frac{R_I}{R'}\right)_{\text{fund}}}{\left(\frac{R_I}{R'}\right)_{\text{dist}}}} = 1 - \sqrt{\frac{(1 + 0.0139 + 0.0047)(1 + 0.09)}{1.2}} = 3.8\%$$

EXAMPLE 7.6[5]

Consider a 2000 kcmil (1010 mm^2) pipe-type cable with the following parameters:

where
$d_{cs} = 1.59$ in
$R' = 5.24\ \mu\Omega/\mathrm{ft}$
$R_s = 7100\ \mu\Omega/\mathrm{ft}$
$D_s = 2.37$ in
$s = 2.59$ in
$k_s = 0.35$
$k_p = 0.33$ for cradled configuration and $k_p = 0.15$ for triangular configuration.

The conductors have enameled strands, and the inside diameter of the pipe is 8.125 in. We will consider two cable arrangements inside the pipe: cradle and triangular. The only difference is the value of the proximity effect coefficient as shown above.

Sample harmonic scenarios are taken from IEEE Standard 519 (1993) and are shown in Table 7.8. The first scenario, designated "A," in the table, is the profile given for a 12-pulse converter as described in Table 13.1 of the Standard. The second scenario takes the values for the first scenario and attenuates all harmonics so that the magnitude of THD limitations of the Standard are met. The third and fourth scenarios are taken from Table 13.7 of the IEEE Standard representing unfiltered and most filtered harmonic distribution. Case "E" is the profile used by Meliopoulos and Martin (1992) as an example applied to a distribution system. For all of these cases, the triple harmonics above the third and all even harmonics are negligible, as is common in power transmission systems. Therefore, only frequencies with nonzero harmonics are shown.

These harmonic scenarios were applied to the cable described above. The cable derating factors were computed for each case taking into account the effect of harmonics on the conductor resistance and losses in screens and pipe. The values of the derating factor for the above cable and harmonic

[4] We assume that the screen loss factor for the fundamental frequency is the same for all three cables.

[5] This example is extracted from Palmer *et al.* (1993).

TABLE 7.8 Harmonics Profile (Percent)

n	A	B	C	D	E
Fundamental	100	100	100	100	100
3	0.00	0.00	0.00	0.00	22.85
5	19.20	5.00	9.59	0.00	3.43
7	13.20	5.00	6.60	2.33	3.43
11	7.30	3.50	3.66	1.70	0.00
13	5.70	3.50	2.85	1.37	0.00
17	3.50	3.00	1.75	0.87	0.00
19	2.70	2.70	1.35	0.68	0.00
23	2.00	1.25	1.0	0.51	0.00
25	1.60	1.25	0.80	0.41	0.00
29	1.40	1.25	0.70	0.36	0.00
31	1.20	1.20	0.60	0.31	0.00
35	1.10	0.70	0.55	0.28	0.00
37	1.00	0.70	0.00	0.00	0.00
41	0.90	0.70	0.00	0.00	0.00
43	0.80	0.70	0.00	0.00	0.00
47	0.80	0.70	0.00	0.00	0.00
49	0.70	0.70	0.00	0.00	0.00
THD	25.7%	10%	12.8%	3.49%	23.3%

scenarios are shown in Table 7.9. These results indicate that harmonics may create a very significant increase in losses. Scenario "A," which was a 12-pulse converter with no filtering, required a derating of as much as 16.25%. Comparison with scenario "B" indicates the very strong influence of even minimal filtering. The benefit of complying with the IEEE Standard are also clearly seen as the derating decreases from 16.25 to 5.29% when the limits are impressed on the harmonic scenario created by the 12-pulse converter.

TABLE 7.9 R_I/R' and Derating Factor for Pipe-Type Cable and Harmonics

		A	B	C	D	E
Cradle	***R/R′***	2.251	1.760	1.738	1.598	1.792
	δ (%)	16.25	5.29	4.71	0.62	6.13
Triangular	***R/R′***	1.811	1.466	1.455	1.356	1.500
	δ (%)	13.90	4.29	1.46	0.50	5.37

A comparison of scenarios "B" and "C" clearly indicates that losses are not dependent on THD alone. While scenario "B" had a THD of 10%, the losses were greater than for scenario "C" which had a THD of 12.8%. The reason for this is the existence of higher harmonics in case "B." Hence, the losses are dependent upon both the individual harmonic level and the total harmonic distortion. Additionally, it is clear that the loss increases are dependent on the cable configuration inside the pipe, resulting in deratings that vary by as much as 3% between the triangular and cradled configurations.

In observing scenario "D," we can see that for systems with a high level of filtering, the effect of harmonics on cable rating is minimal.

Scenario "E" requires a derating of slightly over 6%. This can be compared with the derating of 11% calculated by Meliopoulos and Martin (1992) for a given distribution system.

REFERENCES

AEIC CS4-93 (1993), "Specifications for impregnated-paper-insulated low and medium pressure self contained liquid filled cable," Association of Edison Illuminating Companies, New York.

Arnold, A. H. M. (1936), "The alternating current resistance of tubular conductors," *J. IEE*, no. 78, pp. 580–593.

Arnold, A. H. M. (1941), "Proximity effect in solid and hollow round conductors," *J. IEE*, no. 88 II, pp. 349–359.

Arnold, A. H. M. (1942), "Eddy current losses in single-conductor, paper-insulated, lead-covered unarmoured cables of a single phase system," *J. IEE*, no. 89 II, pp. 639–647.

Arnold, A. H. M. (1946), *The Alternating Current Resistance of Non-Magnetic Conductors*. London: (DSIR) HMSO.

Ball, E. H., and Maschio, G. (1968), "The A.C. resistance of segmental conductors as used in power cables," *IEEE Trans. Power App. Syst.*, vol. PAS-87, no. 4, pp. 1143–1148.

Dwight, H. B. (1922), "Skin effect and proximity effect in tubular conductors," *AIEE Trans.*, no. 41, pp. 189–198.

Goldenberg, H. (1961), "Some approximations to Arnold's formulae for skin and proximity effect factors for circular and shaped conductors," Brit. Elec. Res. Assoc. Report F/T203.

IEC 228 (1978), "Conductors of insulated cables," IEC Publication 228.

IEC 287 (1982), "Calculation of the continuous current rating of cables (100%) load factor," IEC Publication 287.

IEC WG10 (1991), "Skin and proximity effects," Document of WG 10 of IEC SC20A—High Voltage Cables.

IEEE Standard 519 (1993), "IEEE recommended practices and requirements for harmonic control in electric power systems," IEEE Ind. Appl. Soc./Power Eng. Soc., Apr. 1993.

Lewis, W. A., and Tuttle, P. D. (1959), "The resistance and reactance of ACSR," *AIEE Trans.*, vol. 77 III, pp. 1189–1217.

Meliopoulos, A. P. S., and Martin, M. A., Jr. (Apr. 1992), "Calculation of secondary cable losses and ampacity in the presence of harmonics," *IEEE Trans. Power Delivery*, vol. 7, no. 2, pp. 451–457.

Palmer, J. A., Degeneff, R. C., McKernan, T. M., and Halleran, T. M. (Oct. 1993), "Pipe-type cable ampacities in the presence of harmonics," *IEEE Trans. Power Delivery*, vol. 8, no. 4, pp. 1689–1695.

Silver, D. A., and Seman, G. W. (1982), "Investigation of A.C./D.C. resistance ratios of various designs of pipe-type cable systems," *IEEE Trans. Power App. Syst.*, vol. PAS-101, no. 9, pp. 3481–3497.

8

Joule Losses in Screens, Sheaths, Armor, and Pipes

8.1 INTRODUCTION

Dielectric losses and joule losses from conductors have already been considered in Chapters 6 and 7. The third source of losses, considered in this chapter, are those from metallic cable screens and coverings such as sheaths and armoring. These losses are generated by currents induced in these components by the current flowing in the conductors.

Sheath losses are current dependent, and can be divided into two categories according to the type of bonding. These are losses due to circulating currents which flow in the sheaths of single-core cables if the sheaths are bonded together at two points, and losses due to eddy currents, which circulate radially (skin effect) and azimuthally (proximity effect). Eddy current losses occur in both three-core and single-core cables, irrespective of the method of bonding. Eddy current losses in the sheaths of single-core cables which are solidly bonded are considerably smaller than circulating current losses, and are ignored except for cables with large segmental conductors.

Losses in protective armoring also fall into several categories depending on the cable type, the material of the armor, and installation methods. Armored single-core cables without a metallic sheath generally have a nonmagnetic armor because the losses in steel-wire or tape armor would be unacceptably high. For cables with nonmagnetic armor, the armor loss is calculated as if it were a cable sheath, and the calculation method depends on whether the armor is single-point bonded or solidly bonded. For cables having a metallic sheath and nonmagnetic armor, the losses are calculated as for sheath losses, but using the combined resistance of the sheath and armor in parallel and a mean diameter equal to the rms value or the armor and sheath diameters. The same procedure applies to two- and three-core cables having a metallic sheath and nonmagnetic armor. For two- and three-core cables having metallic sheath and magnetic wire armor, eddy current losses in the armor must be considered. For two- and three-core cables having steel tape armor, both eddy current losses and hysteresis losses in the tape must be considered together with the effect of armor on sheath losses.

Submarine cables require special consideration. Single-core ac cables for submarine power connections differ in many respects from underground, buried directly or in ducts; in fact, submarine cables are generally armored, can be manufactured in very long lengths, and are laid with a very large distance between them. For these reasons, calculation methods described in the IEC 287 (1982) must be supplemented and modified in some points. In this chapter, we will introduce required tools to compute sheath and armor losses in single-core cables with metallic sheath and magnetic armor representing common submarine cable construction. However, we will ignore the presence of the sea in our calculations. This implies that the theory presented here is applicable to cables laid in air, buried underground, or in ducts or cables in shallow waters near landing points. The reader interested in calculations involving consideration of the sea impedance is referred to the paper by Bianchi and Luoni (1976).

8.2 SHEATH BONDING ARRANGEMENTS

Sheath losses in single-core cables depend on a number of factors, one of which is the sheath bonding arrangement. In fact, the bonding arrangement is the second most important parameter in cable ampacity computations after the external thermal resistance of the cable. For safety reasons, cable sheaths must be earthed, and hence bonded, at least at one point in a run. There are three basic options for bonding sheaths of single core cables. These are: single-point bonding, solid bonding, and cross bonding (ANSI/IEEE, 1988).

In a single-point-bonded system, the considerable heating effect of circulating currents is avoided, but voltages will be induced along the length of the cable. These voltages are proportional to the conductor current and length of run, and increase as the cable spacing increases. Particular care must be taken to insulate and provide surge protection at the free end of the sheath to avoid danger from the induced voltages.

One way of eliminating the induced voltages is to bond the sheath at both ends of the run (solid bonding). The disadvantage of this is that the circulating currents which then flow in the sheaths reduce the current-carrying capacity of the cable.

Cross bonding of single-core cable sheaths is a method of avoiding circulating currents and excessive sheath voltages while permitting increased cable spacing and long run lengths. The increase in cable spacing increases the thermal independence of each cable, and hence increases its current-carrying capacity. The cross bonding divides the cable run into three sections, and cross connects the sheaths in such a manner that the induced voltages cancel. One disadvantage of this system is that it is very expensive, and therefore is applied mostly in high-voltage installations.

Figure 8.1 gives a diagrammatic representation of the cross connections.

The cable route is divided into three equal lengths, and the sheath continuity is broken at each joint. The induced sheath voltages in each section of each phase are equal in magnitude and 120° out of phase. When the sheaths are cross connected, as shown in Fig. 8.1, each sheath circuit contains one section from each phase such that the total voltage in each sheath circuit sums to zero. If the sheaths are then bonded and earthed at the end of

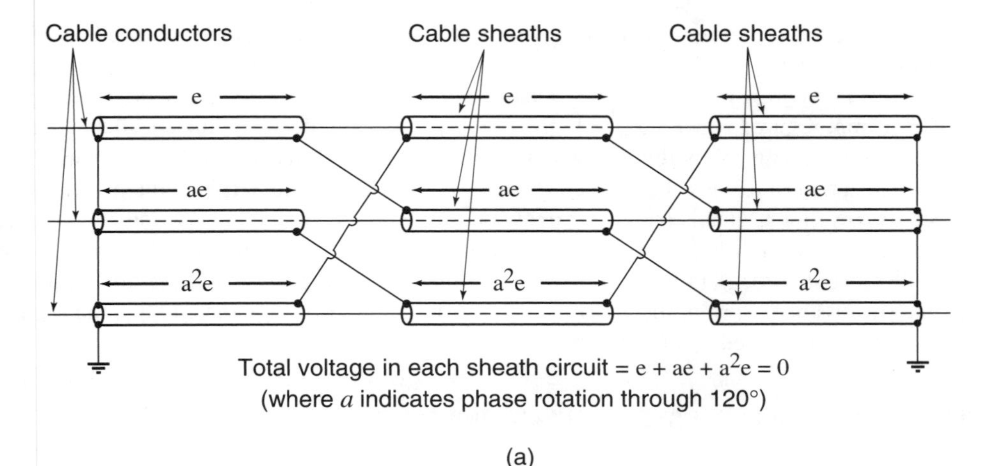

(a)

Sheath sectionalizing insulators
Joints
Transposition of cables
Sheaths solidly bonded and earthed

(b)

Figure 8.1 Diagrammatic representation of a cross-bonded cable system. (a) Cables are not transposed. (b) Cables are transposed.

the run, the net voltage in the loop and the circulating currents will be zero, and the only sheath losses will be these caused by eddy currents.

This method of bonding allows the cables to be spaced to take advantage of improved heat dissipation without incurring the penalty of increased circulating current losses. In practice, the lengths and cable spacings in each section may not be identical, and therefore some circulating currents will be present. The length of each section and cable spacings are limited by the voltages which exist between the sheaths and between the sheaths and earth at each cross-bonding position. For long runs, the route is divided into a number of lengths, each of which is divided into three sections. Cross bonding as described above can be applied to each length independently.

The cross-bonding scheme described above assumes that the cables are arranged symmetrically, that is, in trefoil. It is usual that single-core cables are laid in a flat configuration. In this case, it is a common practice in long-cable circuits or heavily loaded cable lines to transpose the cables as shown in Fig. 8.1b so that each cable occupies each position for a third of the run.

A number of practical points must be considered before adopting cross bonding, the most important of which are the high voltages which can occur on the sheaths and across sheath insulating joints during switching surges or other transients. Experimental work by Gosland (1940) has shown that voltages as high as the full service voltage can appear across insulating glands under transient conditions, even when there are only a few meters of cable in the circuit. These voltages cannot be avoided, but use of suitable surge diverters will prevent damage to the cable system. Other practical points relate to the voltages on the sheath under normal service or fault conditions and the need to ensure that the sheath is effectively insulated from earth for the life of the system.

8.3 CIRCULATING CURRENT LOSSES IN SHEATH AND ARMOR

The basic equations for calculating circulating current losses were developed by a number of authors in the 1920s: Morgan *et al.* (1927), Arnold (1929), and Carter (1927, 1928). In some cases, the effects of eddy currents were included in the equations developed; others concluded that the effects of eddy currents were insignificant compared with circulating current losses, and hence could be ignored. The equations presented in IEC Publication 287 (1982) and in the Neher–McGrath paper (1957) are taken from the work by Arnold (1929), and ignore eddy current losses except for the case of cables with large segmental conductors.

All of the equations for sheath losses given in this section assume that the phase currents are balanced. The equations also require a knowledge of the temperature of the sheath, which cannot be calculated until the cable rating is known, and therefore an iterative process is required. For the first calculation, the sheath temperature must be estimated; this estimate can be checked later after the current rating has been calculated. If necessary, the sheath losses, and hence the current rating, must be recalculated with the revised sheath temperature.

As discussed above, the power loss in the sheath or screen (λ_1) consists of losses caused by circulating currents (λ_1') and eddy currents (λ_1''). Thus,

$$\lambda_1 = \lambda_1' + \lambda_1'' \tag{8.1}$$

The loss factor in armor is also composed of two components: that due to circulating currents (λ_2') and, for magnetic armor, that caused by hysteresis (λ_2''). Thus,

$$\lambda_2 = \lambda_2' + \lambda_2'' \tag{8.2}$$

As mentioned above, for single-core cables with sheaths bonded at both ends of an electrical section, only losses caused by circulating currents are considered. An electrical section is defined as a portion of the route between points at which the sheaths or screens of all cables are solidly bonded. Circulating current losses are much greater than eddy current losses, and they completely dominate the calculations. Of course, there are no circulating currents when the sheaths are isolated or bonded at one point only. The formulas for the circulating current losses in sheath and armor are developed below.

Consider a three-phase cable circuit. The complex currents flowing in the conductor, sheath, and armor are denoted by I_c, I_s, and I_a, respectively. Then, the sheath and armor loss factors due to circulating currents are defined as

$$\lambda_1' = \frac{|I_s|^2 R_s}{|I_c|^2 R} \tag{8.3}$$

$$\lambda_2' = \frac{|I_a|^2 R_a}{|I_c|^2 R} \tag{8.4}$$

where R (Ω/m) denotes the ac resistance of the conductor at operating temperature and the subscripts s and a represent sheath and armor, respectively.

As can be seen from the above equations, in order to compute the loss factors, the sheath and armor currents have to be expressed as functions of the conductor current. In order to compute I_s and I_a, we observe that the voltages along the conductor, sheath, and armor are related to the currents by

$$\begin{pmatrix} V_c \\ V_s \\ V_a \end{pmatrix} = \begin{pmatrix} Z_{cc} & Z_{cs} & Z_{ca} \\ Z_{sc} & Z_{ss} & Z_{sa} \\ Z_{ac} & Z_{as} & Z_{aa} \end{pmatrix} \begin{pmatrix} I_c \\ I_s \\ I_a \end{pmatrix} \tag{8.5}$$

where Z_{ij} is an impedance between element i and j. Calculation of these impedances is discussed in the following subsections. The approach to deriving inductances will be to compute "flux linkages," except for the inductances resulting from magnetic fluxes within the thickness of the armor. In that case, an energy approach turns out to be more suitable.

We will use Fig. 8.2 to illustrate the flux linkage concept. This figure shows three cables in trefoil configuration, where each cable is composed of a conductor, sheath, and magnetic armor. Also illustrated is magnetic flux due to current in the bottom conductor. Consider, for example, the conductor–conductor impedance Z_{cc} which will be evaluated for the bottom conductor in Fig. 8.2. This impedance consists of the conductor resistance, the conductor's self-reactance ($j\omega$ times the self-inductance), and the conductor's mutual reactance with the other two conductors.

The conductor's self-inductance is the linkage of the lower conductor's flux with its own current. The mutual inductance is the linkage of the lower conductor's flux with the current in the other conductors. (Flux linkage, generally, is the spatial integral of flux due to a unit current times the proportion of current that it links.)

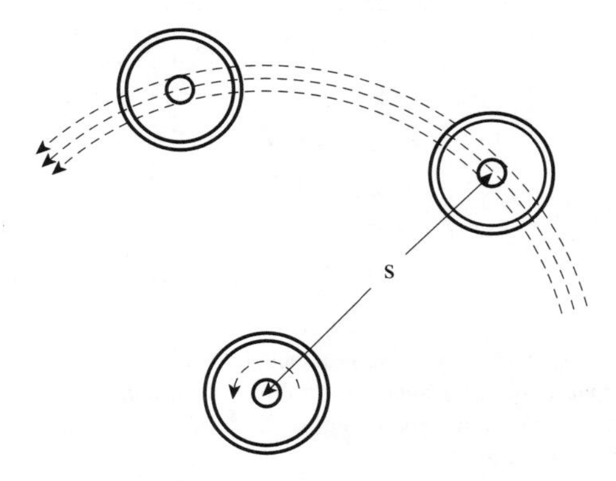

Figure 8.2 Three cables in trefoil configuration.

The evaluation of conductor–conductor inductance, including both the self- and mutual inductances, consists of integrating all of the conductor flux linkage from the center of the conductor to the dashed line passing through the centers of the other two conductors in Fig. 8.2. The reason that this integration ends halfway through the upper two conductors is that flux entering the upper two conductors near their bottom edges, as illustrated by the inner dashed line, links only a small proportion of the current in those conductors. On the other hand, flux at the upper edge, as illustrated by the outer dashed line, links nearly all of the current. The total flux linkage from the lower surface of the two conductors to the upper surface (linking, on average half of the current) is approximately the same as the integral of flux to the halfway point, linking all of the conductor current.

The inductances due to flux within the conductor and within the thicknesses of the sheath and armor walls are derived in the following subsections. We will begin by evaluating all internal cable inductances.

8.3.1 Internal Inductances[1]

8.3.1.1 Hollow Conductor Internal Inductance. The self-inductance of a hollow conductor is computed in the following manner. For unit total current, unit outer radius, and an inner radius of a (see Fig. 8.3), the fraction of conductor current within radius r is given by

$$I_{<r} = \frac{r^2 - a^2}{1 - a^2} \tag{8.6}$$

and the flux within dr is given by

$$d\phi = \frac{\mu_0 I_{<r}}{2\pi r} dr \tag{8.7}$$

The internal inductance of a hollow conductor is thus given by

$$L_{cc-int} = \frac{\mu_0}{2\pi} \int_a^1 \frac{I_{<r}^2 dr}{r} = \frac{\mu_0}{2\pi} \left[\frac{1/4 - a^2 + a^4(3/4 - \ln a)}{(1 - a^2)^2} \right] \tag{8.8}$$

where a is the ratio of inner to outer radius.

For a solid conductor ($a = 0$), this reduces to $(\mu_0/2\pi) \cdot (1/4)$. For a thin shell ($a \to 1$), it converges to $(\mu_0/2\pi) \cdot (1/3)(1 - a) = (\mu_0/2\pi) \cdot (1/3)(t_s/r_s)$

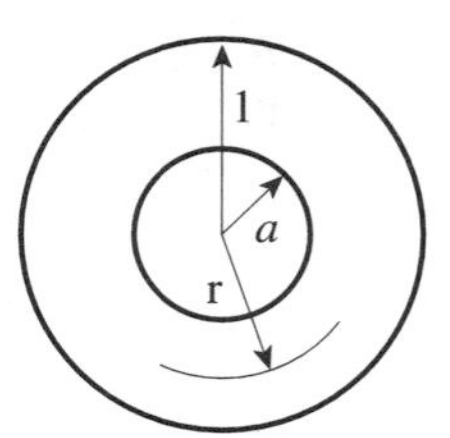

Figure 8.3 Hollow-core conductor with unit outer radius.

[1] In the development of inductances, all linear dimensions should be expressed in meters. However, since the final results contain only the ratios of these dimensions, the formulas contain the symbols for the thicknesses, the spacing, and the armor length of lay without an asterisk, so the dimensions in millimeters can be substituted in the final expressions.

where r_s = mean diameter of the sheath, mm
t_s = mean thickness of the sheath, mm.

8.3.1.2 Conductor–Sheath Internal Inductance. For mean sheath radius r_s and thickness t_s, the conductor flux density for unit conductor current is approximately given by

$$B_c \approx \mu_0 / \left(2\pi r_s^*\right) \tag{8.9}$$

The proportion of sheath current at radius less than r is given approximately by

$$I_{<r} \approx \left(r - r_s + t_s/2\right)/t_s \tag{8.10}$$

The internal inductance is then given by the flux linkage

$$L_{cs-int} \approx \frac{\mu_0}{2\pi r_s} \int_{r_s - t_s/2}^{r_s + t_s/2} \frac{r - r_s + t_s/2}{t_s} dr = \frac{\mu_0}{2\pi r_s} \frac{t_s}{2} \approx \frac{\mu_0}{2\pi} \ln\left(\frac{r_s + t_s/2}{r_s}\right) \tag{8.11}$$

8.3.1.3 Sheath–Sheath Internal Inductance. For mean sheath radius r_s and thickness t_s, and unit total current in the sheath, the fraction of sheath current within a distance x of the inner radius is equal to $I_{<x} = x/t_s$. The flux $d\phi$ within dx encircling this current is given by

$$d\phi = \frac{\mu_0 I_{<x}}{2\pi r_s} dx \tag{8.12}$$

The radius has been approximated by r_s at all values of x. The inductance is given by the integral of the flux times the proportion of current that it surrounds, so that the *internal self-inductance*[2] is approximately given by

$$L_{ss-int} = \frac{\mu_0}{2\pi r_s} \int_0^{t_s} I_{<x}^2 dr = \frac{\mu_0}{2\pi} \cdot \frac{1}{3} \frac{t_s}{r_s} \approx \frac{\mu_0}{2\pi} \ln\left(\frac{r_s + t_s/2}{r_s + t_s/6}\right) \tag{8.13}$$

8.3.1.4 Internal Armor Inductances. Magnetic flux within the thickness of the armor wall affects all of the inductances. The internal inductances resulting from the flux in the armor tapes or wire are fairly complicated. The armor may be treated as a coil surrounded by circular flux as a result of conductor and sheath currents and the armor's own current. In addition, the armor surrounds solenoidal flux as a result of its own current. This solenoidal flux was not considered by Bosone (1931), and is therefore not included in the IEC 287 (1994) Standard. The magnetic fields within the thickness of the armor are a superposition of the axial and circular fields. The circular fields from the conductor and sheath are nearly uniform between the inner and outer surfaces of thin armor and are treated as uniform. The fields from the armor's own current, however, are nonuniform. The axial field is maximum at the inner surface of the armor and decreases to zero at the outer surface

[2] In solving the field equations, the voltage measured at the outer surface of the sheath is equal to ρJ_{surface}. To second order, this is the sum of the resistive voltage due to the total sheath current (or average current density) and the voltage from internal inductance. The latter therefore depends on the difference between the surface current density and average current density. The resulting self-inductance, to second order, however, is the same as that obtained with a uniform-density flux-linkage calculation.

because the axial field at a given radius is caused by current at larger radii. Conversely, the circular field is maximum at the outer surface of the armor and decreases to zero at the inner surface because the circular field at a given radius is caused by current at smaller radii. The situation is complicated by the fact that the magnetic permeability of the wires is different, parallel and perpendicular to the wires. The circular and axial fields in the armor wall must therefore be resolved into directions parallel and perpendicular to the wires (see Fig. 8.4). In deriving inductance formulas, the helical geometry of the currents and fluxes leads to flux linkages that are sometimes difficult to visualize. Therefore, to avoid errors, the derivations presented here are based on the integral of complex power over the volume of the armor material. These integrals are identical to those encountered in flux-linkage calculations. The total electromagnetic power is given by Poynting's Theorem (Jackson, 1962):

$$P + jQ = \frac{d}{dt}\left[\iiint (J \cdot E + B \cdot H)\, dV\right] \tag{8.14}$$

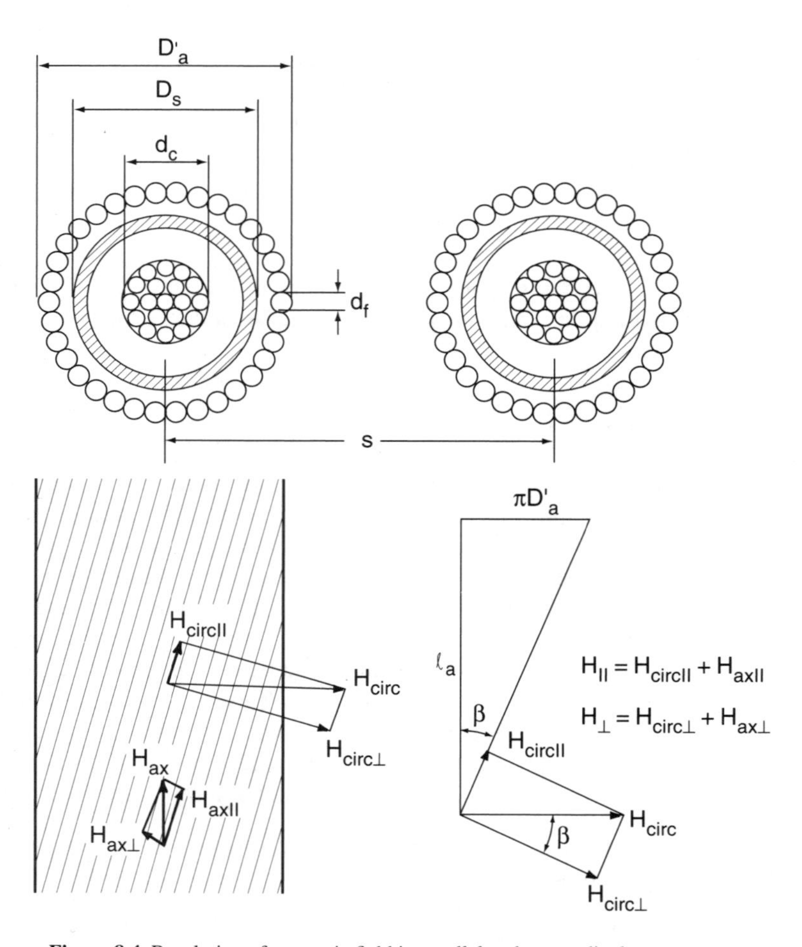

Figure 8.4 Resolution of magnetic field in parallel and perpendicular components.

where J = current density, A/m^2
E = electric field, V/m
B = magnetic flux density, Tesla
H = magnetic field, A-turns/m

and the integral is over all space. The $J \cdot E$ term results in the I^2R power loss. The second term, integrated only over the volume of the armor material, will be used to obtain internal armor inductances, as well as the contribution that internal armor flux makes to all of the other cable inductances.

The complex power (real is dissipated; imaginary is stored) is also given by

$$P + jQ = \sum_{i,j} (I_i \cdot I_j)\, Z_{ij} \tag{8.15}$$

where Z_{ij} = mutual (or self for $i = j$) impedances

$$I_i \cdot I_j = \Re e(I_i)\Re e(I_j) + \Im m(I_i)\Im m(I_j) = (I_i I_j^* + I_i^* I_j)/2 \tag{8.16}$$

Equating these two expressions allows the identification of $j\omega L_{ij}$ (part of Z_{ij}) in the power obtained from the first expression.

All currents, voltages, and fields are rms values. The subscripts may represent the conductor, sheath, or armor. The losses are due to currents in these elements, but take place within the thickness of the armor. Because the longitudinal permeability of the wires is complex and is included in all of the impedances, there is loss corresponding to all of the impedances, not just the diagonal ones.

To simplify the computations, it will be assumed that the armor is composed of tapes rather than wires. The error in the resulting formulas when used for round wires is negligible. The mean radius and the thickness of the armor are denoted by r_a and t_a, respectively. The length of lay of armor wires or tapes is denoted by ℓ_a. The fields within the tape (or wire) are obtained by resolving the solenoidal field $(1 - x)I_a/\ell_a$ and the circular field $(I_c + I_s + xI_a)/(2\pi r_a)$ into components parallel and perpendicular to the tape. The parallel and perpendicular components of the magnetic field at position x within the armor wall (assuming uniform current density in the armor) are given by

$$H_{\parallel} = (1 - x)I_a \frac{\cos(\beta)}{\ell_a} + (I_c + I_s + xI_a)\frac{\sin(\beta)}{2\pi r_a} = (I_c + I_s + I_a)\frac{\cos(\beta)}{\ell_a} \tag{8.17}$$

and

$$\begin{aligned} H_{\perp} &= -(1 - x)I_a \frac{\sin(\beta)}{\ell_a} + (I_c + I_s + xI_a)\frac{\cos(\beta)}{2\pi r_a} \\ &= \left[(I_c + I_s)\cos^2(\beta) + I_a\left(x - \sin^2(\beta)\right)\right]\frac{1}{2\pi r_a \cos(\beta)} \end{aligned} \tag{8.18}$$

where I_c = rms conductor current, A
I_s = rms sheath current, A
I_a = rms armor current, A
x = distance from inner surface/armor thickness ($x = 0$ at inner surface; $x = 1$ at outer surface)
β = helical lay angle with respect to the cable axis
ℓ_a = helical lay length of the armor, mm
r_a = mean radius of the armor, mm.

The identity $2\pi r_a/\ell_a = \sin(\beta)/\cos(\beta)$ was used above and below. The first term of $H_\perp$ has $-(1-x)$ because that term is from the axial field of the armor, and its perpendicular component in the tape is in the opposite direction to the perpendicular components of all the circular fields.

Considering now the second term on the right-hand-side of equation (8.14), the complex parallel power is given by

$$\begin{aligned}(P + jQ)_\parallel &= j\omega\mu_0\mu_e H_\parallel H_\parallel^* && \text{per unit volume}\\ &= \frac{j\omega\mu_0\mu_e}{\ell_a^2}|I_c + I_s + I_a|^2\cos^2(\beta) && \text{per unit volume}\\ &= \frac{j\omega\mu_0\mu_e A_a}{\ell_a^2}|I_c + I_s + I_a|^2\cos^2(\beta) && \text{per unit tape length}\\ &= \frac{j\omega\mu_0\mu_e A_a}{\ell_a^2}|I_c + I_s + I_a|^2\cos(\beta) && \text{per unit cable length}\\ &= \frac{j\omega\mu_0\mu_e A_a}{2\pi r_a \ell_a}|I_c + I_s + I_a|^2\sin(\beta) && \text{per unit cable length}\end{aligned} \tag{8.19}$$

where μ_e = complex relative longitudinal magnetic permeability
A_a = sum of the wire or tape cross-sectional areas (mm^2)
$|I_c + I_s + I_a|^2 = |I_c|^2 + |I_s|^2 + |I_a|^2 + 2I_c \cdot I_s + 2I_c \cdot I_a + 2I_s \cdot I_a$.

From this expression and equations (8.15) and (8.16), it can be seen that $(\mu_0\mu_e A_a/2\pi r_a \ell_a)\sin(\beta)$ contributes equally to all of the cable inductances L_{cc}, L_{cs}, L_{ca}, L_{sc}, L_{ss}, L_{sa}, L_{ac}, L_{as}, and L_{aa}.

The complex perpendicular power is given by

$$\begin{aligned}(P + jQ)_\perp &= j\omega\mu_0\mu_t H_\perp H_\perp^* && \text{per unit volume}\\ &= j\omega\mu_0\mu_t \int_0^1 \left[H_\perp H_\perp^*\right](2\pi r_a t_a)dx && \text{per unit cable length}\\ &= \frac{j\omega\mu_0\mu_t}{2\pi}\frac{t_a}{r_a}\left[|I_c + I_s|^2\cos^2(\beta) + |I_a|^2\left(\frac{1}{3\cos^2(\beta)} - \sin^2(\beta)\right)\right.\\ &\quad \left. + (I_c + I_s)\cdot I_a\left(\frac{1}{2} - \sin^2(\beta)\right)\right] && \text{per unit cable length}\end{aligned} \tag{8.20}$$

where μ_t = complex relative transverse magnetic permeability.

For round wires, μ_t in the IEC 287 Standard has been adjusted so that the transverse inductance for a cylinder may still be used. Although the formula above permits μ_t to be complex, the standard treats it as real. From this point on, we will also treat it as real since it is small compared with the longitudinal permeability, and therefore has little associated hysteresis loss. For touching steel wire armor, $\mu_t = 10$; for nontouching steel wires, $\mu_t = 1$.

From the expression for perpendicular power and equations (8.15) and (8.16), it can be seen that

$[(\mu_0\mu_t)/(2\pi)](t_a/r_a)\cos^2(\beta)$ contributes equally to L_{cc}, L_{cs}, $L_{sc,}$ and L_{ss}

$[(\mu_0\mu_t)/(2\pi)](t_a/r_a)\left[1/\left(3\cos^2(\beta)\right) - \sin^2(\beta)\right]$ contributes to L_{aa}

$[(\mu_0\mu_t)/(2\pi)](t_a/r_a)\left[1/2 - \sin^2(\beta)\right]$ contributes equally to $\left(L_{ca}, L_{sa}, L_{ac,}\right.$ and $\left.L_{as}\right)$.

The total contribution of internal armor flux to each cable inductance is given by the sum of the parallel and perpendicular power contributions.

8.3.2 Total Inductances

With the internal inductances computed in the preceding section, we can now derive the expressions for the total inductances of all cable components.

8.3.2.1 Conductor–Conductor Inductance. Integration of the flux linkage from the center of the conductor to its outer surface yields the internal inductance L_{cc-int} given by equation (8.8). It may be assumed that the permeability of the sheath is μ_0, and so the presence of the sheath makes no difference in the amount of conductor flux within the thickness of the sheath walls. The presence of the sheath does, however, make a difference in the total flux because the conductor flux induces currents in the sheath, which are taken into account by conductor–sheath inductance. If the presence of the armor is temporarily ignored, the external conductor–conductor inductance is given by the integral of conductor flux (linking all of the conductor current) from the conductor surface r_c to distance s, yielding

$$L_{\text{ext-no armor}} = \frac{\mu_0}{2\pi}\int_{r_c}^{s}\frac{1}{r}dr = \frac{\mu_0}{2\pi}\ln\left(\frac{s}{r_c}\right) \tag{8.21}$$

where μ_0 = magnetic permeability of free space, $4\pi \times 10^{-7}$ H/m

r_c = outer radius of the conductor, mm

s = axial distance between conductors, mm.

Magnetic armor makes a difference to the conductor–conductor inductance because of its high magnetic permeability. The inductance from conductor flux within the thickness of the armor is derived in Section 8.3.1.4. However, before adding it to L_{ext} above, it is necessary to subtract the contribution that has been included in equation (8.21) for the space that the armor occupies, which is given by

$$L_{\text{armor space}} = \frac{\mu_0}{2\pi}\int_{r_a-t_a/2}^{r_a+t_a/2}\frac{1}{r}dr \approx \frac{\mu_0}{2\pi}\frac{t_a}{r_a} \tag{8.22}$$

The total *conductor–conductor inductance* is obtained by combining equations (8.8), (8.21), 8.22) and the contribution of the armor as discussed in Section 8.3.1.4:

$$L_{cc} = \frac{\mu_0}{2\pi}[f] + \frac{\mu_0}{2\pi}\ln\left(\frac{s}{r_c}\right) - \frac{\mu_0}{2\pi}\frac{t_a}{r_a} + \frac{\mu_0}{2\pi}\frac{t_a}{r_a}\left[\mu_t \cos^2(\beta)\right] + \frac{\mu_0 \mu_e A_a}{2\pi r_a \ell_a}\sin(\beta) \quad (8.23)$$

where

$$f = \frac{1/4 - a^2 + a^4(3/4 - \ln a)}{(1-a^2)^2} \quad (8.24)$$

and a is the ratio of inner to outer radius of a hollow conductor.

Also,

β = helical lay angle with respect to the cable axis (see Fig. 8.4)

ℓ_a = helical lay length of the armor, mm

r_a = mean radius of the armor, mm

t_a = armor thickness, mm

μ_e = complex relative longitudinal magnetic permeability; the imaginary part of it describes the hysteresis loss of the magnetic material

μ_t = complex relative transverse magnetic permeability

A_a = sum of the wire or tape cross-sectional areas, mm^2.

The first two terms may be combined into one by defining an effective conductor radius αr_c where $\alpha = \exp(-f)$. For a solid conductor, $\alpha = \exp(-1/4) = 0.7788$.

The total conductor–conductor inductance then reduces to

$$L_{cc} = \frac{\mu_0}{2\pi}\ln\left(\frac{s}{\alpha r_c}\right) + \frac{\mu_0}{2\pi}\frac{t_a}{r_a}\left[\mu_t \cos^2(\beta) - 1\right] + \frac{\mu_0 \mu_e A_a}{2\pi r_a \ell_a}\sin(\beta) \quad (8.25)$$

For nonmagnetic armor, the last two terms are ignored. For nontouching magnetic armor wires, the second term is ignored.

8.3.2.2 Conductor–Sheath Inductance. The derivation of conductor–sheath inductance is similar to that of conductor–conductor inductance. The flux linkage is between the flux from the bottom conductor in Fig. 8.2 and all three sheath currents. The internal conductor–sheath inductance L_{cs-int} (the first term below) is given by equation (8.11). The external conductor–sheath flux linkage is obtained by integrating from the outer sheath radius to distance s and then taking into account the presence of the armor as before. The total conductor–sheath inductance is thus given by

$$\begin{aligned} L_{cs} &= \frac{\mu_0}{2\pi}\ln\left(\frac{r_s + t_s/2}{r_s}\right) + \frac{\mu_0}{2\pi}\ln\left(\frac{s}{r_s + t_s/2}\right) + \frac{\mu_0}{2\pi}\frac{t_a}{r_a}\left[\mu_t \cos^2(\beta) - 1\right] \\ &\quad + \frac{\mu_0 \mu_e A_a}{2\pi r_a \ell_a}\sin(\beta) \\ &= \frac{\mu_0}{2\pi}\ln\left(\frac{s}{r_s}\right) + \frac{\mu_0}{2\pi}\frac{t_a}{r_a}\left[\mu_t \cos^2(\beta) - 1\right] + \frac{\mu_0 \mu_e A_a}{2\pi r_a \ell_a}\sin(\beta) \end{aligned} \quad (8.26)$$

where r_s = mean radius of the sheath, mm

t_s = sheath thickness, mm.

It can be seen that, within the thickness of the sheath, the conductor flux links approximately half of the sheath current. The internal and external logarithm terms can thus be combined, to a good approximation, as the integral of flux from the mean radius of the sheath to distance s.

For nonmagnetic armor, the last two terms are ignored. For nontouching magnetic armor wires, the second term is ignored.

8.3.2.3 Sheath–Sheath Inductance. The sheath–sheath inductance is the linkage of the flux from the bottom sheath in Fig. 8.2 with the currents in all three sheaths. The internal sheath–sheath inductance L_{ss-int} (the second term below) is given by equation (8.13).

The external sheath–sheath flux linkage is obtained by integrating from the outer sheath radius to distance s and then taking into account the presence of the armor as before.

$$L_{ss} = \frac{\mu_0}{2\pi}\ln\left(\frac{s}{r_s + t_s/2}\right) + \frac{\mu_0}{2\pi}\ln\left(\frac{r_s + t_s/2}{r_s + t_s/6}\right) + \frac{\mu_0}{2\pi}\frac{t_a}{r_a}\left[\mu_t\cos^2(\beta) - 1\right] + \frac{\mu_0\mu_e A_a}{2\pi r_a \ell_a}\sin(\beta)$$

$$= \frac{\mu_0}{2\pi}\ln\left(\frac{s}{r_s + t_s/6}\right) + \frac{\mu_0}{2\pi}\frac{t_a}{r_a}\left[\mu_t\cos^2(\beta) - 1\right] + \frac{\mu_0\mu_e A_a}{2\pi r_a \ell_a}\sin(\beta) \qquad (8.27)$$

We can observe that, within the thickness of the sheath, the sheath's own flux links approximately one third of the sheath current. The internal and external logarithm terms can thus be combined, to a good approximation, as the integral of flux from $r_s + t_s/6$ of the sheath to distance s. For manual computations, the $t_s/6$ may be ignored with little loss of accuracy.

For nonmagnetic armor, the last two terms are ignored. For nontouching magnetic armor wires, the second term is ignored.

8.3.2.4 Armor Inductances. The conductor–armor, sheath–armor, and armor–armor internal inductances have been derived in Section 8.3.1.4. The external portion of the armor inductances is obtained by integrating from the outer surface of the armor $r_a + t_a/2$ to distance s, yielding

$$L_{aa-ext} = \frac{\mu_0}{2\pi}\ln\left(\frac{s}{r_a + t_a/2}\right) \qquad (8.28)$$

The following armor-related inductances are given as the sum of the external inductance and the internal inductances for both parallel and perpendicular power (power associated with flux parallel and perpendicular to the armor tapes or wires), derived in Section 8.3.1.4.

The *conductor–armor* and *sheath–armor* inductances are given by

$$L_{ca} = L_{sa} = \frac{\mu_0}{2\pi}\ln\left(\frac{s}{r_a + t_a/2}\right) + \frac{\mu_0\mu_t}{2\pi}\frac{t_a}{r_a}\left[\frac{1}{2} - \sin^2(\beta)\right] + \frac{\mu_0\mu_e A_a}{2\pi r_a \ell_a}\sin(\beta) \qquad (8.29)$$

For nonmagnetic armor, the second term becomes $(\mu_0 t_a/2\pi r_a)(1/2)$ and can be absorbed into the first term, to a good approximation, by dropping the "$+t_a/2$" in the denominator in the logarithm term. The last term is ignored for nonmagnetic armor.

For nontouching magnetic armor wires, μ_t is assumed to be 1. If the $\sin^2(\beta)$ term is ignored, the first and second terms may then be combined (by ignoring the "$+t_a/2$" in the denominator in the logarithm term) as before. The last term remains unchanged.

The *armor–armor* inductance is given by

$$L_{aa} = \frac{\mu_0}{2\pi} \ln\left(\frac{s}{r_a + t_a/2}\right) + \frac{\mu_0 \mu_t}{2\pi} \frac{t_a}{r_a} \left[\frac{1}{3\cos^2(\beta)} - \sin^2(\beta)\right] + \frac{\mu_0 \mu_e A_a}{2\pi r_a \ell_a} \sin(\beta) \quad (8.30)$$

For nonmagnetic armor, the second term becomes $(\mu_0 t_a/2\pi r_a)(1/3)$ and can be absorbed into the first term, to a good approximation, by changing the "$r_a + t_2/2$" to "$r_a + t_a/6$" in the denominator in the logarithm term since $t_a/(3r_a) \approx \ln[(r_a + t_a/2)/(r_a + t_a/6)]$. The last term is ignored for nonmagnetic armor.

If the $\sin^2(\beta)$ term is ignored, the first and second terms may the combined (by changing the "$r_a + t_a/2$" to "$r_a + t_a/6$" in the denominator in the logarithm term as before. The last term remains unchanged.

8.3.3 Cable Impedances

In order to compute the loss factors, the impedances will be separated into real and imaginary components. Every cable inductance includes the longitudinal internal armor term given by

$$jX_L = \frac{j\omega\mu_0\mu_e A_a}{2\pi r_a \ell_a} \sin(\beta) \quad (8.31)$$

where ω = angular frequency

μ_e = complex relative longitudinal magnetic permeability; the imaginary part of it describes the hysteresis loss of the magnetic material.

Equation (8.31) may be split into the imaginary and real parts, as shown below.

The *complex longitudinal magnetic permeability* μ_e of the wire may be written as

$$\mu_e = |\mu_e|(\cos\gamma - j\sin\gamma) \quad (8.32)$$

where γ is the angular time delay of the magnetic flux density B w.r.t. the field H.

Real and imaginary components in equation (8.31) are

$$B_0 = \Im m(jX_L) = \frac{\omega\mu_0|\mu_e|A_a}{2\pi r_a \ell_a} \sin(\beta)\cos(\gamma) \quad (8.33)$$

and

$$B_2 = \Re e(jX_L) = \frac{\omega\mu_0|\mu_e|A_a}{2\pi r_a \ell_a} \sin(\beta)\sin(\gamma) \quad (8.34)$$

From equation (8.25), the *conductor–conductor* impedance is given by

$$Z_{cc} = (R + B_2) + \frac{j\omega\mu_0}{2\pi} \ln\left(\frac{s}{\alpha r_c}\right) + jB_0 + \frac{j\omega\mu_0}{2\pi} \frac{t_a}{r_a} \left[\mu_t \cos^2(\beta) - 1\right] \quad (8.35)$$

where R = conductor ac resistance, Ω/m

s = distance between cables in trefoil configuration or the geometric mean of the three spacing between cables in flat formation, mm

αr_c = effective radius of the conductor ($\alpha = 0.7788$ for solid conductor), mm.

From equation (8.26), the *conductor–sheath* impedance is given by

$$Z_{cs} = Z_{sc} = B_2 + jB_1 \tag{8.36}$$

where

$$B_1 = \frac{\omega\mu_0}{2\pi}\ln\left(\frac{s}{r_s}\right) + B_0 + \frac{\omega\mu_0}{2\pi}\frac{t_a}{r_a}\left[\mu_t\cos^2(\beta) - 1\right] \tag{8.37}$$

The *conductor–armor* and *sheath–armor* impedances obtained from equation (8.29), are given by

$$Z_{ca} = Z_{ac} = Z_{sa} = Z_{as} = B_2 + jB_3 \tag{8.38}$$

where

$$B_3 = \frac{\omega\mu_0}{2\pi}\ln\left(\frac{s}{r_a + t_a/2}\right) + B_0 + \frac{\omega\mu_0}{2\pi}\frac{t_a}{r_a}\mu_t\left[\frac{1}{2} - \sin^2(\beta)\right] \tag{8.39}$$

The *sheath–sheath* impedance is obtained from equation (8.27). Without significant loss of accuracy, we can ignore the term $t_s/6$ in the equation, yielding

$$Z_{ss} \approx (R_s + B_2) + jB_1 \tag{8.40}$$

From equation (8.30), the *armor–armor* impedance is given by

$$Z_{aa} \approx (R_a + B_2) + jB_4 \tag{8.41}$$

where

$$B_4 = \frac{\omega\mu_0}{2\pi}\ln\left(\frac{s}{r_a + t_a/2}\right) + B_0 + \frac{\omega\mu_0}{2\pi}\frac{t_a}{r_a}\mu_t\left[\frac{1}{3\cos^2(\beta)} - \sin^2(\beta)\right] \tag{8.42}$$

where R_a = armor resistance, including the stranding increment $1/\cos\beta$, Ω/m

Z_{aa} = the only impedance that would differ for round wires as compared to tapes. For a solid cylinder without helical lay, the internal flux linkage contains the factor $1/3$ (the square bracket of B_4 when $\beta = 0$). For round wires without helical lay, this factor becomes 0.356.

8.3.4 Loss Factors

The voltages V_s and V_a in equation (8.5) are zero when both ends of the sheath and armor are grounded, and so the sheath and armor voltages are described by the following matrix equation:

$$\begin{bmatrix} (R_s + B_2) + jB_1 & B_2 + jB_3 \\ B_2 + jB_3 & (R_a + B_2) + jB_4 \end{bmatrix}\begin{bmatrix} I_s \\ I_a \end{bmatrix} = -\begin{bmatrix} (B_2 + jB_1)I_c \\ (B_2 + jB_3)I_c \end{bmatrix} \tag{8.43}$$

The solutions for I_s and I_a in terms of I_c are given by

$$I_s = -\frac{(B_2 + jB_1)[(R_a + B_2) + jB_4] - (B_2 + jB_3)^2}{[(R_s + B_2) + jB_1][(R_a + B_2) + jB_4] - (B_2 + jB_3)^2} I_c \tag{8.44}$$

and

$$I_a = -\frac{(B_2 + jB_3)R_s}{[(R_s + B_2) + jB_1][(R_a + B_2) + jB_4] - (B_2 + jB_3)^2} I_c \tag{8.45}$$

Define

$$\begin{aligned} Y_1 &= B_2(R_a + B_2) - B_1 B_4 - B_2^2 + B_3^2 \\ Y_2 &= B_1(R_a + B_2) + B_2 B_4 - 2B_2 B_3 \\ Y_3 &= R_s(R_a + B_2) \\ Y_4 &= R_s B_4 \\ Y_5 &= R_s B_2 \\ Y_6 &= R_s B_3. \end{aligned} \tag{8.46}$$

Collecting the real and imaginary parts of equations (8.44) and (8.45) together, the solutions become

$$I_s = -\frac{Y_1 + jY_2}{(Y_1 + Y_3) + j(Y_2 + Y_4)} I_c \tag{8.47}$$

$$I_a = -\frac{Y_5 + jY_6}{(Y_1 + Y_3) + j(Y_2 + Y_4)} I_c \tag{8.48}$$

Define $Y = (Y_1 + Y_3)^2 + (Y_2 + Y_4)^2$. Then

$$\begin{aligned} |I_s|^2 &= \frac{Y_1^2 + Y_2^2}{Y} |I_c|^2 \\ |I_a|^2 &= \frac{Y_5^2 + Y_6^2}{Y} |I_c|^2 \end{aligned} \tag{8.49}$$

The circulating loss factors for the sheath and armor are thus given by

$$\boxed{\lambda_1' = \frac{|I_s|^2 R_s}{|I_c|^2 R} = \frac{R_s}{R} \frac{Y_1^2 + Y_2^2}{Y}} \tag{8.50}$$

and

$$\boxed{\lambda_2' = \frac{|I_a|^2 R_a}{|I_c|^2 R} = \frac{R_a}{R} \frac{Y_5^2 + Y_6^2}{Y}.} \tag{8.51}$$

The hysteresis loss factor for the armor is given by

$$\lambda_2'' = \frac{|I_c + I_s + I_a|^2 B_2}{|I_c|^2 R} = \frac{B_2}{R} \frac{(Y_3 - Y_5)^2 + (Y_4 - Y_6)^2}{Y} \tag{8.52}$$

The armor loss factor is the sum of the loss factors given by equations (8.51) and (8.52).

EXAMPLE 8.1

Consider a cable model No. 4, and assume that the armor is composed of touching steel wires. The parameters of this cable are: 1) lead sheath: $t_s = 3.2$ mm, $D_s = 75.4$ mm; 2) three copper tapes: $D_T = 78$ mm, $t_T = 0.13$ mm, $w_T = 0.13$ mm, 3) steel armor: $\delta_0 = 5.189$ mm, $\ell_a = 121.8$ mm, $D_a' = 98.4$ mm, $n_a = 51$. The circuit is partially located under water and partially underground. In its land portion, the cables are placed in a triangular configuration as shown in Fig. 8.5.

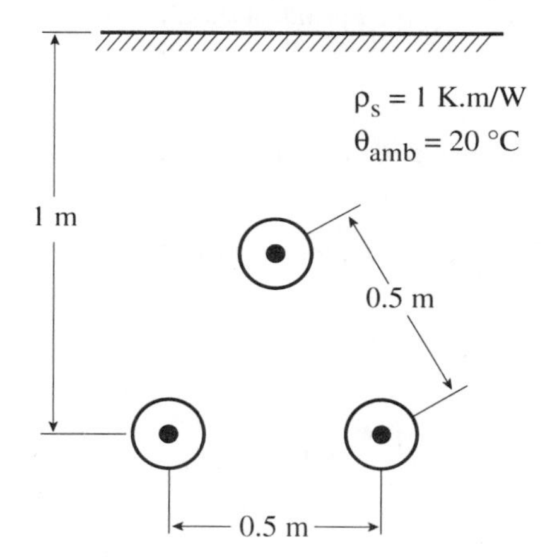

Figure 8.5 Three submarine cables in triangular formation.

All of the dimensions will be expressed in meters. We will begin by computing the resistances of the metallic parts assuming the frequency of the system equal to 50 Hz. The resistances of the conductors, sheaths, and reinforcing copper tapes are slightly different in the upper cable since it operates at slightly lower temperature. However, the differences are very small, and we can assume that the values of the lower cable apply to all cables. We will assume that the sheath operates at 70°C and the armor at 68°C. The conductor ac resistance at the operating temperatures of 85°C is equal to

$$R_c = 0.356 \cdot 10^{-4} \ \Omega/\text{m}$$

The resistance of the lead sheath at $\theta_s = 70$°C is obtained from

$$R_s = \frac{21.4 \cdot 10^{-8}}{\pi(0.0754 - 0.0032) \cdot 0.0032}[1 + 0.004(70 - 20)] = 0.000354 \ \Omega/\text{m}$$

The resistance of the copper reinforcing tape is computed as an equivalent tube resistance. The mean diameter of the tape is equal to

$$d_2^* = 0.078 - 3 \cdot 0.0013 = 0.0741 \text{ m}$$

The cross-sectional area of the tapes is given by

$$A_T = 3 \cdot 0.0013 \cdot 0.025 = 0.975 \cdot 10^{-4} \text{ m}^2$$

The equivalent tube resistance is obtained taking into account the length of lay ℓ_T of the tapes

$$R_T = \frac{\rho_{20}\sqrt{\left(\frac{\pi d_2}{\ell_T}\right)^2 + 1}}{A_T}[1 + \alpha_{20}(\theta_s - 20)]$$

$$= \frac{0.1724 \cdot 10^{-7} \cdot \sqrt{\left(\frac{\pi \cdot 0.0741}{0.1152}\right)^2 + 1}}{0.975 \cdot 10^{-4}}[1 + 0.00393(70 - 20)] = 0.477 \cdot 10^{-3}\ \Omega/\text{m}$$

The resistance of the parallel combination of the sheath and the reinforcing tape at 70°C will continue to be denoted by R_s and is equal to

$$R_s = \frac{0.354 \cdot 10^{-3} \cdot 0.477 \cdot 10^{-3}}{0.354 \cdot 10^{-3} + 0.477 \cdot 10^{-3}} = 0.203 \cdot 10^{-3}\ \Omega/\text{m}$$

The rms diameter of the equivalent sheath is equal to

$$d^* = \sqrt{\frac{0.0754^2 + 0.078^2}{2}} = 0.0767 \text{ m}$$

The cross-sectional area of the armor wires is equal to

$$A_a = \pi \cdot (0.00519)^2 \cdot 51/4 = 0.00108 \text{ m}^2$$

The mean diameter of the armor is given by

$$d_a^* = 0.0984 - 0.00519 = 0.0932 \text{ m}$$

The resistance of armor is obtained taking into account the length of lay of the wires

$$R_a = \frac{0.138 \cdot 10^{-6} \cdot \sqrt{\left(\frac{\pi \cdot 0.0932}{0.1218}\right)^2 + 1}}{0.00108}[1 + 0.0045(68 - 20)] = 0.405 \cdot 10^{-3}\ \Omega/\text{m}$$

Since this is a relatively large cable, the ac resistance of the armor wires is 1.4 times greater than the dc resistance computed above; that is,

$$R_a = 1.4 \cdot 0.405 \cdot 10^{-3} = 0.567 \cdot 10^{-3}\ \Omega/\text{m}$$

In order to compute the remaining impedance, we need to calculate the angle between the axis of each armor wire and the axis of the cable. This angle is obtained from a knowledge of the mean diameter of the armor and the length of lay of the wires (see Fig. 8.4):

$$\beta = \tan^{-1}\frac{\pi d_a}{\ell_a} = 68^\circ$$

The longitudinal relative permeability of steel wires is equal to 400 and $\gamma = 45^\circ$. From equations (8.33) and (8.34), we have

$$B_0 = \frac{\omega\mu_0|\mu_e|A_a}{2\pi r_a \ell_a}\sin(\beta)\cos(\gamma)$$

$$= \left(\frac{2\pi \cdot 50 \cdot 4\pi \cdot 10^{-7} \cdot 400 \cdot 51 \cdot \pi \cdot 0.00519^2}{2\pi \cdot 0.0932/2 \cdot 0.1218 \cdot 4}\right)\sin 68 \cos 45 = 0.00313\ \Omega/\text{m}$$

$$B_2 = \frac{\omega\mu_0|\mu_e|A_a}{2\pi r_a \ell_a}\sin(\beta)\sin(\gamma)$$

$$= \left(\frac{2\pi \cdot 50 \cdot 4\pi \cdot 10^{-7} \cdot 400 \cdot 51 \cdot \pi \cdot 0.00519^2}{2\pi \cdot 0.0932/2 \cdot 0.1218 \cdot 4}\right)\sin 68 \sin 45 = 0.00313\ \Omega/\text{m}$$

The value of B_1 is obtained from equation (8.37) with the value of $\mu_t = 10$

$$B_1 = \frac{\omega\mu_0}{2\pi}\ln\left(\frac{s}{r_s}\right) + B_0 + \frac{\omega\mu_0}{2\pi}\frac{t_a}{r_a}\left[\mu_t\cos^2(\beta) - 1\right]$$

$$= \frac{2\pi\cdot 50\cdot 4\pi\cdot 10^{-7}}{2\pi}\ln\left(\frac{2\cdot 0.5}{0.0767}\right) + 0.00313$$

$$+\frac{2\pi\cdot 50\cdot 4\pi\cdot 10^{-7}}{2\pi}\frac{0.00519\cdot 2}{0.0932}[10\cos^2(68) - 1] = 0.00329\ \Omega/\text{m}$$

The constants B_3 and B_4 from equations (8.39) and (8.42), respectively, are

$$B_3 = \frac{\omega\mu_0}{2\pi}\ln\left(\frac{s}{r_a + t_a/2}\right) + B_0 + \frac{\omega\mu_0}{2\pi}\frac{t_a}{r_a}\mu_t\left[\frac{1}{2} - \sin^2(\beta)\right]$$

$$= \frac{2\pi\cdot 50\cdot 4\pi\cdot 10^{-7}}{2\pi}\ln\left(\frac{0.5}{0.0932/2 + 0.00519/2}\right) + 0.00313$$

$$+\frac{2\pi\cdot 50\cdot 4\pi\cdot 10^{-7}}{2\pi}\frac{0.00519\cdot 2}{0.0932}\cdot 10\left[\frac{1}{2} - \sin^2(68)\right] = 0.00326\ \Omega/\text{m}$$

$$B_4 = \frac{\omega\mu_0}{2\pi}\ln\left(\frac{s}{r_a + t_a/2}\right) + B_0 + \frac{\omega\mu_0}{2\pi}\frac{t_a}{r_a}\mu_t\left[\frac{1}{3\cos^2(\beta)} - \sin^2(\beta)\right]$$

$$= \frac{2\pi\cdot 50\cdot 4\pi\cdot 10^{-7}}{2\pi}\ln\left(\frac{0.5}{0.0932/2 + 0.00519/2}\right) + 0.00313$$

$$+\frac{2\pi\cdot 50\cdot 4\pi\cdot 10^{-7}}{2\pi}\frac{0.00519\cdot 2}{0.0932}\cdot 10\left[\frac{1}{3\cos^2(68)} - \sin^2(68)\right] = 0.00338\ \Omega/\text{m}$$

Constants Y are obtained from equations (8.46)

$$Y_1 = B_2(R_a + B_2) - B_1B_4 - B_2^2 + B_3^2$$
$$= [3.13(0.567 + 3.13) - 3.29\cdot 3.38 - 3.13^2 + 3.26^2]\cdot 10^{-6} = 1.28\cdot 10^{-6}$$
$$Y_2 = B_1(R_a + B_2) + B_2B_4 - 2B_2B_3$$
$$= [3.29(0.567 + 3.13) + 3.13\cdot 3.38 - 2\cdot 3.13\cdot 3.26]\cdot 10^{-6} = 2.33\cdot 10^{-6}$$
$$Y_3 = R_s(R_a + B_2) = [0.203(0.567 + 3.13)]\cdot 10^{-6} = 0.750\cdot 10^{-6}$$
$$Y_4 = R_sB_4 = [0.203\cdot 3.38]\cdot 10^{-6} = 0.686\cdot 10^{-6}$$
$$Y_5 = R_sB_2 = [0.203\cdot 3.13]\cdot 10^{-6} = 0.635\cdot 10^{-6}$$
$$Y_6 = R_sB_3 = [0.203\cdot 3.26]\cdot 10^{-6} = 0.662\cdot 10^{-6}$$
$$Y = (Y_1 + Y_3)^2 + (Y_2 + Y_4)^2 = [(1.28 + 0.750)^2 + (2.33 + 0.686)^2]\cdot 10^{-12} = 13.2\cdot 10^{-12}$$

The loss factors are now computed from equations (8.50)–(8.52)

$$\lambda_1' = \frac{R_s}{R}\frac{Y_1^2 + Y_2^2}{Y} = \frac{0.203}{0.0356}\frac{1.28^2 + 2.33^2}{13.2} = 3.05$$

$$\lambda_2' = \frac{R_a}{R}\frac{Y_5^2 + Y_6^2}{Y} = \frac{0.567}{0.0356}\frac{0.635^2 + 0.662^2}{13.2} = 1.02$$

$$\lambda_2'' = \frac{B_2}{R}\frac{(Y_3 - Y_5)^2 + (Y_4 - Y_6)^2}{Y} = \frac{3.13}{0.0356}\frac{(0.750 - 0.635)^2 + (0.686 - 0.662)^2}{13.2} = 0.08$$

The rating of the hottest (bottom) cable is given by equation (4.3)

$$I = \left[\frac{(\theta_c - \theta_{\text{amb}}) - W_d\,[0.5T_1 + n(T_2 + T_3 + T_4)]}{R_c\,[T_1 + n(1+\lambda_1)T_2 + n(1+\lambda_1+\lambda_2)(T_3+T_4)]}\right]^{0.5}$$

Dielectric losses for the cable considered here are equal to 6.62 W/m. The ambient temperature is equal to 20°C. With the cable construction given in Fig. A.4 and circuit location shown in Fig. 8.5, the thermal resistances of the thermal circuit are as follows (see Chapter 9):

$$T_1 = 0.568 \text{ K}\cdot\text{m/W}, \quad T_2 = 0.082 \text{ K}\cdot\text{m/W}, \quad T_3 = 0.066 \text{ K}\cdot\text{m/W}, \quad T_4 = 0.789 \text{ K}\cdot\text{m/W}$$

where the external thermal resistance is that of the lower cables.

The sheath loss factor is equal to 3.05 since the eddy current losses are neglected and the armor loss factor is equal to 1.1. The rating of the cable is given by

$$I = \left[\frac{85 - 20 - 6.62(0.5\cdot 0.568 + 0.082 + 0.066 + 0.789)}{0.356\cdot 10^{-4}(0.568 + (1+3.05)\cdot 0.082 + (1+3.05+1.1)\cdot(0.066+0.789))}\right]^{0.5} = 549 \text{ A}$$

8.3.5 Circulating Current Losses in the Sheaths—Special Cases

The equations for the sheath and armor circulating currents loss factors derived in this section are very general and are applicable to cables in flat and trefoil configurations with the spacings between cables properly taken into account. The sheath and armor loss factors are computed simultaneously and the calculations are quite involved. The computations can be much simplified, with a small loss of accuracy, if we consider sheath and armor losses separately following the practice presented in the standards. This is illustrated in the following subsections, where we will consider several cable constructions and typical cable arrangements.

8.3.5.1 Two Single-core Cables or Three Single-core Cables in Trefoil Formation, Sheaths Bonded at Both Ends. Since the sheaths are bonded at both ends of an electrical section, $\lambda_1'' = 0$, except for cables having large conductors of segmental construction, in which case λ_1'' is calculated by the method given in Section 8.4.10.

Let us consider only the sheath of a cable. In this case, the last two terms of equation (8.37) will disappear and the reactance B_1, usually denoted by X, takes the form

$$X = B_1 = \frac{\omega\mu_0}{2\pi}\ln\frac{2s}{d} = \omega\cdot 2\cdot 10^{-7}\ln\frac{2s}{d} \tag{8.53}$$

where s = distance between conductor axis in the electrical section being considered, mm

d = mean diameter of the sheath, mm

—for oval-shaped cores d is given by $\sqrt{d_M \cdot d_m}$, where d_M and d_m are the major and minor mean sheath diameters, respectively

—for corrugated sheaths d is given by $(D_{oc} + D_{it})/2$.

Equation (8.49) reduces in this case to

$$I_s = \frac{IX}{\sqrt{R_s^2 + X^2}}$$

and the sheath loss factor is given by:

$$\lambda_1' = \frac{I_s^2 R_s}{I^2 R} = \frac{R_s}{R} \cdot \frac{X^2}{R_s^2 + X^2} = \frac{R_s}{R} \cdot \frac{1}{1 + \left(\frac{R_s}{X}\right)^2} \tag{8.54}$$

Equation (8.54) illustrates that λ_1' is increased by a decrease in the conductor resistance R and an increase in X^2, which means by the use of larger conductors and wider spacing. By computing the first derivative of λ_1' with respect to R_s and equating it to zero, we find that the sheath loss factor attains its maximum when $R_s = X$. Under typical conditions when $R_s > X$, a decrease in R_s increases λ_1'. This explains why, in general, aluminum-sheathed cables have considerably greater sheath losses than lead-sheathed cables.

8.3.5.2 Three Single-core Cables in Flat Formation, with Regular Transposition, Sheaths Bonded at Both Ends. Equations were also developed by Arnold for the calculation of sheath losses of cables laid in a flat formation with the middle cable equidistant from the two outer cables. For a regular transposition condition, the voltage in each sheath at the third transition is equal to the vector average of the voltages induced in each of the three sections. This average voltage V_i is given by

$$V_i = I\left(X + \frac{X_m}{3}\right)$$

where X_m = mutual reactance between the sheath of the outer cable and the conductors of the other two, Ω/m

$= 2\omega 10^{-7} \cdot \ln(2)$, Ω/m.

For convenience, the term $(X + (X_m/3))$ is defined as X_1, and from equation (8.53), it is given by

$$X_1 = 2\omega \cdot 10^{-7} \cdot \ln\left[2 \cdot \sqrt[3]{2}\left(\frac{s}{d}\right)\right] \tag{8.55}$$

The sheath loss factor is given again by equation (8.54) with X replaced by X_1; that is

$$\lambda_1' = \frac{R_s}{R} \cdot \frac{1}{1 + \left(\frac{R_s}{X_1}\right)^2} \tag{8.56}$$

Also, $\lambda_1'' = 0$, except for cables having large conductors of segmental construction when λ_1'' is calculated by the method given in Section 8.4.10.

8.3.5.3 Three Single-core Cables without Transposition, Sheaths Bonded at Both Ends. As before, only balanced currents are considered. When the cables are laid in flat formation with the middle cable equidistant from the two outer ones, the mutual inductance between the middle cable and each of the outer cables is given by twice the value given by equation (8.53). However, since the axial spacing between the two outer cables is $2s$, the mutual reactance between them is given by

$$P = 2\omega \cdot 10^{-7} \ln 2 \cdot \frac{2s}{d} = 2\omega \cdot 10^{-7} \ln 2 + 2\omega \cdot 10^{-7} \ln \frac{2s}{d} = X_m + X$$

Let

$$\begin{aligned} I_1 &= I_2 \left(-\frac{1}{2} + j\frac{\sqrt{3}}{2} \right) \\ I_3 &= I_2 \left(-\frac{1}{2} - j\frac{\sqrt{3}}{2} \right) \end{aligned} \tag{8.57}$$

Clearly,

$$\begin{aligned} I_{s1} + I_{s2} + I_{s3} &= 0 \\ E_{s1} = E_{s2} = E_{s3} &= E_0 \end{aligned} \tag{8.58}$$

where E_0 (V/m) is a residual voltage along the cable sheaths, which usually does not exceed 50 V and could be zero when both ends of the cables are grounded.

Applying equations (8.57) and (8.58), the residual voltage computed for each sheath separately is equal to

$$\begin{aligned} E_{s1} = E_0 &= I_{s1}(R_s + jX) - \frac{1}{2} j I_2 (X - X_m) - \frac{\sqrt{3}}{2} I_2 (X + X_m) - j I_{s3} X_m \\ E_{s2} = E_0 &= I_{s2}(R_s + jX) + j I_2 X \\ E_{s3} = E_0 &= I_{s3}(R_s + jX) - \frac{1}{2} j I_2 (X - X_m) + \frac{\sqrt{3}}{2} I_2 (X + X_m) - j I_{s1} X_m \end{aligned} \tag{8.59}$$

Solving equations (8.58) and (8.59) for the sheath currents, we obtain

$$\begin{aligned} I_{s1} &= \frac{I_2}{2} \left[\frac{Q^2}{R_s^2 + Q^2} + \frac{\sqrt{3} R_s P}{R_s^2 + P^2} + j \left(\frac{R_s Q}{R_s^2 + Q^2} - \frac{\sqrt{3} P^2}{R_s^2 + P^2} \right) \right] \\ I_{s2} &= -I_2 \left(\frac{Q^2}{R_s^2 + Q^2} + j \frac{R_s Q}{R_s^2 + Q^2} \right) \\ I_{s3} &= \frac{I_2}{2} \left[\frac{Q^2}{R_s^2 + Q^2} - \frac{\sqrt{3} R_s P}{R_s^2 + P^2} + j \left(\frac{R_s Q}{R_s^2 + Q^2} + \frac{\sqrt{3} P^2}{R_s^2 + P^2} \right) \right] \end{aligned} \tag{8.60}$$

where P and Q are defined by

$$P = X_m + X, \quad Q = X - X_m/3 \tag{8.61}$$

Taking the magnitudes of the sheath currents in equations (8.60) and remembering that in a balanced system $|I_2| = |I|$, we obtain the following expressions for the sheath loss factors:

$$\lambda'_{11} = \frac{R_s}{R}\left[\frac{\frac{1}{4}Q^2}{R_s^2 + Q^2} + \frac{\frac{3}{4}P^2}{R_s^2 + P^2} - \frac{2R_s P Q X_m}{\sqrt{3}\left(R_s^2 + Q^2\right)\left(R_s^2 + P^2\right)}\right] \quad \text{in the leading phase}$$

$$\lambda'_{1m} = \frac{R_s}{R}\frac{Q^2}{R_s^2 + Q^2} \quad \text{in the middle cable}$$

$$\lambda'_{12} = \frac{R_s}{R}\left[\frac{\frac{1}{4}Q^2}{R_s^2 + Q^2} + \frac{\frac{3}{4}P^2}{R_s^2 + P^2} + \frac{2R_s P Q X_m}{\sqrt{3}\left(R_s^2 + Q^2\right)\left(R_s^2 + P^2\right)}\right] \quad \text{in the lagging phase}$$

(8.62)

Observe from equations (8.62) that the three loss factors are different.

When single-core cables are installed in flat formation without cross bonding or transposition, the sheath losses increase as the cable spacing increases, but not linearly. Also, the external thermal resistance decreases as the cable spacing increases. Therefore, the ideal spacing is a balance between these two factors. For the rating of cables in air, the loss factor for the outer cable carrying the lagging phase should be used since it has the highest loss factor, provided that the spacing is sufficient to ensure thermal independence. For spaced buried cables, the loss for all three cables is used in the calculation of external thermal resistance.

We will illustrate the computation of the loss factors by again examining Example 7.5.

EXAMPLE 8.2

Consider cable model No. 1 with laying conditions specified in Appendix A. We assume that the system load has third harmonic components which constitute 20% of the fundamental current. We compute the loss factors for the fundamental frequency and for the third harmonic for this cable system. The sheath diameter and the cable spacing are $d = 31.2$ mm and $s = 71.6$ mm, respectively. The required resistances were computed in Example 7.5.

The auxiliary quantities required to compute the value of the screen loss factor are obtained from equation (8.61). For the fundamental frequency, we have

$$X = \omega \cdot 2 \cdot 10^{-7} \ln\frac{2s}{d} = 4\pi \cdot 50 \cdot 10^{-7} \cdot \ln\left(\frac{2 \cdot 71.6}{31.2}\right) = 0.957 \cdot 10^{-4}\ \Omega/\text{m}$$

$$X_m = 4\pi \cdot 50 \cdot 10^{-7} \cdot \ln(2) = 0.436 \cdot 10^{-4}\ \Omega/\text{m}$$

$$P = X_m + X = 0.957 \cdot 10^{-4} + 0.436 \cdot 10^{-4} = 0.139 \cdot 10^{-3}\ \Omega/\text{m}$$

$$Q = X - X_m/3 = 0.957 \cdot 10^{-4} - \frac{0.436 \cdot 10^{-4}}{3} = 0.812 \cdot 10^{-4}\ \Omega/\text{m}$$

When the conductor reaches an operating temperature of 90°C, the temperature of the concentric neutral wires is 83°C. Thus, the resistance of the screen is equal to

$$R_s = 0.759 \cdot 10^{-3}(1 + 0.00393 \cdot 63) = 0.947 \cdot 10^{-3}\ \Omega/\text{m}$$

The circulating current loss factor for the outer cable carrying the leading phase is obtained from equation (8.62)

$$R_s^2 + P^2 = (0.947 \cdot 10^{-3})^2 + (0.139 \cdot 10^{-3})^2 = 0.916 \cdot 10^{-6}$$

$$R_s^2 + Q^2 = (0.947 \cdot 10^{-3})^2 + (0.0812 \cdot 10^{-3})^2 = 0.903 \cdot 10^{-6}$$

$$\lambda'_{11} = \frac{R_s}{R}\left[\frac{\frac{1}{4}Q^2}{R_s^2 + Q^2} + \frac{\frac{3}{4}P^2}{R_s^2 + P^2} - \frac{2R_s P Q X_m}{\sqrt{3}\left(R_s^2 + Q^2\right)\left(R_s^2 + P^2\right)}\right]$$

$$= \frac{0.947 \cdot 10^{-3}}{0.0781 \cdot 10^{-3}}\left[\frac{\frac{1}{4} \cdot 0.0812^2 \cdot 10^{-6}}{0.903 \cdot 10^{-6}} + \frac{\frac{3}{4} \cdot 0.139^2 \cdot 10^{-6}}{0.916 \cdot 10^{-6}}\right.$$

$$\left. - \frac{2 \cdot 0.947 \cdot 0.139 \cdot 0.0812 \cdot 0.0436 \cdot 10^{-12}}{\sqrt{3} \cdot 916 \cdot 0.903 \cdot 10^{-12}}\right] = 0.206$$

The loss factor for the other outer cable is also obtained from equation (8.62) and is given by

$$\lambda'_{12} = \frac{R_s}{R}\left[\frac{\frac{1}{4}Q^2}{R_s^2 + Q^2} + \frac{\frac{3}{4}P^2}{R_s^2 + P^2} + \frac{2R_s P Q X_m}{\sqrt{3}\left(R_s^2 + Q^2\right)\left(R_s^2 + P^2\right)}\right]$$

$$= \frac{0.947 \cdot 10^{-3}}{0.0781 \cdot 10^{-3}}\left[\frac{\frac{1}{4} \cdot 0.0812^2 \cdot 10^{-6}}{0.903 \cdot 10^{-6}} + \frac{\frac{3}{4} \cdot 0.139^2 \cdot 10^{-6}}{0.916 \cdot 10^{-6}}\right.$$

$$\left. + \frac{2 \cdot 0.947 \cdot 0.139 \cdot 0.0812 \cdot 0.0436 \cdot 10^{-12}}{\sqrt{3} \cdot 916 \cdot 0.903 \cdot 10^{-12}}\right] = 0.222$$

For the middle cable, the loss factor is equal to

$$\lambda'_{1m} = \frac{R_s}{R}\frac{Q^2}{R_s^2 + Q^2} = \frac{0.947 \cdot 10^{-3}}{0.0781 \cdot 10^{-3}}\frac{0.812^2 \cdot 10^{-8}}{0.903 \cdot 10^{-6}} = 0.088$$

Using a similar approach for the third harmonic, we have

$$(X)_3 = 4\pi \cdot 150 \cdot 10^{-7} \cdot \ln\left(\frac{2 \cdot 71.6}{31.2}\right) = 0.287 \cdot 10^{-3}\ \Omega/\text{m}$$

$$(X_m)_3 = 4\pi \cdot 150 \cdot 10^{-7} \cdot \ln(2) = 0.131 \cdot 10^{-3}\ \Omega/\text{m}$$

$$P = 0.287 \cdot 10^{-3} + 0.131 \cdot 10^{-3} = 0.418 \cdot 10^{-3}\ \Omega/\text{m}$$

$$Q = 0.287 \cdot 10^{-3} - \frac{0.131 \cdot 10^{-3}}{3} = 0.243 \cdot 10^{-3}\ \Omega/\text{m}$$

$$R_s^2 + P^2 = (0.947 \cdot 10^{-3})^2 + (0.418 \cdot 10^{-3})^2 = 1.07 \cdot 10^{-6}$$

$$R_s^2 + Q^2 = (0.947 \cdot 10^{-3})^2 + (0.243 \cdot 10^{-3})^2 = 0.956 \cdot 10^{-6}$$

$$\left(\lambda'_{11}\right)_3 = \frac{0.947 \cdot 10^{-3}}{0.0792 \cdot 10^{-3}}\left[\frac{\frac{1}{4} \cdot 0.243^2 \cdot 10^{-6}}{0.956 \cdot 10^{-6}} + \frac{\frac{3}{4} \cdot 0.418^2 \cdot 10^{-6}}{1.07 \cdot 10^{-6}}\right.$$

$$\left. - \frac{2 \cdot 0.947 \cdot 0.418 \cdot 0.243 \cdot 0.131 \cdot 10^{-12}}{\sqrt{3} \cdot 1.07 \cdot 0.956 \cdot 10^{-12}}\right]$$

$$= 1.48$$

$$(\lambda'_{12})_3 = \frac{0.947 \cdot 10^{-3}}{0.0792 \cdot 10^{-3}} \left[\frac{\frac{1}{4} \cdot 0.243^2 \cdot 10^{-6}}{0.956 \cdot 10^{-6}} + \frac{\frac{3}{4} \cdot 0.418^2 \cdot 10^{-6}}{1.07 \cdot 10^{-6}} + \frac{2 \cdot 0.947 \cdot 0.418 \cdot 0.243 \cdot 0.131 \cdot 10^{-12}}{\sqrt{3} \cdot 1.07 \cdot 0.956 \cdot 10^{-12}} \right] = 1.82$$

$$(\lambda'_{1m})_3 = \frac{0.947 \cdot 10^{-3}}{0.0792 \cdot 10^{-3}} \frac{0.122^2 \cdot 10^{-6}}{0.912 \cdot 10^{-6}} = 0.195$$

8.3.5.4 Variation of Spacing of Single-core Cables Between Sheath Bonding Points. We found that for single-core cables spaced uniformly and with sheaths solidly bonded at both ends and possibly at intermediate points, the circulating currents, and the consequent losses, increase as the spacing increases. However, it is not always possible to install cables with constant spacing along a route. When the distance between cables varies in an electrical section, the mutual reactance will vary, and an average value for this reactance should be used in the formulas in the preceding sections. Note again that a section is defined as a portion of the route between points at which sheaths of all cables are solidly bonded.

When the spacing along a section is not constant but its various values are known, the average value of X is given by

$$X = \frac{l_a X_a + l_b X_b + \cdots + l_n X_n}{l_a + l_b + \cdots + l_n} \tag{8.63}$$

where $l_a, l_b, \ldots, l_n$ = lengths with different spacings along an electrical section

$X_a, X_b, \ldots, X_n$ = the reactances per unit length of cable, where the appropriate spacings $s_a, s_b, \ldots, s_n$ are used.

When, in any section, the spacing between cables and its variation along the route are not known and cannot be anticipated, IEC 287 recommends that the losses in that section, calculated from the design spacing, should be arbitrarily increased by 25%. This increase, which is not applicable to cross-bonded and single point bonded systems, has been determined through experience with lead-sheathed high-voltage cables. Where the section includes a spread-out end, the allowance of 25% may not be sufficient and it is recommended that an estimate of the probable spacing be made and the loss calculated employing formula (8.63).

If the cable ends are widely spaced compared to the main part of the route and form a significant part of the route length, the sheath losses should be based on the average reactance X, as calculated by equation (8.63).

8.3.5.5 Effect of Unequal Section Lengths in Cross-bonded Systems. The ideal cross-bonded system will have equal lengths and spacings in each of the three sections. If the section lengths are different, the induced voltages will not sum to zero and circulating currents will be present. These circulating currents are taken into account by calculating the circulating current loss factor, λ'_1, assuming the cables were not cross-bonded and multiplying this value by a factor to take into account length variations. This factor, F_c, is given by

$$F_c = \left[\frac{p_2 + q_2 - 2}{p_2 + q_2 + 1} \right]^2 \tag{8.64}$$

where $p_2 a$ = length of the longest section

$q_2 a$ = length of the second longest section

a = length of the shortest section.

Where lengths of the minor sections are not known, IEC 287-2-1 (1994) recommends that the value for λ_1' based on experience with carefully installed circuits, be:

$$\lambda_1' = 0.03 \quad \text{for cables laid directly in the ground}$$

$$\lambda_1' = 0.05 \quad \text{for cables installed in ducts.} \tag{8.65}$$

8.3.5.6 Armored Cables with Each Core in a Separate Lead Sheath (SL Type). From the point of view of circulating currents, SL type cables behave like single core lead-sheathed cables in trefoil, except that the magnetic armoring increases the circulating current losses. These losses are now given by the loss factor[3]

$$\boxed{\lambda_1' = \frac{R_S}{R} \frac{1.7}{1 + \left(\frac{R_S}{X}\right)^2}} \tag{8.66}$$

where X = reactance of sheath, $\Omega/\text{m} = 2\omega 10^{-7} \cdot \ln\left(\frac{2s}{d}\right)$, Ω/m

In this case, s = distance between conductor axes, mm.

8.3.5.7 Losses in the Screens and Sheaths of Pipe-Type Cables. Nonmagnetic core screens, copper or lead, of pipe-type cables will carry circulating currents induced in the same manner as in the sheaths of solidly bonded single core cables. As with SL type cables, there is an increase in screen losses due to the presence of the steel pipe. The loss factor is given by [see footnote to equation (8.66)]

$$\boxed{\lambda_1' = \frac{R_S}{R} \frac{1.7}{1 + \left(\frac{R_S}{X}\right)^2}} \tag{8.67}$$

If additional reinforcement is applied over the core screens, the above formula is applied to the combination of sheath and reinforcement. In this case, R_S is replaced by the resistance of the parallel combination of sheath and reinforcement, and the diameter is taken as the rms diameter d', where

$$d' = \sqrt{\frac{d^2 + d_2^2}{2}} \tag{8.68}$$

with d = mean diameter of the screen or sheath, mm

d_2 = mean diameter of the reinforcement, mm.

[3] An amendment to the IEC 287 to be introduced in 1997 will propose to change the factor 1.7 to 1.5 according to the recommendations made by Silver and Seman (1982).

EXAMPLE 8.3

Compute the loss factor λ_1 for a pipe-type cable described by cable model No. 3 in Appendix A.

The parameters of this cable are $D_{is} = 67.26$ mm and $D_s = 67.59$ mm. The combined resistance of the skid wire and the bronze reinforcing tape at operating temperature was computed in Example 7.1 as $R_s = 0.019$ Ω/m. The equivalent diameter is obtained from

$$d' = \sqrt{\frac{D_{is}^2 + D_s^2}{2}} = \sqrt{\frac{67.26^2 + 67.59^2}{2}} = 67.4 \text{ mm}$$

The reactance of the tape and skid wire is

$$X = 2\omega \cdot 10^{-7} \cdot \ln \frac{2 \cdot s}{d'} = 4 \cdot \pi \cdot 60 \cdot 10^{-7} \cdot \ln \frac{2 \cdot 67.59}{67.4} = 0.525 \cdot 10^{-4} \ \Omega/\text{m}$$

Taking the conductor resistance from Table A1 ($R = 0.245 \cdot 10^{-4}$ Ω/m) and applying equation (8.67), we obtain

$$\lambda_1 = \lambda_1' = \frac{R_S}{R} \frac{1.7}{1 + \left(\frac{R_S}{X}\right)^2} = \frac{0.0190}{0.245 \cdot 10^{-4}} \frac{1.7}{1 + \left(\frac{0.0190}{0.525 \cdot 10^{-4}}\right)^2} = 0.010$$

8.3.5.8 Two-core or Three-core Armored Cables Having Extruded Insulation and Copper Tape Screens. By considering the similarity between an SL type cable and an armored cable having copper tape screens around each core, it is clear that the same equations are applicable, that is [see footnote to equation (8.66)],

$$\boxed{\lambda_1' = \frac{R_S}{R} \frac{1.7}{1 + \left(\frac{R_S}{X}\right)^2}} \tag{8.69}$$

In the above equation, it is recommended that the screen resistance is taken as that of the equivalent tube. In practice, the effective resistance will be somewhat higher because there will be contact resistance between each turn of the copper tape and a portion of the induced current will follow the helical path followed by the tape. The use of an equivalent tube resistance will produce a result which errs on the side of safety.

8.3.5.9 Circulating Currents in the Sheaths of Parallel Cables. The equations set out in the preceding sections have been derived by solving a set of simultaneous equations for the three unknown sheath currents. In all of the above cases, it has been reasonable to assume a balanced system in which the conductor currents are equal. When it is necessary to install a number of cables per phase in one circuit, the reactances of the sheaths and conductors are functions of their spacings from all the other sheaths and conductors. Because of this, not only will the impedance of the sheaths vary, but also the impedance of each phase conductor may vary, depending on the relative positions of the cables. Hence, for cables in parallel, the current flowing in each conductor may be different. This leads to the need to solve simultaneous equations for both the conductor and sheath currents. For example, for two cables per phase in a three-phase system, the six conductor currents and the six sheath currents must be found. In this case, a set of 12 simultaneous equations with 12 unknowns has to be solved, each equation having a real and an imaginary component. In general terms, for n cables per phase, $6n$ simultaneous equations must be solved. Additional equations may be required to set up the boundary conditions for voltages and currents. This is not a task for manual calculations.

Equations for the voltage drop can be generalized to represent three-phase multicable installation. In what follows, the term "conductor" refers either to the conductor or to the metallic sheath of a cable. When the cables have nonmagnetic armor, the armor and the sheath are combined into an equivalent sheath as illustrated in Example 8.4 below. Cables with magnetic armor can also be treated by the method discussed in this section; however, the matrix $\mathbf{G}$ would have to be constructed in such a way as to take into account the lay angle of the tapes or wires. In this case, the relative magnetic permeability of the armor material should also be considered.

The longitudinal voltage drop E in a conductor is given by the sum of the resistive and reactive voltage drops (from the fundamental equations set out by Arnold and others):

$$\mathbf{E} = \left[\mathbf{R}_d + j\frac{\omega\mu_0}{2\pi}\mathbf{G}\right] \cdot \mathbf{I} \tag{8.70}$$

where $\mathbf{R}_d = n \times n$ matrix with vector $[R_1, R_2, \cdots, R_n]$ of conductor resistances in the diagonal and zeros outside the diagonal

$\mathbf{E} =$ vector of voltage drops

$\mathbf{I} =$ vector of currents in all conductors

$$\mathbf{G} = \begin{bmatrix} \ln\frac{1}{s_{11}} & \ln\frac{1}{s_{12}} & \cdots & \ln\frac{1}{s_{1n}} \\ \cdot & & & \cdot \\ \cdot & & & \cdot \\ \ln\frac{1}{s_{n1}} & \cdot & \cdots & \ln\frac{1}{s_{nn}} \end{bmatrix}$$

$s_{ii} =$ geometric mean radius of conductor, m

$s_{ij} =$ geometric mean distance between conductors i and j, m.

Since both $\mathbf{E}$ and $\mathbf{I}$ are complex quantities, their separate components must be determined. In effect, therefore, there are $2n$ equations and $4n$ unknown quantities to be found.

The solution of equations (8.70) for conductor currents is based on an application of Kirchhoff's laws: (1) the sum of all the currents flowing into a connection or node is zero, and (2) the potentials between the ends of conductors connected in parallel are equal.

Application of the second law means that the values of the voltage drop for all conductors in parallel in the same phase are equal. It is therefore possible to eliminate the impedance voltages as shown below. The required equations have been developed by Parr (1988) and are set out in IEC 287 (to be published). In the developments presented below, we will consider cables with one additional metallic layer in addition to the conductor, and we will refer to it as the sheath. A more general case will be treated in Section 8.5.

For a three-phase system with p phase conductors per phase, there are a total of n conductors, where $n = 6p$. The phase conductors are assigned even numbers and the sheaths odd numbers so that the phase conductor currents become $I_2, I_4, I_6, \cdots, I_n$ and the sheath currents $I_1, I_3, \cdots, I_{n-1}$.

To set up the equations eliminating the voltage drop in equations (8.70), the parallel conductors are considered in pairs:

For the sheath conductors 1 and 3,

$$0 = I_1(R_s + jX_c - jX_{31}) + I_2(jX_{12} - jX_{32}) + I_3(jX_{13} - R_s - jX_c) + I_4(jX_{14} - jX_{34}) + \cdots + I_n(jX_{1n} - jX_{3n})$$

For the sheath conductors 3 and 5,

$$0 = I_1(jX_{31} - jX_{51}) + I_2(jX_{32} - jX_{52}) + I_3(R_s + jX_c - jX_{53}) + I_4(jX_{34} - jX_{54}) + I_5(jX_{35} - R_s - jX_c) + \cdots + I_n(jX_{3n} - jX_{5n})$$

For the phase conductors 2 and 4,

$$0 = I_1(jX_{21} - jX_{41}) + I_2(R + jX_c - jX_{42}) + I_3(jX_{23} - jX_{43}) + I_4(jX_{24} - R - jX_c) + I_5(jX_{25} - jX_{45}) + \cdots + I_n(jX_{2n} - jX_{4n}) \tag{8.71}$$

In general terms, for conductors f and $f+2$,

$$0 = I_1(jX_{f1} - jX_{f+21}) + \cdots + I_f(R_f + jX_c - jX_{f+2f}) + I_{f+1}(jX_{ff+1} - jX_{f+2f}) + I_{f+2}(jX_{ff+2} - R_f - jX_c) + \cdots + I_n(jX_{fn} - jX_{f+2n}) \tag{8.72}$$

Also, since the sum of the currents in each set of phase conductors must equal the phase currents I_R, I_S, and I_T, we have

$$\begin{aligned} I_R[1 + j0] &= I_2 + I_4 + \cdots + I_{2p} \\ I_S[-0.5 - j0.866] &= I_{2p+2} + I_{2p+4} + \cdots + I_{4p} \\ I_T[-0.5 + j0.866] &= I_{4p+2} + I_{4p+4} + \cdots + I_{6p} \end{aligned} \tag{8.73}$$

and the sum of the sheath currents must equal zero:

$$0 + j0 = I_1 + I_3 + I_5 + \cdots + I_{n-1} \tag{8.74}$$

where, from equation (8.71) and the definition of matrix **G**,

R_f = conductor resistance at operating temperature (sheath or phase), Ω/m

$$X_{fg} = 2\omega 10^{-7} \ln\left(\frac{1}{s_{fg}}\right) \tag{8.75}$$

$X_c = 2\omega 10^{-7} \ln\left(\frac{2}{\alpha d_c^*}\right)$ for phase conductors and $X_c = 2\omega 10^{-7} \ln\left(\frac{2}{d^*}\right)$ for the sheath

d_c^* = conductor diameter, m

d^* = mean diameter of the sheath, m

α = coefficient depending on the conductor construction ($\alpha d_c^*/2$ is often called the geometric mean radius of the conductor).

The terms X_c and X_{fg} are treated as if they were reactances, but they do not represent true physical quantities. In reality, these terms must exist in pairs as a complete loop. Values of α were estimated by Dwight (1923) for multiwire conductors and are given in Table 8.1.

TABLE 8.1 Coefficients α for Computation of Conductor Self-Reactance

Number of Wires	Value of α
1	0.779
3	0.678
7	0.726
19	0.758
37	0.768
61	0.772
91	0.774
127	0.776
Compacted conductors	0.779

The values of α for hollow core conductors are dependent on the inner and outer diameter of the conductor as well as the number of strands, and should be computed separately. The required calculations are described by equation (8.24).

The equations above will give n simultaneous equations with n unknowns I_a. These can be written in a matrix form as

$$\mathbf{Q} = \mathbf{Z} \times \mathbf{I} \tag{8.76}$$

where the matrix **Q** contains the left-hand side of the above equations, **Z** contains the coefficients of I_a from the right-hand side of the above equations, and **I** contains the unknown currents I_a. To solve the equations for I_a, the inverse of the matrix **Z** is found, and equation (8.76) is rewritten as

$$\mathbf{I} = \mathbf{Z}^{-1} \times \mathbf{Q} \tag{8.77}$$

The circulating current loss factor for cable a is then given by

$$\lambda'_a = \left(\frac{I_{sa}}{I_{ca}}\right)^2 \frac{R_s}{R} \tag{8.78}$$

where I_{sa} = circulating current in the sheath of cable a, A

I_{ca} = current in the conductor of cable a, A.

EXAMPLE 8.4

We will consider a circuit composed of model cables No. 4 with two cables per phase in the arrangement shown in Fig. 8.6.

Assuming that the total conductor current is equal to 1600 A per phase, we will compute the loss factors for armor and sheath of each cable and the current split between the two cables in each phase.

The conductor resistance at the operating temperatures of 85°C is equal to $R = 0.356 \cdot 10^{-4}$ Ω/m. Assuming, again, that the sheath and armor temperatures are 70 and 68°C, respectively, we will use

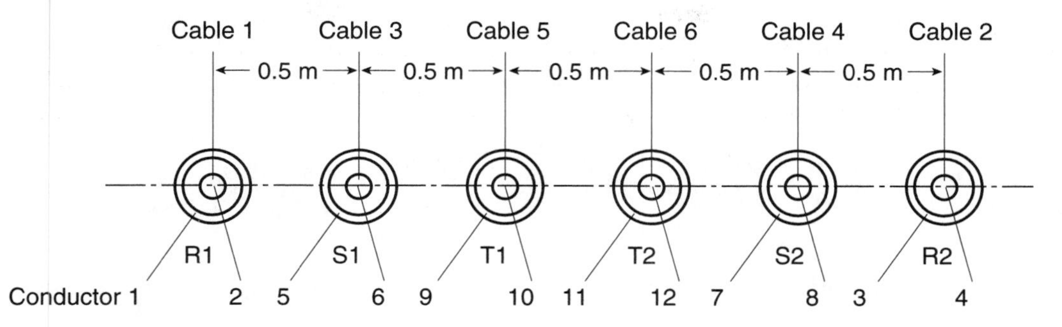

Figure 8.6 Cable arrangement for Example 8.4.

the equivalent resistance of the sheath and the reinforcing tape computed in Example 8.1, that is $R_s = 0.203 \cdot 10^{-3}$ Ω/m. The rms diameter of the equivalent sheath is equal to $d^* = 0.0767$ m. However, the resistance of the copper armor wires has to be recalculated since steel wire armor was used in Example 8.1. The cross-sectional area of the wires and the mean armor diameter were computed in Example 8.1 as 0.00108 m² and 0.0932 m, respectively. Thus,

$$R_a = \frac{0.1724 \cdot 10^{-7} \cdot \sqrt{\left(\frac{\pi \cdot 0.0932}{0.1218}\right)^2 + 1}}{0.00108}[1 + 0.00393(68 - 20)] = 0.0494 \cdot 10^{-3}\ \Omega/\text{m}$$

This multiplied by 1.4 gives $R_a = 0.0692$ Ω/km. Even though equations (8.76) can be set up for both sheath and armor conductors, for the purpose of this example, we will combine sheath and armor into an equivalent sheath. The resistance of the parallel combination of sheath, reinforcing tape, and armor is

$$R_{sa} = \frac{0.203 \cdot 0.0692}{0.203 + 0.0692} \cdot 10^{-3} = 0.516 \cdot 10^{-4}\ \Omega/\text{m}$$

The rms diameter of this combination is equal to

$$d^*_{sa} = \sqrt{\frac{0.0767^2 + 0.0932^2}{2}} = 0.0853 \text{ m}$$

The cable coordinates are entered into the array

$$D = \begin{bmatrix} 0 & 0 \\ 2.5 & 0 \\ 0.5 & 0 \\ 2.0 & 0 \\ 1.0 & 0 \\ 1.5 & 0 \end{bmatrix} \begin{matrix} \text{Cable 1, phase } R \\ \text{Cable 2, phase } R \\ \text{Cable 3, phase } S \\ \text{Cable 4, phase } S \\ \text{Cable 5, phase } T \\ \text{Cable 6, phase } T \end{matrix}$$

(columns: x, y)

The zero coordinates can be fixed at any point; it is convenient to take this point at the center of the leftmost cable.

The cable spacings, in meters, are calculated using the following equation:

$$s_{m,n} = \sqrt{(D_{m,x} - D_{n,x})^2 + (D_{m,y} - D_{n,y})^2}, \qquad m = 1, \cdots, 6; \quad n = 1, \cdots, 6$$

The spacings are given in the following array:

$$s = \begin{bmatrix} 0 & 2.5 & 0.5 & 2.0 & 1.0 & 1.5 \\ 2.5 & 0 & 2.0 & 0.5 & 1.5 & 1.0 \\ 0.5 & 2.0 & 0 & 1.5 & 0.5 & 1.0 \\ 2.0 & 0.5 & 1.5 & 0 & 1.0 & 0.5 \\ 1.0 & 1.5 & 0.5 & 1.0 & 0 & 0.5 \\ 1.5 & 1.0 & 1.0 & 0.5 & 0.5 & 0 \end{bmatrix}$$

Clearly, this array is symmetrical about its diagonal, and it is not necessary to calculate the spacing between cables m and n and again between n and m.

The effective reactances X_c and X_{mn} are calculated using the following equations:

(1) For the phase conductors,

$$F_{2m,2n} = if\left(s_{m,n} > 0, \sqrt{-1}\cdot 2\omega\cdot 10^{-7}\ln\frac{1}{s_{m,n}}, \sqrt{-1}\cdot 2\omega\cdot 10^{-7}\ln\frac{2}{\alpha d_c^*}\right)$$

(2) For the sheath,

$$F_{2m-1,2n} = if\left(s_{m,n} > 0, \sqrt{-1}\cdot 2\omega\cdot 10^{-7}\ln\frac{1}{s_{m,n}}, \sqrt{-1}\cdot 2\omega\cdot 10^{-7}\ln\frac{2}{d^*}\right)$$

$$F_{2m-1,2n-1} = if\left(s_{m,n} > 0, \sqrt{-1}\cdot 2\omega\cdot 10^{-7}\ln\frac{1}{s_{m,n}}, \sqrt{-1}\cdot 2\omega\cdot 10^{-7}\ln\frac{2}{d^*}\right)$$

$$F_{2m,2n-1} = if\left(s_{m,n} > 0, \sqrt{-1}\cdot 2\omega\cdot 10^{-7}\ln\frac{1}{s_{m,n}}, \sqrt{-1}\cdot 2\omega\cdot 10^{-7}\ln\frac{2}{d^*}\right)$$

NOTE: the last three equations are identical; three equations are used only to fill the array. The calculated values are given in the array F with sample computations shown below.

The phase conductors are compacted hollow core. To compute the coefficient α, we apply equation (8.24). Let $a = d_i/d_c = 17.5/33.8 = 0.518$. From equation (8.24),

$$f = \frac{1/4 - a^2 + a^4(3/4 - \ln a)}{(1-a^2)^2} = \frac{1/4 - 0.518^2 + 0.518^4(3/4 - \ln 0.518)}{(1-0.518^2)^2} = 0.1551$$

The coefficient α is then

$$\alpha = e^{-f} = e^{-0.1551} = 0.86$$

The effective self-reactance of the conductor is obtained from equation (8.75):

$$X_c = 2\omega 10^{-7}\ln\left(\frac{2}{\alpha d_c^*}\right) = 4\cdot\pi\cdot 50\cdot 10^{-7}\ln\frac{2}{0.86\cdot 0.0338} = 0.266\cdot 10^{-3}\ \Omega/\mathrm{m}$$

The effective self-reactance of the equivalent sheath, obtained from equation (8.75), is equal to

$$X_s = 2\omega 10^{-7}\ln\left(\frac{2}{d^*}\right) = 4\cdot\pi\cdot 50\cdot 10^{-7}\ln\frac{2}{0.853} = 0.1986\cdot 10^{-3}\ \Omega/\mathrm{m}$$

From equation (8.75), the effective reactance between the sheath of cable 1 and the phase conductor of cable 2 is

$$X_{1,4} = 2\omega 10^{-7}\ln\left(\frac{1}{s_{cs}}\right) = 4\cdot\pi\cdot 50\cdot 10^{-7}\ln\frac{1}{2.5} = -0.576\cdot 10^{-4}\ \Omega/\mathrm{m}$$

Continuing the same way, the following matrix is obtained:

$$
\mathbf{F} = -j \cdot 10^{-4} \left[\begin{array}{ccccccc}
-1.98 & -1.98 & 0.576 & 0.576 & -0.436 & -0.436 & 0.436 \\
-1.98 & -2.66 & 0.576 & 0.576 & -0.436 & -0.436 & 0.436 \\
0.576 & 0.576 & -1.98 & -1.98 & 0.436 & 0.436 & -0.436 \\
0.576 & 0.576 & -1.98 & -2.66 & 0.436 & 0.436 & -0.436 \\
-0.436 & -0.436 & 0.436 & 0.436 & -1.98 & -1.98 & 0.255 \\
-0.436 & -0.436 & 0.436 & 0.436 & -1.98 & -2.66 & 0.255 \\
0.436 & 0.436 & -0.436 & -0.436 & 0.255 & 0.255 & -1.98 \\
0.436 & 0.436 & -0.436 & -0.436 & 0.255 & 0.255 & -1.98 \\
0 & 0 & 0.255 & 0.255 & -0.436 & -0.436 & 0 \\
0 & 0 & 0.255 & 0.255 & -0.436 & -0.436 & 0 \\
0.255 & 0.255 & 0 & 0 & 0 & 0 & -0.436 \\
0.255 & 0.255 & 0 & 0 & 0 & 0 & -0.436
\end{array} \right.
$$

$$
\left. \begin{array}{ccccc}
0.436 & 0 & 0 & 0.255 & 0.255 \\
0.436 & 0 & 0 & 0.255 & 0.255 \\
-0.436 & 0.255 & 0.255 & 0 & 0 \\
-0.436 & 0.255 & 0.255 & 0 & 0 \\
0.255 & -0.436 & -0.436 & 0 & 0 \\
0.255 & -0.436 & -0.436 & 0 & 0 \\
-1.98 & 0 & 0 & -0.436 & -0.436 \\
-2.66 & 0 & 0 & -0.436 & -0.436 \\
0 & -1.98 & -1.98 & -0.436 & -0.436 \\
0 & -1.98 & -2.66 & -0.436 & -0.436 \\
-0.436 & -0.436 & -0.436 & -1.98 & -1.98 \\
-0.436 & -0.436 & -0.436 & -1.98 & -2.66
\end{array} \right]
$$

The components of matrix **Z** in equation (8.76) are calculated as follows:

(1) For the phase conductor loops,

Phase R	Phase S	Phase T	
$Z_{2p} = F_{2,p} - F_{4,p}$	$Z_{6,p} = F_{6,p} - F_{8,p}$	$Z_{10,p} = F_{10,p} - F_{12,p}$	$p = 1, 2, \cdots, 12$
$Z_{2,2} = Z_{2,2} + R$	$Z_{6,6} = Z_{6,6} + R$	$Z_{10,10} = Z_{10,10} + R$	
$Z_{2,4} = Z_{2,4} - R$	$Z_{6,8} = Z_{6,8} - R$	$Z_{10,12} = Z_{10,12} - R.$	

(2) For the sheath loop,

$$Z_{k,p} = F_{k,p} - F_{k+2,p} \quad p = 1, 2, \cdots, 12; \quad k = 1, 3, 5, 7, 9$$

$$Z_{k,k} = Z_{k,k} + R_s$$

$$Z_{k,k+2} = Z_{k,k+2} - R_s$$

We can observe that when constructing matrix **Z** for this example, the rows 4, 8, 11, and 12 will be filled with zeros. In order to save space, we will move the last four rows representing the coefficients of the currents I on the right-hand-side of equations (8.73)–(8.74) into rows 4, 8, 11, and 12. The matrix **H** of the current coefficients in equations (8.73)–(8.74) is given by

$$
\mathbf{H} = \begin{bmatrix}
0 & 1 & 0 & 1 & 0 & 0 & 0 & 0 & 0 & 0 & 0 & 0 \\
0 & 0 & 0 & 0 & 0 & 1 & 0 & 1 & 0 & 0 & 0 & 0 \\
0 & 0 & 0 & 0 & 0 & 0 & 0 & 0 & 0 & 1 & 0 & 1 \\
1 & 0 & 1 & 0 & 1 & 0 & 1 & 0 & 1 & 0 & 1 & 0
\end{bmatrix}
\begin{array}{l}
\text{phase } R \\
\text{phase } S \\
\text{phase } T \\
\text{Sheath}
\end{array}
$$

Substituting numerical values for variables $\mathbf{Z}_{i,j}$ and combining them with the matrix $\mathbf{H}$ in rows 4, 8, 10, and 12, we obtain

$$
\mathbf{Z} = 10^{-4} \left[\begin{array}{cccccccccccc}
0.516 + j2.55 & j2.55 & -0.516 - j2.55 & -j2.55 & j0.876 & j08.76 & -j0.876 & -j0.876 & j0.255 & j0.255 & -j0.255 & -j0.255 \\
j2.55 & 0.356 + j3.2 & -j2.55 & -0.356 - j3.2 & j0.876 & j0.876 & -j0.876 & -j0.876 & j0.255 & j0.255 & -j0.255 & -j0.255 \\
-j1.01 & -j1.01 & 0.516 + j2.42 & j2.42 & -0.516 - j2.42 & -j2.42 & j0.691 & j0.691 & -j0.691 & -j0.691 & 0 & 0 \\
0 & 1 & 0 & 1 & 0 & 0 & 0 & 0 & 0 & 0 & 0 & 0 \\
j0.876 & j0.876 & -j0.876 & -j0.876 & 0.516 + j2.24 & j2.24 & -0.516 - j2.24 & -j2.24 & j0.436 & j0.436 - j0.436 & -j0.436 & j0.436 \\
j0.876 & j0.876 & -j0.876 & -j0.876 & j2.24 & 0.356 + j2.91 & -j2.24 & -0.356 - j2.91 & j0.436 & j0.436 & -j0.436 & -j0.436 \\
-j0.436 & -j0.436 & j0.691 & j0.691 & -j0.691 & -j0.691 & 0.516 + j1.98 & j1.98 & -0.516 - j1.98 & -j1.98 & 0 & 0 \\
0 & 0 & 0 & 0 & 0 & 1 & 0 & 1 & 0 & 0 & 0 & 0 \\
j0.255 & j0.255 & -j0.255 & -j0.255 & j0.436 & j0.436 & -j0.436 & -j0.436 & 0.516 + j1.54 & j1.54 & -0.516 - j1.54 & -j1.54 \\
j0.255 & j0.255 & -j0.255 & -j0.255 & j0.436 & j0.436 & -j0.436 & -j0.436 & j1.54 & 0.356 + j2.22 & -j1.54 & -0.356 - j2.22 \\
0 & 0 & 0 & 0 & 0 & 0 & 0 & 0 & 0 & 1 & 0 & 1 \\
1 & 0 & 1 & 0 & 1 & 0 & 1 & 0 & 1 & 0 & 1 & 0
\end{array} \right]
$$

The array $\mathbf{Q}$ in equation (8.76) is given by

$$\mathbf{Q} = [0\,0\,0\,1\,0\,0\,0\;\; -0.5 - j0.866\;0\;0\;\; -0.5 + j0.866\;0]^t$$

where t denotes transposition ($\mathbf{Q}$ is a column vector).

Inverting matrix $\mathbf{Z}$ and substituting into equation (8.77), we obtain the following vector for the currents and their magnitudes:

$$\mathbf{I} = \begin{bmatrix} -0.4875 - j0.1404 \\ 0.5000 - j0.0000 \\ -0.4875 - j0.1404 \\ 0.5000 + j0.0000 \\ 0.3523 + j0.3014 \\ -0.2500 - j0.4330 \\ 0.3523 + j0.3014 \\ -0.2500 - j0.4330 \\ 0.1305 - j0.4408 \\ -0.2500 + j0.4330 \\ 0.1305 - j0.4408 \\ -0.2500 + j0.4330 \end{bmatrix} \qquad |\mathbf{I}| = \begin{bmatrix} 0.5073 \\ 0.5000 \\ 0.5073 \\ 0.5000 \\ 0.4637 \\ 0.5000 \\ 0.4637 \\ 0.5000 \\ 0.4597 \\ 0.5000 \\ 0.4597 \\ 0.5000 \end{bmatrix} A$$

The magnitudes of the phase conductor and sheath currents are given in Table 8.2, assuming a total phase current of 1600 A. The equivalent sheath loss factors are computed from equation (8.78) and are given in column 4 of Table 8.2.

TABLE 8.2 Results of Example 8.4

	Phase Conductor Current (A)	Sheath Current (A)	Equivalent Sheath Loss Factor	Sheath Loss Factor	Armor Loss Factor
	$\lvert I_{2m}\rvert \cdot 1600$	$\lvert I_{2m-1}\rvert \cdot 1600$	$\lambda_{s,a} = \dfrac{(\lvert I_{2m-1}\rvert \cdot 1600)^2 \cdot R_{sa}}{(\lvert I_{2m}\rvert \cdot 1600)^2 \cdot R}$		
Cable 1, phase R	800	812	1.49	0.379	1.11
Cable 2, phase R	800	813	1.49	0.379	1.11
Cable 3, phase S	800	742	1.25	0.318	0.932
Cable 4, phase S	800	742	1.25	0.318	0.932
Cable 5, phase T	800	736	1.23	0.313	0.917
Cable 6, phase T	800	735	1.23	0.313	0.917

To separate the equivalent sheath loss factor into sheath (with tape) and armor loss factors, we consider each cable separately and observe that

$$I_s = \frac{X_s X_a - X_a^2 - jX_s R_a}{X_a^2 - X_s X_a + j(X_s R_a + X_a R_s)}$$

$$I_a = \frac{-jX_a R_s}{X_a^2 - X_s X_a + j(X_s R_a + X_a R_s)}$$

From the definition of the loss factors, we thus have

$$\frac{\lambda_1'}{\lambda_2} = \frac{(X_s - X_a)^2}{R_s R_a} + \frac{X_a^2 R_a}{X_s^2 R_s} \tag{8.79}$$

When $X_s \approx X_a$, which usually is a good assumption, equation (8.79) leads to

$$\lambda_1' = \frac{R_a}{R_s + R_a} \cdot \lambda_{s,a}, \quad \lambda_2 = \lambda_{s,a} - \lambda_1' \tag{8.80}$$

Using equation (8.80), the results summarized in the last two columns of Table 8.2 are obtained.

8.3.6 Circulating Current Losses in the Armor—Special Cases

8.3.6.1 Armor Materials. Armored single-core cables for general use in ac systems usually have nonmagnetic armor. This is because of the very high losses which would occur in closely spaced single-core cables with magnetic armor. On the other hand, when magnetic armor is used, losses due to eddy currents and due to hysteresis in the steel must be considered. A method of calculating these losses is given by Bosone (1931), and the results agree with those obtained in the limited experimental work reported by Whitehead and Hutchings (1939). The latter work demonstrated that the losses in the sheath and armor combination could be several times the conductor losses, depending on the bonding arrangements of the sheaths and armor (see Fig. 8.7).

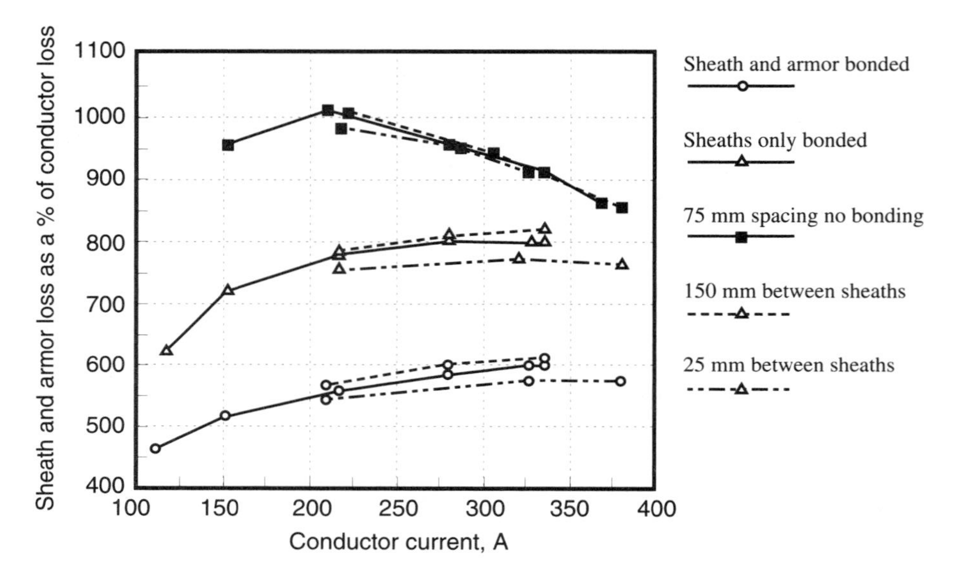

Figure 8.7 Effect of steel wire armor on cable losses.

Several other authors addressed the subject of armor losses. A comprehensive account of this subject is given by Carter (1954), Schurig *et al.* (1929), and Schelkunoff (1934). More recent publications include the works by Bianchi and Luoni (1976), Kawasaki *et al.* (1981), and Weedy (1986).

The armoring or reinforcement on two-core or three-core cables can be either magnetic or nonmagnetic. These cases are treated separately in the next two subsections. Steel wires or tapes are generally used for magnetic armor.

8.3.6.2 Nonmagnetic Armor or Reinforcement. When nonmagnetic armor is used, the losses are calculated as a combination of sheath and armor losses. The equations set out above for sheath losses are applied, but the resistance used is that of the parallel combination of sheath and armor, and the sheath diameter is replaced by the rms value of the mean armor and sheath diameters.

For nonmagnetic tape reinforcement where the tapes do not overlap, the resistance of the reinforcement is a function of the lay length of the tape. The advice given in IEC 287 to deal with this is as follows.

1. If the tapes have a very long lay length, i.e., are almost longitudinal tapes, the resistance taken is that of the equivalent tube, that is, a tube having the same mass per unit length and the same internal diameter as the tapes.
2. If the tapes are wound at about 54° to the axis of the cable, the resistance is taken to be twice the equivalent tube resistance.
3. If the tapes are wound with a very short lay, the resistance is assumed to be infinite; hence, the reinforcement has no effect on the losses.
4. If there are two or more layers of tape in contact with each other and having a very short lay, the resistance is taken to be twice the equivalent tube resistance. This is intended to take account of the effect of the contact resistance between the tapes.

EXAMPLE 8.5

We will compute loss factors and ratings of a three-phase circuit composed of cable model No. 4 with all necessary parameters as given in Appendix A and with the resistances of the metallic parts at operating temperature as computed in Example 8.4. From this example, $R_s = 0.203 \cdot 10^{-3}$ Ω/m and $R_a = 0.0692 \cdot 10^{-3}$ Ω/m. The equivalent sheath/tape/armor resistance is $0.516 \cdot 10^{-4}$ Ω/m and the mean diameter is $d = 85.3$ mm. The conductor ac resistance is given in Table A1 as $R = 0.356 \cdot 10^{-4}$ Ω/m. We will consider two laying conditions: (1) when the cables are transposed, and (2) when they are not transposed. In both cases, the cables are in a flat formation.

(1) Cables regularly transposed

The mutual reactance between one equivalent sheath and a conductor of the neighboring cable is given by equation (8.55):

$$X_1 = 2\omega \cdot 10^{-7} \cdot \ln\left[2 \cdot \sqrt[3]{2}\left(\frac{s}{d}\right)\right] = 4 \cdot \pi \cdot 50 \cdot 10^{-7} \cdot \ln\left[2 \cdot \sqrt[3]{2}\left(\frac{500}{85.3}\right)\right] = 1.69 \cdot 10^{-4}\ \Omega/\text{m}$$

The sheath loss factor is given by equation (8.56):

$$\lambda'_{1sa} = \frac{R_s}{R} \cdot \frac{1}{1 + \left(\frac{R_s}{X_1}\right)^2} = \frac{0.516 \cdot 10^{-4}}{0.356 \cdot 10^{-4}} \cdot \frac{1}{1 + \left(\frac{0.516 \cdot 10^{-4}}{1.69 \cdot 10^{-4}}\right)^2} = 1.33$$

The sheath and armor loss factors are obtained from equation (8.80):

$$\lambda_1 = \lambda'_{1sa}\frac{R_a}{R_s + R_a} = 1.33\frac{0.0692}{0.203 + 0.0692} = 0.338$$

$$\lambda_2 = 1.33 - 0.338 = 0.992$$

The rating of the middle cable is obtained from equation (4.3) with the thermal resistances and the dielectric losses given in Appendix A as $T_1 = 0.568$ K · m/W, $T_2 = 0.082$ K · m/W, $T_3 = 0.066$ K · m/W, and $T_4 = 0.814$ K · m/W. $W_d = 6.62$ W/m. Thus, we have

$$I = \left[\frac{\Delta\theta - W_d[0.5T_1 + n(T_2 + T_3 + T_4)]}{RT_1 + nR(1 + \lambda_1)T_2 + nR(1 + \lambda_1 + \lambda_2)(T_3 + T_4)}\right]^{0.5}$$

$$= \left[\frac{85 - 20 - 6.62(0.5 \cdot 0.568 + 0.082 + 0.066 + 0.814)}{0.356 \cdot 10^{-4}[0.568 + (1 + 0.338) \cdot 0.082 + (1 + 0.338 + 0.992) \cdot (0.066 + 0.814)]}\right]^{0.5}$$

$$= 770\ \text{A}$$

(2) Cables are not transposed

The values of P and Q are given by equation (8.61)

$$X_m = 4 \cdot \pi \cdot 50 \cdot 10^{-7} \ln 2 = 0.436 \cdot 10^{-4}\ \Omega/\text{m}$$

$$X = 2\omega \cdot 10^{-7} \ln \frac{2s}{d} = 4 \cdot \pi \cdot 50 \cdot 10^{-7} \ln \frac{2 \cdot 500}{85.3} = 1.55 \cdot 10^{-4}\ \Omega/\text{m}$$

$$P = X_m + X = 0.436 \cdot 10^{-4} + 1.55 \cdot 10^{-4} = 1.99 \cdot 10^{-4}\ \Omega/\text{m}$$

$$Q = X - X_m/3 = 1.55 \cdot 10^{-4} - 0.436 \cdot 10^{-4}/3 = 1.40 \cdot 10^{-4}\ \Omega/\text{m}$$

The rating of the circuit is computed for the hottest cable. The loss factor for the middle cable is obtained from equations (8.62), with the factor 10^{-4} cancelled in all the fractions

$$\lambda'_{1msa} = \frac{R_s}{R} \frac{Q^2}{R_s^2 + Q^2} = \frac{0.516}{0.356} \cdot \frac{1.40^2}{0.516^2 + 1.40^2} = 1.28$$

The sheath and armor loss factors are obtained from equation (8.80)

$$\lambda_1 = \lambda'_{1sa} \frac{R_a}{R_s + R_a} = 1.28 \frac{0.0692}{0.203 + 0.0692} = 0.325$$

$$\lambda_2 = 1.28 - 0.325 = 0.955$$

The rating of the middle cable is obtained from equation (4.3)

$$I = \left[\frac{85 - 20 - 6.62(0.5 \cdot 0.568 + 0.082 + 0.066 + 0.814)}{0.356 \cdot 10^{-4}[0.568 + (1 + 0.325) \cdot 0.082 + (1 + 0.325 + 0.955) \cdot (0.066 + 0.814)]} \right]^{0.5}$$
$$= 771 \text{ A.}$$

We observe that these loss factors are almost equal to the ones obtained in part (1); hence, the ampacities are almost the same in both cases.

The magnitudes of combined sheath and armor currents can be computed from equations (8.60) or from the loss factors formulas (8.62)

$$I_{s1} = |I| \sqrt{\frac{\frac{1}{4}Q^2}{R_s^2 + Q^2} + \frac{\frac{3}{4}P^2}{R_s^2 + P^2} - \frac{2R_s P Q X_m}{\sqrt{3}(R_s^2 + Q^2)(R_s^2 + P^2)}}$$

$$I_{s2} = |I| \frac{Q}{\sqrt{R_s^2 + Q^2}}$$

$$I_{s3} = |I| \sqrt{\frac{\frac{1}{4}Q^2}{R_s^2 + Q^2} + \frac{\frac{3}{4}P^2}{R_s^2 + P^2} + \frac{2R_s P Q X_m}{\sqrt{3}(R_s^2 + Q^2)(R_s^2 + P^2)}}$$

Substituting numerical values, we have

$$I_{sa1} = 709 \text{ A}, \quad I_{sa2} = 723 \text{ A}, \quad I_{sa3} = 771 \text{ A}$$

$$I_{s1} = \frac{0.0692}{0.203} \cdot 709 = 242 \text{ A}, \quad I_{a1} = 709 - 242 = 467 \text{ A}$$

Similarly,

$$I_{s2} = 246 \text{ A}, \quad I_{a2} = 477 \text{ A}, \quad I_{s3} = 263 \text{ A}, \quad I_{a3} = 508 \text{ A}$$

We can observe that combined sheath and armor current is in the same order of magnitude as the conductor current, and that most of this current flows in the armor.

8.3.6.3 Magnetic Armor or Reinforcement.

8.3.6.3.1 SINGLE-CORE LEAD-SHEATHED CABLES WITH STEEL WIRE ARMOR. The armor losses are lowest when the armor and sheath are bonded together at both ends of a run; thus, this condition is selected for the calculations below. The method, derived by Bosone (1931), gives a combined sheath and armor loss for cables which are very widely spaced (greater than 10 m). It has been applied for submarine cables where the cable spacing may be very wide, and there is a need for the mechanical protection provided by the steel wire armor. The equations contained in the IEC Standard 287 can be also derived by simplifying equations (8.43). The simplification involves replacement of the sheath and armor by a single resistance equal to

$$R_e = \frac{R_s \cdot R_a}{R_s + R_a} \tag{8.81}$$

where R_s = sheath resistance with the cable at its maximum operating temperature, Ω/m

R_a = armor resistance with the cable at its maximum operating temperature, Ω/m.

The ac resistance of the armor wires will vary between about 1.2 and 1.4 times its dc resistance, depending on the wire diameter, but this variation is not critical because the sheath resistance is generally considerably lower than that of the armor wires. The use of the resistance R_e assumes that the current split between sheath and armor is inversely proportional to the resistances. This would be true only if sheath and armor had the same self- and mutual inductances. Even a relatively small difference in inductances can cause the current split (and hence the losses) to be quite different; the actual split will be controlled by inductances.

Two important factors make the sheath and armor inductances different:

1. the gap between the sheath and armor,
2. the longitudinal inductance caused by the lay of the armor.

Next, the impedances B_3 and B_4 appearing in equations (8.50)–(8.52) are both replaced by B_1. Summing now the loss factors given by equations (8.50)–(8.52), we obtain the following expression for the total loss factor

$$\boxed{\lambda_1' + \lambda_2 = \frac{R_e}{R}\left(\frac{B_2^2 + B_1^2 + R_e B_2}{(R_e + B_2)^2 + B_1^2}\right)} \tag{8.82}$$

IEC Standard 287 (1982) suggests that the sheath and armor loss factors have the same value, namely, one half of the value obtained from equation (8.82). Alternatively, equation (8.80) can be used to obtain these loss factors.

Replacing B_3 and B_4 by B_1 in equations (8.50)–(8.52) assumes that sheath and armor have the same inductances, which is approximately correct because the B_0 term of B_3,

B_4, and B_1 completely dominates the other terms. The last terms of B_3 and B_4 must be replaced, incorrectly, with the last term of B_1. The reason that these terms do not converge towards the IEC Standard is that circular fields were assumed to have 100% flux linkage in the armor, and solenoidal fields in the armor were not considered at all in the original paper by Bosone (1931). These errors are fairly minor because of the dominance of the B_0 and B_2 terms resulting from the large longitudinal flux in the steel tapes or wires. Thus, it turns out that the approximation given by equation (8.82) does not introduce any significant error in cable rating (see, for instance, Example 8.6).

EXAMPLE 8.6

We will return to the case examined in Example 8.1, but this time we will apply equation (8.82) to compute the sheath and armor loss factors.

The resistance of the parallel combination of sheath, reinforcing tape, and armor is

$$R_e = \frac{0.203 \cdot 0.567}{0.203 + 0.567} \cdot 10^{-3} = 0.149 \cdot 10^{-3} \ \Omega/\text{m}$$

Applying equation (8.82), sheath and armor loss factors are equal to

$$\lambda_1' = \lambda_2 = \frac{R_e}{2R}\left(\frac{B_2^2 + B_1^2 + R_e B_2}{(R_e + B_2)^2 + B_1^2}\right)$$

$$= \frac{1}{2}\left[\frac{0.149 \cdot 10^{-3}}{0.0356 \cdot 10^{-3}} \frac{0.00313^2 + 0.00329^2 + 0.00313 \cdot 0.149 \cdot 10^{-3}}{(0.149 \cdot 10^{-3} + 0.00313)^2 + 0.00329^2}\right] = 2.05$$

The rating of the hottest (bottom) cable is given by

$$I = \left[\frac{85 - 20 - 6.62(0.5 \cdot 0.568 + 0.082 + 0.066 + 0.789)}{0.356 \cdot 10^{-4}(0.568 + (1 + 2.05) \cdot 0.082 + (1 + 2.05 + 2.05)) \cdot (0.066 + 0.789))}\right]^{0.5}$$
$$= 556 \text{ A}$$

The ampacities computed using equations (8.50)–(8.52) and equation (8.82) are in very good agreement. In numerous studies performed by the author, the difference in the cable ratings obtained using both approaches did not exceed 10%.

8.3.6.3.2 TWO-CORE CABLES WITH STEEL WIRE ARMOR. When calculating armor losses for cables with steel wire, account must be taken of both armor losses and hysteresis losses. Equations for such losses were developed by Arnold (1941) and Whitehead (1939) for both two- and three-core cables. The equation given in IEC 287 for two-core cables is derived from the work by Whitehead, and is intended for cables with shaped conductors, but it is also sufficiently accurate for cables with circular conductors. The relative permeability of the armor wires is assumed to be 300. The armor loss factor is given by

$$\lambda_2 = \frac{0.62\omega^2 10^{-14}}{R R_a} + \frac{3.82 A_a \omega 10^{-5}}{R}\left[\frac{1.48 r_1 + t}{d_a^2 + 95.7 A_a}\right]^2 \tag{8.83}$$

where R_a = ac resistance of the armor at maximum cable operating temperature, Ω/m

d_a = mean diameter of the armor, mm²

A_a = total cross-sectional area of the armor, mm

r_1 = radius of the circle which circumscribes the conductors, mm

t = insulation thickness between conductors, mm.

8.3.6.3.3 THREE-CORE CABLES WITH STEEL WIRE ARMOR AND ROUND CONDUCTORS. The equations generally used for three-core cables are derived from those developed by Arnold:

$$\lambda_2 = 1.23 \frac{R_a}{R} \left(\frac{2c}{d_a}\right)^2 \frac{1}{\left(\frac{2.77 R_a 10^6}{\omega}\right)^2 + 1} \tag{8.84}$$

where c = distance between the axis of the conductor and the center of the cable, mm.

8.3.6.3.4 THREE-CORE AND FOUR-CORE CABLES WITH STEEL WIRE ARMOR AND SECTOR-SHAPED CONDUCTORS. The equations used for this type of cable are derived from those developed by Arnold. For three-core cables,

$$\lambda_2 = 0.358 \frac{R_a}{R} \left(\frac{2r_1}{d_a}\right)^2 \frac{1}{\left(\frac{2.77 R_a 10^6}{\omega}\right)^2 + 1} \tag{8.85}$$

where r_1 = radius of the circle circumscribing the shaped conductors, mm.

For four-core cables, the coefficient 0.358 is replaced by 1.09.

8.3.6.3.5 SL TYPE CABLES. The armor loss in SL type cables is reduced by the screening effect of the sheath currents. It is calculated as for three-core cables with circular conductors and then multiplied by the factor $(1 - \lambda_1')$, where λ_1' is the circulating current loss factor for SL type cables.

8.3.6.3.6 THREE-CORE CABLES WITH STEEL TAPE ARMOR OR REINFORCEMENT. For two- or three-core cables with steel tape armor, it is necessary to consider the hysteresis and eddy current losses separately. Equations for the hysteresis loss in tapes of a thickness between 0.3 and 1 mm have been developed by Whitehead (1939). It is recognized that this equation overestimates the hysteresis loss for tapes less than 0.3 mm in thickness. A suitable approach has been presented in Section 8.3.4.

The equation for hysteresis loss developed by Whitehead has been used to derive the equation given in IEC 287. The armor loss factor λ_2 is given by the sum of the hysteresis loss λ_2' and the eddy current loss factor λ_2''. The hysteresis loss factor is given by

$$\lambda_2' = \frac{s^2 k^2 10^{-7}}{R d_a \delta_0} \tag{8.86}$$

where $k = \dfrac{1}{1 + \dfrac{d_a}{\mu \delta_0}} \times \left(\dfrac{f}{50}\right)$

s = distance between conductor axes, mm

f = supply frequency, Hz

μ = relative permeability of the steel, usually taken as 300

δ_0 = equivalent thickness of the armor = $A_a/\pi d_a$, mm.

The eddy current loss factor is given by

$$\lambda_2'' = \frac{2.25 s^2 k^2 \delta_0 10^{-8}}{R d_a} \tag{8.87}$$

The total armor loss factor is then

$$\lambda_2 = \lambda_2' + \lambda_2'' \tag{8.88}$$

8.3.7 Losses in Steel Pipes

Empirical equations for the losses in steel pipes were developed by Meyerhoff (1952). These equations are used by Neher and McGrath (1957) and in IEC 287. The equations were derived for the size of pipe and type of steel commonly used in the United States at that time, and should be treated with caution for other sizes of pipe or types of steel. Three equations are given by Neher and McGrath (1957) depending on the cable configuration within the pipe.

For a three-core cable,

$$\boxed{\lambda_2 = \left(\frac{0.0199 s - 0.001485 D_d}{R}\right) 10^{-5}} \tag{8.89}$$

where s = axial spacing of adjacent conductors, mm

D_d = internal diameter of the pipe, mm

R = ac conductor resistance at maximum operating temperature, Ω/m.

If the three cores have a flat wire armor applied over them, the pipe losses are ignored and the armor losses calculated as for an SL type cable.

For three single cores in trefoil,

$$\boxed{\lambda_2 = \left(\frac{0.0115 s - 0.001485 D_d}{R}\right) 10^{-5}} \tag{8.90}$$

For three single cores in cradle formation on the bottom of the pipe,

$$\boxed{\lambda_2 = \left(\frac{0.00438 s + 0.00226 D_d}{R}\right) 10^{-5}} \tag{8.91}$$

In practice, the three single cores will lie in a formation somewhere between trefoil and cradle. For practical cases, the losses should be calculated as the mean value between the trefoil and cradle formations using the following empirical equation:

$$\boxed{\lambda_2 = \left(\frac{0.00794 s + 0.00039 D_d}{R}\right) 10^{-5}} \tag{8.92}$$

The above equations apply to systems operating at 60 Hz. For systems operating at 50 Hz, the calculated loss factor should be multiplied by 0.76.

EXAMPLE 8.7

We will compute the pipe loss factor for a pipe-type cable (cable model No. 3). The parameters of this cable obtained from Appendix A are $s = 67.6$ mm, $R = 0.245 \cdot 10^{-4}$ Ω/m, and $D_d = 206.4$ mm. Since the cables are assumed to be in the cradle configuration, we apply equation (8.91) to obtain

$$\lambda_2 = \left(\frac{0.00438s + 0.00226D_d}{R}\right) 10^{-5} = \left(\frac{0.00438 \cdot 67.6 + 0.00226 \cdot 206.4}{0.245 \cdot 10^{-4}}\right) \cdot 10^{-5} = 0.311$$

8.4 SHEATH EDDY CURRENT LOSSES

8.4.1 Overview

Sheath eddy current losses must be included in the equations for calculating current ratings for three-core cables, two-core cables, and single-core cables when circulating current losses are eliminated by choosing a suitable method of bonding.

In this section, we will consider cables with a continuous tubular sheath. Cables with corrugated sheaths are treated as having extruded sheaths with the outer diameter replaced by the geometric mean of inner and outer diameters. Cables with concentric neutral wires when bonded at one end will have no eddy current losses.

Early investigations by Carter (1927), Dwight (1923), and Arnold (1929) form the basis of the formulas used in today's standards. Accurate calculation of eddy current losses for single-core cable sheaths is very complicated, and analytical simplifications leading to semi-empirical equations have been developed by a number of workers. The most notable of these are Carter (1927), Miller (1929), Goldenberg (1958), and Morello (1959). The methods developed by these authors were accurate enough for the lead-sheathed cables in general use at the time. The equations developed by Miller were included in the first edition of IEC 287 (1969).

The increasing use of aluminum sheaths, whose electrical resistance is an order of magnitude lower than the equivalent lead sheath, has led to reconsideration of the equations for eddy current loss. Work by Heyda *et al.* (1973) provided a good general approach to the problem, yet the solutions could only be expressed as complicated mathematical series unsuitable for general use. The work by Heyda *et al.* (1973) was examined by Parr (1979) in order to derive semi-empirical equations for general use. These equations are included in the second edition of IEC 287 (1982). They contain correction factors for eddy current losses in thick sheaths where additional losses due to currents in other conductors and sheath losses due to the conductor current of the cable itself must be included.

The developments presented in this section are based on the work by Jackson (1975). The eddy currents flowing in a tubular sheath are composed of many components. The first-order eddy current in a sheath is caused by the combined effect of its own axial conductor current and the currents of neighboring cables. These two eddy currents can be considered separately, and the resulting losses added (Jackson, 1975). In most practical cases, the self-induced losses are negligible. The second-order eddy current arises from the effect of the magnetic field of first-order eddies in other tubes. The process is repeated until the successive order eddies have a negligible effect on the total eddy current distribution in the tube under consideration.

The original formulas developed by Carter, Dwight, and Arnold were made amenable for hand calculations by considering approximations to the series expressions in the equations. Even though their application may result in an error reaching about 20% in sheath loss factor computations (Jackson, 1975), the convenience of the formulas and a small effect of eddy current losses on cable rating could justify their continued use provided that the range of application is clearly defined. We will consider this subject next.

In the developments presented below we will first assume that the sheath thickness is small compared with its radius, so that eddy currents can be assumed to be uniformly distributed across the wall and to act at the mean sheath radius. The correction to be applied to thick sheaths is considered separately. The resulting geometry and nomenclature for analysis are shown in Fig. 8.8. In this figure, two cables, A and X, separated by a distance s_{AX} are considered. Currents I_A and I_X flow in conductor A and X, respectively. The sheath thickness is denoted by t_s and the outer radius of the sheath is r_A. A magnetic flux at the point P is considered.

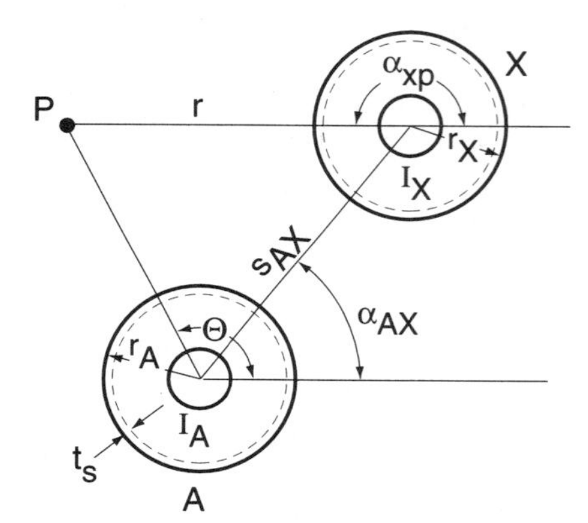

Figure 8.8 Nomenclature for analysis of eddy currents (Jackson, 1975).

8.4.2 Loss Due to External Currents

First-Order Eddy Currents: The eddy current density in any one sheath is determined by first considering the influence of the magnetic field caused by the neighboring conductor currents. A current so induced is the first-order eddy current. The basic induction equation linking the magnetic flux Φ with current density i_A is given by

$$r_A \frac{\partial i_A}{\partial \theta} = j\omega\mu_r\mu_0 \frac{t_s}{\varsigma} \frac{\partial \Phi}{\partial r} \qquad \text{at } r = r_A \tag{8.93}$$

where $\omega = 2\pi f$

$\mu_0 = 4\pi \cdot 10^{-7}$, H/m = permeability of nonmagnetic systems

μ_r = relative permeability, equal to 1 for nonmagnetic materials

t_s = sheath thickness, m

ς = sheath material electrical resistivity, Ω/m

Solving equation (8.93) with the notation in Fig. 8.8, we obtain (Jackson, 1975)

$$i_{A_1} = \frac{I}{2\pi r_A} \sum_{n=1}^{\infty} (F_{An} \cos n\theta + F'_{An} \sin n\theta) \tag{8.94}$$

where

$$F_{An} = (F_n)_A \sum_{X=A+1}^{q} \frac{M_X}{(s_{AX})^n} \exp(j\psi_X) \cos n\alpha_{AX}$$

$$F'_{AX} = (F_n)_A \sum_{X=A+1}^{q} \frac{M_X}{(s_{AX})^n} \exp(j\psi_X) \sin n\alpha_{AX} \tag{8.95}$$

$$(F_n)_A = \frac{j2r_A m_A}{n + jm_A} \tag{8.96}$$

$$m_A = \frac{\mu\mu_0\omega}{4\pi R_s} \tag{8.97}$$

X = a dummy variable and the summation $\sum_{X=A+1}^{q}$ represents the summation of all conductors X exterior to A. The analysis does not assume identical tubes, but each tube may be completely specified by its mean radius and the frequency/resistance ratio m.

M_X and ψ_X = the magnitude and phase angle of the current I_X defined with respect to an arbitrary reference current I; that is,

$$I_X = M_X \exp(j\psi_X) I \tag{8.98}$$

From equation (8.94), we obtain the following expression for the eddy current sheath loss factor:

$$\lambda_1'' = \frac{\frac{\varsigma r_A}{t_s} \int_{\theta=0}^{2\pi} \left| \frac{i_A}{2} \right|^2 d\theta}{I^2 R} = \frac{\varsigma r_A}{t_s} \frac{I^2}{(2\pi r_A)^2} \pi \frac{1}{I^2 R} \sum_{n=1}^{\infty} \left(|F_{An}|^2 + |F'_{An}|^2 \right)$$
$$= \frac{R_s}{2R} \sum_{n=1}^{\infty} \left(|F_{An}|^2 + |F'_{An}|^2 \right) \tag{8.99}$$

Substituting equations (8.95) into (8.99), we obtain

$$\lambda_1'' = \frac{R_s}{R} \sum_{n=1}^{\infty} 2r_A^{2n} \frac{m_A^2}{n + m_A^2} \sum_{X=A+1}^{q} \frac{M_x}{(s_{AX})^n}$$
$$\left[\frac{M_X}{(s_{AX})^n} + \sum_{\substack{Y=X+1 \\ Y \neq A}}^{q} 2 \frac{M_Y}{(s_{AY})^n} \cos n(\alpha_{AX} - \alpha_{AY}) \cos(\psi_{AX} - \psi_{AY}) \right] \tag{8.100}$$

Equation (8.100) gives a general expression for the first-order eddy current loss factor as a function of the system geometry. Computations can become tedious when several cables are involved. For the following common arrangements, further simplifications are possible.

(1) Single-phase go-and-return circuit. Let s be the distance between the cables in millimeters. Since there are only two cables, the second summation term in equation (8.100) is equal to zero. Also, $M_X = 1$; hence, equation (8.100) becomes

$$\lambda_1'' = \frac{R_s}{R} \sum_{n=1}^{\infty} \left[2 \left(\frac{d}{2s} \right)^{2n} \frac{m^2}{n + m^2} \right] \quad (8.101)$$

with $m = m_A$ defined by equation (8.97) and d is the mean sheath diameter, mm.

(2) Cables in flat formation with three-phase balanced currents. For three-phase balanced currents, we have

$$M_A = M_B = M_C = 1, \quad \psi_A = 0, \quad \psi_B = \frac{4\pi}{3}, \quad \psi_C = \frac{2\pi}{3} \quad (8.102)$$

The eddy current losses will be different in all three phases. For the center cable,

$$\lambda_1'' = \frac{R_s}{R} \sum_{n=1}^{\infty} \left[2 \left(\frac{d}{2s} \right)^{2n} \frac{m^2}{n + m^2} (2 - (-1)^n) \right] \quad (8.103)$$

For the outer cables, we have

$$\lambda_1'' = \frac{R_s}{R} \sum_{n=1}^{\infty} \left[2 \left(\frac{d}{2s} \right)^{2n} \frac{m^2}{n + m^2} \left(1 + \frac{1}{2^{2n}} - \frac{1}{2^n} \right) \right] \quad (8.104)$$

(3) Cables in trefoil-touching formation with three-phase balanced currents. Substituting (8.102) into (8.100), we obtain in this case

$$\lambda_1'' = \frac{R_s}{R} \sum_{n=1}^{\infty} \left[2 \left(\frac{d}{2s} \right)^{2n} \frac{m^2}{n + m^2} \left(2 - \cos \frac{n\pi}{3} \right) \right] \quad (8.105)$$

Higher Order Currents: The procedure for determining the second-order eddy current i_{A_2} caused by the magnetic field set up by the first-order eddy currents flowing in adjacent tubes follows the same development as that for finding i_{A_1}. Thus, the total second-order current in sheath A caused by all adjacent q tubes is

$$i_{A_2} = \frac{I}{2\pi r_A} \sum_{n=1}^{\infty} (G_{An} \cos n\theta + G'_{An} \sin n\theta) \quad (8.106)$$

where

$$G_{An} = \left(\frac{F_n}{2}\right)_A \sum_{k=1}^{\infty} \frac{(n+k-1)!(-1)^{k+1}}{(n-1)!k!}$$
$$\cdot \sum_{X=A+1}^{q} \frac{r_X^k}{(s_{AX})^{n+k}} \left[F_{Xk}\cos(n+k)\alpha_{AX} + F'_{Xk}\sin(n+k)\alpha_{AX}\right]$$
$$G'_{AX} = \left(\frac{F_n}{2}\right)_A \sum_{k=1}^{\infty} \frac{(n+k-1)!(-1)^{k+1}}{(n-1)!k!} \tag{8.107}$$
$$\cdot \sum_{X=A+1}^{q} \frac{r_X^k}{(s_{AX})^{n+k}} \left[F_{Xk}\sin(n+k)\alpha_{AX} - F'_{Xk}\cos(n+k)\alpha_{AX}\right]$$

Similar expressions can be written for other sheaths replacing the suffix A by the appropriate legend. Higher order currents follow a similar pattern. For example, the third-order current i_{A_3} is obtained from an equation similar to (8.106) with variable G replaced by variable H. Variables H and H' are computed from equations (8.107) with F_{Xk} and F'_{Xk} replaced by G_{Xk} and G'_{Xk}, respectively. Jackson (1975) has shown that the effect of fourth- and higher order currents can usually be neglected.

Total Eddy Current and Loss Factor Due to External Currents: The total eddy current density in sheath A is

$$i_A = i_{A_1} + i_{A_2} + \cdots = \frac{I}{2\pi r_A}\sum\left[(F_{An} + G_{An} + \cdots)\cos n\theta + \left(F'_{An} + G'_{An} + \cdots\right)\sin n\theta\right] \tag{8.108}$$

Using the notation

$$C_{An} = F_{An} + G_{An} + \cdots$$
$$C'_{An} = F'_{An} + G'_{An} + \cdots \tag{8.109}$$

and repeating the developments which lead to equation (8.99), the following expression for the total eddy current loss factor due to external currents is obtained:

$$\lambda''_1 = \frac{R_s}{2R}\sum_{n=1}^{\infty}\left(|C_{An}|^2 + |C'_{An}|^2\right) \tag{8.110}$$

8.4.3 Loss Due to Internal Current

The eddy current in a tube induced by its own coaxial conductor current circulates in the wall thickness, and is constant for all angular positions θ assuming a uniform density of the current in the coaxial conductor. Thus, there is only a single loss term equivalent to $n = 0$ and the loss can be added directly to the external loss. Since the loss is in general very small, it is permissible to ignore the magnetic field caused by the eddy current itself. The exact expression for eddy current internal loss factor was developed by Imai (1968) and is equal to

$$\lambda''_A = \frac{R_s}{R} M_A^2 \frac{(\beta_1 t_s)^2}{12} \cdot 10^{-12} \tag{8.111}$$

where

$$\beta_1 = \sqrt{\frac{4\pi\omega}{10^7 \varsigma}}$$

and t_s (mm) is the sheath thickness.

The total loss factor is obtained by summing expressions (8.110) and (8.111):

$$\lambda''_1 = \frac{R_s}{2R}\left[\sum_{n=1}^{\infty}\left(|C_{An}|^2 + |C'_{An}|^2\right) + \frac{M_A^2(\beta_1 t_s)^4}{6} \cdot 10^{-12}\right] \tag{8.112}$$

8.4.4 Correction for Wall Thickness

The preceding analysis has assumed that the eddy currents are uniformly distributed across the sheath wall. For thicker than usual sheaths, Jackson (1975) has developed a correction factor by which equation (8.110) is multiplied. This correction factor g_S, is plotted in Fig. 8.9 against the nondimensional parameter $\beta_1 D_s^*$, where D_s^* (m) is the external diameter of the sheath for various ratios of the sheath thickness versus the outside diameter.

Analytical expressions to approximate the curves in Fig. 8.9 have been developed and are given by the following expression:

$$g_s = 1 + \left(\frac{t_s}{D_s}\right)^{1.74} \left(\beta_1 D_s \cdot 10^{-3} - 1.6\right) \tag{8.113}$$

No correction for wall thickness is required for the losses caused by the internal current.

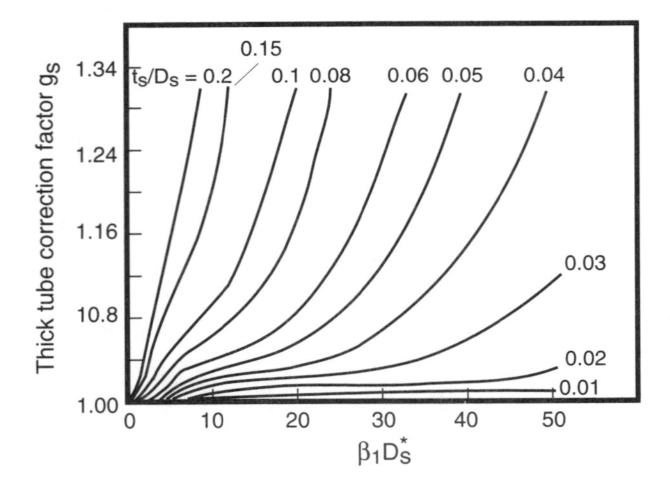

Figure 8.9 Correction to external loss factor (Jackson, 1975).

8.4.5 Simplified Expressions for Three Single-core Cables in Flat and Trefoil Formations

It is evident that the computation of the loss factor from equation (8.112) is fairly complex because it involves a summation of a series within a series. Equation (8.112) can be evaluated by suitable computer programs; however, in applications published in cable rating standards, significant simplifications of this expression are used for the purpose. Two types of simplifications are employed for the external component of the eddy sheath loss factor: 1) the first term in equations (8.101), (8.103)–(8.105) is used to approximate first-order eddies, and 2) an analytical expression is used for approximating the remaining terms in equation (8.112). The relevant expressions are as follows. Equation (8.112) is approximated by

$$\lambda_1'' = \frac{R_s}{R}\left[g_s\lambda_0(1+\Delta_1+\Delta_2) + \frac{(\beta_1 t_s)^4}{12}\cdot 10^{-12}\right] \tag{8.114}$$

where g_s is given by equation (8.113). The formulas for λ_0 (the first term of the first-order eddy current loss) Δ_1 and Δ_2 are given below.

(1) Three single-core cables in trefoil formation.

$$\begin{aligned} \lambda_0 &= 3\left(\frac{d}{2s}\right)^2 \frac{m^2}{1+m^2} \quad \text{obtained by substituting } n = 1 \text{ in (8.105)} \\ \Delta_1 &= \left(1.14m^{2.45}+0.33\right)\left(\frac{d}{2s}\right)^{0.92m+1.66} \\ \Delta_2 &= 0 \end{aligned} \tag{8.115}$$

where m is defined by the same equation as m_A in (8.97).

(2) Three single-core cables, flat formation.

(a) center cable:

$$\begin{aligned} \lambda_0 &= 6\left(\frac{d}{2s}\right)^2 \frac{m^2}{1+m^2} \quad \text{obtained by substituting } n = 1 \text{ in (8.103)} \\ \Delta_1 &= 0.86m^{3.08}\left(\frac{d}{2s}\right)^{1.4m+0.7} \\ \Delta_2 &= 0 \end{aligned} \tag{8.116}$$

(b) outer cable leading phase:

$$\begin{aligned} \lambda_0 &= 1.5\left(\frac{d}{2s}\right)^2 \frac{m^2}{1+m^2} \\ \Delta_1 &= 4.7m^{0.7}\left(\frac{d}{2s}\right)^{0.16m+2} \\ \Delta_2 &= 21m^{3.3}\left(\frac{d}{2s}\right)^{1.47m+5.06} \end{aligned} \tag{8.117}$$

(c) outer cable lagging phase:

$$\lambda_0 = 1.5\left(\frac{d}{2s}\right)^2 \frac{m^2}{1+m^2} \text{ obtained by substituting } n = 1 \text{ in (8.104)}$$

$$\Delta_1 = -\frac{0.74(m+2)m^{0.5}}{2+(m-0.3)^2}\left(\frac{d}{2s}\right)^{m+1} \tag{8.118}$$

$$\Delta_2 = 0.92m^{3.7}\left(\frac{d}{2s}\right)^{m+2}$$

In many practical cases, only the first-order eddy currents will be of interest. Jackson (1975) has shown that the approximate formulas (8.101), (8.103)–(8.105) are good approximations (maximum error of 10%) for all cases where $d/2s \leq 0.2, m \leq 1.0$. For the majority of power cables, m values are typically in the range 0.3–0.7 with spacing factors less than 0.2, and here the use of the first-order solution is permissible. In addition, for lead-sheathed cables, g_s can be taken as unity, and the term $(\beta_1 t_s)^4/12 \cdot 10^{-12}$ can be neglected. For aluminum-sheathed cables, both terms may have to be evaluated when the sheath diameter is greater than about 70 mm or the sheath is thicker than usual.

EXAMPLE 8.8

We will find the sheath loss factor for the center cable in the circuit composed of three cables (model No. 5) in flat formation. The sheaths in this cable circuit are cross bonded and the lengths of minor sections are not known. The spacing between cable centers is 0.5 m. The required parameters of this circuit are taken from Appendix A: $D_{oc} = 113$ mm, $D_{it} = 102$ mm, and $t_s = 6$ mm.

To compute the cross-sectional area of the sheath, we will take the diameter of the equivalent noncorrugated sheath and multiply it by the sheath thickness and by π. The mean diameter of the equivalent solid sheath is equal to

$$d = \frac{D_{oc} + D_{it}}{2} = \frac{102 + 113}{2} = 107.5 \text{ mm}$$

Assuming sheath operating temperature of 75°C, the sheath resistance is given by

$$R_s = \frac{2.84 \cdot 10^{-8}}{\pi 107.5 \cdot 6 \cdot 10^{-6}}[1 + 0.00403(75 - 20)] = 1.71 \cdot 10^{-5} \ \Omega/\text{m}$$

The frequency-to-resistance ratio is given by equation (8.97):

$$m = \frac{\mu\mu_0\omega}{4\pi R_s} = \frac{2\pi \cdot 60 \cdot 10^{-7}}{1.71 \cdot 10^{-5}} = 2.2$$

With the sheath temperature assumed to be 75°C, the component of the loss factor caused by the conductor current in the same cable is obtained from equation (8.111) as

$$\beta_1 = \sqrt{\frac{4\pi\omega}{10^7 \varsigma}} = \sqrt{\frac{4\pi \cdot 2\pi \cdot 60}{10^7 \cdot 2.84(1 + 0.00403 \cdot 55) \cdot 10^{-8}}} = 116.8$$

$$\lambda_A'' = \frac{R_s}{R} M_A^2 \frac{(\beta_1 t_s)^4}{12} \cdot 10^{-12} = \frac{1.71 \cdot 10^{-5}}{1.26 \cdot 10^{-5}} \cdot 1 \cdot \frac{(116.8 \cdot 6)^4}{12} \cdot 10^{-12} = 0.027$$

The outside diameter of the equivalent sheath is given by

$$D_s = \frac{D_{oc} + D_{it}}{2} + t_s = \frac{102 + 113}{2} + 6 = 113.5 \text{ mm}$$

The correction factor for thick sheaths is given by equation (8.113) as

$$g_s = 1 + \left(\frac{t_s}{D_s}\right)^{1.74} (\beta_1 D_s \cdot 10^{-3} - 1.6) = 1 + \left(\frac{6}{113.5}\right)^{1.74} (129.2 \cdot 113.5 \cdot 10^{-3} - 1.6) = 1.08$$

The approximate equations (8.116) for eddy current loss factor yield

$$\lambda_0 = 6\left(\frac{d}{2s}\right)^2 \frac{m^2}{1+m^2} = 6\left(\frac{107.5}{2 \cdot 500}\right)^2 \frac{2.2^2}{1+2.2^2} = 0.0575$$

$$\Delta_1 = 0.86 m^{3.08} \left(\frac{d}{2s}\right)^{1.4m+0.7} = 0.86 \cdot 2.2^{3.08} \left(\frac{107.5}{2 \cdot 500}\right)^{1.4 \cdot 2.2+0.7} = 0.002$$

$$\Delta_2 = 0$$

$$\lambda_1'' = \frac{R_s}{R}\left[g_s \lambda_0 (1 + \Delta_1 + \Delta_2) + \frac{(\beta_1 t_s)^4}{12} \cdot 10^{-12}\right]$$

$$= \frac{1.71 \cdot 10^{-5}}{1.26 \cdot 10^{-5}} \cdot [1.08 \cdot 0.0575 \cdot (1 + 0.002)] + 0.027 = 0.11$$

Since the lengths of the minor sections are unknown, the value of the circulating current loss is assigned by equation (8.65), that is, $\lambda_1' = 0.03$, and the total sheath loss factor is equal to

$$\lambda_1 = 0.11 + 0.03 = 0.14.$$

8.4.6 Two-core Unarmored Cables with Common Sheath

The equations used for the calculation of eddy current losses in two-core cable with round or oval conductors are approximations made by Whitehead (1939) of the equations developed by Carter (1927). The equations for sector-shaped conductors are further developments of Carter's work.

For two-core cables with round or oval conductors, the loss factor is given by

$$\lambda_1'' = \frac{16\omega^2 10^{-14}}{RR_s}\left(\frac{c}{d}\right)^2\left[1 + \left(\frac{c}{d}\right)^2\right] \tag{8.119}$$

and for sector-shaped conductors,

$$\lambda_1'' = \frac{10.8\omega^2 \cdot 10^{-16}}{RR_s}\left(\frac{1.48 r_1 + t}{d}\right)^2\left[1 + \left(\frac{1.48 r_1 + t}{d}\right)^2\right] \tag{8.120}$$

where c = distance between the axis of one conductor and the axis of the cable, mm

t = insulation thickness between the conductors, mm

r_1 = radius of the circle circumscribing the two sector-shaped conductors, mm

d = mean diameter of the sheath, mm; for oval sheaths, $d = \sqrt{d_M \cdot d_m}$ where d_M and d_m denote the major and minor diameters of the sheath; for corrugated sheaths, $d = (D_{oc} + D_{it}/2)$ where D_{oc} is the external diameter over the crests of the sheath and D_{it} is the internal diameter of the roots of the corrugations.

8.4.7 Three-core Unarmored Cables with a Common Sheath

The equations used for the calculation of eddy current losses in three-core cable with round or oval conductors are based on the equations developed by Carter (1927). For higher resistance sheaths, the equations are those developed by Whitehead (1939). For lower resistance sheaths, the equation has been extended by one term in the series to improve accuracy. The equations for sector-shaped conductors are further developments of Carter's work.

For three-core cables with round or oval conductors and with a sheath resistance $\leq 100\mu\Omega$/m, the loss factor is given by

$$\lambda_1'' = \frac{3R_s}{R}\left[\left(\frac{2c}{d}\right)^2 \frac{1}{1+\left(\frac{R_s 10^7}{\omega}\right)^2} + \left(\frac{2c}{d}\right)^4 \frac{1}{1+4\left(\frac{R_s 10^7}{\omega}\right)^2}\right] \tag{8.121}$$

For round or oval conductors and with a sheath resistance greater than $100\mu\Omega$/m, the loss factor is given by

$$\lambda_1'' = \frac{3.2\omega^2}{RR_s}\left(\frac{2c}{d}\right)^2 10^{-14} \tag{8.122}$$

For sector-shaped conductors at any value of R_s,

$$\lambda_1'' = 0.94\frac{R_s}{R}\left(\frac{2r_1+t}{d}\right)^2 \frac{1}{1+\left(\frac{R_s 10^7}{\omega}\right)^2} \tag{8.123}$$

where r_1 = radius of the circle circumscribing the three sector-shaped conductors, mm
t = insulation thickness between the conductors, mm
d = mean diameter of the sheath, mm.

8.4.8 Two-core and Three-core Cables with Steel Tape Armor

The addition of a magnetic tape armor to multicore cables will increase the eddy current losses in the sheath. The equation which has been developed for the increase is only known to apply to tapes between 0.3 and 1.0 mm in thickness and installed in a single layer. The factor F_t by which the eddy current loss factor calculated from the above formulas must be multiplied is given by

$$F_t = \left[1+\left(\frac{d}{d_a}\right)^2 \frac{1}{1+\frac{d_a}{\mu\delta_0}}\right]^2$$

so that

$$\lambda''^{*}_{1} = \lambda''_{1} F_t \tag{8.124}$$

where d_a = mean diameter of armor, mm

μ = relative permeability of the steel tape (usually taken as 300)

δ_0 = equivalent thickness of the armor = $A_a/\pi d_a$, mm

A_a = cross-sectional area of armor, mm^2.

8.4.9 Armored Three-core Cables with Each Core Having a Separate Lead Sheath (SL Type)

For this type of cable, the eddy current losses are zero and sheath losses are restricted to circulating current losses. The calculation of circulating current losses is covered in Section 8.3.2.6.

8.4.10 Effect of Large Segmental-type Conductors

In this section, we consider cables with large conductors of insulated segmental construction bonded at both ends. Where the conductor is designed to have a reduced proximity effect, such as in the cables considered in this section, sheath eddy current losses must be considered in addition to circulating current losses. When cable sheaths are single-point bonded, eddy currents losses are considered to be due only to the electromagnetic effects of the conductor currents. In solidly bonded systems, sheath eddy current losses are a function of both conductor currents and sheath circulating currents. Accurate calculation of eddy current losses under these conditions is extremely complex and unnecessary when it is considered that the eddy current losses have only a small effect on the current rating of solidly bonded single-core cables. A suitable approximation for the manner in which circulating currents modify the eddy current loss was developed by Miller (1929). This approximation is used in IEC 287 where the eddy current losses are calculated by assuming only conductor currents, and then are multiplied by a factor F given by the formula (IEC 287, 1982)

$$F = \frac{4M^2N^2 + (M+N)^2}{4(M^2+1)(N^2+1)} \tag{8.125}$$

where

$$M = N = \frac{R_s}{X} \qquad \text{for cables in trefoil formation}$$

and

$$\left.\begin{aligned} M &= \frac{R_s}{X + X_m} \\ N &= \frac{R_s}{X - \dfrac{X_m}{3}} \end{aligned}\right\} \qquad \text{for cables in flat formation with equidistant spacing}$$

If the spacing along a section is not constant, the value of X is calculated from equation (8.63).

8.5 GENERAL METHOD OF COMPUTATION OF JOULE LOSS FACTORS USING FILAMENT HEAT SOURCE SIMULATION METHOD

A general configuration for a closely spaced cable system with random separations is illustrated in Fig. 8.10.

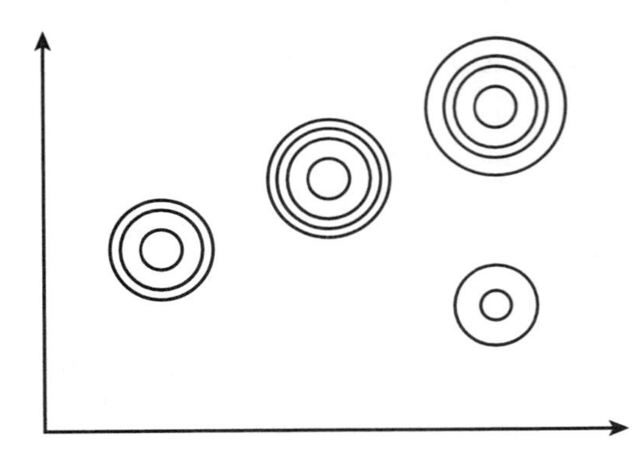

Figure 8.10 General system of n parallel conductors.

Such systems can conveniently be analyzed with the application of the filament heat source simulation (f.h.s.s.) method (Ford, 1970). We will use, again, the term conductor to denote any metallic component of the cable. Applying this method, the conductors are replaced by a large number of smaller cylindrical subconductors or filaments. The number of filaments should be large enough so that the current density can be assumed uniform throughout each filament cross section. The size of the filaments is calculated such that the sum total cross-sectional area of the filaments equals the total conductor cross-sectional area. Helically wound wires are replaced by tubes with equivalent resistances. The longitudinal magnetic fluxes are not indicated in the presentation below, but are included in the computer software mentioned earlier (Anders *et al.*, 1990).

The expressions describing the electrical connections of the filaments are as follows:

$$\mathbf{I}^c = \mathbf{B} \cdot \mathbf{I} \tag{8.126}$$

$$\mathbf{E} = \mathbf{B}^{\mathbf{t}} \cdot \mathbf{E}^{\mathbf{c}} \tag{8.127}$$

where the superscript t denotes transpositions and

$$\mathbf{I}^{\mathbf{c}} = [I_1^c, I_2^c, \cdots, I_{NC}^c]^t = \text{column vector of total conductor currents}$$

$$\mathbf{I} = [I_1, I_2, \cdots, I_n]^t = \text{column vector of filament currents}$$

$$\mathbf{E}^{\mathbf{c}} = [E_1^c, E_2^c, \cdots, E_{NC}^c]^t = \text{column vector of conductor longitudinal voltage drop}$$

$$\mathbf{E} = [E_1, E_2, \cdots, E_n]^t = \text{column vector of filament longitudinal voltage drop}$$

$$\mathbf{B} = \begin{bmatrix} 1 & 1 & . & . & . & 1 & 0 & 0 & . & . & . & 0 & 0 & . & 0 \\ 0 & 0 & . & . & . & . & 1 & 1 & . & . & . & 1 & 0 & . & 0 \\ 0 & . & . & . & . & . & . & . & . & . & . & . & 1 & . & . \\ . & & & & & & & & & & & & & & . \\ . & & & & & & & & & & & & & & . \\ . & & & & & & & & & & & & & & . \\ . & & & & & & & & & & & & & & 0 \\ 0 & & & & & & & & & & & & & & 1 \end{bmatrix} = \text{connection matrix}$$

where NC = number of conductors
n = number of filaments
c = superscript referring to conductor quantities

The value of NC will depend on the number of cables per phase and on whether or not the cables have metallic armor or sheath. The presence of neutral cables and earth conductors will further increase NC.

We can see from equations (8.126) and (8.127), the connection matrix **B** is such that the sum of the filament currents in each conductor equals the total conductor current, and the longitudinal voltage drops in each filament of a conductor are equal to the conductor longitudinal voltage drop.

Our aim is to express filament currents as a function of phase conductor currents and the geometry of the system. The longitudinal voltage drop in a filament is given by equation (8.70), and is repeated here with an inclusion of the relative permeability of the material and the assumption of tubular sheath and armor:

$$\mathbf{E} = \left[\mathbf{R}_d + j\frac{\omega\mu_0\mu_r}{2\pi}\mathbf{G}\right] \cdot \mathbf{I} \tag{8.128}$$

where $\mathbf{R}_d = n \times n$ matrix with vector $[R_1, R_2, \cdots, R_n]$ of filament resistances in the diagonal and zeros outside the diagonal.
$\mu_0 = 4\pi \cdot 10^{-7}$ = permeability of air, H/m
μ_r = relative permeability of the material ($\mu_r = 1$ for nonmagnetic materials)

$$\mathbf{G} = \begin{bmatrix} \ln\frac{1}{s_{11}} & \ln\frac{1}{s_{12}} & . \; . \; . & \ln\frac{1}{s_{1n}} \\ . & & & . \\ . & & & . \\ \ln\frac{1}{s_{n1}} & . & . \; . \; . & \ln\frac{1}{s_{nn}} \end{bmatrix}$$

where s_{ii} = geometric mean radius of filament, m
s_{ij} = geometric mean distance between filaments i and j, m

Since both **E** and **I** are complex quantities, their separate components must be determined. In effect, therefore, there are $2n$ equations and $4n$ unknown quantities to be found.

By equating **E** in equations (8.128) and (8.127), we obtain

$$\mathbf{B^tE^c} = \left[\mathbf{R}_d + j\frac{\omega\mu}{2\pi}\mathbf{G}\right] \cdot \mathbf{I}$$

or

$$\mathbf{I} = \left[\mathbf{R}_d + j\frac{\omega\mu}{2\pi}\mathbf{G}\right]^{-1} \mathbf{B^tE^c} \tag{8.129}$$

where $\mu = \mu_0\mu_r$. Premultiplying by **B** and substituting for $\mathbf{B} \cdot \mathbf{I}$ in equation (8.128), we obtain

$$\mathbf{I}^c = \mathbf{B}\left[\mathbf{R}_d + j\frac{\omega\mu}{2\pi}\mathbf{G}\right]^{-1} \mathbf{B^tE^c}$$

or (8.130)

$$\mathbf{E^c} = \left[\mathbf{B}\left[\mathbf{R}_d + j\frac{\omega\mu}{2\pi}\mathbf{G}\right]^{-1} \mathbf{B^t}\right]^{-1} \mathbf{I}^c$$

Premultiplying the last equation by $\mathbf{B^t}$ and substituting for $\mathbf{B^tE^c}$ in equation (8.129), we obtain

$$\mathbf{I} = \left[\mathbf{R}_d + j\frac{\omega\mu}{2\pi}G\right]^{-1} \mathbf{B^t}\left[\mathbf{B}\left[\mathbf{R}_d + j\frac{\omega\mu}{2\pi}\mathbf{G}\right]^{-1} \mathbf{B^t}\right]^{-1} \mathbf{I}^c \tag{8.131}$$

For systems where all the conductor currents are known, evaluation of the above equations represents the required solution. For systems in which total conductor currents are not known, calculations must be performed to determine the unknown values of the currents.

Equations (8.131) can now be used to determine the sheath and armor loss factors by suitably specifying the matrix boundary conditions. If the sheaths are solidly bonded, the sheath and armor filaments are solidly bonded. Equation (8.131) yields both the circulating and approximate eddy currents after observing the boundary conditions that the voltage drops in all sheath filaments are equal and the sum of all sheath filament currents is zero. If the sheaths are bonded at one end only, the filaments representing the sheath of each cable are bonded together, but not those belonging to different cables. The boundary conditions now require the sum of sheath filament currents in each cable be equal to zero. Thus, the eddy currents and standing voltages are computed from equations (8.131) and (8.128).

For solidly bonded systems, for example, we proceed as follows. Let us suppose that the first $i-1$ entries in the vector $\mathbf{I^c}$ represent known cable conductor currents. From Kirchhoff's first and second laws, we have

$$\begin{aligned} I_1^c + I_2^c + \cdots + I_{i-1}^c + \cdots + I_{NC}^c &= 0 \\ E_i^c = E_{i+1}^c = \cdots = E_{NC}^c = E_0 &= \text{ a constant} \end{aligned} \tag{8.132}$$

where E_0 is the sheath longitudinal voltage drop. Combining equations (8.130) and (8.132) and letting

$$\mathbf{C} = \left[\mathbf{B}\left[\mathbf{R}_d + j\frac{\omega\mu}{2\pi}\mathbf{G}\right]^{-1} \mathbf{B^t}\right]^{-1}$$

we obtain

$$\begin{bmatrix} E_1^c \\ \cdot \\ \cdot \\ \cdot \\ E_{i-1}^c \\ E_0 \\ \cdot \\ \cdot \\ E_0 \\ 0 \end{bmatrix} = \begin{bmatrix} C_{11} & C_{12} & \cdots & C_{1,NC} \\ \cdot & & & \\ \cdot & & & \\ \cdot & & & \\ \cdot & & & \\ \cdot & & & \\ \cdot & & & \\ \cdot & & & \\ C_{NC,1} & & & C_{NC,NC} \\ 1 & 1 & \cdots & 1 \end{bmatrix} \cdot \begin{bmatrix} I_1^c \\ I_2^c \\ \cdot \\ \cdot \\ I_{i-1}^c \\ \cdot \\ \cdot \\ \cdot \\ \cdot \\ I_{NC}^c \end{bmatrix} \tag{8.133}$$

This constitutes a set of $NC + 1$ equations in $NC + 1$ unknowns. Some reduction in computational effort can be obtained by noting that the longitudinal voltage drops of the central conductors are not of interest (see Section 8.3.5.9).

Equations similar to (8.133) can be set up for single-point bonded systems. The sheath voltages in this case are different and are to be computed.

The loss factor for a particular conductor (a sheath, armor, or pipe) composed of filaments k to m is equal to

$$\lambda = \frac{\sum_{i=k}^{m} |I|_i^2 R_i}{\sum_{j} |I|_j^2 R_j} \tag{8.134}$$

where j is the index of central conductor filaments belonging to the same cable as the sheath or armor. The currents represent the rms values.

REFERENCES

ANSI/IEEE Standard 575 (1988), "Application of sheath-bonding methods for single conductor cables and the calculation of induced voltages and currents in cable sheaths."

Anders, G. J., Moshref, A., and Roiz, J. (July 1990), "Advanced computer programs for power cable ampacity calculations," *IEEE Comput. Appl. in Power*, vol. 3, no. 3, pp. 42–45 (1996 revision).

Arnold, A. H. M. (1929), "Theory of sheath losses in single-conductor lead-covered cables," *J. IEE*, no. 67, pp. 69–89.

Arnold, A. H. M. (1941), "Eddy current losses in multi-core paper-insulated lead-covered cables, armored and unarmored, carrying balanced 3-phase current," *J. IEE*, no. 88, part II, pp. 52–63.

Bianchi, G. and Luoni, G. (1976), "Induced currents and losses in single-core submarine cables," *IEEE Trans. Power App. Syst.*, vol. PAS-95, no. 1, pp. 49–58.

Bosone, L. (1931), "Contributo allo studio delle perdite e dell'autoinduzione dei cavi unipolari armati con fili di ferro," *l'Elettrotecnica*, vol. 18, pp. 2–8.

Carter, F. W. (1927), "Eddy currents in thin cylinders of uniform conductivity due to periodically changing magnetic fields, in two dimensions," *Proc. Cambridge Philosophical Soc.*, vol. 23, pp. 901–906.

Carter, F. W. (1928), "Note on losses in cable sheaths," *Proc. Cambridge Philosophical Soc.*, vol. 24, pp. 65–73.

Dwight, H. B. (1923), "Proximity effect in wires and thin tubes," *AIEE Trans.*, no. 42, pp. 850–859.

Ford, G. L. (1970), "Calculation and measurement of current distributions in closely spaced bus system," Ontario Hydro Technologies, Private Communication.

Goldenberg, H. (1958), "The calculation of continuous current ratings and rating factors for transmission and distribution cables," ERA Report F/T187.

Gosland, L. (1940), "Transient sparkover voltages at insulating glands on lead-sheathed single-core cables—Suggestions for their reduction or elimination," ERA Report G/T113.

Heyda, P. G., Kitchie, G. E., and Taylor, J. E. (1973), "Computation of eddy current losses in cable sheaths and bus enclosures," *Proc. IEE*, vol. 120, no. 4, pp. 447–452.

IEC 287 (1969), "Calculation of the continuous current rating of cables (100% load factor)," 1st ed., IEC Publication 287.

IEC 287 (1982), "Calculation of the continuous current rating of cables (100% load factor)," 2nd ed., IEC Publication 287.

IEC 287-1-1 (1994), "Electric cables—calculation of the current rating—Part 1: Current rating equations (100% load factor) and calculation of losses—Section 1: General," IEC Publication 287.

Imai, T. (1968), "Exact equation for calculation of sheath proximity loss of single-conductor cables," *Proc. IEE*, vol. 56, no. 7, pp. 1172–1181.

Jackson, R. L. (1975), "Eddy-current losses in unbonded tubes," *Proc. IEE*, vol. 122, no. 5, pp. 551–557.

Jackson, W. D. (1962), *Classical Electrodynamics*. New York: John Wiley & Sons Inc., 1962.

Kaniuk, G. (1984), "Calculation of sheath eddy current losses for a double circuit installation," ERA Report 84-0189.

Kawasaki, K., Inami, M., and Ishikawa, T. (1981), "Theoretical considerations of eddy current losses in non-magnetic and magnetic pipes for power transmission systems," *IEEE Trans. Power App. Syst.*, vol. PAS-100, no. 2, pp. 474–484.

King, S. Y., and Halfter, N. A. (1983), *Underground Power Cables*. London: Longman.

Meyerhoff, L. (1952), "A.C. resistance of pipe-cable systems with segmental conductors," *AIEE Trans.*, vol. 71, part III, pp. 393–414.

Miller, K. W. (1929), "Sheath currents, sheath losses, induced sheath voltages and apparent conductor impedances of metal sheathed cables carrying alternating currents," Thesis in Electrical Engineering, University of Illinois, Urbana, IL.

Morgan, P. D., Wedmore, S., and Whitehead, E. B. (1927), "A critical study of a three-phase system of unarmored single-conductor cables, from the standpoint of the power losses, line constants and interference with communication circuits," ERA Report F/T22.

Morello, A. (1959), "Calculation of the current ratings for power cables," *l'Elettrotecnica*, vol. 46, pp. 2–17.

Neher, J. H., and McGrath, M. H. (1957), "The calculation of the temperature rise and load capability of cable systems," *AIEE Trans.*, vol. 76, part III, pp. 752–772.

Parr, R. G. (1979), "Formulae for eddy current loss factors in single-point or cross-bonded cable sheaths," ERA Report 79-97.

Parr, R. G., and Coates, M. W. (1988), "Current sharing between armored single core cables in parallel," ERA Report 88-0393.

Schelkunoff, S.A. (1934), "The electromagnetic theory of coaxial transmission lines and cylindrical shields," *Bell Syst. Tech. J.*, pp. 532–575.

Schurig, O. R., Kuehni, H. P., and Buller, F. H. (1929), "Losses in armored single-conductor lead-covered A.C. cables," *AIEE Trans.*, vol. 48, pp. 417–435.

Silver, D. A., and Seman, G. W. (1982), "Investigation of A.C./D.C. resistance ratios of various designs of pipe-type cable systems," *IEEE Trans. Power App. Syst.*, vol. PAS-101, no. 9, pp. 3481–3497.

Weedy, B. M. (1986), "Prediction of return currents and losses in underwater single-core armored AC cables with large spacings," *Elec. Power Syst. Res.*, vol. 10, pp. 77–85.

Whitehead, S., and Hutchings, E. E. (1939), "Current rating of cables for transmission and distribution," ERA Report F/T131.

9

Thermal Resistances and Capacitances

9.1 INTRODUCTION

In this chapter, we will discuss the computation of thermal resistances and capacitances associated with the components and the environment of cables. The internal thermal resistances and capacitances are characteristics of a given cable construction and were defined in Section 3.2. Without loss of accuracy, we will assume that these quantities are constant and independent of the component temperature. Where screening layers are present, we will also assume that for thermal calculations, metallic tapes are part of the conductor or sheath, while semiconducting layers (including metallized carbon paper tapes) are part of insulation.

We recall from Chapter 4 that the current rating of a cable is a function of the following thermal resistances: 1) T_1 = thermal resistance between conductor and sheath, 2) T_2 = thermal resistance between sheath and armor, 3) T_3 = thermal resistance of external covering, and 4) T_4 = thermal resistance of cable external environment. The units of the thermal resistance are K/W for a specified length. Since the length considered here is 1 m, the thermal resistance of a cable component is expressed in K/W per meter, which is most often written as $K \cdot m/W$. This should be distinguished from the unit of thermal resistivity which is also expressed as $K \cdot m/W$. In transient computations discussed in Chapter 5, thermal capacitances associated with the same parts of the cable were identified. The unit of thermal capacitance is $J/K \cdot m$ and the unit of the specific heat of the material is $J/K \cdot m^3$. The thermal resistivities and specific heats of materials used for insulation and for protective coverings are given in Table 9.1.

The thermal resistances of the insulation and the external environment of a cable have the greatest influence on cable rating. In fact, for the majority of buried cables, the external

TABLE 9.1 Thermal Resistivities and Capacities of Materials (Based on IEC 287-2-1, 1994)

Material	Thermal Resistivity (ρ) (K·m/W)	Specific Heat ($c \cdot 10^6$) [J/(m^3·K)]
Insulating materials[*]		
Paper insulation in solid type cables	6.0	2.0
Paper insulation in fluid-filled cables	5.0	2.0
Paper insulation in cables with external gas pressure	5.5	2.0
Paper insulation in cables with internal gas pressure		
preimpregnated	6.5	2.0
mass impregnated	6.0	2.0
PE	3.5	2.4
XLPE	3.5	2.4
Polyvinyl chloride		
up to and including 3 kV cables	5.0	1.7
greater than 3 kV cables	6.0	1.7
EPR		
up to and including 3 kV cables	3.5	2.0
greater than 3 kV cables	5.0	2.0
Butyl rubber	5.0	2.0
Rubber	5.0	2.0
Paper-polypropylene-paper (PPL)	6.5	2.0
Protective coverings		
Compounded jute and fibrous materials	6.0	2.0
Rubber sandwich protection	6.0	2.0
Polychloroprene	5.5	2.0
PVC		
up to and including 35 kV cables	5.0	1.7
greater than 35 kV cables	6.0	1.7
PVC/bitumen on corrugated aluminum sheaths	6.0	1.7
PE	3.5	2.4
Material for duct installations		
Concrete	1.0	1.9
Fiber	4.8	2.0
Asbestos	2.0	2.0
Earthenware	1.2	1.7
PVC	6.0	1.7
PE	3.5	2.4

[*]For the purpose of current rating computations, semiconducting screening materials are assumed to have the same thermal properties as the adjacent dielectric materials.

thermal resistance accounts for more than 70% of the temperature rise of the conductor. For cables in air, the external thermal resistance has a smaller effect on cable rating than in the case of buried cables.

The calculation of thermal resistances of the internal components of cables for single-core cables, whether based on rigorous mathematical computations or empirical investigations, is straightforward. For three-core cables, the calculations are somewhat more involved. Also, the calculation of the external thermal resistance requires particular attention. These topics are discussed next.

9.2 THERMAL RESISTANCE BETWEEN ONE CONDUCTOR AND SHEATH T_1

9.2.1 Single-core Cables

The thermal resistance between one conductor and the sheath is computed from equation (3.3):

$$T_1 = \frac{\rho}{2\pi} \ln\left(1 + \frac{2t_1}{d_c}\right) \tag{9.1}$$

where ρ = thermal resistivity of insulation, K · m/W

d_c = diameter of conductor, mm

t_1 = thickness of insulation between conductor and sheath, mm

T_1 = thermal resistance of the insulation, K · m/W.

For corrugated sheaths, t_1 is based on the mean internal diameter of the sheath which is equal to

$$\left(\frac{D_{it} + D_{oc}}{2} - t_s\right)$$

The dimensions of the cable occur in

$$\ln\left(1 + \frac{2t_1}{d_c}\right),$$

and therefore this expression plays the role of a geometric factor or shape modulus, and has been defined as the *geometric factor.*

EXAMPLE 9.1

We will compute the value of T_1 for model cable No. 5. The required parameters of this circuit are taken from Appendix A: $d_c = 58$ mm, $D_{oc} = 113$ mm, $D_{it} = 102$ mm, and $t_s = 6$ mm.

First, we compute the mean internal diameter of the sheath:

$$d = \frac{D_{it} + D_{oc}}{2} - t_s = \frac{113 + 102}{2} - 6 = 101.5 \text{ mm}$$

From equation (9.1), we obtain

$$T_1 = \frac{\rho}{2\pi} \ln\left(1 + \frac{2t_1}{d_c}\right) = \frac{6.5}{2\pi} = \ln\left(\frac{101.5}{58}\right) = 0.579 \text{ K} \cdot \text{m/W}$$

9.2.2 Three-core Cables

9.2.2.1 Overview. The computation of the internal thermal resistance of three-core cables is more complicated than for the single-core case. Rigorous mathematical formulas cannot be determined, although mathematical expressions to fit the conditions have been derived either experimentally or numerically. The general method of computation employs geometric factor (G) in place of the logarithmic term in equation (9.1); that is,

$$T_1 = \frac{\rho}{2\pi} G \tag{9.2}$$

Several methods of determining such factors have been devised. The first paper on this subject dates from 1905 (Mie, 1905). Further work on determining the value of T_1 for various types of three-core cables was done by Russel (1914), Atkinson (1924), and Simmons (1923, 1932). The values of the geometric factor published in IEC 287 are based on empirical investigations carried out at E.R.A. in the United Kingdom in the 1930s. These values are, in turn, based on measurements of electrical resistance performed in a series of tests on models comprising copper-tube electrodes soldered to resistance-alloy sheets to represent the dielectric. The results were further checked using the graphical method devised by Wedmore (E.R.A., 1923) and Simmons (1923, 1932).

The difficulty in analytically solving the problem of a three-core cable has been overcome by employing numerical methods such as the integral-equation method, the filament heat source simulation (f.h.s.s.) method (King and Halfter, 1982), and the finite-element method (Van Geertruyden, 1994; Anders, *et al.* 1997). By the integral-equation method, the actual conductor surfaces, including the sheath, are regarded as boundaries. The thermal field within the domain defined by the boundaries, which, in this case, is the region of cable insulation, is then formulated and solved by numerical means.

In the f.h.s.s. method, instead of taking the conductor surfaces as boundaries, the surfaces are simulated with the aid of filament heat sources (see also Section 8.5). This approach is similar to Mie's method (Mie, 1905), but a larger number of heat sources is employed. Once the positions and magnitudes of the heat sources are fixed, the thermal field can be determined.

In the finite-element method, the thermal resistance of the insulation is computed directly, assuming that the conductor and sheath boundaries are isothermal. With given losses dissipated by each conductor per unit length W_c (W/m) and the temperature θ_s of the sheath, the finite-element method (see Chapter 11) is used to compute the temperature of the conductors θ_c. The value of T_1 is then obtained from

$$T_1 = \frac{\theta_c - \theta_s}{W_c} \tag{9.3}$$

The values of the geometric factor for various cable constructions as standardized in IEC Publication 287-2-1 (1994), as well as those obtained by numerical methods, are discussed in the following sections.

9.2.2.2 Two-core Belted Cables with Circular Conductors. The curves defining the value of the geometric factor are shown in Fig. 9.1 and analytical expressions fitting the curves are given in Appendix C1.

9.2.2.3 Three-core Belted Cables with Circular and Oval Conductors. The curves defining the value of the geometric factor are shown in Fig. 9.2 and analytical expressions fitting the curves are given in Appendix C2.

Simmons (1923) also proposed the following empirical formula to evaluate the geometric factor for each core of a three-core belted cable:

$$G = \left[0.85 + 0.2\left(\frac{2t_1}{t} - 1\right)\right] \ln\left[\left(8.3 - 2.2\left(\frac{2t_1}{t} - 1\right)\right)\left(\frac{t_1}{d_c}\right) + 1\right] \tag{9.4}$$

where t and t_1 (mm) are the thicknesses of the insulation between conductors and between one conductor and the sheath, respectively.

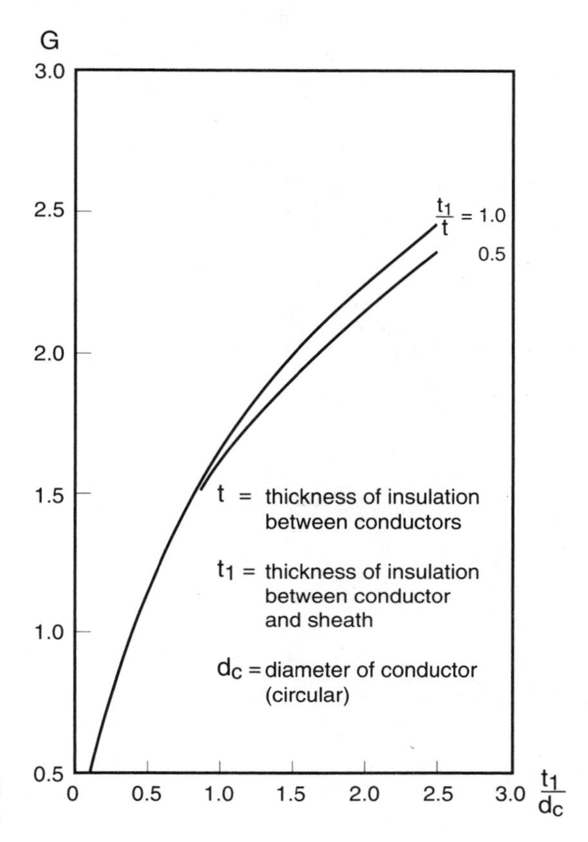

Figure 9.1 Geometric factor for two-core belted cables with circular conductors (IEC 287-2-1, 1994).

King and Halfter (1982) developed the empirical formulas given in Table 9.2 using an integral-equation approach.

Cables with oval conductors are treated as cables with an equivalent circular conductor with an equivalent diameter $d_c = \sqrt{d_{cM} d_{cm}}$ (mm) where d_{cM} and d_{cm} are the major and minor diameters of the oval conductor, respectively.

9.2.2.4 Three-core Cables with Circular Conductors and Extruded Insulation. Today's standards are based on work performed by Simmons (1932) and Whitehead and Hutchings (1938). Over 60 years have passed since the method for computation of T_1 used in today's standards was developed. Since then, many new insulating materials have appeared in three-core cable constructions; cross-linked polyethylene (XLPE) insulation in particular has been a material of choice in newer three-core cable designs. All three-core cables require fillers to fill the space between insulated cores and the belt insulation or a sheath. In the past, when impregnated paper was used to insulate the conductors, the resistivity of the filler material very closely matched that of the paper (around 6 K · m/W). With polyethylene insulation having much lower thermal resistivity (3.5 K · m/W), the higher thermal resistivity of the filler may have a significant influence on the overall value of T_1, and hence on the cable rating. To give an indication of the effect of the resistivity on the internal thermal resistance of the cable, finite-element studies were conducted for a range of values of the thermal resistivities of the insulation and the filler. From these investigations,

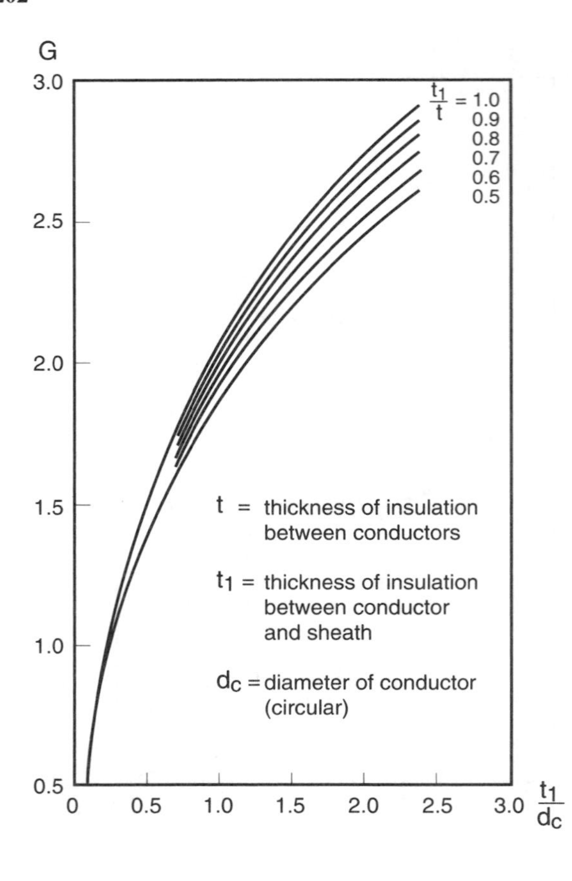

Figure 9.2 Geometric factor for three-core belted cables with circular conductors (IEC 287-2-1, 1994).

TABLE 9.2 Geometric Factor for Three-Core Unscreened Cables

Cable Configuration $(2t_1-t)/t$	Geometric Factor—Empirical Formula for $0.5 \leq x \leq 2$ with $t_1/d_c = x$
1	$0.332 + 3.029x - 1.954x^2 + 0.774x^3 - 0.126x^4$
0.5	$0.344 + 2.985x - 2.028x^2 + 0.827x^3 - 0.137x^4$
0	$0.358 + 2.914x - 2.13x^2 + 0.913x^3 - 0.157x^4$

the following approximating formula emerged (Anders *et al.*, 1997):

$$T_1^{\text{filler}} = \frac{\rho_i}{2\pi} G + 0.031 \left(\rho_f - \rho_i\right) e^{0.67 t_1/d_c} \tag{9.5}$$

where ρ_f and ρ_i are the thermal resistivities of filler and insulation, respectively, and G is the geometric factor obtained from Fig. 9.2 assuming $\rho_f = \rho_i$.

When the values of T_1, computed from equation (9.5) and obtained using the finite-element method for a wide range of cable designs were plotted and regression analysis was performed, the regression curve had a slope of 0.998 with a standard error of 0.8% (Anders *et al.*, 1997).

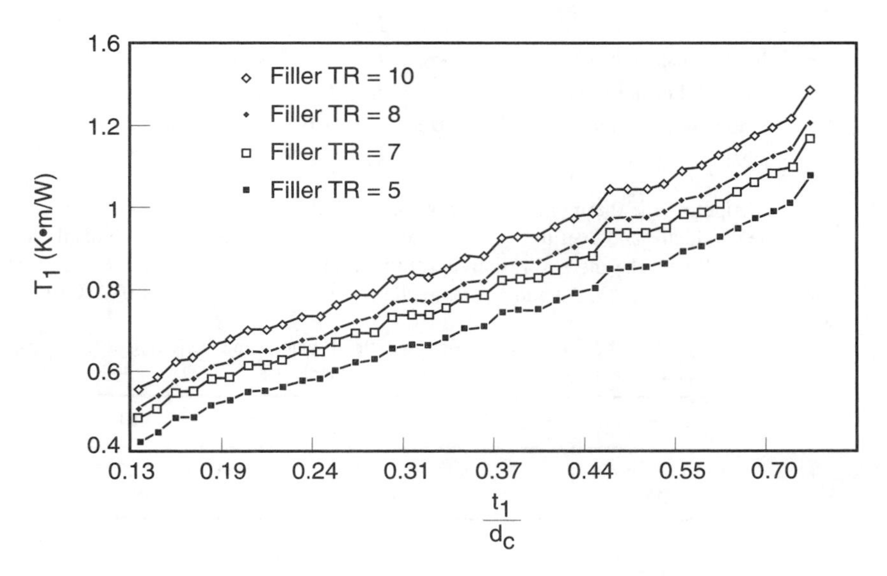

Figure 9.3 T_1 as a function of filler thermal resistivity for various cable parameters.

Figure 9.3 shows the relationships between T_1 and the ratio t_1/d_c for various values of ρ_f.

Equation (9.5) was developed for unscreened cable constructions. Modern three-core cables usually have a metallic screen around each core. However, in ampacity computations for such cables, a two-step procedure is followed: first, the value of T_1 for unscreened cable is computed [e.g., using equation (9.2) or (9.5)]; then, this resistance is modified as described in Section 9.2.2.6.

EXAMPLE 9.2

In order to demonstrate the effect of the thermal resistivity of the filler on cable ampacity, we will consider modified cable model No. 2, ignoring the presence of the metallic screen around each core and adding a metallic sheath over the laid-up cores. The parameters of this cable are as follows:

conductor diameter	$d_c = 20.5$ mm
diameter over conductor shield	$d_{cs} = 21.7$ mm
insulation thickness	$t_1 = 3.4$ mm
diameter over semiconducting insulation screen	$D_{is} = 30.1$ mm
diameter over the three cores	$D_{3c} = 64.7$ mm
diameter over the lead sheath	$D_s = 68.7$ mm
diameter over the PVC jacket	$D_e = 74.7$ mm

The cable is installed in air, away from a wall, with no direct solar radiation. The ambient air temperature is 25°C.

With the above parameters, the following values of thermal resistances and loss factors are obtained using standard calculation procedures:

thermal resistance of the insulation	$T_1 = 0.518$ K · m/W
thermal resistance of the jacket	$T_3 = 0.067$ K · m/W
external thermal resistance	$T_4 = 0.346$ K · m/W
conductor resistance at 90°C (60 Hz)	$R = 0.798 \cdot 10^{-4}\ \Omega/\text{m}$
sheath loss factor	$\lambda_1 = 0.026$

Computing the thermal resistance of the insulation, the thicknesses of semiconducting screens over the conductor and over the insulation are added to the thickness of the insulation.

We will vary the thermal resistivity of the filler between 3.5 and 10 K · m/W. The values of T_1 are summarized in Table 9.3 and a sample computation for $\rho_f = 6$ K · m/W is given below.

TABLE 9.3 Thermal Resistance of the Insulation and the Rating of the Three-Core Cable Examined in Example 9.2

ρ_f (K·m/W)	3.5	6	8	10
T_1(K·m/W)	0.518	0.609	0.666	0.733
***I* (A)**	675	658	648	638

From Fig. 9.2, $G = 0.93$ and the thermal resistance of the insulation is obtained from equation (9.5) as

$$T_1 = \frac{\rho_i}{2\pi}G + 0.031\left(\rho_f - \rho_i\right)e^{0.67t_1/d_c} = \frac{3.5}{2\pi}0.93 + 0.031(6 - 3.5)e^{0.67 \cdot 4.8/20.5} = 0.609 \text{ K} \cdot \text{m/W}$$

The rating of the cable is obtained from equation (4.3) as

$$I = \left[\frac{(\theta_c - \theta_{\text{amb}}) - W_d\left[0.5T_1 + n\left(T_2 + T_3 + T_4\right)\right]}{RT_1 + nR\left(1 + \lambda_1\right)T_2 + nR\left(1 + \lambda_1 + \lambda_2\right)\left(T_3 + T_4\right)}\right]^{0.5}$$

For the cable under consideration, we have values of $T_2 = 0$, $W_d = 0$, and $\lambda_2 = 0$. Hence,

$$I = \left[\frac{90 - 25}{0.798 \cdot 10^{-4}[0.609 + 3 \cdot 1.026 \cdot (0.067 + 0.346)]}\right]^{0.5} = 658 \text{ A}$$

The effect of the filler resistivity is quite noticeable. In this example, for the highest value of ρ_f, the ampacity is reduced by about 6% in comparison with the value computed from equation (9.2).

Finally, we will compare the value of the geometric factor obtained from Fig. 9.2 with the values given by Simmons [equation (9.4)] and by King and Halfter (Table 9.2). With the ratio of the insulation thickness between conductor and sheath to the insulation thickness between conductors equal to $t_1/t = 0.5$, the geometric factor obtained from equation (9.4) is equal to

$$G = \left[0.85 + 0.2\left(\frac{2t_1}{t} - 1\right)\right]\ln\left[\left(8.3 - 2.2\left(\frac{2t_1}{t} - 1\right)\right)\left(\frac{t_1}{d_c}\right) + 1\right]$$
$$= [0.85 + 0.2(2 \cdot 0.5 - 1)]\ln\left[(8.3 - 2.2(2 \cdot 0.5 - 1))\left(\frac{4.8}{20.5}\right) + 1\right] = 0.92$$

From Table 9.2, for $x = 4.8/20.5 = 0.234$, we obtain

$$G = 0.358 + 2.914x - 2.13x^2 + 0.913x^3 - 0.157x^4$$
$$= 0.358 + 2.914 \cdot 0.234 - 2.13 \cdot 0.234^2 + 0.913 \cdot 0.234^3 - 0.157 \cdot 0.234^4 = 0.93$$

The values obtained from equation (9.4), Table 9.2, and Fig. 9.2 are identical, and they also agree with the computations performed using the finite-element method.

9.2.2.5 Shaped Conductors. The use of shaped conductors reduces the thermal resistance of a cable, the precise effect depending on the configuration of the conductor. Some tests on this aspect were carried by Atkinson, and sector correction factors published (Atkinson, 1924). Later tests carried out by E.R.A. covered twin-, three-, and four-core models having a wide range of dimensions. The results of these tests are used in today's standards.

9.2.2.5.1 TWO-CORE BELTED CABLES WITH SECTOR-SHAPED CONDUCTORS. The geometric factor is given by

$$G = F_1 \ln\left(\frac{d_a}{2r_1}\right) \tag{9.6}$$

where $F_1 = 2 + \dfrac{4.4t}{2\pi(d_x + t) - t}$

d_a = external diameter of the belt insulation, mm

r_1 = radius of the circle circumscribing the conductors, mm

d_x = diameter of a circular conductor having the same cross-sectional area and degree of compaction as the shaped one, mm

t = insulation thickness between conductors, mm

9.2.2.5.2 THREE-CORE BELTED CABLES WITH SECTOR-SHAPED CONDUCTORS. The geometric factor is given by equation (9.6) with the coefficient F_1 defined as

$$F_1 = 3 + \frac{9t}{2\pi(d_x + t) - t} \tag{9.7}$$

King and Halfter (1982) developed the equations for the geometric factor using the f.h.s.s. method. These are shown in Table 9.4.

When compared with the case of round conductors, a reduction of about 15% in geometric factor in the case of $(2t_1 - t)/t = 0.5$ is observed.

TABLE 9.4 Geometric Factor for Three-Core Unscreened Cables with Sector-Shaped Conductors

Cable Configuration $(2t_1 - t)/t$	Geometric Factor—Empirical Formula for $0.5 \le x \le 2$ with $t_1/d_c = x$
1	$0.2 + 2.53x - 1.032x^2 + 0.183x^3$
0.5	$0.2 + 2.58x - 1.248x^2 + 0.3x^3 + 0.0163x^4$
0	$0.23 + 2.529x - 1.547x^2 + 0.599x^3 - 0.098x^4$

9.2.2.6 Three-core Cables with Metal Screens Around Each Core.

9.2.2.6.1 CIRCULAR OR OVAL CONDUCTORS. Screening reduces the thermal resistance of a cable by providing additional heat paths along the screening material of high thermal conductivity, in parallel with the path through the dielectric. The thermal resistance of the insulation is thus obtained in two steps. First, the cables of this type are considered as belted cables for which $t_1/t = 0.5$. Then, in order to take account of the thermal conductivity of the metallic screens, the results are multiplied by a factor K, called the *screening factor*, values of which are obtained from Fig. 9.4, and the analytical expressions fitting the curves are given in Appendix C3. Thus, we have

$$\boxed{T_1 = K\frac{\rho}{2\pi}G} \tag{9.8}$$

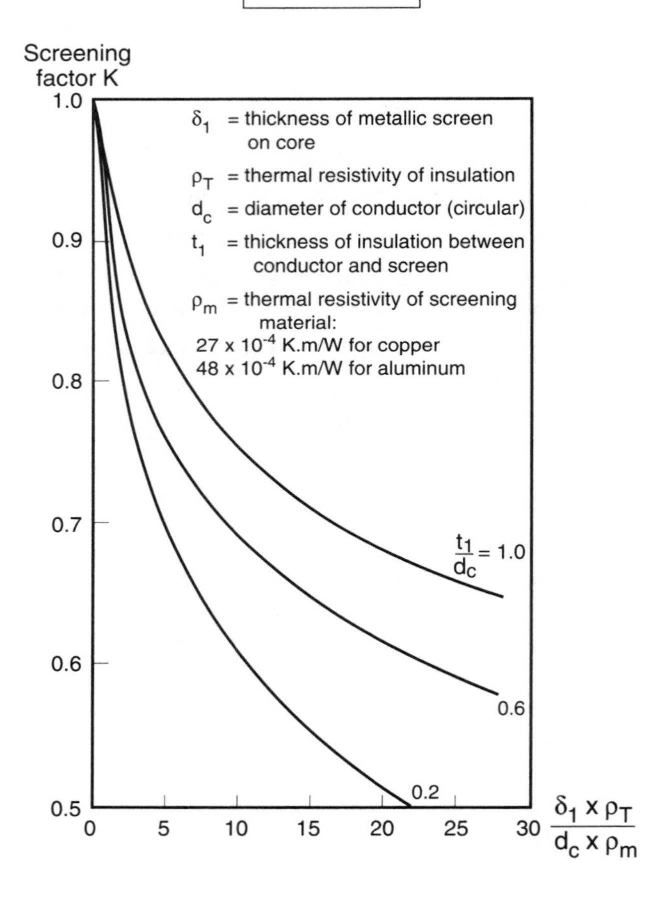

Figure 9.4 Thermal resistance of three-core screened cables with circular conductors as compare to that of a corresponding unscreened cable (IEC 287-2-1, 1994).

The values of factor K were obtained in the experiments carried out at E.R.A. (Whitehead and Hutchings, 1938). The models previously described were provided with an additional copper strip soldered to the resistance alloy and surrounding the core electrode to simulate the screen. The screen was so dimensioned as to have the correct relative conductance as determined by the parameter $t\rho_i/(2\pi\rho_{sc})$, where t is the thickness of the screen, r the radius or equivalent radius of conductor, ρ_i the thermal resistivity of the dielectric, and ρ_{sc} the thermal resistivity of the screen material.

As discussed in Section 9.2.2.4, filler resistivity may have a significant influence on the value of T_1 for plastic-insulated cables. At present, an equation similar to (9.5) is being developed for screened cables with fillers. However, until this work is completed, it is recommended that the thermal resistance of such cables be first computed from equation (9.5) and then multiplied by the screening factor K obtained from Fig. 9.4.

EXAMPLE 9.3

We will compute the value of T_1 for model cable No. 2.

First, we determine the value of the screening factor. For this cable, $t_1/d_c = 4.8/20.5 = 0.234$ and the ratio $\delta_1 \cdot \rho_t/(d_c \cdot \rho_m) = 0.2 \cdot 3.5/(20.5 \cdot 27 \cdot 10^{-4}) = 12.65$. From Fig. 9.4, we obtain $K = 0.59$.

To obtain the geometric factor G, we first assume $t_1/t = 0.5$, and from Fig. 9.2, $G = 0.93$. The thermal resistance of the insulation is obtained from equation (9.8):

$$T_1 = K\frac{\rho}{2\pi}G = 0.59\frac{3.5}{2\pi}0.94 = 0.306 \text{ K} \cdot \text{m/W}$$

This value is somewhat different from the one given in Table A1 because the latter was computed with the aid of formulas in Appendix C.

Cables with oval conductors are treated as an equivalent circular conductor with an equivalent diameter $d_c = \sqrt{d_{cM}d_{cm}}$.

9.2.2.6.2 SHAPED CONDUCTORS. For these cables, T_1 is calculated the same way as for belted cables with sector-shaped conductors, but d_a is taken as the diameter of a circle which circumscribes the core assembly. The result is multiplied by a screening factor given in Fig. 9.5 or computed as described in Appendix C4.

9.2.2.7 Fluid-filled Cables.

9.2.2.7.1 THREE-CORE CABLES WITH CIRCULAR CONDUCTORS AND METALLIZED PAPER CORE SCREENS AND CIRCULAR OIL DUCTS BETWEEN THE CORES. In cables of this construction, oil ducts are provided by laying up an open spiral duct of metal strip in each filler space as shown in Fig. 9.6.

The expression for the thermal resistance between one core and the sheath was obtained experimentally:

$$T_1 = 0.358\rho\left(\frac{2t_i}{d_c + 2t_i}\right) \tag{9.9}$$

where t_i (mm) is the thickness of core insulation including carbon black and metallized paper tapes plus half of any nonmetallic tapes over the three laid-up cores. Equation (9.9)

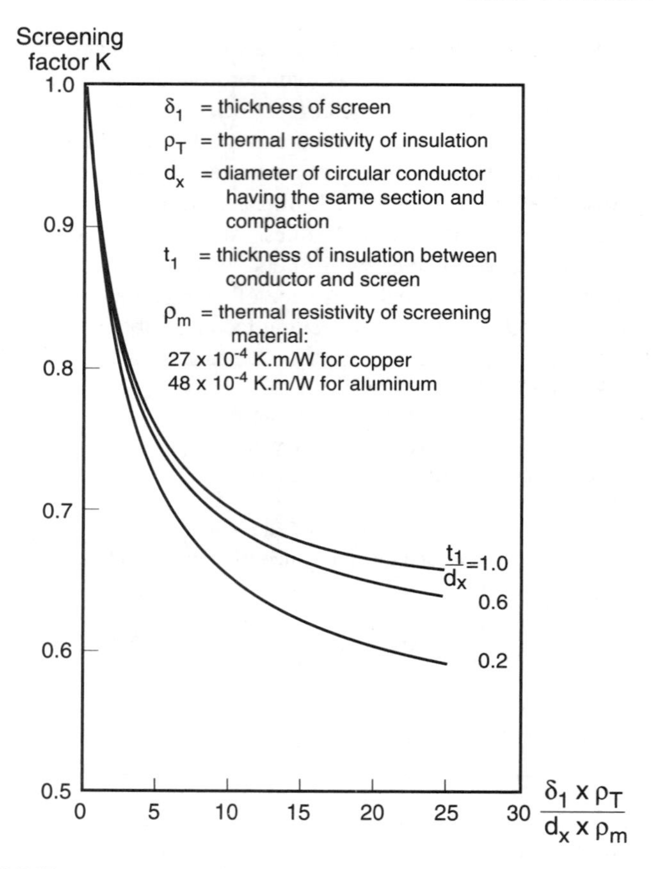

Figure 9.5 Thermal resistance of three-core screened cables with sector-shaped conductors as compared to that of a corresponding unscreened cable (IEC 287-2-1, 1994).

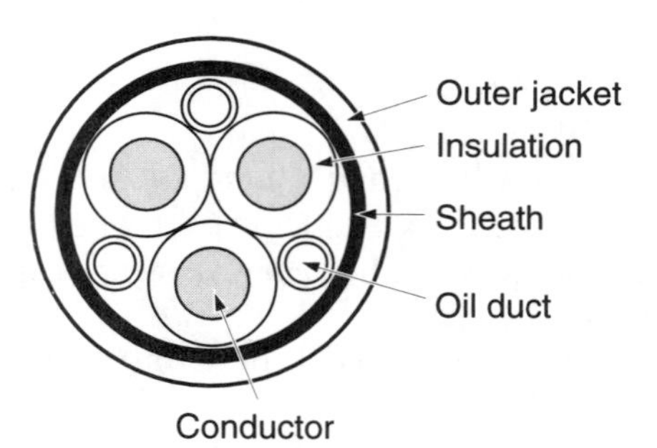

Figure 9.6 Three-core oil-filled cable.

assumes that the space occupied by the metal ducts and the oil inside them has a very high thermal conductance compared with that of the insulation; the equation therefore applies irrespective of the metal used to form the duct or its thickness.

9.2.2.7.2 THREE-CORE CABLES WITH CIRCULAR CONDUCTORS AND METAL TAPE CORE SCREENS AND CIRCULAR OIL DUCTS BETWEEN THE CORES. The thermal resistance between one conductor and the sheath is

$$T_1 = 0.35\rho\left(0.923 - \frac{2t_i}{d_c + 2t_i}\right) \tag{9.10}$$

where t_i (mm) is the thickness of core insulation including the metal screening tapes plus half of any nonmetallic tapes over the three laid-up cores.

9.2.2.7.3 THREE-CORE CABLES WITH CIRCULAR CONDUCTORS, METAL TAPE CORE SCREENS, WITHOUT FILLERS AND OIL DUCTS, HAVING A COPPER WOVEN FABRIC TAPE BINDING THE CORES TOGETHER AND A CORRUGATED ALUMINUM SHEATH. Studies carried out in Germany (Brakelmann *et al.*, 1991) have shown that the air gap between the cable insulation and corrugated sheath forms additional thermal resistance when the corrugated sheath is not in direct contact with the underlying layer. The thermal resistance T_1 of these cables was obtained recently in experiments carried out in Britain, and is given by the following equation (IEC 287-2-1, 1994):

$$T_1 = \frac{475}{D_c^{1.74}}\left(\frac{t_g}{D_c}\right)^{0.62} + \frac{\rho}{2\pi}\ln\left(\frac{D_c - 2\delta_1}{d_c}\right) \tag{9.11}$$

where $t_g = 0.5\left[\left(\frac{D_{it} + D_{ic}}{2}\right) - 2.16D_c\right]$

D_c = diameter of a core over its metallic screen tapes, mm

δ_1 = average nominal clearance between the core metallic screen tapes and the average inside diameter of the sheath, mm.

Equation (9.11) is independent of the metal used for the screen tapes.

9.2.2.8 SL Type Cables. In SL type cables, the lead sheath around each core may be assumed isothermal. The thermal resistance T_1 is calculated from equation (9.1) the same way as for single-core cables.

9.3 THERMAL RESISTANCE BETWEEN SHEATH AND ARMOR T_2

9.3.1 Single-core, Two-core, and Three-core Cables Having a Common Metallic Sheath

The thermal resistance between sheath and armor is obtained from equation (3.3) representing thermal resistance of any concentric layer. With the notation applicable to this part of the cable, we have

$$\boxed{T_2 = \frac{\rho}{2\pi}\ln\left(1 + \frac{2t_2}{D_s}\right)} \tag{9.12}$$

where ρ = thermal resistivity of the armor bedding, K · m/W

D_s = external diameter of the sheath, mm

t_2 = thickness of the bedding, mm

9.3.2 SL Type Cables

In these cables, the thermal resistance of the fillers between sheaths and armoring has been obtained using the graphical method of Wedmore (1929). The thermal resistance of fillers and bedding under the armor is given by

$$T_2 = \frac{\rho}{6\pi}\bar{G} \tag{9.13}$$

where $\bar{G}$ is the geometric factor given in Fig. 9.7 and provided analytically in Appendix C5.

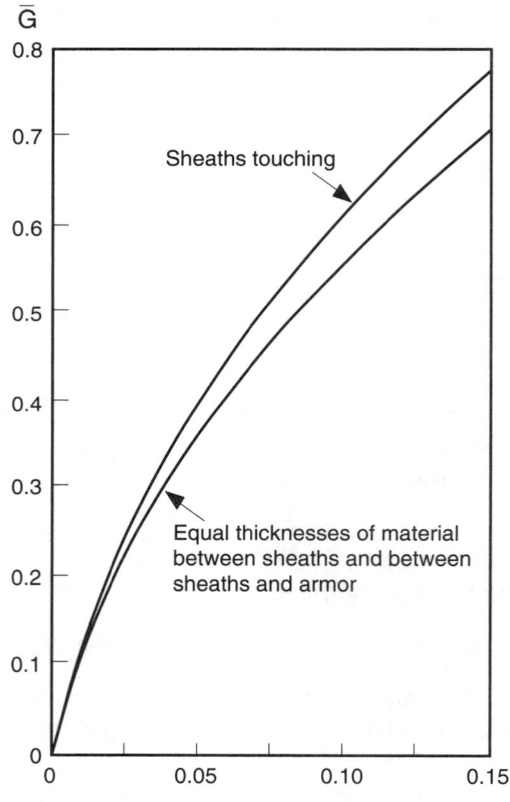

Figure 9.7 Geometric factor for obtaining the thermal resistances of the filling material between the sheaths and armor of SL-type cables (IEC 287-1-1, 1994).

9.4 THERMAL RESISTANCE OF OUTER COVERING (SERVING) T_3

The external servings are generally in the form of concentric layers, and the thermal resistance T_3 is given by

$$T_3 = \frac{\rho}{2\pi} \ln\left(1 + \frac{2t_3}{D'_a}\right) \tag{9.14}$$

where ρ = thermal resistivity of the serving, K · m/W

D'_a = external diameter of the armor, mm; for unarmored cables, D'_a is taken as the external diameter of the component immediately beneath it, that is, sheath, screen or bedding

t_3 = thickness of serving, mm

For corrugated sheaths,

$$T_3 = \frac{\rho}{2\pi} \ln\left[\frac{D_{oc} + 2t_3}{\left(\frac{D_{oc} + D_{it}}{2}\right) + t_s}\right] \tag{9.15}$$

9.5 PIPE-TYPE CABLES

A basic construction of a pipe-type cable is shown in Fig. 1.6.

For these three-core cables, the following computational rules apply:

1. The thermal resistance T_1 of the insulation of each core between the conductor and the screen is calculated from equation (9.1) for single-core cables.
2. The thermal resistance T_2 is made up of two parts:
   (a) The thermal resistance of any serving over the screen or sheath of each core. The value to be substituted for part of T_2 in the rating equation (4.3) is the value per cable, that is, the value for a three-core cable is one third of the value of a single core. The value per core is calculated by the method given in Section 9.3.1 for the bedding of single-core cables. For oval cores, the geometric mean of the major and minor diameters $\sqrt{d_M d_m}$ is used in place of the diameter for a circular core assembly.
   (b) The thermal resistance of the gas or liquid between the surface of the cores and the pipe. This resistance is calculated in the same way as that part T_4 which is between a cable and the internal surface of a duct, as given in Section 9.6.4.1. The value calculated will be per cable and should be added to the quantity calculated in a) above before substituting for T_2 in the rating equation (4.3).
3. The thermal resistance T_3 of any external covering on the pipe is dealt with as in Section 9.4. The thermal resistance of the metallic pipe itself is negligible.

We will illustrate computation of the thermal resistances of the pipe-type cables in Example 9.8 after we review computations for cables in ducts.

9.6 EXTERNAL THERMAL RESISTANCE

The current-carrying capability of cables depends to a large extent on the thermal resistance of the medium surrounding the cable. For a cable laid underground, this resistance accounts for more than 70% of the temperature rise of the conductor. For underground installations, the external thermal resistance depends on the thermal characteristics of the soil, the diameter of the cable, the depth of laying, mode of installation (e.g., directly buried, in thermal backfill, in pipe or duct, etc.), and on the thermal field generated by neighboring cables. For cables in air, the external thermal resistance has a smaller effect on the cable rating. For aerial cables, the effect of installation conditions (e.g., indoors or outdoors, proximity of walls and other cables, etc.) is an important factor in the computation of the external thermal resistance. In the following sections, we will describe how the external thermal resistance of buried and aerial cables is computed.

9.6.1 Single Buried Cable

When an analytical expression is sought for the value of the external thermal resistance, it is necessary to consider the thermal resistivity of the soil as being unaffected by the temperature which may be attained at various points in the general field. We further idealize the conditions by assuming that the thermal resistivity is constant throughout the field. Under this condition, the superposition theorem becomes applicable, that is, the temperature change existing at any point in the general heat field becomes equal to the sum of the temperature changes produced at that point by each of the heat fields by itself.

We will first consider a single cable laid directly in a uniform soil. If the diameter of the cable is small compared with the depth of burial, it will be reasonable to represent the cable as a filament heat source laid in an infinite medium. Under steady-state conditions, equation (2.15) now simplifies to

$$\frac{d\theta}{dr} + \frac{\rho_s}{2\pi r} W_t = 0 \tag{9.16}$$

The temperature rise at any point M located at a distance d from the center of the cable is obtained by integrating equation (9.16) between the limits $r = \infty$ and $r = d$. Thus,

$$\Delta\theta = \int_{\infty}^{d} -\frac{\rho_s}{2\pi r} W_t dr = -\frac{\rho_s}{2\pi} W_t \ln d \tag{9.17}$$

As explained in Section 5.2.2, in order to avoid the assumption of an infinite uniform medium, we have to use another hypothesis, namely, the hypothesis of Kennelly referenced in Chapter 5, which requires the assumption that the earth surface is an isotherm. Under this hypothesis, the temperature rise at any point M in the soil is, at any time, the sum of the temperature rises caused by the heat source W_t and by its fictitious image placed symmetrically with respect to the earth surface and emitting heat $-W_t$ (see Fig. 9.8). If these two heat flows operate simultaneously, the resulting temperature is obtained by the superposition theorem, adding to equation (9.17) a term corresponding to the fictitious heat source located at the distance d' from point M (see Fig. 9.8):

$$\Delta\theta = -\frac{\rho_s}{2\pi} W_t \ln d + \frac{\rho_s}{2\pi} W_t \ln d' = \frac{\rho_s}{2\pi} W_t \ln \frac{d'}{d} \tag{9.18}$$

If the point M is placed at the surface of the cable, and expressing the values of d and d' in terms of L and D_e, the depth of the center of the cable and its diameter, equation (9.18) can be written as

$$\Delta\theta = \frac{\rho_s}{2\pi} W_t \ln(2u) \tag{9.19}$$

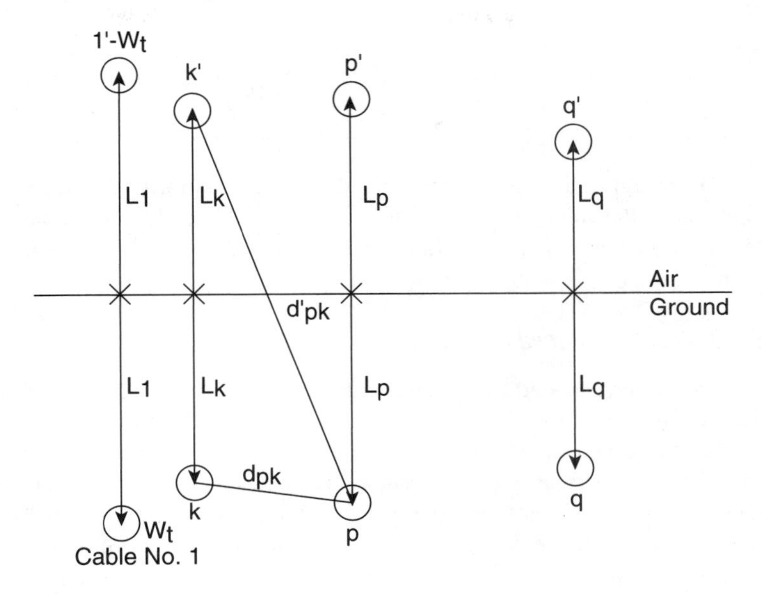

Figure 9.8 Illustration of the development of an equation for the external thermal resistance of a single cable buried under an isothermal plane.

where $u = 2L/D_e$ and

ρ_s = soil thermal resistivity, K · m/W

D_e = external diameter of the cable, mm

L = depth of burial of the center of the cable, mm

W_t = total losses inside the cable (W/m).

Equation (9.19) assumes that the heat flow lines emerge from the geometric center of the cable. Strictly speaking, the heat flow lines terminate at a small distance vertically above the geometric center of the heat source. The magnitude of this displacement or eccentricity is given by

$$e = D_e/2\left(u + \sqrt{u^2 - 1}\right)$$

Thus, equation (9.19) takes the form (known as a long form of the Kennelly formula)

$$\Delta\theta = \frac{\rho_s}{2\pi} W_t \ln\left(u + \sqrt{u^2 - 1}\right) \tag{9.19a}$$

From equations (9.19a) and (3.6), the external thermal resistance can be obtained as

$$\boxed{T_4 = \frac{\rho_s}{2\pi} \ln\left(u + \sqrt{u^2 - 1}\right)} \tag{9.20}$$

When the depth of burial is much greater than the cable diameter, u becomes much greater than one, and equation (9.20) reduces to

$$\boxed{T_4 = \frac{\rho_s}{2\pi} \ln \frac{4L}{D_e}} \tag{9.21}$$

The magnitude of the error involved in using equation (9.21) in place of equation (9.20) depends on the value of u and reaches the high value of 15% for $u = 1.5$. In practice, u is seldom less than 10, and the short form of the Kennelly formula [equation (9.21)] can be used.

9.6.2 Groups of Buried Cables (Not Touching)

For several loaded cables placed underground, we must deal with superimposed heat fields. The principle of superposition is applicable if we assume that each cable acts as a line source and does not distort the heat field of the other cables. Therefore, in the following subsections, we will assume that the cables are spaced sufficiently apart so that this assumption is approximately valid. The axial separation of the cables should be at least two cable diameters. The case when the superposition principle is not applicable is discussed in Section 9.6.3.

9.6.2.1 Unequally Loaded Cables. The method suggested for the calculation for ratings of a group of cables set apart is to calculate the temperature rise at the surface of the cable under consideration caused by the other cables of the group, and to subtract this rise from the value of $\Delta\theta$ used in the equation (4.3) for the rated current. An estimate of the power dissipated per unit length of each cable must be made beforehand, and this can be subsequently amended as a result of the calculation where it becomes necessary.

From equation (9.18), the temperature rise $\Delta\theta_{kp}$ at the surface of the cable p produced by the power W_k watt per unit length dissipated in cable k is equal to

$$\Delta\theta_{kp} = \frac{\rho_s}{2\pi} W_k \ln \frac{d'_{pk}}{d_{pk}} \tag{9.22}$$

The distances of d_{pk} and d'_{pk} are measured from the center of the pth cable to the center of cable k, and the center of the reflection of cable k in the ground–air surface, respectively (see Fig. 9.8). Thus, the temperature rise $\Delta\theta_p$ above ambient at the surface of the pth cable, whose rating is being determined, caused by the power dissipated by the other $(q - 1)$ cables in the group, is given by

$$\Delta\theta_p = \Delta\theta_{1p} + \Delta\theta_{2p} + \cdots + \Delta\theta_{kp} + \cdots + \Delta\theta_{qp} \tag{9.23}$$

with the term $\Delta\theta_{pp}$ excluded from the summation.

The value of $\Delta\theta$ in equation (4.3) for the rated current is then reduced by the amount of $\Delta\theta_p$, and the rating of the pth cable is determined using a value of T_4 corresponding to an isolated cable at position p. This calculation is performed for all cables in the group, and is repeated where necessary to avoid the possibility of overheating any of the cables.

Let θ_{ep} denote the external temperature of cable p in isolation. Substituting equation (9.22) into the right-hand side of equation (9.23) and applying equation (9.20), the following general expression for the external thermal resistance of cable p is obtained:

$$T_4^p = \frac{\theta_{ep} + \Delta\theta_p - \theta_{amb}}{W_p} = \frac{\rho_s}{2\pi}\left(\ln\left(u + \sqrt{u^2 - 1}\right) + \frac{1}{W_p}\sum_{\substack{k=1\\k\neq p}}^{q} W_k \ln\frac{d'_{pk}}{d_{pk}}\right) \quad (9.24)$$

EXAMPLE 9.4

Consider a cable model No. 2 and a cable model No. 3 located horizontally 1 m below the ground and the centers spaced 50 cm apart. The thermal resistivity of the soil is 1 K · m/W and the ambient temperature is 15°C. We will compute the rating of cable model No. 2 assuming that the pipe-type cable (cable model No. 3) carries the rated current as specified in Table A1.

From Table A1, the parameters of the three-core cable are

thermal resistance of the insulation	$T_1 = 0.307$ K · m/W
thermal resistance of the jacket	$T_3 = 0.078$ K · m/W
conductor resistance at 90°C (50 Hz)	$R = 0.798 \cdot 10^{-4}$ Ω/m
sheath loss factor	$\lambda_1 = 0.0218$
external diameter	$D_e = 72.9$ mm

The external thermal resistance of the three-core distribution cable is obtained from equation (9.20):

$$u = \frac{2L}{D_e} = \frac{2 \cdot 1000}{72.9} = 27.4,$$

$$T_4 = \frac{\rho_s}{2\pi}\ln\left(u + \sqrt{u^2 - 1}\right) = \frac{1}{2\pi}\ln\left(27.4 + \sqrt{27.4^2 - 1}\right) = 0.637 \text{ K} \cdot \text{m/W}$$

The temperature rise of the three-core cable due to the heat dissipated in the pipe-type cable (3 · 31.15 W) is obtained from equation (9.22):

$$\Delta\theta_{kp} = \frac{\rho_s}{2\pi} W_k \ln\frac{d'_{pk}}{d_{pk}} = \frac{1}{2\pi} 3 \cdot 31.15 \ln\frac{\sqrt{2^2 + 0.5^2}}{0.5} = 21.1°\text{C}$$

From equation (4.3), the rating of the three-core cable is

$$I = \left[\frac{90 - 15 - 21.1}{0.798 \cdot 10^{-4}(0.307 + 3 \cdot 1.0218 \cdot (0.078 + 0.637))}\right]^{0.5} = 520 \text{ A}$$

The single three-core cable without the presence of the pipe-type cable would have an ampacity of 613 A.

9.6.2.2 Equally Loaded Identical Cables. When a group of identical, equally loaded cables is considered, the computations can be much simplified. In this type of grouping, the rating of the group is determined by the ampacity of the hottest cable. It is usually possible to decide from the configuration of the installation which cable will be the hottest, and to calculate the rating for this one. In cases of difficulty, a further calculation for another cable may be necessary. The method is to calculate a modified value of T_4 which takes into account the mutual heating of the group and to leave unaltered the value of $\Delta\theta$ used in the rating equation (4.3).

When the losses in the group of cables are equal, equation (9.24) simplifies to

$$T_4 = \frac{\rho_s}{2\pi} \ln \left\{ \left(u + \sqrt{u^2 - 1}\right) \cdot \left[\left(\frac{d'_{p1}}{d_{p1}}\right) \left(\frac{d'_{p2}}{d_{p2}}\right) \cdots \left(\frac{d'_{pk}}{d_{pk}}\right) \cdots \left(\frac{d'_{pq}}{d_{pq}}\right) \right] \right\} \tag{9.25}$$

There are $(q - 1)$ factors in square brackets, with the term d'_{pp}/d_{pp} excluded.

EXAMPLE 9.5

We will compute the rating of a circuit composed of three single-core cables of the model cable No. 4. The location of the cables is shown in Fig. A5 with the backfill removed. The native soil thermal resistivity is 1 K · m/W and the ambient temperature is 20°C. The external diameter of this cable is $D_e = 105$ mm. The loss factors for this cable were computed in Example 8.5. They are $\lambda_1 = 0.325$ and $\lambda_2 = 0.955$. From Appendix A, $T_1 = 0.568$ K · m/W, $T_2 = 0.082$ K · m/W, and $T_3 = 0.066$ K · m/W.

The external thermal resistance of the middle cable, which is usually the hottest, is obtained from equation (9.25):

$$\begin{aligned} T_4 &= \frac{\rho_s}{2\pi} \ln \left\{ \left[u + \sqrt{u^2 - 1}\right] \cdot \left[\left(\frac{\sqrt{s_1^2 + L^2}}{s_1}\right) \left(\frac{\sqrt{s_1^2 + L^2}}{s_1}\right) \right] \right\} \\ &= \frac{\rho_s}{2\pi} \ln \left\{ \left[u + \sqrt{u^2 - 1}\right] \cdot \left[1 + \left(\frac{2L}{s_1}\right)^2 \right] \right\} \end{aligned} \tag{9.26}$$

where s_1 is the axial spacing between two adjacent cables (mm). With the factor $\left[u + \sqrt{u^2 - 1}\right]$ approximated by $2u$ and substituting numerical values, we obtain

$$T_4 = \frac{1}{2\pi} \ln \left\{ \left(\frac{4 \cdot 1000}{105}\right) \left[1 + \left(\frac{2 \cdot 1000}{500}\right)^2 \right] \right\} = 1.03 \text{ K} \cdot \text{m/W}$$

The rating of this circuit is therefore equal to

$$I = \left[\frac{85 - 20 - 6.62(0.5 \cdot 0.568 + 0.082 + 0.066 + 1.03)}{0.356 \cdot 10^{-4} (0.568 + (1 + 0.325) \cdot 0.082 + (1 + 0.325 + 0.955) \cdot (0.066 + 1.03))} \right]^{0.5} = 700 \text{ A}$$

When the losses in the sheaths of single-core cables laid in a horizontal plane are appreciable, and the sheaths are laid without transposition and/or sheaths are bonded at all joints, their inequality affects the external thermal resistance of the hottest cable. In such cases, the value of T_4 to be used in the numerator of the rating equation (4.3) is as given by equation (9.26), but a modified value of T_4 must be used in the denominator. This value is obtained from equation (9.24) remembering that the conductor losses are assumed to be equal, but the sheath losses are different. The total joule losses in the cable are expressed as $W_I = I^2 R \left(1 + \lambda'_{1i}\right)$ where the subscript i equals 1 or 2 for the outer cables and equals m for the middle cable. Substituting this in equation (9.24), we obtain

$$T_4 = \frac{\rho_s}{2\pi} \left\{ \ln \left[u + \sqrt{u^2 - 1}\right] + \left[\frac{1 + 0.5\left(\lambda'_{11} + \lambda'_{12}\right)}{1 + \lambda'_{1m}} \right] \cdot \ln \left[1 + \left(\frac{2L}{s_1}\right)^2 \right] \right\} \tag{9.27}$$

This assumes that the center cable is the hottest cable. The value of λ_1 to be used in rating equation (4.3) is that for the center cable. Equation (9.27) is modified slightly if the cable has both sheath and armor as illustrated in Example 9.6 in Section 9.6.3.3.

9.6.3 Groups of Buried Cables (Touching) Equally Loaded

9.6.3.1 Overview. When the cables are touching or are laid in a close proximity to each other, the thermal field of a cable will be distorted by the thermal fields of the cables located nearby. The principle of superposition is not applicable in this case. Goldenberg (1969a, 1969b) has shown that the minimal axial separation of the cables when equation (9.25) can be safely used is equal to two cable diameters.

The derivation of the formulas for the external thermal resistance of cables in flat and trefoil touching formations was carried out by Symm (1969) and Goldenberg (1969a, 1969b). Symm considered the general case of touching cables and used the integral-equation method for potential theory. The equations governing the distribution of an electric field around a current-carrying conductor are very similar to the heat conduction equations discussed in Chapter 2. Goldenberg solved the heat conduction problem by developing several formulas for the external thermal resistance of cables in flat and trefoil formations. In the case of two buried cables, Poritsky's (1931) formula was applied to give the potential distribution due to two parallel conducting cylinders of equal radius, in infinite space, with equal charges of the same sign each. For cables in trefoil-touching formation, Goldenberg employed the technique of conformal transformation, and the formulas were derived using restricted application of the principle of superposition.

Other authors also have tried to solve the problem. Cronin and Conangla (1971) examined the relationship between heat flux and cable sheath temperature for closely spaced buried power cables using both a numerical method and an electrolytic tank analog. King and Halfter (1977) used the method of successive images to determine the external thermal resistance of groups of equally or unequally loaded touching cables. More recently, numerical studies for three cables in flat and trefoil formations and two cables in flat formation were performed by Van Geertruyden (1992, 1993) using the finite-element method.

All authors using analytical techniques assumed that the cable surfaces are isothermal. In all cases, the derivations are very involved and will not be repeated here. Instead, we will quote the approximations used in today's standards and discuss their validity. The formulas developed by Van Geertruyden are also reported below.

9.6.3.2 Two Single-core Cables in Flat Formation. Using Poritsky's (1931) expression for the potential distribution due to two parallel conducting cylinders of equal radius, Goldenberg (1969a) developed analytically the following formula for the external thermal resistance of two cables touching in flat formation:

$$T_4 = \frac{\rho_s}{\pi}\left[\ln\left(\coth\frac{\pi}{4u}\right)\right] \tag{9.28}$$

where u is defined in equation (9.19). This formula can be simplified by using a series expansion for $\coth x$. If $u \geq 5$, equation (9.28) can be replaced by

$$T_4 = \frac{\rho_s}{\pi}\left[\ln\left(\frac{4u}{\pi} + \frac{\pi}{12u}\right)\right] \tag{9.29}$$

with an error of less than 0.01%.

Van Geertruyden (1993), using finite-element analysis, developed the following formulas for the external thermal resistance of two cables touching in flat formation.

For metallic sheathed cables, with the sheath assumed to have sufficient thermal conductance to provide an isotherm at the cable surface,

$$T_4 = \frac{\rho_s}{\pi}[\ln(2u) - 0.451] \tag{9.30}$$

When the external surfaces of the cables cannot be assumed to be isothermal, we have

$$T_4 = \frac{\rho_s}{\pi}[\ln(2u) - 0.295] \tag{9.31}$$

This formula applies for nonmetallic sheathed cables having a copper wire screen and for the external thermal resistance of touching ducts. The value of T_4 computed form equation (9.31) differs by less than 0.4% from the value computed by the finite-element method.

9.6.3.3 Three Single-core Cables in Flat Formation. The expression used in today's standards for the external thermal resistance of three touching cables in flat formation is based on Symm's (1969) paper. The following equation, not in the quoted paper, was derived empirically by the members of WG10 of SC20A of the IEC, to fit the calculated results given in Table 2 of the paper:

$$T_4 = \rho_s\,[0.475\ln(2u) - 0.346] \qquad \text{for } u \geq 5 \tag{9.32}$$

Performing numerical studies using finite-element analysis, Van Geertruyden (1992) concluded that equation (9.32) gives temperature values corresponding to the mean temperature of the outer cables. This means that when using this equation, the external temperature of the central cable, and thus the conductor temperature, will be underestimated. On the other hand, using equation (9.25) for the same case, she found that the temperatures will be slightly overestimated. She concluded that equation (9.25) is better suited for the computation of the rating of the hottest cable in this configuration than is equation (9.32). In addition, she proposed the following equation, derived from numerical studies, when the cable surfaces are nonisothermal (e.g., for cables in ducts):

$$T_4 = \frac{1.5\rho_s}{\pi}[\ln(2u) - 0.297] \tag{9.33}$$

EXAMPLE 9.6

We will reconsider Example 9.5, but will place the cables in flat touching formation. We will compute the rating of the hottest cable applying equations (9.25), (9.32), and (9.33). We will recall that the cables are located 1 m underground and the external diameter of this cable is 105 mm.

(1) Isothermal cable surface; principle of superposition assumed.

Since the cable has both the sheath and an armor, the total joule losses are obtained from

$$W_I = I^2R\left(1 + \lambda_1' + \lambda_2\right)$$

Substituting this into equation (9.24) and remembering that sheath and armor are combined in cable No. 4, we obtain

$$T_4 = \frac{\rho_s}{2\pi}\left\{\ln\left[u + \sqrt{u^2 - 1}\right] + \left[\frac{1 + 0.5\,(\lambda_{11sA} + \lambda_{12sA})}{1 + \lambda_{1msA}'}\right]\cdot\ln\left[1 + \left(\frac{2L}{s_1}\right)^2\right]\right\}$$

The loss coefficients for this new arrangement have to be recomputed because the spacing of the cables has changed. Employing the same procedure as used in part 2 of Example 8.5, we obtain the following values of the loss factors: $\lambda'_{11} = 0.159$, $\lambda'_{12} = 0.295$, $\lambda'_{1m} = 0.089$, and $\lambda_2 = 0.865$. Since $u = 2 \cdot 1000/105 = 19.05$, we have

$$T_4 = \frac{1}{2\pi}\left\{\ln\left[19.05 + \sqrt{19.05^2 - 1}\right] + \frac{1 + 0.865 + 0.5 \cdot (0.159 + 0.295)}{1 + 0.089 + 0.865} \cdot \ln\left[1 + \left(\frac{2 \cdot 1000}{105}\right)^2\right]\right\}$$
$$= 1.58 \text{ K} \cdot \text{m/W}$$

Since the sheath loss factors for the outer cables are almost the same as for the center cable, the coefficient in front of the second logarithm is almost equal to one. Therefore, the same value of T_4 will be used in the numerator and denominator of equation (4.3). The rating of the center cable becomes

$$I = \left[\frac{85 - 20 - 6.62(0.5 \cdot 0.568 + 0.082 + 0.066 + 1.58)}{0.356 \cdot 10^{-4}[0.568 + (1 + 0.089) \cdot 0.082 + (1 + 0.089 + 0.865) \cdot (0.066 + 1.58)]}\right]^{0.5}$$
$$= 612 \text{ A}$$

(2) Isothermal cable surface; Symm's equation (9.32).

$$T_4 = \rho_s[0.475\ln(2u) - 0.346] = 1 \cdot [0.475\ln(2 \cdot 19.05) - 0.346] = 1.38 \text{ K} \cdot \text{m/W}$$

$$I = \left[\frac{(\theta_c - \theta_{amb}) - W_d\,[0.5T_1 + n(T_2 + T_3 + T_4)]}{RT_1 + nR\,(1 + \lambda_1)\,T_2 + nR\,(1 + \lambda_1 + \lambda_2)\,(T_3 + T_4)}\right]^{0.5}$$
$$= \left[\frac{85 - 20 - 6.62(0.5 \cdot 0.568 + 0.082 + 0.066 + 1.38)}{0.356 \cdot 10^{-4}[0.568 + (1 + 0.089) \cdot 0.082 + (1 + 0.089 + 0.865) \cdot (0.066 + 1.38)]}\right]^{0.5}$$
$$= 654 \text{ A}$$

(3) Nonisothermal cable surface; finite-element equation (9.33).

$$T_4 = \frac{1.5\rho_s}{\pi}[\ln(2u) - 0.297] = \frac{1.5 \cdot 1}{\pi}[\ln(2 \cdot 19.05) - 0.297] = 1.60 \text{ K} \cdot \text{m/W}$$

$$I = \left[\frac{85 - 20 - 6.62(0.5 \cdot 0.568 + 0.082 + 0.066 + 1.60)}{0.356 \cdot 10^{-4}\,(0.568 + (1 + 0.089) \cdot 0.082 + (1 + 0.089 + 0.865) \cdot (0.066 + 1.60))}\right]^{0.5}$$
$$= 608 \text{ A}$$

We can observe that, in this example, the external thermal resistance increases by about 16% and the rating is reduced by 7% between two extreme cases. If we accept the fact that the finite-element equation (9.33) best represents the present situation, Van Geertruyden's conclusion that equation (9.25) is better suited for the computation of T_4 than equation given in the IEC Standard 287 (1982) can be seen to be justified in this case.

9.6.3.4 Three Single-core Cables in Trefoil Formation. For this configuration, L is measured to the center of the trefoil group and D_e is the diameter of one cable. The external thermal resistance T_4 is that of any of the cables, and the configuration may be with the apex either at the top or at the bottom of the group.

As in the case of touching cables in flat formation, the early work used in today's standards was performed by Goldenberg (1969b) and Symm (1969). Goldenberg used restricted application of superposition to derive the following equation in the case when the external surfaces of the cables can be assumed to be isothermal (Goldenberg, 1969b; IEC 287-2-1, 1994):

$$\boxed{T_4 = \frac{1.5\rho_s}{\pi}[\ln(2u) - 0.630]} \tag{9.34}$$

Symm (1969), using the integral-equation method of potential theory, confirmed that equation (9.34) is valid for $u \geq 4$.

If we neglect the effect of circumferential heat conduction of metallic layers for touching cables in trefoil formation, the thermal resistance of the insulation and the outer covering is increased because dissipation of the heat is obstructed. Thus,

$$T_1^* = f_\varphi T_1, \qquad T_3^* = f_\varphi T_3 \tag{9.35}$$

where T_1 and T_3 are calculated by the methods in Sections 9.2 and 9.4, respectively. The value of f_φ represents the fraction of the cable circumference which is obstructed by the neighboring cables. This fraction will depend on the angle φ shown in Fig. 9.9. Referring to this figure,

$$f_\varphi = \frac{\pi}{\pi - \varphi} \tag{9.36}$$

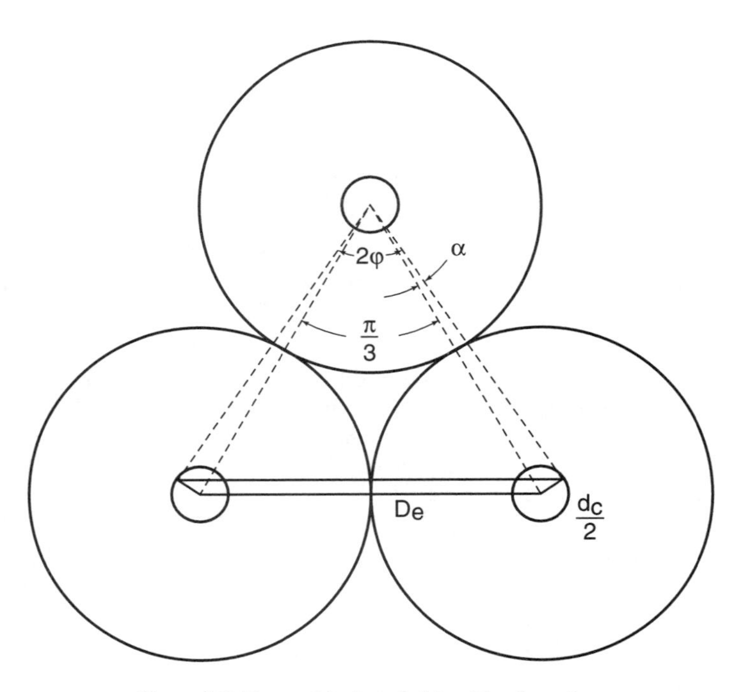

Figure 9.9 Three cables in trefoil-touching formation.

We can also see that

$$\varphi = \frac{\pi}{6} + \alpha = \frac{\pi}{6} + \frac{\pi}{180}\sin^{-1}\left(\frac{d_c}{2D_e}\right) \tag{9.37}$$

For the standard range of cable dimensions, equations (9.36) and (9.37) give the value of f_φ between 1.27 and 1.42. IEC 287 recommends that the value of $f_\varphi = 1.6$ be used in the computation of the thermal resistance of the outer covering T_3. Van Geertruyden (1995) has performed several numerical studies using the finite-element method to empirically determine the values of f_φ for cables in touching trefoil and flat formations. She concluded that the value of $f_\varphi = 1.6$ suggested in IEC 287 (1982) is too large, but its magnitude has a negligible effect on the cable rating (see also Example 9.7).

When the cable is only partially metal-covered (where helically laid armor or screen wires cover from 20 to 50% of the cable circumference), equation (9.34) applies for wires having a long lay (15 times the diameter under the wire screen), 0.7 mm diameter individual copper wires having a total cross-sectional area of between 15 and 25 mm^2. In this case, the factor f_φ will take the following values for the standard cable dimensions:

$$\begin{aligned} T_1: \quad & f_\varphi = 1.07 \text{ for cables up to 35 kV} \\ & f_\varphi = 1.16 \text{ for cables from 35 to 110 kV} \\ T_3: \quad & f_\varphi = 1.6. \end{aligned} \tag{9.38}$$

Cables for voltage above 110 kV are usually not buried in trefoil-touching formation. But if this were the case, equations (9.36) and (9.37) should be used in conjunction with equations (9.35).

In the nonisothermal case, the IEC 287 (1982) document proposes the following formula:

$$T_4 = \frac{\rho_s}{2\pi}[\ln(2u) + 2\ln(u)] \tag{9.39}$$

Performing finite-element studies, Van Geertruyden (1992) concluded that equations (9.34) and (9.39) are well suited for cables in trefoil formation.

EXAMPLE 9.7

Even though high-voltage cables are not normally placed in touching formation, we will reconsider Example 9.6 and assume that the cables are in trefoil. We will consider both cases where the cable surface is isothermal or nonisothermal.

(1) Isothermal cable surface; equation (9.34)

The external thermal resistance is equal to

$$T_4 = \frac{1.5\rho_s}{\pi}[\ln(2u) - 0.630] = \frac{1.5}{\pi}[\ln(2 \cdot 19.05) - 0.630] = 1.437 \text{ K} \cdot \text{m/W}$$

Since cable model No. 4 used in this example has a metallic sheath, there is no reduction in the thermal resistance of nonmetallic layers. However, in order to illustrate the computational procedure, we will compute the factor f_φ given by equations (9.36) and (9.37). Since the same considerations apply for the armor bedding, the value of T_2 is also increased by the factor f_φ. The reduction factor is given by

$$\varphi = \frac{\pi}{6} + \alpha = \frac{\pi}{6} + \frac{\pi}{180}\sin^{-1}\left(\frac{d_c}{2D_e}\right) = \frac{\pi}{6} + \frac{\pi}{180}\sin^{-1}\frac{33.8}{2 \cdot 105} = 0.685$$

$$f_\varphi = \frac{\pi}{\pi - \varphi} = \frac{\pi}{\pi - 0.685} = 1.28$$

We observe that this value is lower than that recommended by IEC 287 (1982). The internal thermal resistances are now obtained from equation (9.35):

$$T_1^* = 1.28 \cdot 0.568 = 0.727, \quad T_2^* = 1.28 \cdot 0.082 = 0.105, \quad T_3^* = 1.28 \cdot 0.066 = 0.0845$$

The sheath and armor loss factors are computed from equations (8.56) and (8.80) and are equal to $\lambda_1' = 0.206$ and $\lambda_2 = 0.604$. The cable rating becomes

$$I = \left[\frac{85 - 20 - 6.62(0.5 \cdot 0.727 + 0.105 + 0.0845 + 1.437)}{0.356 \cdot 10^{-4}[0.727 + (1 + 0.206) \cdot 0.105 + (1 + 0.206 + 0.604) \cdot (0.0845 + 1.437)]}\right]^{0.5}$$

$$= 635 \text{ A}$$

The rating is comparable to the case of the flat touching formation examined in Example 9.6, but is much smaller in comparison with the case when the cables are separated as examined in Example 9.5.

9.6.4 Cables in Ducts and Pipes

This section deals with the external thermal resistance of cables in ducts or pipes filled with air or a liquid. Cables in ducts which have been completely filled with a pumpable material having a thermal resistivity not exceeding that of the surrounding soil, either in the dry state or when sealed to preserve the moisture content of the filling material, may be treated as directly buried cables.

The external thermal resistance of a cable in duct or pipe consists of three parts:

1. The thermal resistance of the air or liquid between the cable surface and the duct internal surface, T_4'.
2. The thermal resistance of the duct itself, T_4''. The thermal resistance of a metal pipe is negligible.
3. The external thermal resistance of the duct, T_4'''.

The value of T_4 to be substituted in the current rating equation (4.3) will be the sum of the individual parts; that is,

$$T_4 = T_4' + T_4'' + T_4''' \tag{9.40}$$

9.6.4.1 Thermal Resistance Between Cable and Duct (or Pipe) T_4'. The development of a rigorous equation for this thermal resistance is quite involved, and the expression depends on the cable surface temperature. This equation, even though amenable to computer implementation, is not suitable for standardization. Therefore, after developing a general expression for T_4', we will simplify it in several stages until we arrive at the equation given by Neher and McGrath (1957) and in IEC 287-2-1 (1994).

In the development presented below, we will assume that the inner surface of the duct or pipe is isothermal. This assumption is usually valid for metallic conduits. For ducts made from materials having poor heat transfer properties, the average temperature inside the duct will be assumed.

Considering the outside surface of the jacket under steady-state conditions, the conduction heat flux from its inner surface is equal to the heat loss through conduction, free convection, and thermal radiation. The energy balance equation (2.8) takes the form

$$W_t = W_{\text{conv},s} + W_{\text{cond}} + W_{\text{rad},s-w} \tag{9.41}$$

where $W_{conv,s}$ = natural convection heat transfer rate between the cable outside surface and the surrounding medium per unit length, W/m

W_{cond} = conductive heat transfer rate in the medium surrounding the cable, W/m

$W_{rad,s-w}$ = thermal radiation heat transfer rate between the duct (pipe) inner surface and the cable outside surface, per unit length, W/m

W_t = total energy per unit length generated within the cable, W/m. Its value is given by equation (4.6).

Free convection heat transfer in the annular space between long, horizontal concentric cylinders has been considered by Raithby and Hollands (1975). The heat transfer rate per unit length of the duct may be obtained from equation (2.2):

$$W_{conv,s} = h_s(\theta_s - \theta_w)A_s \tag{9.42}$$

where h_s = natural convection coefficient at the surface of the cable, $W/K \cdot m^2$

θ_s = average temperature of the cable outside surface, °C

θ_w = temperature of the duct/pipe inner surface,[1] °C

A_s = area effective for convective heat transfer, (m^2), for unit length.

The value of A_s reflects the series connection of two thermal resistances corresponding to the outer surface of the cable and the inner surface of the duct wall and is equal to (Incropera and de Witt, 1990)[2]

$$A_s = \frac{2\pi}{\ln \dfrac{D_d^*}{D_e^*}} \tag{9.43}$$

The convective heat transfer coefficient is obtained by assuming that the cable and the conduit are concentric cylinders. This assumption is almost always violated in practical installations since cables are usually placed at the bottom of the conduit. The topic is discussed in detail in Anders *et al.* (1987) where the heat flux emanating from various parts of the duct is examined. However, since the thermal resistance of the gas/liquid surrounding a cable in duct or pipe constitutes a small portion of the total external thermal resistance of the cable, the proposed simplifications have a very small effect on the accuracy of the final results. The heat transfer coefficient represents in this case the *effective thermal conductivity* of the fluid (gas or oil). The empirical correlation is given by Raithby and Hollands (1985):

$$\begin{aligned} h_s &= 0.386\frac{1}{\rho}\left(\frac{\mathrm{Pr}}{0.861 + \mathrm{Pr}}\right)^{1/4}(\mathrm{Ra})^{1/4} \\ \mathrm{Ra} &= \frac{\left[\ln\left(D_d^*/D_e^*\right)\right]^4}{\left(D_d^{*-3/5} + D_e^{*-3/5}\right)^5} \cdot \frac{g\beta(\theta_s - \theta_w)d^2\rho c_p}{\mu} \end{aligned} \tag{9.44}$$

[1] To be more precise, the average temperature of the medium surrounding the cable should be used in equation (9.42). However, in order to facilitate further simplifications, the temperature of the inner wall of the duct is used here with only a small loss of accuracy.

[2] Diameters with an asterisk denote dimensions in meters.

where Ra = Rayleigh Number

β = volumetric thermal expansion coefficient, K^{-1}

c_p = specific heat at constant pressure, J/kg · K

d = mass density, kg/m^3

g = acceleration due to gravity mm/s^2

μ = viscosity, kg/s · m

ρ = thermal resistivity of the fluid, K · m/W

Pr = Prandtl Number

D_d^* = inside diameter of the conduit, m

D_e^* = external diameter of the cable, m

When the formula is used for a group of cables in a conduit, D_e^* becomes the equivalent diameter of the group as follows:

- two cables: $D_e^* = 1.65$ times the outside diameter of one cable, m
- three cables: $D_e^* = 2.15$ times the outside diameter of one cable, m
- four cables: $D_e^* = 2.50$ times the outside diameter of one cable, m

Equation (9.44) may be used in the range $10^2 \leq \text{Ra} \leq 10^7$. For Ra < 100, $h_s = 1/\rho$. Denoting by D_f^* the factor representing the cable-duct geometry and substituting (9.43) and (9.44) into (9.42), we obtain

$$W_{\text{conv},s} = 2\pi \cdot 0.386 \left(\frac{\text{Pr}}{0.861 + \text{Pr}}\right)^{1/4} D_f^{*3/4} \cdot \left[\frac{g\beta d^2 c_p}{\mu \rho^3}\right]^{1/4} (\theta_s - \theta_w)^{5/4} \tag{9.45}$$

with

$$D_f^* = \left(D_d^{*-3/5} + D_e^{*-3/5}\right)^{-5/3} \tag{9.46}$$

If the medium between the cable and the enclosure wall is air at atmospheric pressure, which is usually the case for cables in ducts, the values of the physical constants can be obtained from the formulas in Appendix D. For other gases and fluids, the material properties can be obtained from the tables found in most of the books on heat transfer, for example, in Incropera and de Witt (1990).

From the theoretical standpoint, the expression for the conduction component should take into account any eccentricity between cylindrical radiation and the enveloping isothermal enclosure. But as in the case of convection, we can assume without much loss of accuracy that the cable and the conduit are concentric cylinders.[3] The conductive heat loss is thus given by the formula

$$W_{\text{cond}} = \frac{2\pi(\theta_s - \theta_w)}{\rho \ln \dfrac{D_d^*}{D_e^*}} \tag{9.47}$$

[3] Whitehead and Hutchings (1939) give an expression to account for the eccentricity of the cable and conduit.

The net radiation heat transfer rate between the cable and the inside surface of the conduit is based on the radiative exchange between two surfaces:[4]

$$W_{\text{rad},s-w} = A_{sr} F_{s,w} \sigma_B \left(\theta_s^{*4} - \theta_w^{*4}\right) \tag{9.48}$$

where σ_B = Stefan–Boltzmann constant, equal to 5.67×10^{-8} W/(m^2· K^4)

$F_{s,w}$ = thermal radiation shape factor, its value depending on the geometry of the system

A_{sr} = area of the cable surface effective for heat radiation, (m^2), for unit length of the cable.

This equation is applicable to the case of gas-filled ducts or pipes. Computation of the radiation shape factor $F_{s,w}$ and the effective radiation area is discussed in detail in Chapter 10. The thermal properties which occur in the above equations are temperature dependent. However, for approximate calculations, they can be assumed constant over the small range of expected operating temperatures. For some of the parameters in equation (9.45), Sellers and Black (1995) proposed the values listed in Table 9.5.

TABLE 9.5 Thermal Properties for the Intervening Media

Quantity	Air (1 atm)	Gas (200 psi)	Oil
ρ—thermal resistivity (K·m/W)	35.5	39.0	7.15
β—thermal coefficient of expansion (1/K)	0.00308	0.00310	0.00068
Pr—Prandtl Number	0.715	0.7567	1126.1
ν—kinematic viscosity* (m^2/s)	$1.88 \cdot 10^{-5}$	$1.303 \cdot 10^{-6}$	$8.278 \cdot 10^{-6}$

* The following relation holds between kinematic viscosity ν and viscosity μ: $\nu = \mu / d$, where d is the mass density of the liquid.

The thermal resistance between the cable surface and the inner surface of the duct (pipe) is obtained by dividing the temperature drop across the duct (pipe) gap by the total heat emanating from the cable surface. Therefore, by equation (9.41),

$$T_4' = \frac{\theta_s - \theta_w}{W_t} = \frac{\theta_s - \theta_w}{W_{\text{conv},s} + W_{\text{cond}} + W_{\text{rad},s-w}} \tag{9.49}$$

When the first attempts to determine the thermal resistance between the cable and the duct wall were made (Whitehead and Hutchings, 1939; Buller and Neher, 1950), finding the solution of equation (9.49) appeared to be a formidable task. The value of T_4' depends on the unknown cable surface and inner wall temperatures, and the material parameters are also dependent on the mean temperature of the medium. Several iterations are required to compute T_4'. In the absence of digital computers, it became apparent that several simplifications were required. The approach proposed by Buller and Neher (1950) provided such

[4] We recall that the temperature symbols with an asterisk denote absolute temperatures.

simplifications. The same approach was later adopted by Neher and McGrath (1957), and became the basis for North American and IEC standards.

The first approximation concerns the representation of cable/conduit geometry. The effective diameter given in equation (9.46) is approximated by

$$\left(D_f^*\right)^{3/4} = \frac{\left(D_e^*\right)^{3/4}}{1.39 + \dfrac{D_e^*}{D_d^*}} \tag{9.50}$$

Next, the following assumptions were made.

1. In the case of inert gas, the physical properties of the medium were assumed to be substantially independent of temperature over the working range, but it was noted that the density is a direct function of the pressure. Thus, if P represents the pressure (in atmospheres), with appropriate numerical values for the material constants, we obtain from equations (9.45) and (9.50)

$$\frac{W_{\mathrm{conv},s}}{\Delta\theta_{sw}}(\mathrm{gas}) = 4.744 \cdot \frac{\left(D_e^*\right)^{3/4}}{1.39 + \dfrac{D_e^*}{D_d^*}} \cdot P^{1/2} \cdot \Delta\theta_{sw}^{1/4} \tag{9.51}$$

where $\Delta\theta_{sw} = \theta_s - \theta_w$.

2. When mineral oil is employed as the pipe-filling medium, it was assumed that the physical constants are substantially independent of pressure with the exception of viscosity which, for the type of oil commonly employed, was taken as varying inversely as the cube of the temperature. This leads to the following form of equation (9.45):

$$\frac{W_{\mathrm{conv},s}}{\Delta\theta_{sw}}(\mathrm{oil}) = 2.733 \cdot \frac{\left(D_e^*\right)^{3/4}}{1.39 + \dfrac{D_e^*}{D_d^*}} \cdot \theta_m^{3/4} \cdot \Delta\theta_{sw}^{1/4} \tag{9.52}$$

where θ_m is the mean oil temperature in °C.

3. The radiation component with gas as the medium is assumed to be given with sufficient accuracy by the following expression:

$$\frac{W_{\mathrm{rad},s-w}}{\Delta\theta_{sw}}(\mathrm{gas}) = 13.21 \cdot D_e^* \cdot \varepsilon_s \cdot (1 + 0.0167 \cdot \theta_m) \tag{9.53}$$

where ε_s is the emissivity of the cable outside surface. The radiation component is ignored for pipes filled with oil. Substituting (9.47) and (9.51)–(9.53) into equation (9.49) with appropriate values of the thermal resistivities, we obtain

$$\frac{1}{T_4'}(\mathrm{gas}) = 4.744 \cdot \frac{\left(D_e^*\right)^{3/4}}{1.39 + \dfrac{D_e^*}{D_d^*}} \cdot P^{1/2} \cdot \Delta\theta_{sw}^{1/4} + \frac{0.5279}{\ln \dfrac{D_d^*}{D_e^*}} + 13.21 \cdot D_e^* \cdot \varepsilon_s \cdot (1 + 0.0167 \cdot \theta_m) \tag{9.54}$$

$$\frac{1}{T_4'}(\text{oil}) = 2.733 \cdot \frac{(D_e^*)^{3/4}}{1.39 + \dfrac{D_e^*}{D_d^*}} \cdot \theta_m^{3/4} \cdot \Delta\theta_{sw}^{1/4} + \frac{0.8763}{\ln \dfrac{D_d^*}{D_e^*}} \tag{9.55}$$

Next, Buller and Neher (1950) proposed to linearize equations (9.54) and (9.55). First, they assumed that the second term in both equations and the radiation term in equation (9.54) are constant. Considering equation (9.54), the conduction term constitutes about 14% of the total in the case of a typical cable in duct installation, and about 8% for a typical gas-filled pipe-type installation at 200 psi. The corresponding values for the radiation term are 63 and 43%, respectively. Normal variation of D_e/D_d may produce considerable variation in the conduction term, but the overall effect is small because conduction is such a small part of the total heat flow. In addition, the variation of this ratio has opposite effects on the convection and conduction terms. Buller and Neher concluded that a minimum error should therefore prevail when the conduction term is treated as a constant if the denominator of the convection term is also treated as a constant.

Variation of θ_m can affect the radiation term by as much as 20% over a sufficiently wide operating range; however, when calculating a cable rating with fixed conductor temperature on the order of 70–80°C, the range of this variable is very small, and an inaccuracy on the order of 3–5% may be expected.

In the case of equation (9.55), the conduction term constitutes about 24% of the total for a typical oil-filled pipe installation. Variation of θ_m is more important than in the case of gas-filled pipe cables, but is still within tolerable limits.

Under the above assumptions, equations (9.54) and (9.55) can be rewritten as (Neher and McGrath, 1957)

$$T_4'(\text{gas}) = \frac{1}{D_e^* \left[a \left(\dfrac{\Delta\theta_{sw} \cdot P^2}{D_e^*} \right)^{1/4} + b + c \cdot \theta_m \right]} \tag{9.56}$$

$$T_4'(\text{oil}) = \frac{1}{0.120 \cdot \left(D_e^{*3} \theta_m^3 \Delta\theta_{sw} \right)^{1/4} + 0.183} \tag{9.57}$$

The constants a, b, and c in equation (9.56) were established empirically from data in Buller and Neher (1950) for cables in pipe, and from data in Greebler and Barnett (1950) for cables in fiber and transite ducts. Their values are given in Table 9.6. The constants in equation (9.57) were also determined empirically.

By further restricting the value of $\Delta\theta_{sw}$ to 20°C for cables in ducts and to 10°C for gas-filled pipe-type cables, and restricting the range of D_e to 25–100 mm for cables in ducts and to 75–125 mm for three cores in pipe, equation (9.56) can be reduced to[5]

$$\boxed{T_4' = \frac{U}{1 + 0.1(V + Y\theta_m)D_e}} \tag{9.58}$$

[5] It should be noted that in equation (9.58), the cable external diameter D_e is expressed in mm.

TABLE 9.6 Values of Constants a, b, c and U, V, Y

Installation Condition	a	b	c	U	V	Y
In metallic conduit	11.41	15.63	0.2196	5.2	1.4	0.011
In fiber duct in air	11.41	4.65	0.1163	5.2	0.83	0.006
In fiber duct in concrete	11.41	5.55	0.1806	5.2	0.91	0.010
In asbestos cement						
duct in air	11.41	11.11	0.1033	5.2	1.2	0.006
duct in concrete	11.41	10.20	0.2067	5.2	1.1	0.011
Gas pressure cable in pipe	11.41	15.63	0.2196	0.95	0.46	0.0021
Oil pressure pipe-type cable	—	—	—	0.26	0.0	0.0026
Earthenware ducts	—	—	—	1.87	0.28	0.0036

in which the values of the constants U, V, and Y depend on the installation and are given in Table 9.6.

In the case of oil-filled pipe cable, assuming the average value of $\Delta\theta_{sw} = 7°\text{C}$ and a range of 3810–8890 mm ·°C for $D_e\theta_m$, equation (9.57) reduces to equation (9.58) with the values of U, V, and Y given in Table 9.6.

9.6.4.2 Thermal Resistance of the Duct (or Pipe) Itself T_4''. This resistance is obtained by a direct application of equation (3.3):

$$\boxed{T_4'' = \frac{\rho}{2\pi} \ln \frac{D_o}{D_d}} \tag{9.59}$$

where ρ is the thermal resistivity of the material and D_o (mm) is the outside diameter of the duct.

9.6.4.3 External Thermal Resistance of the Duct (or Pipe) T_4'''. For a single-way duct not embedded in concrete, this thermal resistance is calculated in the same way as for a cable, using the appropriate formulas from Sections 9.6.1, 9.6.2, or 9.6.3 with the external radius of the duct or pipe including any protective covering thereon replacing the external radius of the cable. Ducts embedded in concrete are treated as described in Section 9.6.5.

The external thermal resistance of buried pipes for pipe-type cables is calculated as in the case of ordinary cables, using equation (9.20). In this case, the depth of laying L is measured to the center of the pipe and D_e is the external diameter of the pipe, including anti-corrosion coating.

EXAMPLE 9.8

We will determine the value of the thermal resistances of the oil, pipe covering, and the external thermal resistance for the cable model No. 3 (the pipe-type cable from the Neher–McGrath (1957) paper) and compute the percentage of the heat transfer rate in the oil attributed to the two modes of heat transfer. The parameters of this cable are given in Appendix A as $D_i = 67.26$ mm, $d_c = 41.45$ mm, $D_s = 67.59$ mm, $D_e = 244.48$ mm, $D_o = 219.08$ mm, $D_d = 206.38$ mm, $\lambda_1 = 0.010$, $\lambda_2 = 0.311$, $T_1 = 0.422$ K · m/W, $T_2 = 0.082$ K · m/W, and $T_3 = 0.017$ K · m/W, $W_c = 19.93$ W/m, $W_d = 4.83$ W/m.

Since the oil temperature is unknown, we will initially assume that its average value is 60°C. The external diameter appearing in equation (9.58) is that over the skid wire D_s. Applying equation (9.58) with the constants taken from Table 9.6 and remembering that we have three cores in a pipe, we obtain (Note: for pipe-type cables $T_4' = T_2$ and $T_4'' = T_3$: see Section 9.5)

$$T_4' = \frac{U}{1 + 0.1(V + Y\theta_m)D_s} = \frac{0.26}{1 + 0.1(0. + 0.0026 \cdot 60)2.15 \cdot 67.59} = 0.080 \text{ K} \cdot \text{m/W}$$

Let us employ the computed parameters for this cable as given in Table A1. The temperatures of the cable surface and inner wall are equal to

$$\theta_s = \theta_c - 3 \cdot (W_c \cdot T_1/3 + W_d \cdot T_1/6) = 70 - 3 \cdot (19.93 \cdot 0.422/3 + 4.83 \cdot 0.422/6) = 60.6°\text{C}$$

$$\theta_w = \theta_s - 3\,[W_c(1 + \lambda_1) + W_d]\,T_2 = 60.6 - 3[19.93(1 + 0.01) + 4.83] \cdot 0.082 = 54.5°\text{C}$$

Thus, the mean temperature of the oil is equal to $(60.6 + 54.5)/2 = 57.6°$C. With this temperature, T_4' is recalculated. Equation (9.58) yields $T_4' = 0.082$ K · m/W, and the temperature of the inner wall of the pipe is equal to 54.4°C. The resulting value of $T_4' = 0.082$ K · m/W, and the process has converged.

The heat transferred by convection is obtained from equation (9.52):

$$W_{\text{conv},s}(\text{oil}) = 2.733 \cdot \frac{\left(D_e^*\right)^{3/4}}{1.39 + \dfrac{D_e^*}{D_d^*}} \cdot \theta_m^{3/4} \cdot \Delta\theta_{sw}^{1/4} \cdot \Delta\theta_{sw}$$

$$W_{\text{conv},s}(\text{oil}) = 2.733 \cdot \frac{(2.15 \cdot 0.06759)^{3/4}}{1.39 + \dfrac{2.15 \cdot 0.06759}{0.20638}} \cdot 57.5^{3/4} \cdot (60.6 - 54.4)^{5/4} = 62.75 \text{ W/m}$$

The conduction heat transfer is obtained from the second term in equation (9.55):

$$W_{\text{cond}}(\text{oil}) = \frac{\Delta\theta_{sw}}{T_4'} = \frac{0.8763 \cdot \Delta\theta_{sw}}{\ln \dfrac{D_d^*}{D_e^*}} = \frac{0.8763 \cdot (60.6 - 54.4)}{\ln \dfrac{0.20638}{2.15 \cdot 0.06759}} = 15.49 \text{ W/m}$$

Thus, the conductive heat transfer constitutes $15.49/(62.75 + 15.49) = 0.20$ or 20% of the total heat flow. Using the computed values of temperatures and heat flow rates, the thermal resistance of the oil becomes

$$T_4' = \frac{\theta_s - \theta_w}{W_{\text{conv},s} + W_{\text{cond}}} = \frac{60.6 - 54.4}{62.75 + 15.49} = 0.079 \text{ K} \cdot \text{m/W}$$

which gives an error of about 3% in comparison with the value obtained using equation (9.58).

The thermal resistance of the pipe can be neglected, but the losses in it must be taken into account. The relevant loss factor was computed in Example 8.7 and is equal to 0.311. The thermal resistance of the pipe coating is calculated from equation (9.59) and is equal to

$$T_4'' = \frac{\rho}{2\pi} \ln \frac{D_e}{D_o} = \frac{1}{2\pi} \ln \frac{0.24448}{0.21908} = 0.017 \text{ K} \cdot \text{m/W}$$

The external thermal resistance of the pipe is computed from equation (9.20):

$$T_4''' = \frac{\rho_s}{2\pi} \ln\left(u + \sqrt{u^2 - 1}\right) = \frac{0.8}{2\pi} \ln\left(\frac{2 \cdot 0.91}{0.24448} + \sqrt{\left(\frac{2 \cdot 0.91}{0.24448}\right)^2 - 1}\right) = 0.343 \text{ K} \cdot \text{m/W}$$

9.6.5 Cables in Backfills and Duct Banks

In many North American cities, medium- and low-voltage cables are often located in duct banks in order to allow a large number of circuits to be laid in the same trench.

The ducts are first installed in layers with the aid of distance pieces, and then a bedding of filler material is compacted after each layer is positioned. Concrete is the material most often used as a filler. High- and extra-high-voltage cables are, on the other hand, often placed in an envelope of well-conducting backfill to improve heat dissipation. What both methods of installation have in common is the presence of a material which has a different thermal resistivity from that of the native soil. The first attempt to model the presence of a duct bank or a backfill was presented by Neher and McGrath (1957) and later adopted in IEC Standard 287 (1982). In later works by El-Kady and others (El-Kady and Horrocks, 1985; El-Kady *et al.* 1988, Tarasiewicz *et al.* 1987; Sellers and Black, 1995), the basic method of Neher and McGrath was extended to take into account backfills and duct banks of elongated rectangular shapes, and to remove the assumption that the external perimeter of the rectangle is isothermal.

In the following subsections, we will review the computation of the external thermal resistances of cables laid in backfills and duct banks, neglecting the effect of moisture migration. The subject of moisture migration in the vicinity of backfills and duct banks is treated briefly in Heinhold (1990).

9.6.5.1 The Neher–McGrath Approach. When the cable system is contained within an envelope of thermal resistivity ρ_c, the effect of thermal resistivity of the concrete or backfill envelope being different from that of the surrounding soil ρ_e is handled by first assuming that the thermal resistivity of the medium is ρ_c throughout. A correction is then added algebraically to account for the difference in the thermal resistivities of the envelope and the native soil. The correction to the thermal resistance is given by

$$T_4^{\text{corr}} = \frac{N}{2\pi}(\rho_e - \rho_c)\ln\left(u + \sqrt{u^2 - 1}\right) \tag{9.60}$$

where N = number of loaded cables in the envelope

$u = \dfrac{L_G}{r_b}$

L_G = depth of laying to center of duct bank or backfill, mm

r_b = equivalent radius of the envelope, mm.

The equivalent radius of the thermal envelope is obtained as follows. Considering the surface of the duct bank to be an isothermal circle of radius r_b, the thermal resistance between the duct bank and the earth's surface will be a logarithmic function of L_G and r_b. In order to evaluate r_b in terms of the dimensions x and y ($x \leq y$) of a rectangular thermal envelope, consider two circles: one inscribed and outside the envelope. The radius of the circle inscribed in the envelope is

$$r_1 = x/2$$

and the radius of a larger circle touching the four corners is

$$r_2 = 0.5\sqrt{x^2 + y^2}$$

Let us assume that the circle of radius r_b lies between these circles, and that the magnitude of r_b is such that it divides the thermal resistance between r_1 and r_2 in the ratio of the portions of the heat field between r_1 and r_2 occupied and unoccupied by the envelope. Thus,

$$\ln\frac{r_b}{r_1} = \frac{xy - \pi r_1^2}{\pi\left(r_2^2 - r_1^2\right)}\left(\ln\frac{r_2}{r_1}\right) \quad \text{or} \quad \ln\frac{r_2}{r_b} = \frac{\pi r_2^2 - xy}{\pi\left(r_2^2 - r_1^2\right)}\left(\ln\frac{r_2}{r_1}\right)$$

from which

$$\ln r_b = \frac{1}{2}\frac{x}{y}\left(\frac{4}{\pi} - \frac{x}{y}\right)\ln\left(1 + \frac{y^2}{x^2}\right) + \ln\frac{x}{2}$$

or

$$r_b = \exp\left[\frac{1}{2}\frac{x}{y}\left(\frac{4}{\pi} - \frac{x}{y}\right)\ln\left(1 + \frac{y^2}{x^2}\right) + \ln\frac{x}{2}\right] \tag{9.61}$$

Equation (9.61) is only valid for ratios of y/x less than 3.

EXAMPLE 9.9

We will reconsider Example 9.5 with the cables located in the thermal backfill as shown in Fig. 9.10.

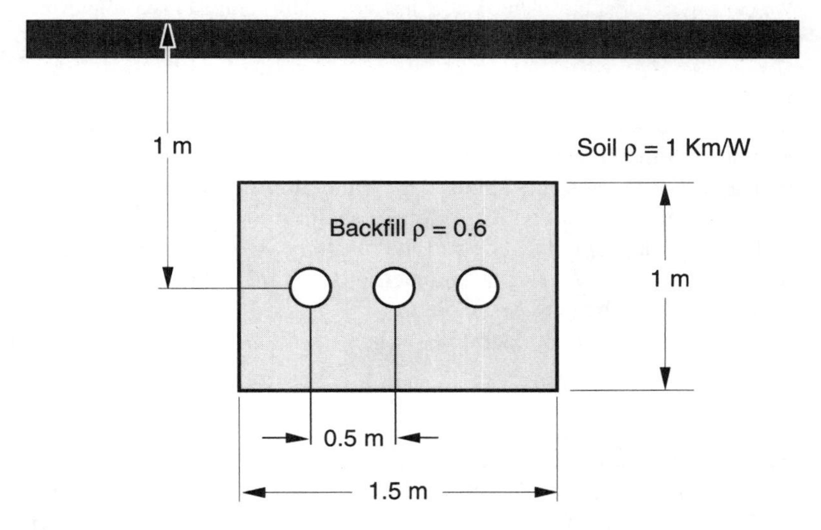

Figure 9.10 Cable model No. 4 in a thermal backfill.

The external thermal resistance of the middle cable, assuming a uniform soil thermal resistivity of 0.6 K · m/W, is obtained from the short form of equation (9.26):

$$T_4 = \frac{\rho_s}{2\pi}\ln\left\{2u\cdot\left[1 + \left(\frac{2L}{s_1}\right)^2\right]\right\} = \frac{0.6}{2\pi}\ln\left\{\left(\frac{4\cdot 1000}{500}\right)\left[1 + \left(\frac{2\cdot 1000}{500}\right)^2\right]\right\} = 0.618 \text{ K}\cdot\text{M/W}$$

For the installation in Fig. 9.10, $x = 1000$ mm, $y = 1500$ mm, and $L_G = 1000$ mm. From equation (9.61), we have

$$\begin{aligned} r_b &= \exp\left[\frac{1}{2}\frac{x}{y}\left(\frac{4}{\pi} - \frac{x}{y}\right)\ln\left(1 + \frac{y^2}{x^2}\right) + \ln\frac{x}{2}\right] \\ &= \exp\left[\frac{1}{2}\frac{1000}{1500}\left(\frac{4}{\pi} - \frac{1000}{1500}\right)\ln\left(1 + \frac{1500^2}{1000^2}\right) + \ln\frac{1000}{2}\right] = 634.6 \text{ mm.} \end{aligned}$$

Thus, $u = 1000/634.6 = 1.576$, and from equation (9.60) the correction factor becomes

$$T_4^{\mathrm{corr}} = \frac{N}{2\pi}(\rho_e - \rho_c)\ln\left(u + \sqrt{u^2 - 1}\right) = \frac{3}{2\pi}(1 - 0.6)\ln\left(1.576 + \sqrt{1.576^2 - 1}\right) = 0.196\ \mathrm{K \cdot m/W}$$

The external thermal resistance of the middle cable is

$$T_4 = 0.618 + 0.196 = 0.814\ \mathrm{K \cdot m/W}$$

The loss factors were computed in Example 8.5 and are equal to $\lambda_1 = 0.325$ and $\lambda_2 = 0.955$. The rating of the hottest cable is obtained from equation (4.3):

$$I = \left[\frac{(\theta_c - \theta_{\mathrm{amb}}) - W_d\,[0.5T_1 + n(T_2 + T_3 + T_4)]}{RT_1 + nR(1 + \lambda_1)T_2 + nR(1 + \lambda_1 + \lambda_2)(T_3 + T_4)}\right]^{0.5}$$

$$= \left[\frac{85 - 20 - 6.62(0.5 \cdot 0.568 + 0.082 + 0.066 + 0.814)}{0.356 \cdot 10^{-4}[0.568 + (1 + 0.325) \cdot 0.082 + (1 + 0.325 + 0.955) \cdot (0.066 + 0.814)]}\right]^{0.5}$$

$$= 771\ \mathrm{A}$$

We can observe that the addition of a thermal backfill resulted in a 10% increase in cable rating in comparison with the results in Example 9.5.

9.6.5.2 Extended Values of the Geometric Factor. The approximation of a rectangular duct bank or backfill by an isothermal circle with a radius given by equation (9.61) is only valid if the height/width ratio is in the range of 1/3 to 3. To overcome this restriction, El-Kady and Horrocks (1985) used the finite-element method to develop values for the geometric factor if the thermal envelopes have ratios beyond the above range. Their method is described in Example 11.5, and the resulting values of the geometric factor are shown in Table 9.7. The values of the geometric factor are displayed in terms of the height/width (h/w) and depth/height (L_G/h) ratios.

Since the geometric factor $G_b = \ln 2L_G/r_b$, the equivalent radius is equal to

$$r_b = \frac{2L_G}{e^{G_b}} \tag{9.62}$$

where G_b is obtained from Table 9.7.

El-Kady and Horrocks (1985) compared the values in Table 9.7 with these computed by the Neher–McGrath formula [the logarithmic factor in equation (9.60)]. The comparison had shown that for a given value of (L_G/h) and in the range of $h/w > 1.0$, the geometric factor calculated using the Neher–McGrath method tends to decrease as the ratio h/w increases. This contradicts the more exact results obtained by the finite-element method (Table 9.7). This contradiction can be explained by the fact that the Neher–McGrath method does not differentiate between cases where $h > w$ and cases where $h < w$. On the other hand, the finite-element method recognizes the difference between the heat flux patterns associated with these two cases.

Sellers and Black (1995) used conformal transformation to calculate the external thermal resistance of cables in duct banks and backfills. However, as explained in Section 9.6.6, this method is not easy to use in a general case.

9.6.5.3 Geometric Factor for Transient Computations. The method of Neher and McGrath to calculate the steady-state temperature rise of cables in duct banks and backfills,

TABLE 9.7 Extended Values of the Geometric Factor for Duct Banks and Backfills

h/w \ L_G/h	0.6	1.0	2.0	3.0	4.0	5.0	6.0	7.0	8.0	9.0	10.0	11.0	12.0	13.0	14.0	15.0	16.0	17.0	18.0	20.0
0.05	0.08	0.32	0.39	0.59	0.77	0.93	1.08	1.21	1.34	1.45	1.56	1.67	1.77	1.87	1.96	2.05	2.14	2.23	2.31	2.47
0.1	0.10	0.36	0.65	0.94	1.18	1.39	1.57	1.72	1.87	2.00	2.13	2.25	2.37	2.47	2.57	2.66	2.76	2.85	2.94	3.12
0.2	1.14	0.45	1.00	1.37	1.68	1.93	2.12	2.24	2.39	2.53	2.66	2.79	2.90	3.01	3.12	3.21	3.31	3.41	3.51	3.69
0.3	0.18	0.56	1.26	1.68	2.02	2.29	2.48	2.60	2.75	2.89	3.02	3.15	3.27	3.38	3.49	3.59	3.69	3.9	3.89	4.08
0.4	0.22	0.68	1.43	1.86	2.19	2.45	2.66	2.80	2.95	3.09	3.22	3.35	3.47	3.58	3.69	3.79	2.88	3.95	4.02	4.12
0.5	0.25	0.81	1.51	1.92	2.21	2.46	2.67	2.83	2.99	3.13	3.25	3.38	3.50	3.61	3.71	3.81	3.91	4.01	4.11	4.29
0.6	0.29	0.90	1.62	2.04	2.34	2.69	2.81	2.98	3.15	3.29	3.42	3.55	3.68	3.80	3.91	4.02	4.13	4.24	4.35	4.56
0.7	0.32	0.97	1.71	2.14	2.44	3.70	2.92	3.10	3.27	3.43	3.57	3.72	3.86	3.99	4.12	4.24	4.37	4.49	4.62	4.86
0.8	0.35	1.04	1.81	2.26	2.58	2.87	3.12	3.34	3.55	3.74	3.92	4.11	4.29	4.47	4.64	4.81	5.00	5.19	5.39	5.79
0.9	0.39	1.11	1.90	2.39	2.74	3.07	3.37	3.64	3.91	4.16	4.40	4.65	4.90	5.15	5.39	5.63	5.89	6.14	6.41	6.94
1.0	0.42	1.17	2.00	2.52	2.93	3.31	3.67	4.01	4.35	4.66	5.01	5.34	5.68	6.01	6.35	6.68	7.01	7.34	7.67	8.33
1.2	0.47	1.24	2.06	2.58	2.98	3.35	3.70	4.03	4.36	4.68	5.02	5.34	5.67	5.98	6.30	6.61	6.93	7.25	7.57	8.21
1.4	0.52	1.31	2.12	2.64	3.03	3.40	3.75	4.08	4.41	4.73	5.05	5.37	5.69	6.00	6.31	6.62	6.92	7.25	7.57	8.20
1.6	0.56	1.37	2.18	2.70	3.10	3.47	3.82	4.15	4.48	4.81	5.14	5.46	5.78	6.09	6.40	6.71	7.03	7.34	7.66	8.29
1.8	0.60	1.43	2.24	2.76	3.17	3.55	3.91	4.24	4.58	4.92	5.26	5.59	5.92	6.24	6.56	6.87	7.19	7.52	7.85	8.50
2.0	0.64	1.48	2.31	2.83	3.25	3.64	4.01	4.36	4.72	5.07	5.43	5.78	6.12	6.45	6.78	7.11	7.45	7.79	8.13	8.82
2.2	0.67	1.52	2.39	2.90	3.35	3.77	4.17	4.55	4.94	5.32	5.71	6.09	6.47	6.84	7.21	7.58	7.96	8.33	8.71	9.46
2.4	0.70	1.56	2.46	2.98	3.44	3.89	4.32	4.74	5.16	5.58	6.00	6.42	6.83	7.24	7.65	8.05	8.46	8.87	9.28	10.11
2.6	0.73	1.59	2.53	3.05	3.54	4.02	4.49	4.94	5.39	5.84	6.29	6.74	7.19	7.63	8.08	8.52	8.97	9.41	9.86	10.75
2.8	0.76	1.62	2.60	3.13	3.65	4.15	4.65	5.13	5.62	6.10	6.58	7.06	7.55	8.03	8.51	8.99	9.47	9.96	10.4	11.41
3.0	0.79	1.64	2.66	3.2-	3.74	4.28	4.81	5.33	5.85	6.37	6.88	7.40	7.92	8.43	8.95	9.47	9.99	10.51	11.0	12.06
3.2	0.82	1.67	2.72	3.27	3.84	4.41	4.97	5.53	6.08	6.63	7.18	7.73	8.29	8.84	9.39	9.95	10.50	11.06	11.6	12.72
3.4	0.84	1.70	2.77	3.35	3.95	4.55	5.14	5.73	6.32	6.90	7.48	8.07	8.66	9.25	9.84	10.43	11.02	11.61	12.2	13.38
3.6	0.86	1.72	2.81	3.42	4.05	4.68	5.31	5.94	6.56	7.10	7.79	8.41	9.04	9.66	10.29	10.92	11.54	12.17	12.7	14.04
3.8	0.88	1.75	2.85	3.49	4.16	4.82	5.48	6.14	6.80	7.45	8.10	8.76	9.42	10.08	10.74	11.41	12.07	12.73	13.3	14.71
4.0	0.90	1.77	2.89	3.56	4.26	4.96	5.66	6.35	7.04	7.73	8.42	9.11	9.81	10.50	11.20	11.90	12.60	13.29	13.9	15.38
4.5	0.94	1.83	2.96	3.74	4.53	5.31	6.10	6.88	7.66	8.44	9.22	10.0	10.79	11.57	12.35	13.14	13.93	14.71	15.5	17.08
5.0	0.97	1.88	3.00	3.91	4.79	5.67	6.55	7.42	8.29	9.17	10.04	10.9	11.79	12.66	13.53	14.40	15.28	16.15	17.0	18.79

given in Section 9.6.5.1, permits the introduction of a thermal resistivity for the duct bank which may differ from that used for the earth. This is compensated for by a correction factor. Recently, El-Kady and Affolter (Affolter, 1987) proposed a simple correction factor to take into account the presence of a duct bank or backfill in transient calculations.

In analogy to equation (9.20), the external thermal resistance in transient analysis is obtained from equation (5.10) as

$$T_4(t) = \frac{\rho_c}{4\pi}\left[-Ei\left(-\frac{D_e^{*2}}{16\delta t}\right) + Ei\left(-\frac{L^{*2}}{\delta t}\right)\right] \tag{9.63}$$

where δ = soil thermal diffusivity (m^2/s) and t is the time in seconds.

In analogy to equation (9.60), the following formula can be applied for the correction term:

$$T_4^{\text{corr}}(t) = \frac{\rho_e - \rho_c}{4\pi}\left[-Ei\left(-\frac{r_b^{*2}}{4\delta t}\right) + Ei\left(-\frac{L_b^{*2}}{\delta t}\right)\right] \tag{9.64}$$

This correction is then added to (9.63) in the same way as was done in the steady-state analysis.

9.6.6 External Thermal Resistance of Cables Laid in Materials Having Different Thermal Resistivities

The approach for the computation of the external thermal resistance discussed so far assumes that the heat path between the cable and the ground surface is composed of a region having uniform thermal resistivity. Even in the case of a backfill or duct bank, the same assumption was made initially, and a correction factor was applied later to account for different thermal resistivities. In practice, several layers of different thermal resistivities may be present between the cable surface and the ground/air interface. To deal with a general case of varying thermal resistivities of the soil, CIGRE WG 02 proposed a method using conformal transformation to compute the value of the external thermal resistance (CIGRE, 1985). This method is briefly described in the following.

We will start by writing the temperature rise of the surface of cable i in a group of N cables in the following form:

$$\Delta\theta_i = \sum_{k=1}^{N} R_{ik} W_{tk} \tag{9.65}$$

where W_{tk} = total power loss per unit length of cable k, W/m

R_{ii} = self thermal resistance of ith cable, K · m/W

R_{ik} $(i \neq k)$ = mutual thermal resistance between cable i and cable k, K · m/W.

The matrix with elements R_{ik} plays the same role as the external thermal resistance T_4 for one cable as given in IEC Publication 287 (1982), but covers the case where there are several cables, each of which may have different losses.

To explain the concept of the particular conformal transformation, we will consider a single cable in a uniform soil. The field can be drawn for this situation, resulting in the isothermal lines taking the form of eccentric circles as shown in Fig. 9.11, one of which is the cable surface.

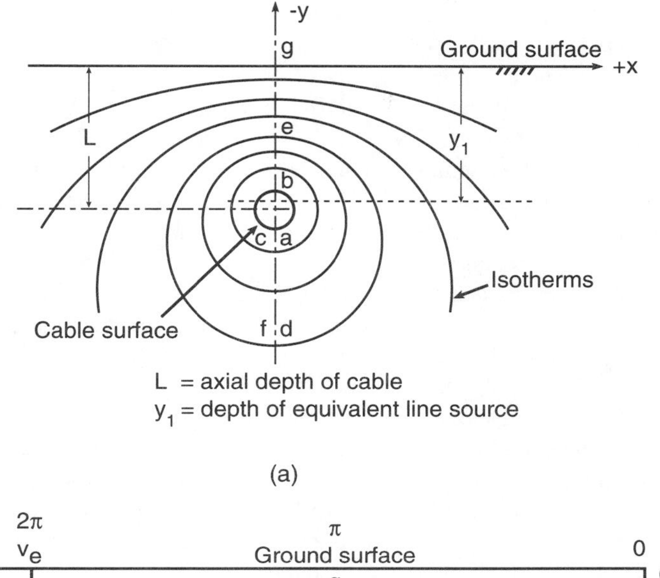

(a)

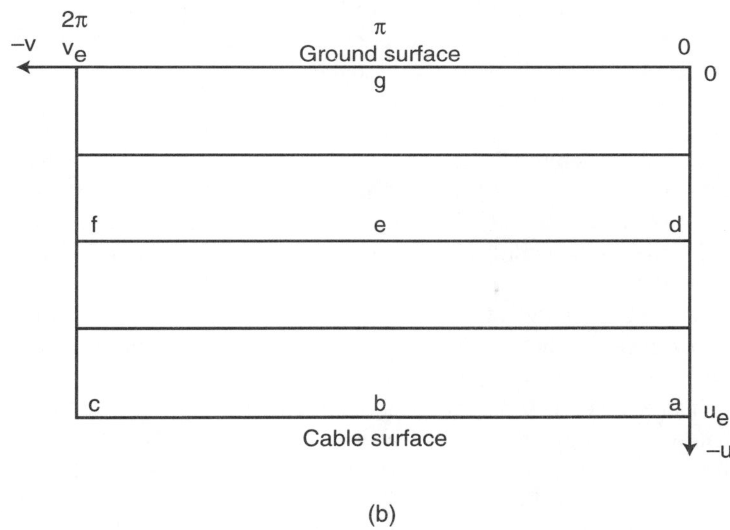

(b)

Figure 9.11 (a) Kennelly assumption for thermal field around a cable; (b) conformal map of field around a cable. (CIGRE, 1985)

The conformal transform used to redraw this simple case is given by the function

$$w = \Phi(z) = \ln\left[\frac{z + j|y_1|}{z - j|y_1|}\right] \tag{9.66}$$

where $z = x + jy =$ complex coordinates of a point being considered in the original plane of Fig. 9.11a

$w = u + jv =$ complex coordinates of the corresponding point in the transformed plane represented in Fig. 9.11b

$0 + jy_1 =$ complex coordinates of the line source corresponding to an isothermal surface of the reference cable in Fig. 9.11a.

The function Φ has been chosen so that the new diagram (Fig. 9.11b) has rectangular finite boundaries, the upper and lower ones representing the ground and cable surfaces, respectively. The two side boundaries both correspond to a particular flux line; in the case of Fig. 9.11a, it is the flux line *cr.ad* on the y axis passing through the center of the cable.

The differential equation (2.12) of heat conduction is invariant under a conformal transform, so that the temperatures at corresponding locations in the w and in the z planes are the same when the boundary conditions are the same. The advantage of the transformed plane is that the isotherms become straight lines parallel to the line representing the isothermal ground surface, and the flux lines can be represented by a second set of parallel lines perpendicular to the isotherms. The thermal resistance along any tube of flux lines is the same in both planes, but is more readily computed in the rectangular geometry of the w plane. The external thermal resistance is given directly by the height u of the transformed domain.

If we now consider an installation where several cables are present, similar transformations may be applied, using each cable in turn as the reference cable, so as to obtain the external thermal resistance R_{ii} of each cable separately. In so doing, the presence of other cables is ignored in accord with the restricted application of superposition. The determination of the temperatures at the location of the centers of the remaining cables in the w plane will yield the "mutual resistance" terms R_{ik} $(i \neq k)$.

If, finally, we include variations in soil resistivity, the boundaries between regions of different resistivity may be mapped, point by point, on the w plane using the same function Φ. The only difference from the case with uniform soil in Fig. 9.11b is that, when imposing a temperature difference between the lines representing the soil surface ($u = 0$) and the cable surface ($u = u_c$), the resulting flux lines are not necessarily parallel to the u axis due to refraction at the boundaries between regions of different resistivity.

Evaluation of the temperatures in the thermal field of the w plane is achieved by approximating the composite medium represented in the w plane with an analogous resistance network covering the same rectangular area. The equations of the network may be solved by any standard method of numerical analysis.

There are several drawbacks to the method outlined above. The major one is that the equations describing the transformed network are equivalent to finite-difference equations obtained by discretizing the heat equation in the w plane, and hence the complexity of a numerical solution of the heat conduction problem is not avoided. Another drawback is that both the earth and cable surfaces are assumed to be isothermal. In addition, transformation of the boundaries between regions with different resistivities point by point is very laborious, and the resulting computer software cannot efficiently handle more than four cables in one installation.

All of these limitations can easily be overcome by the application of the finite-element method. The nonuniform soil conditions and nonisothermal boundaries are handled naturally by this method. The computational efficiency of this approach is also quite satisfying. With presently available personal computers, calculations involving networks with several thousand nodes can be performed in a matter of minutes.[6] The finite-element method is described in Chapter 11.

[6] The author performed computations on a network of 7000 nodes in less than 5 min using a 66 MHz 486 PC.

9.6.7 The Neher–McGrath Modification of T_4 to Account for Cyclic Loading

As discussed in Section 5.6, the computation of a cyclic rating factor requires an evaluation of the cable capacitances and conductor temperatures at several time points. Neher and McGrath (1957) proposed an alternative, simple method to account for the cyclic loading of a cable. Their approach requires a modification of the external thermal resistance of the cable. The underlying principles are discussed below.

In order to evaluate the effect of a cyclic load upon the maximum temperature rise of a cable system, Neher (1953) observed that one can look upon a heating effect of a cyclical load as a wave front which progresses alternately outwardly and inwardly in respect to the conductor during the cycle. He further assumed that the heat flow during the loss cycle is represented by a steady component of magnitude μW_t plus a transient component $(1-\mu)W_t$ which operates for a period of time τ during each cycle. The transient component of the heat flow will penetrate the earth only to a limited distance from the cable, thus the corresponding thermal resistance T_{4et} will be smaller than its counterpart T_{4ss} which pertains during steady-state conditions.

Assuming that the temperature rise over the internal thermal cable resistance is complete by the end of the transient period τ, the maximum temperature rise at the conductor may be expressed as

$$\Delta\theta = W_t\,[T + \mu T_{4ss} + (1-\mu)T_{4et}] \tag{9.67}$$

where T is the internal thermal resistance of the cable, T_{4ss} is the external thermal resistance with constant load, and T_{4et} is the effective transient thermal resistance in the earth. Further, Neher (1953) assumed that the last thermal resistance may be represented with sufficient accuracy by an expression of the general form

$$T_{4et} = A\rho_s \ln \frac{B\sqrt{\delta\cdot\tau}}{D_e}$$

in which constants A and B were evaluated empirically to best fit the temperature rises calculated over a range of cable sizes. Using measured data, Neher (1953) obtained the following values for the constants $A = 1/2\pi$ and $B = 61\,200$ when τ is expressed in hours and δ is expressed in m^2/s.

Introducing the notation

$$D_x = 61\,200\sqrt{\delta(\text{length of cycle in hours})} \tag{9.68}$$

the external thermal resistance in equation (9.67) can be written as

$$\begin{aligned} T_4 = \mu T_{4ss} + (1-\mu)T_{4et} &= \frac{\rho_s}{2\pi}\left[\mu \ln\frac{4L}{D_e} + (1-\mu)\ln\frac{D_x}{D_e}\right] \\ &= \frac{\rho_s}{2\pi}\left[\ln\frac{D_x}{D_e} + \mu\ln\frac{4L}{D_x}\right] \end{aligned} \tag{9.69a}$$

The last equation can be generalized for a group of equally loaded identical cables taking into account equation (9.25)

$$T_4 = \frac{\rho_s}{2\pi}\left\langle \ln\frac{D_x}{D_e} + \mu \cdot \ln\left\{\left[u + \sqrt{u^2 - 1}\right] \cdot \left[\left(\frac{d'_{p1}}{d_{p1}}\right)\left(\frac{d'_{p2}}{d_{p2}}\right)\cdots\left(\frac{d'_{pk}}{d_{pk}}\right)\cdots\left(\frac{d'_{pq}}{d_{pq}}\right)\right]\right\}\right\rangle \tag{9.69b}$$

where $u = 4L/D_x$. When the cables are laid in a flat formation without transpositions, and when the sheaths are solidly bonded, equation (9.27) rather than (9.25) is substituted into (9.69a) (see Example 9.10).

From equation (9.68), we observe that the fictitious diameter D_x (mm) at which the effect of loss factor commences is a function of the diffusivity of the medium δ and the length of the loss cycle.

In the majority of cases, the soil diffusivity will not be known. In such cases, a value of $0.5 \cdot 10^{-6}$ m²/s can be used. This value is based on a soil thermal resistivity of 1.0 K · m/W and a moisture content of about 7% of dry weight (see Section 5.4 for more details on this subject). The value of D_x for a load cycle lasting 24 h and with a representative soil diffusivity of $0.5 \cdot 10^{-6}$ m²/s is 211 mm (or 8.3 in).

Alternative expressions for D_x are given by Heinhold (1990):

$$D_x = \frac{205}{\sqrt{w}\rho_s^{0.4}} \quad \text{for sinusoidal load variation} \tag{9.70}$$

$$D_x = \frac{493\sqrt{\mu}}{\sqrt{w}\rho_s^{0.4}} \quad \text{for rectilinear load variation} \tag{9.71}$$

$$D_x = \frac{103 + 246\sqrt{\mu}}{\sqrt{w}\rho_s^{0.4}} \quad \text{for other load variations} \tag{9.72}$$

where w is the number of load cycles in a 24 h period.

EXAMPLE 9.10

We will compute the rating of the cable circuit studied in Example 5.6 using the Neher–McGrath approach. The parameters of this circuit are $\lambda'_{11} = 0.206$, $\lambda'_{12} = 222$, $\lambda'_{1m} = 0.089$, $T_1 = 0.214$ K · m/W, and $T_3 = 0.104$ K · m/W. Cables are spaced two cable diameters apart.

The fictitious diameter D_x is first obtained from equation (9.68) and is equal to 211 mm. The external thermal resistance of the middle cable is obtained from equation (9.69b) with the loss factor computed in Example 5.5 equal to 0.504. In this case, equation (9.27) is used for the external thermal resistance with unity load factor, and with the sheath loss factors computed in Example 8.2:

$$T_4 = \frac{\rho_s}{2\pi}\left\langle \ln\frac{D_x}{D_e} + \mu\left\{\ln\frac{4L}{D_x} + \left[\frac{1 + 0.5\left(\lambda'_{11} + \lambda'_{12}\right)}{1 + \lambda'_{1m}}\right] \cdot \ln\left[1 + \left(\frac{2L}{s_1}\right)^2\right]\right\}\right\rangle$$

$$= \frac{1}{2\pi}\left\langle \ln\frac{211}{35.8} + 0.504 \cdot \left\{\ln\frac{4 \cdot 1000}{211} + \frac{1 + 0.5(0.206 + 0.222)}{1 + 0.088}\ln \cdot \left[1 + \left(\frac{2 \cdot 1000}{2 \cdot 35.8}\right)^2\right]\right\}\right\rangle$$

$$= 1.11 \text{ K} \cdot \text{m/W}$$

Assuming the sheath loss factor for the middle cable, the rating is equal to

$$I = \left[\frac{90 - 15}{0.781 \cdot 10^{-4}[0.214 + (1 + 0.09) \cdot (0.104 + 1.11)]}\right]^{0.5} = 790 \text{ A}$$

If we use Heinhold's equation (9.72), the fictitious diameter D_x is equal to

$$D_x = \frac{103 + 246\sqrt{0.504}}{\sqrt{1} \cdot 1^{0.4}} = 277.6 \text{ mm}$$

The external thermal resistance and the rating of the cable become

$$T_4 = \frac{1}{2\pi}\left\langle \ln\frac{277.6}{35.8} + 0.504 \cdot \left\{ \ln\frac{4 \cdot 1000}{277.6} + \frac{1 + 0.5(0.206 + 0.222)}{1 + 0.089} \ln\left[1 + \left(\frac{2 \cdot 1000}{2 \cdot 35.8}\right)^2\right]\right\}\right\rangle$$

$$= 1.14 \text{ K} \cdot \text{m/W}$$

$$I = \left[\frac{90 - 15}{0.781 \cdot 10^{-4}[0.214 + (1 + 0.09) \cdot (0.104 + 1.14)]}\right]^{0.5} = 782 \text{ A}$$

We can observe that these values are not very different form the rating obtained with the more precise method in Example 5.6, with the Heinhold approximation giving the more accurate result.

9.6.8 Cables in Air

9.6.8.1 General Equation for the External Thermal Resistance. In this section, we will consider cables either installed horizontally on noncontinuous brackets, ladder supports, or cleats or clipped to a vertical wall. Heat transfer phenomena are more complex for cables installed in free air than for those located underground, and proper handling of these situations requires the solution of a set of energy balance equations. The relevant energy balance equation for the surface of the cable was formulated in Section 2.3.2 [equation (2.20)] and, for a cable with n conductors, is given by

$$nI^2R(1 + \lambda_1 + \lambda_2) + W_d + \sigma D_e^* H - \pi D_e^* h_{\text{conv}}\left(\theta_e^* - \theta_{\text{amb}}^*\right) - \pi D_e^* \varepsilon_s \sigma_B \left(\theta_e^{*4} - \theta_{\text{amb}}^{*4}\right) = 0 \tag{9.73}$$

where
h_{conv} = convective heat transfer coefficient, $\text{W/m}^2 \cdot \text{K}^{-1}$
θ_e^* = cable surface temperature, K
σ = solar absorption coefficient (see Table 9.10)
H = intensity of solar radiation, W/m^2
σ_B = Stefan–Boltzmann constant, equal to $5.67 \cdot 10^{-8}$ $\text{W/m}^2 \cdot \text{K}^4$
ε_s = emissivity of the cable outer covering
D_e^* = cable external diameter, m
θ_{amb}^* = ambient temperature, K

There are two unknown variables in equation (9.73), the conductor current I and the cable surface temperature θ_e^*. The second equation linking these two variables is obtained from the following relationship between conductor and cable surface temperatures [see equation (4.5)]:

$$\theta_c^* - \theta_e^* = n\left(I^2 R \cdot T + \Delta\theta_d\right) \tag{9.74}$$

where T, the internal thermal resistance of the cable, is defined in equation (4.4) and the temperature rise caused by dielectric losses, $\Delta\theta_d$, is given by equation (4.7). The constant n represents the number of the conductors in the cable.

To obtain an expression for the external thermal resistance of a cable in air, we first rewrite equation (9.73) using the notation in equation (3.10a),

$$W_t = \pi D_e^* h_t \Delta\theta_s \tag{9.75}$$

where $\Delta\theta_s = \theta_e^* - \theta_{\text{amb}}^*$ is the cable surface temperature rise above ambient, h_t is the total heat transfer coefficient defined in equation (3.10), and W_t includes the heat gained by solar radiation. From equation (9.75), the external thermal resistance of the cable is given by

$$\boxed{T_4 = \frac{\Delta\theta_s}{W_t} = \frac{1}{\pi D_e^* h_t}} \tag{9.76}$$

Before the external thermal resistance is computed from equation (9.76), the value of the heat transfer coefficient h_t has to be determined first. This coefficient is a nonlinear function of the cable surface temperature (see Table 10.2 in Section 10.3.4.2). However, for standardization purposes, a simpler method is required for the determination of T_4. Such a method is discussed next.

9.6.8.2 IEC Standard 287—Simple Configurations. From a large number of tests on various cables in various configurations, carried out in the United Kingdom in the 1930s, Whitehead and Hutchings (1939) deduced that the total thermal dissipation from the surface of cable in air may conveniently be expressed as

$$W_t = \pi D_e^* h (\Delta\theta_s)^{5/4} \tag{9.77}$$

where h $\left(\text{W/m}^2 \cdot \text{K}^{5/4}\right)$ is the heat transfer coefficient embodying convection, radiation, conduction, and mutual heating. From equation (9.77), the following expression for the external thermal resistance of cables in free air is obtained:

$$\boxed{T_4 = \frac{1}{\pi D_e^* h (\Delta\theta_s)^{1/4}}} \tag{9.78}$$

From equations (9.76) and (9.78), we can observe that for a single cable,

$$h_t = h \Delta\theta_s^{1/4} \tag{9.79}$$

The values of the heat transfer coefficient h were obtained experimentally and plotted as a function of the cable diameter for various cable arrangements (Whitehead and Hutchings, 1939; IEC 287-2-1, 1994) The curves of Whitehead and Hutchings were later fitted with the following analytical expression (IEC 287, 1982):

$$\boxed{h = \frac{Z}{(D_e^*)^g} + E} \tag{9.80}$$

where constants Z, E, and g are given in Table 9.8.

TABLE 9.8 Values for Constants Z, E, and g for Black Surfaces of Cables in Free Air

	Installation	Z	E	g	Mode
1	Single cable*	0.21	3.94	0.60	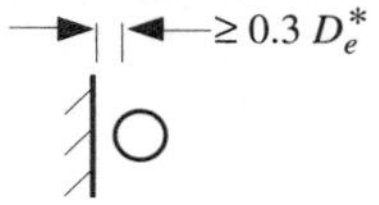 $\geq 0.3\ D_e^*$
2	Two cables touching, horizontal	0.21	2.35	0.50	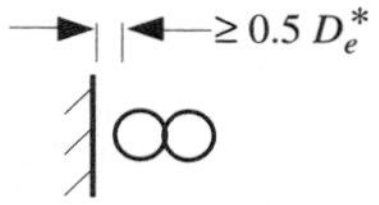 $\geq 0.5\ D_e^*$
3	Three cables in trefoil	0.96	1.25	0.20	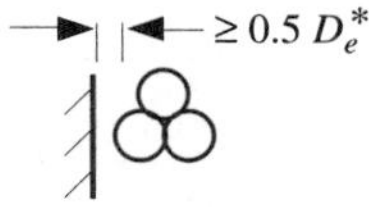 $\geq 0.5\ D_e^*$
4	Three cables touching, horizontal	0.62	1.95	0.25	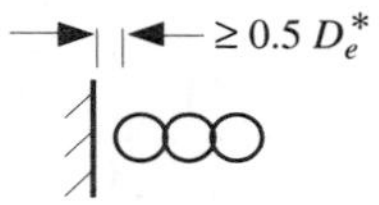 $\geq 0.5\ D_e^*$
5	Two cables touching, vertical	1.42	0.86	0.25	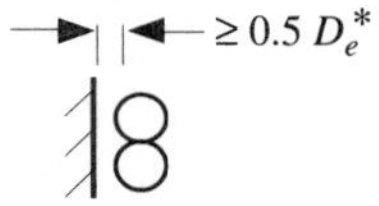 $\geq 0.5\ D_e^*$
6	Two cables spaced D_e^*, vertical	0.75	2.80	0.30	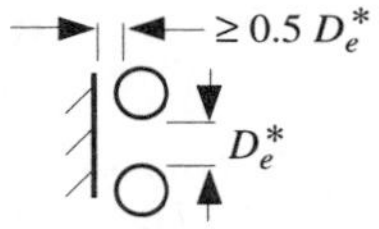 $\geq 0.5\ D_e^*$; D_e^*
7	Three cables touching, vertical	1.61	0.43	0.20	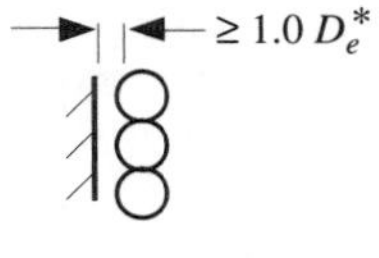 $\geq 1.0\ D_e^*$
8	Three cables spaced D_e^*, vertical	1.31	2.00	0.20	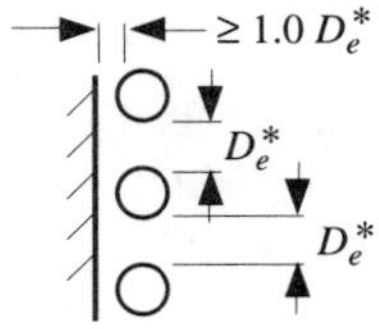 $\geq 1.0\ D_e^*$; D_e^*; D_e^*
9	Single cable	1.69	0.63	0.25	
10	Three cables in trefoil	0.94	0.79	0.20	

*Values for a "single cable" also apply to each cable of a group when they are spaced horizontally with a clearance between cables of at least 0.75 times the cable overall diameter.

Served cables and cables having a nonmetallic surface are considered to have a black surface. Unserved cables, either plain lead or armored, should be given a value of h equal to 88% of the value for the black surface.

Morgan (1994) has shown that the exponent 1/4 and heat transfer coefficient h in equation (9.78) vary with the temperature rise of the surface of the cable. By performing numerous calculations, he determined the values of constants Z, g, and E listed in Table 9.9, as well as of the exponent $q-1$ to which the cable surface temperature rise is raised [1/4 in (9.78)]. The values in Table 9.9 are displayed for various temperature ranges and as a function of the number n of bundled cables.

TABLE 9.9 Values for Constants Z, E, and g for Black Surfaces of Cables in Free Air as a Function of the Cable Surface Temperature Rise (Morgan, 1994)

	$n=1$			$n=2$			$n=3$			$n=4$		
	W	X	Y	W	X	Y	W	X	Y	W	X	Y
Z	0.880	0.825	0.835	0.522	0.325	0.142	0.860	0.605	0.460	0.823	0.870	0.395
g	0.350	0.300	0.263	0.447	0.479	0.521	0.335	0.349	0.316	0.337	0.322	0.345
E	4.50	3.30	2.00	4.00	3.00	1.85	4.00	3.00	1.80	4.00	2.80	1.90
$q-1$	0.165	0.230	0.320	0.120	0.210	0.340	0.135	0.220	0.330	0.135	0.226	0.330

W:10 K $\leq \Delta\theta_s \leq$ 33 K; X:33 K $< \Delta\theta_s \leq$ 66 K; Y:66 K $< \Delta\theta_s \leq$ 100 K; $\theta_{amb} = 25°C$, $\varepsilon = 0.9$

As can be seen from Table 9.9, the IEC value of $q-1 = 0.25$ (for $n=1$) is approximately correct only for the case $X(\Delta\theta_s = 33 \div 66\text{ K})$, which is the usual operating range of cables.

The external thermal resistance given by equation (9.78) depends on the unknown cable surface temperature rise $\Delta\theta_s$. This temperature rise can be obtained iteratively by considering the energy balance equation at the surface of the cable. The energy conservation principle states that the heat dissipated by conduction inside the cable is equal to the heat dissipated outside the cable by convection, radiation, and conduction. Thus, the energy balance equation for the cable surface can be written as

$$\theta_c - \theta_{\text{amb}} - \Delta\theta_s + \Delta\theta'_d = \frac{\pi D_e^* h(\Delta\theta_s)^{5/4} T}{1+\lambda_1+\lambda_2} \tag{9.81}$$

where T is the internal thermal resistance of the cable given by equation (4.4) and $\Delta\theta'_d$ is the temperature rise caused by dielectric losses. The latter is obtained using equations (4.4) and (4.7):

$$\Delta\theta'_d = W_d\left[\left(\frac{1}{1+\lambda_1+\lambda_2} - \frac{1}{2}\right)T_1 - \frac{n\lambda_2 T_2}{1+\lambda_1+\lambda_2}\right] \tag{9.82}$$

If the dielectric losses are neglected, $\Delta\theta_d = 0$.

Denoting

$$K_A = \frac{\pi D_e^* h T}{1+\lambda_1+\lambda_2}, \tag{9.83}$$

equation (9.81) can be written as

$$(\Delta\theta_s)^{1/4} = \left[\frac{\Delta\theta + \Delta\theta_d'}{1 + K_A(\Delta\theta_s)^{1/4}}\right]^{1/4} \tag{9.84}$$

where $\Delta\theta = \theta_c - \theta_{amb}$ is the permissible conductor temperature rise above ambient. Equation (9.84) suggests the following iterative procedure for determination of $(\Delta\theta_s)^{1/4}$. Set the initial value of $(\Delta\theta_s)^{1/4} = 2$ and compute consecutive values from

$$(\Delta\theta_s)_{n+1}^{1/4} = \left[\frac{\Delta\theta + \Delta\theta_d'}{1 + K_A(\Delta\theta_s)_n^{1/4}}\right]^{1/4} \tag{9.85}$$

Iterate until $(\Delta\theta_s)_{n+1}^{1/4} - (\Delta\theta_s)_n^{1/4} \leq 0.001$.

If the cables are directly exposed to solar radiation, the effect of solar heating must be taken into account by modifying slightly equation (9.85):

$$(\Delta\theta_s)_{n+1}^{1/4} = \left[\frac{\Delta\theta + \Delta\theta_d' + \sigma H K_A/\pi h}{1 + K_A\,(\Delta\theta_s)_n^{1/4}}\right]^{1/4} \tag{9.86}$$

In (9.86), σ is the absorption coefficient for the cable surface (see Table 9.10) and H is the intensity of solar radiation. IEC Standard 287 recommends values for H for various countries. When the actual value of H is unknown, this value should be taken as 1000 W/m^2 for most latitudes.

TABLE 9.10 Absorption Coefficient of Solar Radiation and Emissivity Values for Cable Surfaces

Material	σ	ε_s
Bitumen/jute serving	0.8	0.95
Polychloroprene	0.8	0.9
PVC	0.6	0.9
PE	0.4	0.9
Lead	0.6	0.63

EXAMPLE 9.11

Consider cable model No. 2 installed in free air with air temperature of 25°C. The parameters of this cable taken from Appendix A are: D_e = 72.9 mm, λ_1 = 0.0218, T_1 = 0.307 K · m/W, and T_3 = 0.078 K · m/W.

We will determine the external thermal resistance and cable rating for the following conditions: (1) single cable not clipped to a vertical wall and shaded from solar radiation, (2) the same case with the cable being exposed to solar radiation with intensity of 1000 W/m^2, and (3) the cable is clipped to a vertical wall and shaded from solar radiation.

(1) Cable shaded from solar radiation

The heat transfer coefficient is obtained from equation (9.80) with the constants taken from Table 9.8:

$$h = \frac{Z}{(D_e^*)^g} + E = \frac{0.21}{0.0729^{0.6}} + 3.94 = 4.95 \text{ W/m}^2 \cdot \text{K}^{3/4}$$

Constant K_A is obtained from equation (9.83) with the required cable parameters listed above

$$K_A = \frac{\pi D_e^* hT}{1+\lambda_1+\lambda_2} = \frac{\pi \cdot 0.0729 \cdot 4.95}{1+0.0218}\left[\frac{0.307}{3} + (1+0.0218)\cdot 0.078\right] = 0.202$$

The cable surface temperature rise is obtained from equation (9.85) with $\Delta\theta_d' = 0$:

$$(\Delta\theta_s)_{n+1}^{1/4} = \left[\frac{\Delta\theta + \Delta\theta_d'}{1 + K_A\,(\Delta\theta_s)_n^{1/4}}\right]^{1/4}$$

$$(\Delta\theta_s)_1^{1/4} = \left[\frac{90-25}{1+0.202\cdot 2}\right]^{1/4} = 2.6085$$

$$(\Delta\theta_s)_2^{1/4} = \left[\frac{90-25}{1+0.202\cdot 2.6085}\right]^{1/4} = 2.5543$$

$$(\Delta\theta_s)_3^{1/4} = \left[\frac{90-25}{1+0.202\cdot 2.5543}\right]^{1/4} = 2.5589$$

$$(\Delta\theta_s)_4^{1/4} = \left[\frac{90-25}{1+0.202\cdot 2.5589}\right]^{1/4} = 2.5585$$

The iterations are stopped since $(\Delta\theta_s)_4^{1/4} - (\Delta\theta_s)_3^{1/4} \le 0.001$. The external thermal resistance is now obtained from equation (9.78):

$$T_4 = \frac{1}{\pi D_e^* h\,(\Delta\theta_s)^{1/4}} = \frac{1}{\pi \cdot 0.0729 \cdot 4.95 \cdot 2.558} = 0.345\ \text{K}\cdot\text{m/W}$$

The rating of this cable is obtained from equation (4.3):

$$I = \left[\frac{90-25}{0.798\cdot 10^{-4}\,(0.307 + 3(1+0.0218)\cdot(0.078+0.345))}\right]^{0.5} = 713\ \text{A}$$

(2) Cable exposed to solar radiation

The first iteration of equation (9.86) is

$$(\Delta\theta_s)_1^{1/4} = \left[\frac{\Delta\theta + \Delta\theta_d + \sigma H K_A/\pi h}{1 + K_A\,(\Delta\theta_s)_0^{1/4}}\right]^{1/4} = \left[\frac{90 - 25 + 0.60\cdot 1000\cdot 0.202/(\pi\cdot 4.95)}{1+0.202\cdot 2}\right]^{1/4}$$
$$= 2.6834$$

Repeating the application of equation (9.86), convergence is achieved in the third iteration with $(\Delta\theta_s)_3^{1/4} = 2.626$. The external thermal resistance is obtained from equation (9.78):

$$T_4 = \frac{1}{\pi\cdot 0.0729\cdot 4.95\cdot 2.626} = 0.336\ \text{K}\cdot\text{m/W}$$

The rating of this cable is obtained from equation (4.15):

$$I = \left[\frac{90-25-0.6\cdot 0.0729\cdot 1000\cdot 0.336}{0.798\cdot 10^{-4}[0.307 + 3(1+0.0218)\cdot(0.078+0.336)]}\right]^{0.5} = 632\ \text{A}$$

In this case, solar heating of the cable surface decreases the cable rating by about 11%.

(3) Cable clipped to a vertical wall and shaded from solar radiation

The heat transfer coefficient is equal to

$$h = \frac{1.69}{0.0729^{0.25}} + 0.63 = 3.88 \text{ W/m}^2 \cdot \text{K}^{5/4}$$

Constant K_A is obtained from equation (9.83) and is equal to 0.158. The cable surface temperature rise is obtained after three iterations of equation (9.69) and is equal to $(\Delta\theta_s)_3^{1/4} = 2.605$. The external thermal resistance is obtained from equation (9.78):

$$T_4 = \frac{1}{\pi \cdot 0.0729 \cdot 3.88 \cdot 2.605} = 0.432 \text{ K} \cdot \text{m/W}$$

The rating of this cable is obtained from equation (4.3):

$$I = \left[\frac{90 - 25}{0.798 \cdot 10^{-4} (0.307 + (1 + 0.0218) \cdot (0.078 + 0.432))}\right]^{0.5} = 660 \text{ A}$$

The ampacity of the cable is smaller than in case 1) because heat dissipation is now more difficult with the cable clipped to a vertical wall.

9.6.8.3 IEC Standard 287—Derating Factors for Groups of Cables. Cable installations discussed in Section 9.6.8.2 were limited to a maximum of three cables touching, either in trefoil or in flat formation. In this section, we will consider installations of groups of cables in air belonging to one of the configurations shown in Fig. 9.12.

The discussion in this section is limited to:

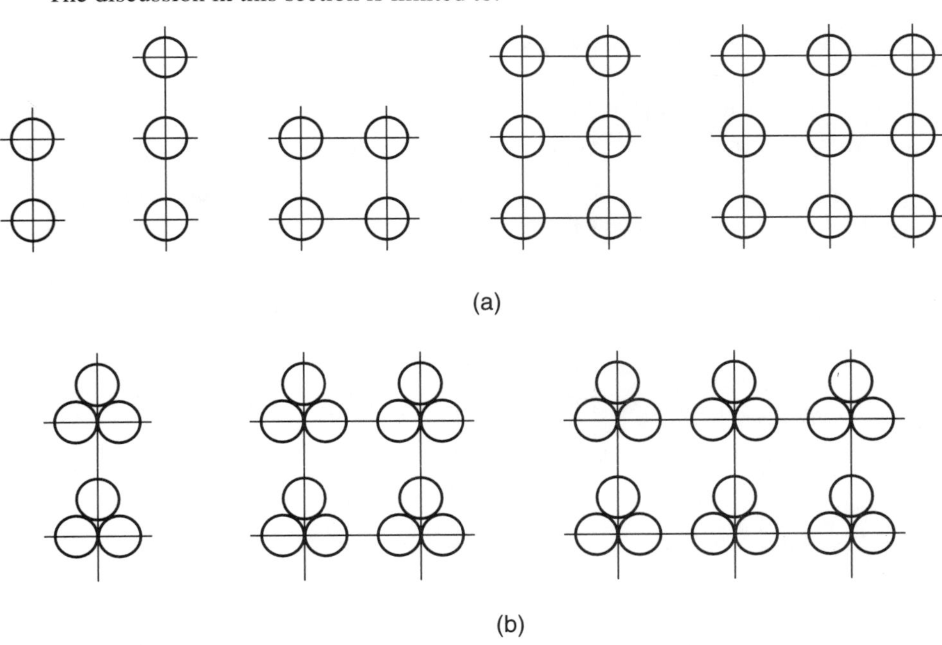

Figure 9.12 Typical groups of (a) multicore cables, and (b) trefoil circuits.

1. a maximum of nine cables in a square formation (the last arrangement in Fig. 9.12a),
2. a maximum of six circuits, each comprising three cables mounted in trefoil, with up to three circuits placed side by side or two circuits placed one above the other (the last arrangement in Fig. 12b),
3. cables for which dielectric losses can be neglected (usually, only lower voltage polymeric cables are installed in groups).

When cables are installed in groups as shown in Fig. 9.12, the rating of the hottest cable will be lower than in the case when the same cable is installed in isolation. This reduction is caused by mutual heating. A simple method to account for this mutual heating effect is to calculate the rating of a single cable or circuit using the method described in the previous section and apply a reduction factor. This is defined as follows.

$$I_g = F_g \cdot I_1 \tag{9.87}$$

where I_g = rating of the hottest cable in the group, A

I_1 = rating of the same cable or circuit isolated, A

F_g = group reduction factor

The reduction in current is a result of an increase in the external thermal resistance of a cable or circuit in a group as compared to the case when the cable or circuit is installed in isolation. A basic criterion is that the temperature rise above ambient of the conductor be the same for the grouped and isolated cases. Since we neglect dielectric losses, equation (4.5) yields

$$I_1^2 (RT + R_t T_{4I}) = F_g^2 I_1^2 \left(RT + R_t T_{4g}\right) \tag{9.88}$$

where $R_t = R(1 + \lambda_1 + \lambda_2)$

T_{4g} = external thermal resistance of the hottest cable in the group, K · m/W

T_{4I} = external thermal resistance of one cable, assumed to be isolated, used to compute the rated current I, K · m/W

Defining k_1 as the cable surface temperature rise factor of one multicore cable or one single-core cable mounted in trefoil and assumed isolated in free air, we have

$$k_1 = \frac{\text{cable surface temperature rise}}{\text{conductor temperature rise}} = \frac{I_1^2 R_t T_{4I}}{I_1^2 (RT + R_t T_{4I})}$$

Substituting this into (9.88), we obtain

$$F_g = \sqrt{\frac{1}{k_1 \left(\dfrac{RT}{R_t T_{4I}} + \dfrac{T_{4g}}{T_{4I}}\right)}}. \tag{9.89}$$

Also,

$$\frac{RT}{R_t T_{4I}} = \frac{1 - k_1}{k_1}.$$

Substituting this value into equation (9.89), yields

$$F_g = \sqrt{\frac{1}{1 - k_1 + k_1 (T_{4g}/T_{4I})}} \tag{9.90}$$

Since the cable surface temperature rise equals $W_t \cdot T_{4I}$, we have

$$k_1 = \frac{W_t \cdot T_{4I}}{\theta_c - \theta_{\text{amb}}} \tag{9.91}$$

where W_t (W/m) is the power loss from one multicore cable or one single-core cable mounted in trefoil, assumed to be isolated, when carrying the current I_1.

To compute the term T_{4g}/T_{4I}, we use equations (9.78) and (9.87):

$$\frac{T_{4g}}{T_{4I}} = \frac{h_I}{h_g}\left(\frac{\Delta\theta_{sI}}{\Delta\theta_{sg}}\right)^{0.25} = \frac{h_I}{h_g}\left(\frac{I_1^2 R_t T_{4I}}{F_g^2 I_1^2 R_t T_{4g}}\right)^{0.25} = \frac{h_I}{h_g}\left[\frac{1-k_1}{T_{4g}/T_{4I}} + k_1\right]^{0.25}$$

Thus, the term T_{4g}/T_{4I} can be calculated from the ratio (h_I/h_g) by using the iterative relationship

$$\left(\frac{T_{4g}}{T_{4I}}\right)_{n+1} = \frac{h_I}{h_g}\left[\frac{1-k_1}{(T_{4g}/T_{4I})_n} + k_1\right]^{0.25} \tag{9.92}$$

and starting with $(T_{4g}/T_{4I})_1 = (h_I/h_g)$.

Equation (9.92) converges quickly; one iteration with $(T_{4g}/T_{4I})_1 = (h_I/h_g)$ is usually sufficient. Alternatively, when (h_I/h_g) is less than 1.4, it is sufficient to substitute (h_I/h_g) for (T_{4g}/T_{4I}) in equation (9.90).

Values for the ratio (h_I/h_g) have been obtained empirically and are given in Table 9.11 (IEC 1042, 1991).

If a clearance exceeding the appropriate value given in column 2 of Table 9.11 cannot be maintained with confidence throughout the length of the cable, the reduction coefficient is determined as follows.

1. For horizontal clearances, we assume that the cables are touching each other or the vertical surface. Appropriate values are given in column 4 of Table 9.11.
2. For vertical clearances, the reduction coefficient due to grouping is derived according to the value of the expected clearance:

   a. Where the clearance is less than the appropriate value given in column 2 of Table 9.11, but can be maintained at a value equal or exceeding the minimum given in column 3, the appropriate value of (h_I/h_g) is obtained from the formula in column 4.
   b. Where the clearance is less than the minimum given in column 3, we assume that the cables are touching each other. Suitable values of (h_I/h_g) are provided in column 4 of Table 9.11.

The values in column 4 are the average values for cables having diameters from 13 to 76 mm. More precise values for multicore cables may be evaluated for a specific cable diameter, both inside and outside this range, by consulting Table 9.8.

TABLE 9.11 Data for Calculating Reduction Factors for Grouped Cables

Arrangement of Cable 1	Thermal Proximity Effect Is Negligible if e/D_e Is Greater Than or Equal To: 2	Thermal Proximity Effect Is Not Negligible If e/D_e Is Less Than 3	Average Value of h_1/h_g * 4
Side by side			
2 multicore	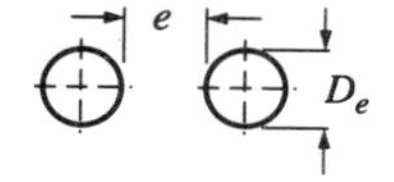 0.5	0.5	1.41
3 multicore	 0.75	0.75	1.65
2 trefoils	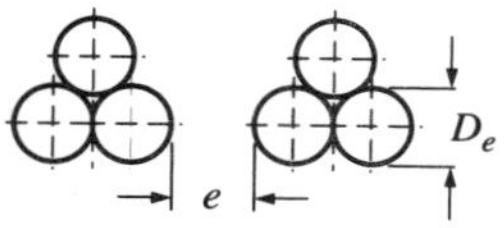 1.0	1.0	1.2
3 trefoils	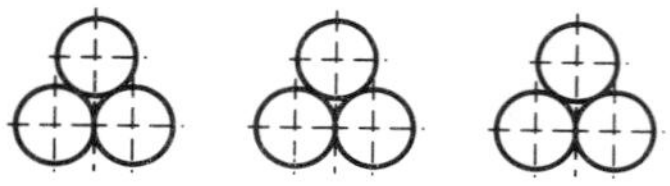 1.5	1.5	1.25
One above the other			
2 multicore	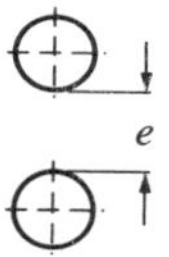 2 or 0.5	2	$1.085\,(e/D_e)^{-0.128}$ or 1.35
3 multicore	 4 or 0.5	4	$1.19\,(e/D_e)^{-0.135}$ or 1.57
2 trefoils	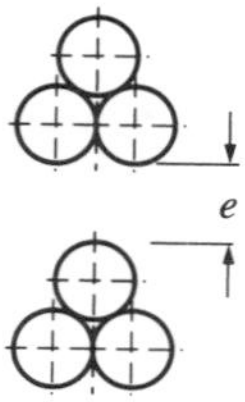 4 or 0.5	4	$1.106(e/D_e)^{-0.078}$ or 1.39
Near to a vertical surface or to a horizontal surface below the cable	 0.5	0.5	1.23

* The formulas for (h_1/h_g) given in column 4 of this table shall not be used for values of (e/D_e) less than 0.5 or greater than the appropriate values given in column 2.

The reduction factor given by equation (9.90) is used when the rating of a single cable or a single circuit is known. When the ampacity of the hottest cable in a group has to be determined from the beginning, an alternative way to do so is to use equation (9.78) to obtain the external thermal resistance with the heat transfer coefficient h_g substituted for the coefficient h. For the group configurations covered in Table 9.11, values of the heat transfer coefficient h_g are derived from

$$h_g = \frac{h}{(h_I/h_g)} \tag{9.93}$$

where the parameter h is given in equation (9.80) for one multicore cable or for a single cable mounted in a trefoil group, assumed to be isolated, and the ratio (h_I/h_g) is obtained from Table 9.11.

When the cables are installed in more than one plane, factors for current-carrying capacities for the hottest cable in a group are evaluated by using the appropriate value of (h_I/h_g) for the vertical clearance and ensuring that the horizontal clearance between cables, e, is not less than the value given in Table 9.11 for neglecting the side-by-side thermal proximity effect.

EXAMPLE 9.12

We will again consider cable model No. 2 with the installation examined in part 1 of Example 9.11, but this time the cable arrangement is as shown in Fig. 9.13.

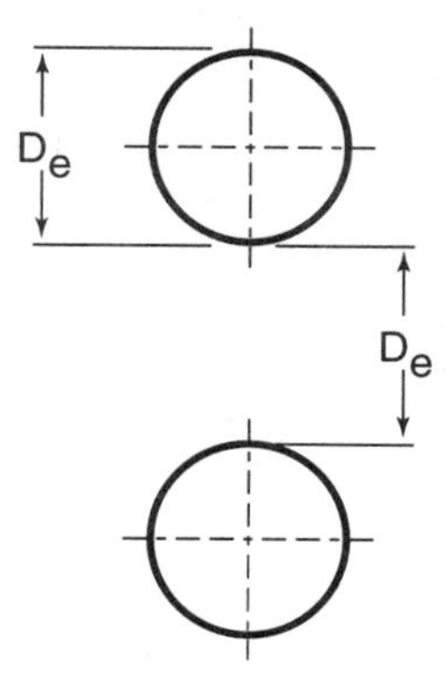

Figure 9.13 Installation of cables in air for Example 9.12.

The rating of a single cable computed in part 1 of Example 9.11 was 713 A. The ratio (h_I/h_g) is calculated from the following equation, obtained from column 4 of Table 9.11:

$$(h_I/h_g) = 1.085\left(\frac{e}{D_e}\right)^{-0.128} = 1.085\left(\frac{D_e}{D_e}\right)^{-0.128} = 1.085$$

The cable surface attainment factor is obtained from equation (9.91) as

$$k_1 = \frac{W_t \cdot T_{4I}}{\theta_c - \theta_{\text{amb}}} = \frac{3I^2R(1+\lambda_1)\cdot T_{4I}}{\theta_c - \theta_{\text{amb}}} = \frac{3\cdot 713^2 \cdot 0.798 \cdot 10^{-4}(1+0.0218)\cdot 0.345}{90-25} = 0.66$$

To compute the reduction factor F_g, we first apply equation (9.92). Since the ratio (h_I/h_g) is smaller than 1.4, we substitute (h_l/h_g) for (T_{4g}/T_{4l}). Thus, from equation (9.90), we have

$$F_g = \sqrt{\frac{1}{1 - k_1 + k_1\left(T_{4g}/T_{4l}\right)}} = \sqrt{\frac{1}{1 - 0.66 + 0.66 \cdot 1.085}} = 0.973$$

The ampacity of the hottest cable is therefore equal to

$$I_g = 0.973 \cdot 713 = 694 \text{ A}$$

9.6.8.4 The Effect of Wind Velocity and Mixed Convection. Equation (9.78) was developed under the assumption that the cable is subjected to natural cooling only and that h is constant for a fixed value of D_e^*. Cables installed outdoors may be subjected to wind, resulting in forced convective cooling. Morgan has presented forced convective heat transfer data for a single-core cable (Morgan, 1982) and bundles of two, three, and four single-core cables (Morgan, 1993). With low wind speed, natural convective cooling and forced convective cooling may occur at the same time. We will briefly review the issues involved in computation of the heat transfer coefficient when both modes of cooling are present.

Neglecting conductive heat transfer, which is very small in free air, the heat transfer coefficient is given by equation (3.10) and is equal to

$$h_t = h_{\text{conv}} + h_{\text{rad}}. \tag{9.94}$$

The radiative heat transfer coefficient h_{rad} can be computed from equation (3.9), and can then be approximated by (Black and Rehberg, 1985) with not-too-high accuracy:

$$h_{\text{rad}} = \frac{\sigma\varepsilon\left[(\theta_e + 273)^4 - (\theta_{\text{amb}} + 273)^4\right]}{\Delta\theta_s} \approx \sigma_B\varepsilon\left(1.38 \cdot 10^8 + 1.39 \cdot 10^6\theta_{\text{amb}}\right) \tag{9.95}$$

In the following, we concentrate on the computation of the convective heat transfer coefficient h_{conv}. We denote the natural convective heat transfer coefficient as $h_{o,n}$. Its value is obtained from standard correlations (Morgan, 1982; Incropera and de Witt, 1990):

$$h_{o,n} = \frac{k_{\text{air}} \cdot c(\text{Gr} \cdot \text{Pr})^m}{D_e^*} \tag{9.96}$$

where Pr = Prandtl Number (a property of air); its computation is described in Appendix D

$\text{Gr} = \dfrac{D_e^{*3} g\beta\Delta\theta_s}{\nu^2}$ is the Grashoff Number; all of the symbols were defined in Section 9.6.4.1.

The values of coefficients c and m are given in Table 10.4 ($m = n$) for various ranges of the Rayleigh Number (the product of Prandtl and Grashoff Numbers). In the temperature ranges $-10°\text{C} \le \theta_{\text{amb}} \le 40°\text{C}$ and $10°\text{C} \le \theta_e \le 100°\text{C}$, and for $10^4 < (\text{Gr} \cdot \text{Pr}) \le 10^7$, the following approximation may be made (Morgan, 1993):

$$0.48 k_{\text{air}}\left(\frac{\beta g \text{Pr}}{\nu^2}\right)^{0.25} \approx 1.23(\pm 4\%) \tag{9.97}$$

In the case of forced convection, the convective heat transfer coefficient is correlated with the Reynolds Number, Re, which is defined by

$$\mathrm{Re} = \frac{U D_e^*}{\nu} \tag{9.98}$$

where U (m/s) is the wind velocity.

We denote the forced convective heat transfer coefficient as $h_{o,f}$. Its value is obtained from the following correlations (Morgan, 1982; Incropera and de Witt, 1990):

$$h_{o,f} = \frac{k_{\mathrm{air}} b (\mathrm{Re})^p}{D_e^*} \tag{9.99}$$

Values of the coefficients b and p for various ranges of the Reynolds Number are given in Table 10.5.

As mentioned before, both natural and forced convection are present at low wind velocities. For this mixed convection, an equivalent Reynolds Number $\mathrm{Re}_{\mathrm{eq}}$, is first calculated (Morgan, 1992):

$$\mathrm{Re}_{\mathrm{eq}} = \left[\frac{c(\mathrm{Gr} \cdot \mathrm{Pr})^m}{b}\right]^{1/p} \tag{9.100}$$

Next, assuming that forced and natural flows are perpendicular to each other, we compute the effective Reynolds Number from

$$\mathrm{Re}_{\mathrm{eff}} = \left(\mathrm{Re}^2 + \mathrm{Re}_{\mathrm{eq}}^2\right)^{0.5} \tag{9.101}$$

The effective Reynolds Number is then inserted in place of Re in equation (9.99) to obtain the mixed convection heat transfer coefficient $h_{o,m}$:

$$h_{o,m} = \frac{k_{\mathrm{air}} b (\mathrm{Re}_{\mathrm{eff}})^p}{D_e^*} \tag{9.102}$$

Morgan (1992) has shown that for cables with the outside surface temperature rise above ambient not exceeding 20 K, as is often the case in outdoor cable installations, the natural convective cooling can be ignored when the wind velocity exceeds 1 m/s. By substituting for h_{conv} either $h_{o,n}$ or $h_{o,f}$ or $h_{o,m}$, as the case may be, we can solve equations (9.73) and (9.74) to determine the cable rating in free air. For groups of cables, we can apply derating factors as described in Section 9.6.8.2. An alternative, simpler approach to account for the effect of wind velocity was proposed by Neher and McGrath (1957) and is discussed in the next section.

EXAMPLE 9.13

Consider the cable installation described in part 1 of Example 9.11, and assume that the wind velocity is 3 m/s. We will determine the external thermal resistance and cable rating. We will consider two approaches: (1) by solving the energy conservation equations, and (2) by applying the approximations described in this section.

The loss factors and convection coefficient are temperature dependent. Therefore, an iterative procedure is required for the determination of the cable rating. However, for the purpose of illustrating the procedure, we will take the values of R and λ_1 from Table A1, that is, $R = 0.798 \cdot 10^{-4}\ \Omega/\text{m}$ and $\lambda_1 = 0.0218$. Other required parameters of this cable are: $D_e^* = 0.0729$ m, $\varepsilon = 0.9$, $T_1 = 0.307$ K · m/W, and $T_3 = 0.078$ K · m/W.

The Reynolds Number is obtained from equation (9.98) with the air viscosity computed from equation (D1) in Appendix D. We will assume an average cable surface temperature of 55°C. The air properties are evaluated at the average film temperature = (55 + 25)/2 = 40°C:

$$v = 1.7 \cdot 10^{-5}\ \text{m}^2/\text{s}, \quad k_{\text{air}} = 0.0272\ \text{W/K} \cdot \text{m}$$

$$\text{Re} = \frac{U D_e^*}{v} = \frac{3 \cdot 0.0729}{1.7 \cdot 10^{-5}} = 1.29 \cdot 10^4$$

The convective heat transfer coefficient is obtained from equation (9.99). From Table 10.5, $b = 0.193$ and $p = 0.618$. Thus,

$$h_{o,f} = \frac{k_{\text{air}} b (\text{Re})^p}{D_e^*} = \frac{0.0272 \cdot 0.193 \left(1.29 \cdot 10^4\right)^{0.618}}{0.0729} = 25.0\ \text{W/m}^2 \cdot \text{K}$$

(1) Exact solution

Since the cables are shaded from solar radiation, there is no solar heat gain term in equation (9.73). Also, dielectric losses for this cable are neglected. Substituting the numerical values into equation (9.73) and (9.74), we obtain

$$\begin{aligned} n I^2 R (1 + \lambda_1 + \lambda_2) + W_d + \sigma D_e^* H - \pi D_e^* h_{\text{conv}} \left(\theta_e^* - \theta_{\text{amb}}^*\right) - \pi D_e^* \varepsilon \sigma_B \left(\theta_e^{*4} - \theta_{\text{amb}}^{*4}\right) \\ = 3 \cdot I^2 \cdot 0.798 \cdot 10^{-4} \cdot 1.0218 - \pi \cdot 0.0729 \cdot 25 \left(\theta_e^* - 298\right) \\ - \pi \cdot 0.0729 \cdot 0.9 \cdot 5.667 \cdot 10^{-8} \left[\left(\theta_e^*\right)^4 - (298)^4\right] = 0 \end{aligned}$$

$$\theta_c^* - \theta_e^* = n \left(I^2 R \cdot T + \Delta\theta_d\right)$$

$$363 - \theta_e^* = 3 \cdot \left[I^2 \cdot 0.798 \cdot 10^{-4} \left(\frac{0.0307}{3} + 1.0218 \cdot 0.078\right)\right]$$

Solving these equations, we obtain $I = 910$ A and $\theta_e^* = 327$ K or 54°C. Since the outside cable temperature is very close to the assumed value, no further iterations are required, and the above values represent the solution.

We can observe that the presence of a moderate wind increases the cable rating by almost 28%.

(2) Approximate solution

With a jacket emissivity of 0.90 (Table 9.10), the radiative heat transfer coefficient is obtained from equation (9.95):

$$\begin{aligned} h_{\text{rad}} &= \sigma_B \varepsilon_s \left(1.38 \cdot 10^8 + 1.39 \cdot 10^6 \theta_{\text{amb}}\right) = 5.667 \cdot 10^{-8} \cdot 0.9 \cdot \left(1.38 \cdot 10^8 + 1.39 \cdot 10^6 \cdot 25\right) \\ &= 8.8\ \text{W/m}^2 \cdot \text{K} \end{aligned}$$

The total heat transfer coefficient is the sum of the convective and radiative coefficients:

$$h_t = 25.0 + 8.8 = 33.8\ \text{W/m}^2 \cdot \text{K}$$

The external thermal resistance is now obtained from equation (9.76) as

$$T_4 = \frac{1}{\pi D_e^* h_t} = \frac{1}{\pi \cdot 0.0729 \cdot 33.8} = 0.129\ \text{K} \cdot \text{m/W}$$

The rating of the cable is thus

$$I = \left[\frac{90 - 25}{0.798 \cdot 10^{-4}[0.307 + 3(1 + 0.0218) \cdot (0.078 + 0.129)]}\right]^{0.5} = 930 \text{ A}$$

The approximate method gives, in this case, a somewhat optimistic cable rating with the ampacity less than 3% higher than that provided by the exact solution.

9.6.8.5 Neher–McGrath Approach. The approximation in equation (9.95) is rather crude, except for $\Delta\theta_s = 0$ K and $\Delta\theta_s = 100$ K (Morgan, 1993). The approach proposed by Neher and McGrath (1957) is based on the work performed by Buller and Neher (1950), and involves approximating the convective and radiative heat losses as functions of the cable surface temperature rise.

The approximation adopted by Neher and McGrath for the radiative heat transfer coefficient is to select a temperature θ_m about midway between maximum and minimum expected temperatures for the surface of the cable and use this value in the following equation [see also equation (9.53)]:

$$\frac{W_{\text{rad}}}{\Delta\theta_s} = 4.2\pi D_e^* \varepsilon_s (1 + 0.0167\theta_m) \tag{9.103}$$

Substituting equation (9.103) into (9.76), we obtain

$$T_4 = \frac{1}{\pi D_e^* [h_{\text{conv}} + 4.2 \cdot \varepsilon_s \cdot (1 + 0.0167\theta_m)]} \tag{9.104}$$

Further, Neher and McGrath (1957) proposed the following simplified version of equation (9.80):

$$h_{o,n} = \frac{1.05}{\left(D_e^*\right)^{1/4}} \tag{9.105}$$

The constant in equation (9.105) was selected to best fit experimental data for 1.3, 3.5, and 10.8 in (33.02, 88.9, and 274.32 mm) black pipes.

The forced convection heat transfer coefficient was approximated by Neher and McGrath (1957) by

$$h_{o,f} = 2.87\sqrt{\frac{U}{D_e^*}} \tag{9.106}$$

EXAMPLE 9.14

Consider again part 1 of Example 9.11 and Example 9.13. This time we will apply the equations developed by Neher and McGrath and presented in this section.

Assuming a mean cable surface temperature of 50°C, the radiative heat coefficient obtained from equation (9.103) is equal to

$$h_{\text{rad}} = 4.2\varepsilon_s (1 + 0.0167\theta_m) = 4.2 \cdot 0.9 \cdot (1 + 0.0167 \cdot 50) = 6.94 \text{ W/m}^2 \cdot \text{K}$$

(1) Free convection

The convection heat transfer coefficient is given by equation (9.105) and is equal to

$$h_{o,n} = \frac{1.05}{\left(D_e^*\right)^{1/4}} = \frac{1.05}{(0.0729)^{1/4}} = 2.02 \text{ W/m}^2 \cdot \text{K}$$

The external thermal resistance and the rating are obtained from equations (9.104) and (4.3), respectively, as

$$T_4 = \frac{1}{\pi D_e^* \left[h_{\text{conv}} + h_{\text{rad}}\right]} = \frac{1}{\pi \cdot 0.0729[2.02 + 6.94]} = 0.487 \text{ K} \cdot \text{m/W}$$

$$I = \left[\frac{90 - 25}{0.798 \cdot 10^{-4}\,(0.307 + 3(1 + 0.0218) \cdot (0.078 + 0.487))}\right]^{0.5} = 632 \text{ A}$$

(2) Forced convection

From equation (9.106), we have

$$h_{o,f} = 2.87\sqrt{\frac{U}{D_e^*}} = 2.87\sqrt{\frac{3}{0.0729}} = 18.4 \text{ W/m}^2 \cdot \text{K}$$

The external thermal resistance and the rating of the cable are equal to

$$T_4 = \frac{1}{\pi D_e^* \left[h_{\text{conv}} + h_{\text{rad}}\right]} = \frac{1}{\pi \cdot 0.0729 \cdot (18.4 + 6.94)} = 0.17 \text{ K} \cdot \text{m/W}$$

$$I = \left[\frac{90 - 25}{0.798 \cdot 10^{-4}\,(0.307 + 3(1 + 0.0218) \cdot (0.078 + 0.17))}\right]^{0.5} = 874 \text{ A}$$

When equation (9.103) is used to approximate radiation losses in Example 9.13, the following results are obtained: $T_4 = 0.136$ K · m/W, $I = 920$ A. Comparing the approximate results with the exact solution obtained in part 1 of Example 9.13, it appears that, in this case, the best approximation is provided by equation (9.99) for the convective heat transfer calculations and by the Neher–McGrath approximation (equation 9.103) for the radiative heat transfer.

9.7 THERMAL CAPACITANCES

In Section 3.2.2, thermal capacitance was defined as the product of the volume of the material and its specific heat [equation (3.13)]. In the following section, formulas for special and concentric layers are presented. In all formulas, Q is the thermal capacitance (J/K · m) and c is the specific heat of the material (J/K · m^3).

9.7.1 Oil in the Conductor

The thermal capacitance of oil inside the conductor is obtained from

$$Q_o = \left(\frac{\pi d_i^{*2}}{4} + S \cdot F\right) \cdot c \tag{9.107}$$

where d_i^* = conductor internal diameter, m

S = conductor cross section, m^2

F = factor representing unfilled cross section (usually 0.36).

9.7.2 Conductor

The thermal capacitance of the conductor is given by

$$Q_c = S \cdot c \tag{9.108}$$

9.7.3 Insulation

Representation of the capacitance of the insulation depends on the duration of the transient. For transients lasting longer than about 1 h, the capacitance of the insulation is divided between the conductor and the sheath positions according to a method given by Van Wormer (1955) (see Section 3.3.1). This assumes logarithmic temperature distribution across the dielectric during the transient, as in the steady state. The total thermal capacitance is obtained from equation (3.13) as

$$Q_i = \frac{\pi}{4}\left(D_i^{*2} - d_c^{*2}\right) \cdot c \tag{9.109}$$

The portion pQ_i is placed at the conductor and the portion $(1 - p)Q_i$ at the sheath, where p is the Van Wormer coefficient defined in equation (3.19) as

$$p = \frac{1}{2\ln\left(\dfrac{D_i}{d_c}\right)} - \frac{1}{\left(\dfrac{D_i}{d_c}\right)^2 - 1} \tag{9.110}$$

From the thermal point of view, the thickness of the dielectric includes any nonmetallic semiconducting layer either on the conductor or on the insulation.

For shorter durations (less than 1 h), it has been found necessary to divide the insulation into two portions having equal thermal resistances. The thermal capacitance of each portion of the insulation is then assumed to be located at its boundaries, using the Van Wormer coefficient to split the capacitances as shown in Fig. 3.4. The thermal capacitance of the first part of the insulation is given by

$$Q_{I1} = \frac{\pi}{4}\left(D_i^* \cdot d_c^* - d_c^{*2}\right) \cdot c \tag{9.111}$$

This capacitance is split into two parts using the Van Wormer coefficient as follows:

$$Q_{i1} = p^* Q_{I1}, \quad Q_{i2} = (1 - p^*)Q_{I1} \tag{9.112}$$

$$p^* = \frac{1}{\ln \dfrac{D_i}{d_c}} - \frac{1}{\dfrac{D_i}{d_c} - 1} \tag{9.113}$$

The total thermal capacitance of the second part is given by

$$Q_{I2} = \frac{\pi}{4}\left(D_i^{*2} - D_i^* \cdot d_c^*\right) c \tag{9.114}$$

which leads to the third and fourth part of the thermal capacitance of the insulation, defined as

$$Q_{i3} = p^* Q_{I2}, \quad Q_{i4} = (1 - p^*)Q_{I2} \tag{9.115}$$

In the case of dielectric losses, the cable thermal circuit is the same as shown in Fig. 3.4, with the Van Wormer coefficient apportioning to the conductor a fraction of the dielectric thermal capacitance given by equation (9.110). For cables at voltages higher than 275 kV, the Van Wormer coefficient is given by equation (3.23).

9.7.4 Metallic Sheath or Any Other Concentric Layer

The thermal capacitance of all other concentric layers of cable components such as sheath, armor, jacket, armor bedding, or serving is computed by using equation (3.13). However, one should remember that the thermal capacitances of nonmetallic layers have to be divided into two parts using the Van Wormer factor given by equation (3.20). The appropriate dimensions for the inner and outer diameters must be used in order to attain sufficient accuracy for short-duration transients.

9.7.5 Reinforcing Tapes

The thermal capacitance of the reinforcing tapes over the sheath is

$$Q_T = n_t \left[w_t^* t_t^* \sqrt{\ell_t^{*2} + \left(\pi d_2^*\right)^2} \right] c \tag{9.116}$$

where n_t = number of tapes
w_t^* = tape width, m
t_t^* = tape thickness, m
ℓ_t^* = length of lay, m
d_2^* = mean diameter of tapes, m.

9.7.6 Armor

The thermal capacitance of the armor is obtained from

$$Q_a = n_1 \frac{\pi d_f^{*2}}{4} \sqrt{\ell_a^* d_a^*} \cdot c \tag{9.117}$$

where n_1 = number of armor wires
d_f^* = mean diameter of armor wire, m
ℓ_a^* = length of lay, m
d_a^* = mean diameter of armor, m

9.7.7 Pipe-type Cables

The thermal capacitances of cable components are computed as described above, and the thermal capacitance of the oil in the pipe is obtained from

$$Q_{op} = \frac{\pi}{4} \left(D_d^{*2} - 3D_s^{*2}\right) c \tag{9.118}$$

where D_d^* = internal diameter of the pipe, m
D_s^* = external diameter of one cable, m

The capacitance of the oil is divided into two equal parts.

The thermal capacitance of the skid wires is generally neglected, but may be computed using equation (9.117) if needed. The thermal capacitance of the metallic pipe and of the external covering are computed using equation (3.13).

EXAMPLE 9.15

We will compute the insulation thermal capacitances of cable model No. 2 located in a PVC duct for short-duration transients (capacitances for long-duration transients were computed in Example 3.8).

We recall that for short-duration transients, the mutual heating of the cores is neglected. The time limit for the short-duration transient is equal to $\Sigma T \cdot \Sigma Q$ where ΣT and ΣQ refer to one core only (see Section 3.3.1). The short-duration transients for cable No. 2 located in duct in air last less then 30 min. The numerical values required for this case are as follows: $D_i = 30.1$ mm, $d_c = 20.5$ mm, and $c = 2.4 \cdot 10^{-6}$ J/(m$^3 \cdot$ K).

From equations (9.111) and (9.114), the thermal resistance of the first and second part of the insulation is equal to

$$Q_{I1} = \frac{\pi}{4}\left(D_i^* \cdot d_c^* - d_c^{*2}\right) \cdot c = \frac{\pi}{4}\left(30.1 \cdot 20.5 - 20.5^2\right) \cdot 10^6 \cdot 2.4 \cdot 10^{-6} = 371.0 \text{ J/K} \cdot \text{m}$$

$$Q_{I2} = \frac{\pi}{4}\left(D_i^{*2} - D_i^* \cdot d_c^*\right) c = \frac{\pi}{4}\left(30.1^2 - 30.1 \cdot 20.5\right) \cdot 10^6 \cdot 2.4 \cdot 10^{-6} = 544.7 \text{ J/K} \cdot \text{m}$$

The Van Wormer coefficient is obtained from equation (9.113) as

$$p^* = \frac{1}{\ln \dfrac{D_i}{d_c}} - \frac{1}{\dfrac{D_i}{d_c} - 1} = \frac{1}{\ln \dfrac{30.1}{20.5}} - \frac{1}{\dfrac{30.1}{20.5} - 1} = 0.468$$

Thus, the capacitances per core of the four parts of the insulation, obtained from equations (9.112) and (9.115), are equal to

$$\begin{aligned}
Q_{i1} &= p^* Q_{I1} = 0.468 \cdot 371.0 = 173.6 \text{ J/K} \cdot \text{m}, \\
Q_{i2} &= (1 - p^*) Q_{I1} = (1 - 0.468) \cdot 371.0 = 197.4 \text{ J/K} \cdot \text{m}, \\
Q_{i3} &= p^* Q_{I2} = 0.468 \cdot 544.7 = 254.9 \text{ J/K} \cdot \text{m}, \\
Q_{i4} &= (1 - p^*) Q_{I2} = (1 - 0.468) \cdot 544.7 = 289.8 \text{ J/K} \cdot \text{m}
\end{aligned}$$

REFERENCES

Affolter, J. F. (1987), "An improved methodology for transient analysis of underground power cables using electrical network analogy," M.E. Thesis, McMaster University, Hamilton, Canada.

Anders, G. J., Napieralski, A., and Zamojski, W. (1997), "Internal thermal resistance of 3-core cables with fillers," submitted for publication to *J. IEE*.

Anders, G. J., Bedard, N., Chaaban, M., and Ganton, R. W. (Oct. 1987), "New approach to ampacity evaluation of cables in ducts using finite element technique," *IEEE Trans. Power Delivery*, vol. PWRD-2, No. 4, pp. 969–975.

Atkinson, R. W. (1924), "Dielectric field in an electric power cable," *Trans. AIEE*, vol. 43, pp. 966–988.

Black, Z. W., and Rehberg, R. L. (1985), "Simplified model for steady state and real-time capacity of overhead conductors," *IEEE Trans. Power App. Syst.*, vol. PAS-104, pp. 2942–2963.

Brakelmann, H., Honerla, J., and Rasquin, W. (1991), "Thermal resistance of cables with corrugated sheaths," *ETAP*, vol. 1, no. 6, pp. 341–346.

Buller, F. H., and Neher, J. H. (1950), "The thermal resistance between cables and a surrounding pipe or duct wall," *AIEE Trans.*, vol. 69, part 1, pp. 342–349.

CIGRE (1985), "The calculation of the effective thermal resistance of cables laid in materials having different thermal resistivities," *Electra*, no. 98, pp. 19–42.

El-Kady, M. A., and Horrocks, D. J. (1985), "Extended values of geometric factor of external thermal resistance of cables in duct banks," *IEEE Trans. Power App. Syst.*, vol. PAS-104, pp. 1958–1962.

El-Kady, M. A., Anders, G. J., Horrocks, D.J., and Motlis, J. (Oct. 1988), "Modified values for geometric factor of external thermal resistance of cables in ducts," *IEEE Trans. Power Delivery*, vol. 3, no. 4, pp. 1303–1309.

Goldenberg, H. (1969a), "External thermal resistance of two buried cables. Restricted application of superposition," *Proc. IEE*, vol. 116, no. 5, pp. 822–826.

Goldenberg, H. (1969b), "External thermal resistance of three buried cables in trefoil touching formation. Restricted application of superposition," *Proc. IEE*, vol. 116, no. 11, pp. 1885–1890.

Greebler, P., and Barnett, G. F. (1950), "Heat transfer study on power cable ducts and duct assemblies," *AIEE Trans.*, vol. 69, part 1, pp. 357–367.

Heinhold, L. (1990), *Power Cables and their Application. Part 1*, 3rd ed., Siemens Aktiengesellschaft.

Holman, J. P. (1990), *Heat Transfer*. New York: McGraw-Hill.

IEC (1982), "Calculation of the continuous current rating of cables (100% load factor)," IEC Standard Publication 287, 2nd ed., 1982.

IEC Standard 287-2-1 (1994), "Electric cables—calculation of the current rating. Part 2: Thermal resistance—Section 1: calculation of thermal resistance," 1st ed.

Incropera, F. P., and De Witt, D. P. (1990), *Introduction to Heat Transfer*. New York: Wiley.

King, S. Y., and Halfter, N. A. (1982), *Underground Power Cables*. London: Longman.

Mie, G. (1905), "Über die Warmeleitung in einem verseilten Kable," *Electrotechnische Zeitschrift*, pp. 137.

Morgan, V. T. (1982), "The thermal rating of overhead-line conductors, Part I. The steady-state thermal model," *Elec. Power Syst. Res.*, vol. 5, pp. 119–139.

Morgan, V. T. (Mar. 1992), "Effect of mixed convection on the external resistance of single-core and multicore bundled cables in air," *Proc. IEE*, vol. 139, part C, no. 2, pp. 109–116.

Morgan, V. T. (Mar. 1993), "External thermal resistance of aerial bundled cables," *Proc. IEE*, vol. 140, part C, no. 2, pp. 65–62.

Morgan, V. T. (1994), "Effect of surface temperature rise on external thermal resistance of single-core and multi-core bundled cables in air," *Proc. IEE*, vol. 141, part C, no. 3, pp. 215–218.

Neher, J. H. (June 1953), "Procedures for calculating the temperature rise of pipe cable and buried cables for sinusoidal and rectangular loss cycles," *AIEE Tran.*, vol. 72, part 3, pp. 541–545.

Neher, J. H., and McGrath, M. H. (Oct. 1957), "The calculation of the temperature rise and load capability of cable systems," *AIEE Trans.*, vol. 76, part 3, pp. 752–772.

Poritsky, H. (1931), "The field due to two equally charged parallel conducting cylinders," *J. Math. Phys.*, vol. 32, no. 11, pp. 213–217.

Raithby, G. D., and Hollands, K. G. T. (1985), "Natural convection," in *Handbook of Heat Transfer*, W. M. Rohsenow and J. P. Hartnett, Eds. New York: McGraw-Hill.

Russel, A. (1914), *Theory of Alternating Currents.* Cambridge: Cambridge University Press.

Sellers, S. M., and Black, W. Z. (1995), "Refinements to the Neher-McGrath model for calculating the ampacity of underground cables," IEEE PES Winter Meeting paper 95 WM 0115-8 PWRD; to be published in *IEEE Trans. Power Delivery.*

Simmons, D. M. (1923), "Cable geometry and the calculation of current-carrying capacity," *Trans. AIEE*, vol. 42, pp. 600 – 615.

Simmons, D. M. (1932), "Calculation of the electrical problems of underground cables," *Elec. J.*, vol. 29, no. 9. pp. 395– 426.

Symm, G. T. (1969), "External thermal resistance of buried cables and troughs," *Proc. IEE*, vol. 166, no. 10, pp. 1696 –1698.

Tarasiewicz, E., El-Kady, M. A., and Anders, G. J. (Jan. 1987), "Generalized coefficients of external thermal resistance for ampacity evaluation of underground multiple cable systems," *IEEE Trans. Power Delivery*, vol. PWRD-2, no. 1, pp. 15–20.

Van Geertruyden, A. (1992), "External thermal resistance of three buried single-core cables in flat and in trefoil formation," Laborelec Report No. DMO-RD - 92-003/AVG.

Van Geertruyden, A. (1993), "External thermal resistance of two buried single-core cables in flat formation," Laborelec Report No. DMO-RD - 93-002/AVG.

Van Geertruyden, A. (1994), "Internal thermal resistance of extruded cables," Laborelec Report No. SMI-RD - 94-002/AVG.

Van Geertruyden, A. (1995), "Thermal resistance of the external serving for three single-core touching cables in trefoil and in flat formations," Laborelec Report No. SMI-APP-95-007/E/AVG.

Wedmore, E. B. (1923), "Permissible current loading of British standard impregnated paper-insulated electric cables," *J. IEE*, vol. 61, p. 517.

Weedy, B. M. (1988), *Thermal Design of Underground Systems.* Chichester, England: Wiley.

Whitehead, S., and Hutchings, E. E. (1938), "Current ratings of cables for transmission and distribution," *J. IEE*, vol. 38, pp. 517–557.

PART III ADVANCED TOPICS

10

Special Cable Installations

10.1 INTRODUCTION

The majority of power cables are installed underground or in free air, and the rating of cables in such installations has been discussed in earlier chapters. However, there exists a large number of installations for which the rating techniques discussed earlier do not apply directly. Examples of such installations are: 1) cables on riser poles, 2) cables in open and covered trays, 3) cables in tunnels and shafts, and 4) cables in shallow troughs. What these installations have in common is the heat transfer mechanism from the cable surface to the outside environment. In general, we can divide cable installations considered in this chapter as either located within protective walls (e.g., protective riser, tray cover, tunnel walls) or located in trays without cover. In the former case, we will talk about heat transfer between the cable surface and the protective wall, as well as the heat transfer from the outside surface of the wall. In the latter case, we will consider heat transfer between the cable surface and the surrounding air.

For cables installed in air, the significant modes of heat transfer are as follows:

1. By natural or free convection when no longitudinal induced flow is present
2. By forced convection by air flow along the cables
3. By radiation of the heat from the cable surface to the ambient air, walls, or covers.

Since conduction in air accounts for a small fraction of heat transfer in the installations under consideration in this chapter, we will, in agreement with common practice, ignore this mode of heat transfer in further analysis.

Rating of cables in air is often based on the assumption that only natural convection and radiation are present. Forced convection, if present, will result in a lower cable temperature for the same current compared to the natural convection case only.

The convective heat transfer process is determined by the following factors:

1. The length, diameter, and thickness of the jacket and wall
2. Venting conditions at the ends
3. The heat flux generated in the cable, which depends on the electric current and cable type
4. The environmental conditions such as temperature, humidity, solar radiation, wind speed, and others.

Thermal radiation is an important heat transfer mode in air-filled cable-wall systems. Thermal radiation transfers energy from the cable surface to the wall inside surface. This is different from convective heat transfer. Thermal radiation from the cable surface accounts for 40–60% of the total heat transfer (Keyhani and Kulacki, 1985). Thus, with free convection and air flow at low velocities, the proportion of heat removed by radiation is substantial and must be accounted for in calculations. The amount of heat transferred by radiation depends upon a number of factors, including surface temperatures and emissivities.

To compute the rating of cables in air considered in this chapter, the temperatures at various points of the thermal circuit are required. To obtain the required temperatures, a set of energy conservation equations has to be solved. In all of the cases considered here, the same system of equations can be solved, with the differences arising in the selection of coefficients and constants. In the next section, we will develop a general set of energy balance equations for a cable system surrounded by a wall, and the selection of appropriate coefficients will be discussed in detail in the sections discussing specific installations. Since, for cables installed on riser poles, in trays, in tunnels and in shafts, the same set of energy conservation equations is solved, we will limit the number of numerical examples presented in this chapter.

10.2 ENERGY CONSERVATION EQUATIONS

The following assumptions are introduced to simplify the thermal resistance calculation:

1. The process is steady state.
2. The length of the wall and the cable are large, so the heat transfer can be considered as one dimensional.
3. The wall is opaque and the cable jacket material is radiatively gray and opaque; the air inside the protective wall is radiatively transparent.
4. The physical properties of all materials in the cable system are temperature dependent. The model takes into account the variation of physical properties with temperature.

10.2.1 Energy Conservation Equation for the Cable Outside Surface

Considering the outside surface of the jacket under steady-state conditions, the conduction heat flux from its inner surface is equal to the heat loss through free convection and thermal radiation in the air between the cable surface and the wall. The energy balance equation (2.8) takes the form

$$W_t = W_{\text{conv},s} + W_{\text{rad},s-w} \tag{10.1}$$

where $W_{\text{conv},s}$ = natural convection heat transfer rate between the cable outside surface and the air per unit length, W/m

$W_{\text{rad},s-w}$ = thermal radiation heat transfer rate between the wall inner surface and the cable outside surface, per unit length, W/m

W_t = total energy per unit length generated within the cable. Its value is given by equation (4.6):

$$W_t = W_I + W_d \tag{10.2}$$

The joule losses W_I are defined in equation (4.1):

$$W_I = n \cdot W_c(1 + \lambda_1 + \lambda_2) = n \cdot I^2 R(1 + \lambda_1 + \lambda_2) \tag{10.3}$$

where I is the conductor current and R is the conductor ac resistance at operating temperature. λ_1 and λ_2 are the sheath and armor loss factors and n is the number of conductors in the cable.

The dielectric losses are obtained form equation (6.4):

$$W_d = \omega C U_0^2 \tan\delta \tag{10.4}$$

The dielectric loss formula is developed in Chapter 6 and the variables are defined there. Computation of all of these quantities is discussed in detail in this book.

Newton's law of cooling [equation (2.2)] is used to determine the convection heat transfer rate between the outside surface of the cable and the air inside the walls:

$$W_{\text{conv},s} = h_s(\theta_s - \theta_{\text{gas}})A_s \tag{10.5}$$

where h_s = natural convection coefficient at the surface of the cable, W/K·m^2

θ_s = average temperature of the cable outside surface, °C

θ_{gas} = air temperature in the gap, °C

A_s = area of the cable (or cable mass) outside surface, (m^2), for unit length.

The net radiation heat transfer rate between the cable and the inside surface of the wall is based on the radiative exchange between two surfaces [equation (2.6)]:

$$W_{\text{rad},s-w} = A_{sr} F_{s,w} \sigma_B (\theta_s^{*4} - \theta_w^{*4}) \tag{10.6}$$

where σ_B = Boltzmann constant, equal to 5.67×10^{-8}W/(m$^2 \cdot$ K^4)

θ_w^* = average temperature of the wall inner surface, K

$F_{s,w}$ = thermal radiation shape factor; its value depends on the geometry of the system

A_{sr} = area of the cable surface effective for heat radiation, (m^2), for unit length of the cable.

10.2.2 Energy Conservation Equation for the Wall Inside Surface

For the wall inside surface, the energy transferred by conduction through the wall material is equal to the energy transferred through convection and radiation on the inner surface of the wall. Thus, the energy conservation equation under steady-state conditions is

$$W_{\text{cond},w-o} = W_{\text{conv},w} + W_{\text{rad},s-w} \tag{10.7}$$

where $W_{\text{cond},w-o}$ = heat conduction rate from the wall inner surface to its outside surface per unit length, W/m

$W_{\text{conv},w}$ = natural convection between the wall inside surface and the air per unit length, W/m.

Similar to the heat transfer at the cable surface, the free convection heat transfer in the wall inner surface is given by

$$W_{\text{conv},w} = h_w(\theta_{\text{gas}} - \theta_w)A_w \tag{10.8}$$

where h_w is the natural convection coefficient at this surface, W/K·m^2 and A_w is the area of the wall per unit length, (m^2).

The conduction part of the heat transfer is simply

$$W_{\text{cond},w-o} = \frac{\theta_w - \theta_o}{T_4''} \tag{10.9}$$

where θ_o = average temperature at the wall outside surface, °C

T_4'' = thermal resistance of the wall per unit length, K·m/W

10.2.3 Energy Conservation Equation for the Wall Outside Surface

In this section, we will consider cable installations which have air as a medium outside the wall. At the outside surface of the wall, the energy transferred through the wall material by conduction and the energy gain due to solar radiation are balanced by the convective and radiative energy losses to the atmosphere. Thus, the energy conservation equation is

$$W_{\text{cond},w-o} + W_{\text{sol}} = W_{\text{conv},o} + W_{\text{rad},o-\text{sur}} \tag{10.10}$$

where $W_{\text{conv},o}$ = natural convection heat transfer between the wall outside surface and atmosphere air, per unit length, W/m

$W_{\text{rad},o-\text{sur}}$ = thermal radiation heat transfer rate between wall surface and surrounding objects, per unit length, W/m

W_{sol} = solar radiation absorbed by the wall surface, per unit length, W/m. This quantity is only considered for installations exposed to solar radiation.

The convection heat transfer between the wall and the surroundings is

$$W_{\text{conv},o} = h_o(\theta_o - \theta_{\text{amb}})A_o \tag{10.11}$$

where h_o = natural or forced convection coefficient at this surface, W/K·m2
θ_{amb} = air temperature in atmosphere, °C
A_o = area of the outside wall surface, per unit length, (m^2).

The thermal radiation heat transfer rate between the outside surface of the wall and the surroundings is based on the assumption that the wall is a small body enclosed by the surroundings. The net radiative exchange between these surfaces is

$$W_{rad,o-sur} = A_o \varepsilon_o \sigma_B (\theta_o^{*4} - \theta_{amb}^{*4}) \tag{10.12}$$

where ε_o = emissivity of the wall outside surface.

If the solar flux incident on one square meter of the surface is H, then the solar radiation absorbed by the wall is

$$W_{sol} = A_{os} \alpha_o H \tag{10.13}$$

where α_o = wall surface absorptivity to solar radiation, which is generally different from the emissivity of the wall
A_{os} = equivalent area of the wall per unit length perpendicular to sun rays, m^2

10.2.4 Energy Conservation Equations

Equations (10.3)–(10.13) are the basic energy conservation equations for the cable-wall system. There are three unknown temperatures, θ_s, θ_w, and θ_o. These temperatures are thus computed from the following three equations:

$$\begin{aligned}
&n \cdot I^2 R(1 + \lambda_1 + \lambda_2) + W_d = h_s(\theta_s - \theta_{gas})A_s + A_{sr} F_{s,w} \sigma_B (\theta_s^{*4} - \theta_w^{*4}) \\
&\frac{\theta_w - \theta_o}{T_4''} = h_w(\theta_{gas} - \theta_w)A_w + A_{sr} F_{s,w} \sigma_B (\theta_s^{*4} - \theta_w^{*4}) \\
&\frac{\theta_w - \theta_o}{T_4''} + A_{os}\alpha_o H = h_o(\theta_o - \theta_{amb})A_o + A_o \varepsilon_o \sigma_B (\theta_o^{*4} - \theta_{amb}^{*4})
\end{aligned} \tag{10.14}$$

In the above equations, the equivalent conductor electric resistance and the convection coefficients are temperature dependent. Before equations (10.14) can be solved, the values of the intensity of solar radiation and the convection coefficients have to be selected. These subjects are discussed in the following sections.

10.3 CABLES ON RISER POLES

10.3.1 Introduction

Power delivery systems frequently consist of a combination of overhead lines and underground cables. In most cases, the underground cable system is connected to the overhead line through a short section of cable located in a protective riser. Figure 10.1 shows a cross section of a submarine cable installed on a riser pole with a protective guard.

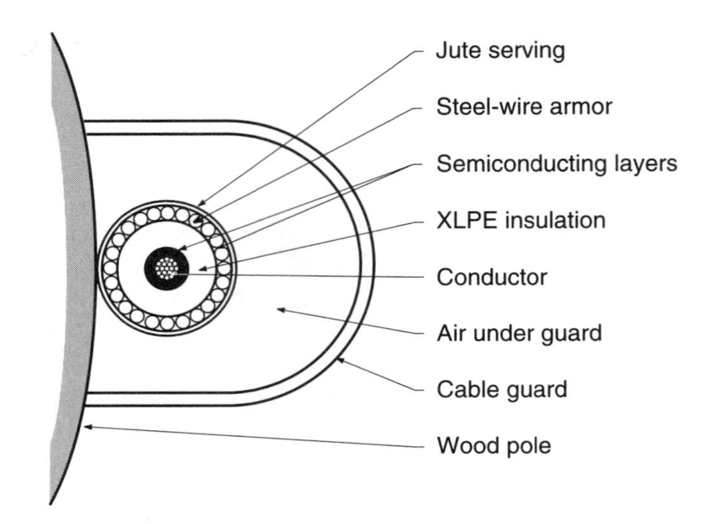

Figure 10.1 Cross section of a submarine cable on riser pole (Cress and Motlis, 1991).

The protective guard is often simply referred to as a riser. The current-carrying capacity of the composite system is limited by that segment of the system that operates at the maximum temperature. Very often, the riser-pole portion of the cable system will be the limiting segment.

Considering the importance of accurately rating power cable systems consisting of cables on riser poles, Hartlein and Black (1983) introduced a mathematical model to represent such systems. The model is based on a modified thermal circuit consisting of thermal resistances separated by local system temperatures. The analysis results in a number of algebraic equations, similar to equations (10.14), that are simultaneously solved for the system temperatures for a given cable ampacity. The theoretical developments were substantiated by experimental evidence for cables in protective risers located indoors without solar radiation and wind.

The pioneering work by Hartlein and Black suffered from gaps in knowledge (no formulas were given for the computation of heat transfer coefficients under certain conditions) and, in several cases, required assumptions which are incompatible with typical cable-riser geometry. Much new experimental work has been reported during the 12-year period since the publication of their paper. Anders and Gu (Anders, 1996) have updated the work of Hartlein and Black (1983) by redefining the mathematical model and supplementing information lacking in their work. A comparison of both models was offered. The new model was tested against Hartlein's and Black's experimental data, as well as the data for outdoor tests reported by Cress and Motlis (1991). The new model is implemented in the CEA's Cable Ampacity Program (CAP) (Anders *et al.*, 1990).

10.3.2 Thermal Model

The assumption used in developing the mathematical model for the cable-riser system is that the cable and the riser are concentric bodies with their length much greater than their diameters. Equations (10.14) can be used to determine the required temperatures. The radiation shape factor is obtained considering two long concentric cylinders. In the case of a single cable in the riser, we have

$$F_{s,w} = (1 + \sigma_s/\varepsilon_s + A_s\sigma_w/A_w\varepsilon_w)^{-1} \tag{10.15}$$

where σ_s = reflectivity of the cable outside surface
ε_s = emissivity of the cable outside surface
σ_w = reflectivity of the wall inner surface
ε_w = emissivity of the wall inner surface
$A_s = \pi D_e^*$, $A_w = \pi D_d^*$, $A_o = \pi D_o^*$.

D_e^*, D_d^*, and D_o^*, (m) are the cable outside diameter and riser inside and outside diameters, respectively. The maximum area exposed to solar radiation is LD_o^*. When several cables are present, the mutual radian area between them must be subtracted from the area radiating to the riser inner surface. The most common installations have either one or three cables inside the guard. The effective radiating area to the guard walls is obtained as follows (Weedy, 1988).

Elastic bands are imagined stretched around two arbitrary concave surfaces (see Fig. 10.2). The lengths of the bands stretched between the two surfaces are found as follows. The length of the internal band is $l_{\text{int}} = AD + BC'C$ where C' is the point of intersection of BC with surface CD. The length of the external bands is $l_{\text{ext}} = AC + BD$. The mutual radiation area A_m per unit axial length is equal to half the difference in lengths; that is,

$$A_m = \frac{l_{\text{int}} - l_{\text{ext}}}{2} = \frac{AD + BC'C}{2} - \frac{AC + BD}{2} \tag{10.16}$$

Application of the method to three cables in touching trefoil formation yields (Weedy, 1988)

$$A_{sr} = 3\pi D_e^* - 3 \cdot 0.618 D_e^* \tag{10.17}$$

In this case, the radiation shape factor has the form

$$F_{s,w} = (1 + \sigma_s/\varepsilon_s + \zeta A_s\sigma_w/A_w\varepsilon_w)^{-1}$$
$$\zeta = \frac{3}{1 - \dfrac{6\pi A_s}{A_w}\dfrac{\varepsilon_s\sigma_w}{\varepsilon_w}} \tag{10.18}$$

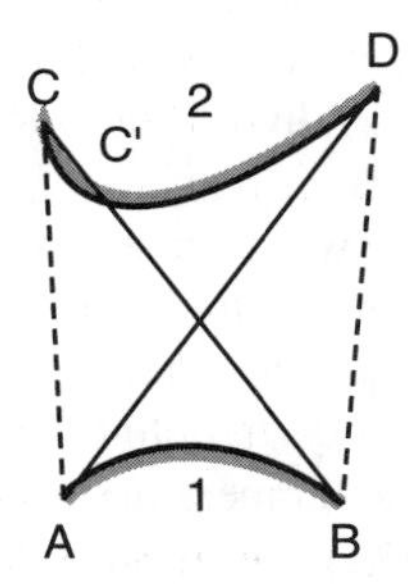

Figure 10.2 Calculation of mutual radiation area. (Weedy, 1988)

Also,

$$T_4'' = \frac{\rho_c}{2\pi} \ln \frac{D_o}{D_d}$$

10.3.3 Intensity of Solar Radiation

The solar radiation absorbed by the riser depends on the solar altitude angle α, the angle of the solar beam with respect to the axis of the conductor, the diffuse sky radiation, and the albedo (reflectance) of the surface of the ground beneath the conductor (Morgan, 1982). Assuming that the reflected-beam component hits the guard at the same angle α as the direct beam, and neglecting the diffuse radiation, a simplified formula for the solar intensity is obtained as

$$H = H_B \cos\alpha \tag{10.19}$$

where H_B is the intensity of the direct solar beam on a surface normal to the beam. Thus, if the intensity of solar radiation is obtained from measurements, the intensity of solar radiation incident on the guard is obtained by multiplying the measured value by $\cos\alpha$.

The angle α depends on the altitude angle ϕ, the declination δ_s, and the hour angle of the sun ω_n, with solar noon being zero, mornings being positive, and afternoons negative (Morgan, 1982; Cress and Motlis, 1991):

$$\sin\alpha = \sin\delta_s \sin\phi + \cos\delta_s \cos\phi \cos\omega_n \tag{10.20}$$

The declination angle depends on the day of the year N_d as

$$\delta_s = 23.45 \sin[360(284 + N_d)/365] \tag{10.21}$$

The hour angle ω_n increases by 15° for every hour from zero at solar noon, and is positive before noon. The hour angle at sunrise ω_0 is found from equation (10.20), putting $\alpha = 0$:

$$\omega_0 = \cos^{-1}(-\tan\delta_s \tan\phi)$$

hence, the hour angle at any time t_n after sunrise is given by

$$\omega_n = \cos^{-1}(-\tan\delta_s \tan\phi) - 15t_n \tag{10.22}$$

To obtain solar time, add 4 min/degree of longitude east of standard time, or subtract 4 min/degree west of standard time.

The conductor temperature will normally show significant dependence on the preceding as well as the immediate solar radiation intensity. However, only a single value of H is used in the ampacity computations. Cress and Motlis (1991) introduced the concept of equivalent solar radiation intensity $\hat{H}$ defined as a constant value that, at time t_n, would bring the conductor to the same temperature as would exposure to the time-varying solar radiation values H_i. They derived the following equations to compute the value of $\hat{H}$:

$$\hat{H} = \frac{\sum_{i=1}^{n} w_i H_i}{\sum_{i=1}^{n} w_i} \tag{10.23}$$

$$w_i = \frac{b \cdot a^{n-i}}{1 - \alpha^n}, \qquad a = e^{-1/\tau}, \qquad b = 1 - a$$

where τ is the time constant of the cable/guard system.

Another factor to consider is the movement of the sun from east to west during the day. This movement results in only partial exposure of the guard to the sun's rays. Considering an azimuth angle and a usual shape of the riser (see Fig. 10.3), Cress and Motlis (1991) developed a computational method to determine the maximum amount of guard surface that could be exposed to incident solar radiation at any one time.

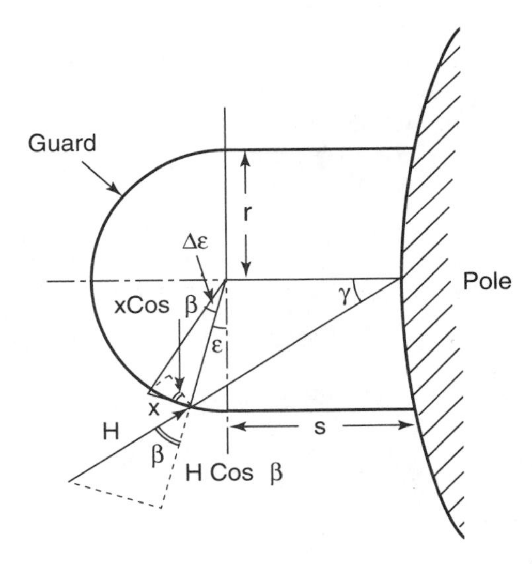

Figure 10.3 Angular solar radiation on riser surface. (Cress and Motlic, 1991)

Considering Fig. 10.3, we note that the solar power, per meter of riser, on the short tangent of length x is

$$H \cos \beta(x) = H(x \cos \beta)$$

Thus, the portion of the semi-circular guard length that is exposed to H can be calculated as the sum of the incremental projections similar to $x \cos \beta$. Including the straight and curved portions of the riser, the length L_e exposed to H can be expressed as

$$L_e = s \cdot \cos(90 - \gamma) + \sum_{\varepsilon=0}^{180} r \cdot \tan(\Delta\varepsilon) \cdot \cos(90 - \gamma - \varepsilon) \tag{10.24}$$

where all the quantities are defined in Fig. 10.3. For a given riser geometry and azimuth angle, the exposed length is computed from equation (10.24) with the aid of a computer program. Numerical studies performed by Cress and Motlis (1991) have shown that, for a typical aluminum guard, the maximum surface that could be exposed to solar radiation was about 32% of the total riser surface. This occurred when $\gamma = 63°$.

10.3.4 Convection Coefficients

The cable and the riser form a vertical annulus. If the temperature of the cable or riser is different from the air temperature in atmosphere, natural convection occurs in the annulus gap. This natural convection makes the heat transfer processes in cable-riser systems very complicated.

Convection coefficients required in equations (10.14) are summarized in Tables 10.1–10.3. The basis for the selection of these coefficients is discussed in Anders (1995) and briefly described below.

TABLE 10.1 Convection Correlations for the Cable Outside Surface

Case	Model Correlation	Comments	References
I. Closed at top and bottom	$h_s = \mathrm{Nu} \cdot k_{air} / \delta$ $\mathrm{Nu} = 0.797\mathrm{Ra}^{0.077} G^{-0.052} K^{0.505}$ for $\mathrm{Ra} < 363K^{025} G^{0.76}$ $\mathrm{Nu} = 0.188\mathrm{Ra}^{0.322} G^{-0.238} K^{0.442}$ for $363K^{025} G^{0.76} < \mathrm{Ra} < 2.3 \cdot 10^6$ $\mathrm{Ra} = \mathrm{Pr} \cdot g\beta(\theta_s - \theta_w)\, \delta^3 / v^2$	1. Correlations apply for vertical annulus. 2. Air properties evaluated at film temperature: $\theta_f = (\theta_s + \theta_w) / 2$	1. Keyhani and Kulacki (1985) 2. Keyhani *et al.* (1985) 3. Keyhani *et al.* (1986)
II. Open at both ends	$h_s = C_v \cdot \mathrm{Nu} \cdot k_{air} / \delta;\ K < 5$ $\mathrm{Nu} = \mathrm{Gr} \cdot \mathrm{Pr} / 9.07$ if $\mathrm{Gr} \cdot \mathrm{Pr} < 10$ $\mathrm{Nu} = 0.62(\mathrm{Gr} \cdot \mathrm{Pr})^{0.25}$ $\mathrm{Gr} = \dfrac{g\beta(\theta_s - \theta_{amb})\delta^4}{Lv^2}$ $C_v = 0.46K + 0.54$	1. C_v is applicable for fully developed flow. It may vary in other flow regimes. 2. C_v is fitted when $1 < D_d / D_e < 5$ 4. $\theta_f = (\theta_s + \theta_{amb}) / 2$	1. Raithby and Hollands (1985) 2. El-Shaarawi *et al.* (1981) 3. Al-Nimr (1993)
III. Open at the top and closed at bottom	$h_s = \mathrm{Nu} \cdot k_{air} / L$ $\mathrm{Nu} = c(\mathrm{Gr} \cdot \mathrm{Pr})^n$ $\mathrm{Gr} = g\beta(\theta_s - \theta_w)L^3 / v^2$	1. Applies to vertical cylinder, not to vertical annulus. Annular effects are not considered. $\theta_f = (\theta_s + \theta_w + 2\theta_{amb}) / 4$	Morgan (1982, 1992, 1993)

10.3.4.1 Riser Outside Surface. The convection heat transfer on the outside surface of the riser includes natural and forced convection. Normally, the forced convection is much stronger than free convection. In the case of natural convection, the vertical cylinder free convection correlation (Morgan, 1982) can be used. In the case of forced convection, the heat transfer coefficient can be calculated as described in Holman (1990), Incropera and De Witt (1990), or Burmeister (1983).

Mixed natural and forced convection may exist with very low wind speed conditions only. The natural convective flow is vertically upwards if the cable is warmer than the air. If the wind direction is perpendicular to that of the natural convective flow, the total convective

TABLE 10.2 Convection Correlations for the Riser Inside Surface

Case	Model Correlations	Comments	References
I. Closed at top and bottom	Same as the cable outside surface $h_w = h_s \frac{D_e}{D_d}$	Same as case I in Table 10.2.	1. Keyhani *et al.* (1985, 1985, 1986)
II. Open at both ends	$h_w = C_v \mathrm{Nu} \cdot k_{air} / \delta;\ K < 5$ $\mathrm{Nu} = \mathrm{Gr} \cdot \mathrm{Pr} / 9.07$ if $\mathrm{Gr} \cdot \mathrm{Pr} < 10$ $\mathrm{Nu} = 0.62 (\mathrm{Gr} \cdot \mathrm{Pr})^{0.25}$ $\mathrm{Gr} = \frac{g\beta(\theta_w - \theta_{amb})\delta^4}{Lv^2}$ $C_v = 0.6 / K + 0.4$	1. C_v is applicable for fully developed flow. It may vary in other flow regimes. 2. C_v is fitted when $1 < D_d / D_e < 5$ 3. $\theta_f = (\theta_w + \theta_{amb}) / 2$	1. Raithby and Hollands (1985) 2. El-Shaarawi *et al.* (1981) 3. Al-Nimr (1993)
III. Open at top and closed at bottom	$h_w = \mathrm{Nu} \cdot k_{air} / (D_d / 2)$ $\mathrm{Nu} = (\mathrm{Gr} \cdot \mathrm{Pr}) / 400$ if $\mathrm{Gr} \cdot \mathrm{Pr} < 200$ $\mathrm{Nu} = 0.35 (\mathrm{Gr} \cdot \mathrm{Pr})^{0.28}$ $\mathrm{Gr} = g\beta(\theta_{gas} - \theta_w)(D_d^* / 2) / (Lv^2)$	1. Developed for a vertical circular duct closed at its lower end, not for annular layer. 2. $\theta_f = \theta_w$	1. Dryer (1985) 2. Martin and Cohen (1954) 3. Martin (1955)

TABLE 10.3 Convection Correlations for the Riser Outside Surface

	Correlations in Proposed Model	Comments	References
No wind, natural convection	$h_{o,n} = \mathrm{Nu} \cdot k_{air} / L$ $\mathrm{Nu} = c(\mathrm{Gr} \cdot \mathrm{Pr})^n$ $\mathrm{Gr} = g\beta(\theta_o - \theta_{amb}) L^3 / v^2$	$\theta_f = (\theta_o + \theta_{amb}) / 2$	Morgan (1982)
Forced convection	$h_{o,f} = \mathrm{Nu} \cdot k_{air} / D_o^*$ $\mathrm{Nu} = b(\mathrm{Re})^p \mathrm{Pr}^{1/3}$ $(\mathrm{Re}) = V_{air} D_o^* / v$	$\theta_f = (\theta_o + \theta_{amb}) / 2$	Holman (1990)

heat transfer coefficient on the riser outside surface can be calculated by (Morgan, 1982, 1992, 1993)

$$h_o = (h_{o,n}^2 + h_{o,f}^2)^{1/2} \tag{10.25}$$

We will assume in the proposed model that the wind direction is perpendicular to the natural convective flow.

10.3.4.2 Convection in the Air Gap.

Case I: Riser Closed at the Top and Bottom

An increasing number of applications in microelectronic packaging, nuclear engineering, and solar systems resulted in remarkable attention being focused on natural convection heat transfer inside confined spaces. The correlations for this case are taken from Keyhani and Kulacki (1985). The gas temperature is equal to θ_w in equation (10.5) and θ_s in equation (10.8).

Case II: Riser Open at the Top and Bottom

Considerable research has been done in recent years on the laminar natural convection in open-ended vertical concentric annuli. Joshi (1987, 1988) studied the natural convection in an isothermal vertical annular duct, and discovered that the heat transfer strongly depends on the radius ratio $K = D_d/D_e$. If K is less than 1.2, the solutions for annular ducts coincide with those for parallel plates. Natural convection in the air gap formed by two vertical parallel plates has been studied for many years, and corresponding heat transfer correlations can be found in many publications. Al-Nimr (1993) studied free convection in vertical concentric annuli and the dependence of heat transfer rate on the radius ratio K. If $K = 1$, the Nusselt Number at the cable outside surface is equal to that on the riser inside surface. As K grows, the Nusselt Number increases by a factor of about $(0.46K + 0.54)$ on the cable outside surface, and decreases by a factor of about $(0.6/K + 0.4)$ on the riser inside surface. In the proposed model, the convective heat transfer coefficients are calculated based on these results and the convection correlations for vertical slot formed by two parallel plates. The gas temperature is equal to θ_{amb} in equations (10.5) and (10.8).

Case III: Riser Open at the Top and Closed at the Bottom

There are very few published data available for the heat transfer in this particular configuration. Hartlein and Black (1983) used the natural convection correlation for vertical plates to calculate the convection coefficient at the cable outside surface. They also applied the glycerin free convection correlation in open thermosyphons to calculate the heat transfer coefficient of air at the inside surface of the riser. In the model presented in Table 10.1, the natural convection heat transfer correlation for vertical cylinders is used for the outside surface of the cable (Morgan, 1982).

The correlation for air natural convection in open thermosyphons is used for the inside surface of the riser. An open thermosyphon is a device to transfer heat from a high-temperature region to a low-temperature reservoir or atmosphere. The simplest open thermosyphon is a vertical tube sealed at its lower end. The tube is heated by the high-temperature heat source from which energy is transferred to the outside by the natural convection of the fluid in the tube. In the proposed model, the natural convection correlation for open thermosyphons is adopted to calculate the heat transfer between the riser inside surface and the gas (Dryer, 1978; Martin and Cohen, 1954; Martin, 1955). The gas temperature is equal to θ_w in equation (10.5) and $(\theta_s + \theta_w + 2\theta_{amb})/4$ in equation (10.8).

As an alternative to the correlations given in Table 10.3 for this case, Morgan (1995) suggested to use the correlation for water given by Seki *et al.* (1980), $\mathrm{Nu} = 0.204(\mathrm{Gr}\cdot\mathrm{Pr})^{0.25}$, in the range $4 \cdot 10^4 \leq \mathrm{Ra} \leq 4 \cdot 10^6$, where the Nusselt and Rayleigh Numbers are based on $(D_w - D_s)/2D_s$.

In the above tables, θ_f is the temperature of the film at which gas properties are evaluated (see Appendix D). Also,

$$G = \text{aspect ratio}, \ G = L/\delta$$
$$K = \text{diameter ratio}, \ K = D_d/D_e$$

The constants c, n, b, and p are given in Tables 10.4 and 10.5.

EXAMPLE 10.1

The sample calculations presented in this example illustrate the steps required to determine the temperature profile for a given cable and riser system. We will examine model cable No. 1 located

TABLE 10.4 Constants for Nusselt Number in Natural Convection (Morgan, 1982)

c	n	Restrictions
0.675	0.058	$10^{-10} < GR \cdot PR < 10^{-2}$
1.02	0.148	$10^{-2} < GR \cdot PR < 10^{2}$
0.850	0.188	$10^{2} < GR \cdot PR < 10^{4}$
0.480	0.250	$10^{4} < GR \cdot PR < 10^{7}$
0.125	0.333	$10^{7} < GR \cdot PR < 10^{12}$

TABLE 10.5 Constants for Nusselt Number in Forced Convection (Holman, 1990)

Re	b	p
0.4–4	0.989	0.330
4–40	0.911	0.385
40–4000	0.683	0.466
4000–40 000	0.193	0.618
40 000–400 000	0.0266	0.805

in a riser as specified in Hartlein and Black (1983). Riser dimensions are: $L = 4.87$ m, $D_d^* = 0.1016$ m, and $D_o^* = 0.1144$ m, thermal resistivity $\rho_C = 3.33$ K·m/W, emissivity $\varepsilon_w = \varepsilon_o = 0.7$, and solar absorption coefficient $\alpha_o = 0.4$. Assume that the riser is closed at the bottom and vented at the top. The wind velocity is equal to 3 m/s and the intensity of solar radiation is 500 W/m^2. Each of the three single-core cables is located in a separate riser. The air ambient temperature is 20°C. The computed parameters listed in Appendix A are: $W_c = 30.85$ W/m, $\lambda_1 = 0.214$, and $T_3 = 0.104$ K·m/W.

We will take the values of the skin and proximity effects as well as the loss factors from Table A1. For the installation conditions examined in this example, these values will be somewhat different from those listed in Table A1. However, the error will be very small and the computations much shorter.

The solution of equations (10.14) is iterative in nature. To illustrate the method of calculation without resorting to an iterative process, the correct conductor temperature is initially assumed. Assume the conductor temperature θ_c is 66°C. The conductor losses are 30.82 W/m and the total losses are 33.38 W/m. Dielectric losses are neglected.

The temperature at the outside surface of the cable is obtained from equation (4.2) and is equal to

$$\theta_s = \theta_c - W_c T_1 - [W_c(1 + \lambda_1)](T_3) = 66 - 30.85 \cdot 0.214 - 33.59 \cdot 0.104 = 55.9°\text{C}$$

The temperature at the inside surface of the riser can be calculated from the first equation in (10.14). The air thermophysical properties are evaluated at a film temperature $\theta_f = (\theta_s + \theta_w + 2\theta_{\text{amb}})/4$. At this point, θ_w is not known; however, it can be approximated for the purpose of performing the calculations. If $\theta_w = 26$°C, then

$$\theta_f = (\theta_s + \theta_w + 2\theta_{\text{amb}})/4 = (55.9 + 26.0 + 2 \cdot 20)/4 = 30.5°\text{C}$$

At this temperature, the air thermophysical properties are (see Appendix D)

$$\beta = 3.3 \cdot 10^{-3} \text{ 1/K}$$
$$v = 16.1 \cdot 10^{-6} \text{ m}^2/\text{s}$$
$$\text{Pr} = 0.71$$
$$k_{\text{air}} = 0.0265 \text{ W/K·m}$$

Because the cable and riser are opaque,

$$\sigma_s = 1 - \varepsilon_s = 0.1, \qquad \sigma_w = 1 - \varepsilon_w = 0.3$$

The thermal radiation shape factor is obtained from equation (10.15) and is equal to

$$F_{s,w} = (1 + \sigma_s/\varepsilon_s + A_s\sigma_w/A_w\varepsilon_w)^{-1} = [1 + 0.1/0.9 + \pi \cdot 0.0358 \cdot 0.3/(\pi \cdot 0.1144 \cdot 0.7)]^{-1} = 0.8$$

The convection coefficient is obtained from Table 10.1. For the computation of the Grashoff Number, we assume the inside riser temperature to be 26°C:

$$\mathrm{Gr} = g\beta(\theta_s - \theta_w)L^3/v^2 = 9.81 \cdot 3.3 \cdot 10^{-3}(55.9 - 26)4.87^3/(16.1 \cdot 10^{-6})^2 = 4.3 \cdot 10^{11}$$

$$\mathrm{Gr{\cdot}Pr} = 4.3 \cdot 10^{11} \cdot 0.71 = 3.1 \cdot 10^{11}$$

$$\mathrm{Nu} = c(\mathrm{Gr{\cdot}Pr})^n = 0.125(3.1 \cdot 10^{11})^{0.333} = 838.6$$

$$h_s = \mathrm{Nu} \cdot k_{\mathrm{air}}/L = 838.6 \cdot 0.0265/4.87 = 4.56 \ \mathrm{W/m^2{\cdot}K}$$

The first equation in (10.14) thus becomes

$$W_t = h_s(\theta_s - \theta_{\mathrm{gas}})A_s + A_{sr}F_{s,w}\sigma_B(\theta_s^{*4} - \theta_w^{*4})$$

$$33.38 = 4.56 \cdot (55.9 - \theta_w)\pi \cdot 0.0358$$

$$+ \pi \cdot 0.0358 \cdot 0.8 \cdot 5.667 \cdot 10^{-8}[(55.9 + 273)^4 - (\theta_w + 273)^4]$$

The solution to this equation is $\theta_w^* = 299.7$ K or $\theta_w = 26.7$°C.

The value of the convective heat transfer from the surface of the cable is given by the first term in the last equation:

$$W_{\mathrm{conv},s} = 13 \ \mathrm{W/m}$$

Next, we will use the second equation in (10.14) to compute the riser outside temperature. The air properties are evaluated at the riser inside surface temperature. At 27°C, the air properties are as follows:

$$\beta = 3.3 \cdot 10^{-3} \ 1/\mathrm{K}$$

$$v = 15.8 \cdot 10^{-6} \ \mathrm{m^2/s}$$

$$\mathrm{Pr} = 0.71$$

$$k_{\mathrm{air}} = 0.0262 \ \mathrm{W/K{\cdot}m}$$

The heat convection coefficient for the inside surface of the riser is obtained from Table 10.2:

$$\mathrm{Gr} = g\beta(\theta_{\mathrm{gas}} - \theta_w)(D_d^*/2)^4/(Lv^2)$$

$$= \frac{9.81 \cdot 3.3 \cdot 10^{-3}\left(\dfrac{55.9 + 26.7 + 2.20}{4} - 26.7\right)0.0508^4}{4.87 \cdot (15.8 \cdot 10^{-6})^2} = 700$$

$$\mathrm{Gr{\cdot}Pr} = 700 \cdot 0.71 = 497$$

$$\mathrm{Nu} = 0.35(\mathrm{Gr{\cdot}Pr})^{0.28} = 0.35(497)^{0.28} = 1.99$$

$$h_w = \mathrm{Nu} \cdot k_{\mathrm{air}}/(D_d/2) = 1.99 \cdot 0.0262/0.0508 = 1.03 \ \mathrm{W/m^2{\cdot}K}$$

The thermal resistance per unit length of the riser is equal to

$$T_4'' = \frac{\rho}{2\pi}\ln\frac{D_o}{D_d} = \frac{3.33}{2\pi}\ln\frac{0.1144}{0.1016} = 0.063 \ \mathrm{K{\cdot}m/W}$$

The energy balance equation (10.7) [second equation in (10.14)] becomes

$$\frac{\theta_w - \theta_o}{T_4''} = h_w(\theta_{gas} - \theta_w)A_w + A_{sr}F_{s,w}\sigma_B(\theta_s^{*4} - \theta_w^{*4})$$

$$\frac{26.7 - \theta_o}{0.063} = 1.03(30.6 - 26.7)\pi \cdot 0.1016_\pi \cdot 0.0358 \cdot 0.8 \cdot 5.667 \cdot 10^{-8}$$
$$[(55.9 + 273)^4 - (26.7 + 273)^4]$$

The solution to this equation gives $\theta_o = 25.4$°C.

Finally, equation (10.10) is used to compute the ambient temperature. The air properties are evaluated at the outside riser film temperature:

$$\theta_f = 0.5(25.4 + 20) = 22.7°\text{C}$$

At this temperature, the air properties are as follows:

$$\beta = 3.38 \cdot 10^{-3}\ 1/\text{K}$$
$$v = 15.4 \cdot 10^{-6}\ \text{m}^2/\text{s}$$
$$\text{Pr} = 0.71$$
$$k_{air} = 0.0259\ \text{W/K}\cdot\text{m}$$

And the heat convection coefficient is obtained from Table 10.3:

$$\text{Re} = V_{air}D_o^*/v = 3.0 \cdot 0.1144/(15.4 \cdot 10^{-6}) = 2.2 \cdot 10^4$$
$$\text{Nu} = b(\text{Re})^p\text{Pr}^{1/3} = 0.193(2.2 \cdot 10^4)^{0.618} \cdot 0.71^{0.333} = 83.1$$
$$h_{o,f} = \text{Nu} \cdot k_{air}/D_o^* = 83.1 \cdot 0.0259/0.1144 = 18.8\ \text{W/m}^2\cdot\text{K}$$

Before we apply equation (10.10), we have to determine the area of the riser exposed to solar radiation. Considering the results obtained by Cress and Motlis (1991) discussed in Section 10.3.3, we assume that $A_{os} = 0.32 \cdot 4.87 \cdot 0.1144 = 0.178$ m^2 per unit length. The energy balance equation (10.10) [third equation in (10.14)] becomes

$$\frac{\theta_w - \theta_o}{T_4''} + A_{os}\alpha_o H = A_o h_o(\theta_o - \theta_{amb}) + A_o\varepsilon_o\sigma_B(\theta_o^{*4} - \theta_{amb}^{*4})$$

$$\frac{26.7 - 25.4}{0.063} + 0.178 \cdot 0.4 \cdot 500$$
$$= \pi \cdot 0.1144[18.8 \cdot (25.4 + 273 - \theta_{amb}^*) + 0.7 \cdot 5.667 \cdot 10^{-8}$$
$$(25.4 + 273)^4 - \theta_{amb}^{*4})]$$

Again, a numerical technique was used to find $\theta_{amb} = 18.5$°C This value is close to the actual ambient temperature; hence, the assumed conductor temperature is correct. If the calculated ambient temperature is significantly different from the given value, the equations should be solved again with different values for the initial conditions. In a computer program, equations (10.14) should be solved simultaneously using iterative techniques.

10.4 CABLES IN TRAYS

A typical cable tray installation which is found in the electric power generation and distribution industry can be visualized as a 3 in deep, 24 in wide metal trough containing anywhere from to 20 to 400 randomly arranged cables ranging in size from #12 AWG to 750 kcmil. This array of cables is usually secured along the cable tray to prevent it from shifting if additional cables should be pulled into the tray. In many cases, especially in nuclear power plants, the trays are covered with fire-protection wrap around the raceway. Because of the very strong mutual heating effects, the ampacity of cables in trays is usually lower than computed with the formulas given in earlier chapters. In the following sections, we will develop models for the ampacity computations of cables in open-top and enclosed cable trays. As a special case of the enclosed tray, we will consider cables in fire-protection wrapped cable trays.

10.4.1 Cables in Single Open-Top Cable Tray

10.4.1.1 Introduction. Since 1975, the accepted technique in North America for calculating the ampacity of power cables in trays has been the use of tables provided by ICEA/NEMA Standard P54-440 (ICEA, 1986). These tables are based upon a thermal model originally proposed by Stolpe (1971) that assumes that every cable in the cable tray carries the maximum current producing the maximum cable bundle temperature. The ICEA Standard also assumes that the heat generated in the cables is uniformly distributed throughout the cross section of the cable mass.

In order to remove the conservatism in the thermal model based on the above assumptions, Harshe and Black (1994) proposed a simple thermal model that can be used to determine the maximum temperature of a mass of cables in a single, open-top horizontal cable tray. The model accounts for load diversity within the tray by providing two different loading options. The thermal model has been verified by comparing predicted cable bundle temperatures with temperatures measured at an actual installation. The model is a subset of equations (10.14), and is described in the following section.

10.4.1.2 Thermal Model. The basis of the thermal model that relates the current in the conductors and the temperature distribution within the cable bundle is the conservation of energy; that is, all of the heat that is generated in the cable bundle must be transferred through the cables to the surface of the cable bundle and from the surface to the air by radiation and convection. In the following model, the cable bundle may consist of any number of cables with arbitrary conductor sizes. Each cable in the bundle is assumed to have a known current, and the thermal model calculates the temperature profile throughout the cable bundle. The model accounts for nonuniform heating within the cable tray by providing two different loading options. The first option assumes that the cables generate heat uniformly across the cable-tray cross section; Harshe and Black called this the Uniform Model. The second option assumes that the heavily loaded cables are concentrated at the cable tray centerline, and that they are surrounded by more lightly loaded cables. This model is referred to as the Hot-Spot Model, and is schematically shown in Fig. 10.4. The mathematical formulation assumes that the cables are located indoors in still air.

The energy balance for the bundle of N cables in the cable tray is a combination of equations (10.2) and (10.10) with the cover removed, and can be expressed as

$$W_{\text{total}} = \sum_{i=1}^{N} n_i R_i I_i^2 = h_s A_s(\theta_s - \theta_{\text{amb}}) + \sigma \varepsilon A_s(\theta_s^{*4} - \theta_{\text{amb}}^{*4}) \tag{10.26}$$

where n_i is the number of cables in the subgroup of cables of the same size, "total" refers to the total value for the entire cable bundle, and the subscript s denotes the surface value. The

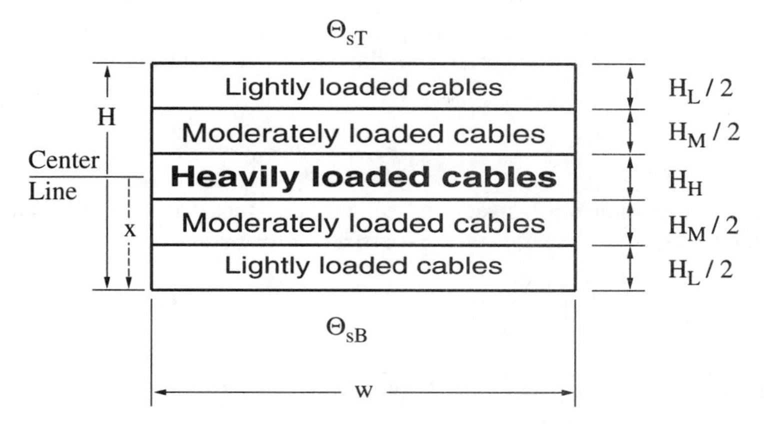

Figure 10.4 Geometry of the Hot-Spot Model assumption (Harshe and Black 1994).

remaining quantities are defined in Section 10.2. The temperature at the bottom surface of the bundle θ_{sB} will, in general, be different from the temperature θ_{sT} at the top of the cable mass because the free convection coefficients are different. The convective heat transfer coefficient for the bottom surface of the cable trays is given by (Harshe and Black, 1994)

$$h_B = 0.248 k_{\text{air}} w^{-0.25} \left[\left(\frac{g\beta}{\nu^2} \right) (\theta_{sB} - \theta_{\text{amb}}) \right]^{0.25} \tag{10.27}$$

and the convective coefficient for the top surface of the cable tray depends upon the value for the Rayleigh Number, Ra:

$$h_T = 0.496 k_{\text{air}} w^{-0.25} \left[\left(\frac{g\beta}{\nu^2} \right) (\theta_{sT} - \theta_{\text{amb}}) \right]^{0.25} \quad \text{for} \quad 10^5 < \text{Ra} < 10^7$$
$$h_T = 0.134 k_{\text{air}} \left[\left(\frac{g\beta}{\nu^2} \right) (\theta_{sT} - \theta_{\text{amb}}) \right]^{0.333} \quad \text{for} \quad 10^7 < \text{Ra} < 10^{10} \tag{10.28}$$

The thermal properties in equations (10.27)–(10.28) are evaluated at an average film temperature $\theta_f = (\theta_s + \theta_{\text{amb}})/2$ and their computation is described in Appendix D, with the value of θ equal to θ_{sB} or θ_{sT}, depending on which surface of the cable mass is considered. The Rayleigh Number is given by

$$\text{Ra} = 0.71 w^3 \left(\frac{g\beta}{\nu^2} \right) (\theta_s - \theta_{\text{amb}}) \tag{10.29}$$

The relationship between the centerline, or maximum temperature within the cable bundle, and the temperature of the surface of the bundle is obtained by applying equation (3.6):

$$\theta_{\max} = \theta_{sB} + \frac{\rho}{8w} \left[W_H \left(\frac{H_H}{2} + H_L + H_M \right) + W_M \left(\frac{H_M}{2} + H_L \right) + W_L \frac{H_L}{2} \right] \tag{10.30}$$

where ρ is the thermal resistivity of the cable bundle including conductor, insulation, and encapsulated air, K·m/W. The W_H, W_M, and W_L are the watts generated per unit length of cable tray in the three layers.

If we make a simplification proposed by Harshe and Black (1994), and assume that $\theta_{sB} = \theta_{sT} = \theta_s$, then the product of the convective heat transfer coefficient and the surface area is

$$h_s A_s = w(h_B + h_T) \tag{10.31}$$

and the relationship between the centerline, or maximum temperature within the cable bundle, and the temperature of the surface of the bundle is

$$\theta_{\text{max}} = \theta_s + \frac{\rho}{4w}\left[W_H\left(\frac{H_H}{2} + H_L + H_M\right) + W_M\left(\frac{H_M}{2} + H_L\right) + W_L\frac{H_L}{2}\right] \tag{10.32}$$

If the heat is generated uniformly throughout the cable bundle, equation (10.32) reduces to

$$\theta_{\text{max}} = \theta_s + \frac{\rho H}{4w}\left(\frac{W_{\text{total}}}{2}\right) \tag{10.33}$$

If the current in each cable is known, equation (10.32) can be used to calculate the cable bundle surface temperature. The maximum cable temperature within the bundle occurs at the centerline, and is a function of the distribution of energy generated throughout the cable bundle. Referring to Fig. 10.4, the depths of the three cable layers are

$$H_x = \frac{\xi}{w}\sum_{i=1}^{n_x} n_i d_i^2; \qquad x = H, M, L \tag{10.34}$$

where w = width of the cable tray, m

H, M, L = subscripts representing the value of heavily, moderately, and lightly loaded cables, respectively

ξ = geometric parameter that accounts for the packing between the cables in the bundle and

ξ = 1.0 for cable packing used in ICEA Standard P54-440

ξ = 0.786 for cable packing using circular cross-section area.

The thermal model given by equations (10.26)–(10.30) is solved iteratively because the heat transfer coefficients and electrical resistances of the metallic parts of the cables are both functions of the cable temperature. In the computational algorithms, the resistances of the cables outside the centerline can conservatively be assumed at the maximum or centerline temperature for the given cable loading. Harshe and Black (1994) report that selecting the initial surface and centerline temperatures at 10 and 20°C, respectively, above the ambient temperature results in convergence within five iterations. The computational procedure is very similar to that described in Example 10.1, and therefore will not be repeated here.

10.4.2 Cables in Covered Trays

Raceway systems in electric generating stations are often enclosed. Engmann (1984) presented a method for the calculation of ampacity of cables in a covered tray. He later extended the model to trays with raised covers (Engmann, 1986). In 1989, Save and Engmann further extended the above techniques to fire-protection wrapped cable trays.

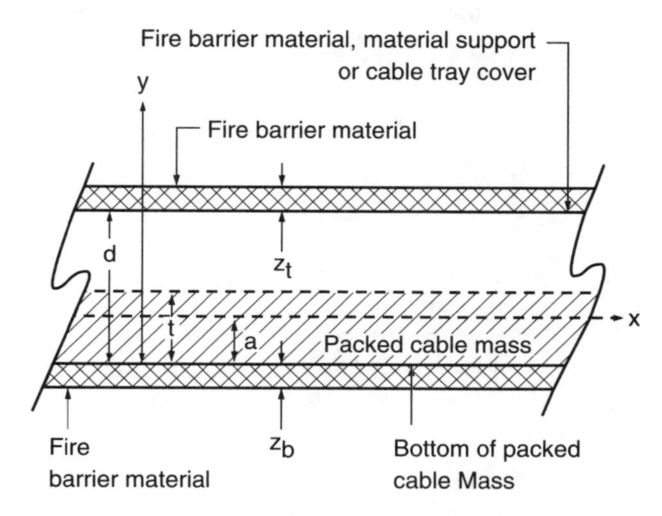

Figure 10.5 Cables in fire-protection wrapped cable tray (Save and Engmann, 1989).

The thermal model of the covered tray cable system is a combination of the models describing cables in riser poles and cables in open-top trays. The geometry of a typical thermal system is shown in Fig. 10.5.

The energy balance equations (10.14) can be used to describe the heat transfer process from the top of the covered tray to the environment. With the absence of solar radiation, the modified equations (10.14) are as follows:

$$\begin{aligned} W_{\text{total}} &= h_s(\theta_{sT} - \theta_{\text{gas}})A_s + A_{sr}F_{s,w}\sigma_B(\theta_{sT}^{*4} - \theta_w^{*4}) \\ \frac{\theta_w - \theta_o}{T_4''} &= h_w(\theta_{\text{gas}} - \theta_w)A_w + A_{sr}F_{s,w}\sigma_B(\theta_{sT}^{*4} - \theta_w^{*4}) \\ \frac{\theta_w - \theta_o}{T_4''} &= h_o(\theta_o - \theta_{\text{amb}})A_o + A_o\varepsilon_o\sigma_B(\theta_o^{*4} - \theta_{\text{amb}}^{*4}) \end{aligned} \tag{10.35}$$

Subscripts s, w, and o denote the top surface of the cable bundle, the inside surface of the cover, and the outside surface of the cover, respectively. The heat transfer coefficients h_s and h_w are defined as follows.

Coefficient h_o for the outside surface of the tray cover is the same as $h_{o,n}$ given in Table 10.3 with the length L replaced by $w/2$ and constant $c = 0.57$. Even though turbulent flow conditions normally do not apply in this case ($10^7 < \text{Ra} < 10^{11}$), the following correlations proposed by Raithby and Hollands (1985) can be used for a wide range of the Rayleigh Number ($\text{Ra} > 1$):

$$\begin{aligned} \text{Nu} &= [(\text{Nu}_l)^{10} + (\text{Nu}_t)^{10}]^{0.1} \\ \text{Nu}_l &= 1.4/\ln(1 + 0.602\text{Ra}^{0.25}), \quad \text{Nu}_t = 0.14\text{Ra}^{0.333} \end{aligned} \tag{10.36}$$

The coefficient h_s is given by

$$
\begin{aligned}
&h_s = h_w = \mathrm{Nu} \cdot k_{\mathrm{air}}/\delta \\
&\mathrm{Nu} = 1 \qquad \text{if} \quad \mathrm{Gr} < 1708 \\
&\mathrm{Nu} = 0.195\mathrm{Gr}^{0.25} \qquad \text{if} \quad 10^4 < \mathrm{GrPr} < 4 \cdot 10^5 \\
&\mathrm{Nu} = 0.069(\mathrm{Gr{\cdot}Pr})^{0.333} \qquad \text{if} \quad 4 \cdot 10^5 < \mathrm{GrPr} < 10^9 \\
&\mathrm{Gr} = g\beta(\theta_s - \theta_w)\delta^3/v^2
\end{aligned}
\tag{10.37}
$$

where δ (m) is the thickness of the air gap in the tray. Air thermal properties are evaluated at the average film temperature.

Equation (10.26) can be applied for the bottom of the tray. Subscript s pertains here to the bottom surface of the tray, and the convection coefficient is given by equation (10.27).

Non-uniformly heated bundle can also be considered in this case by applying equation (10.30) [Harshe and Black, (1996)].

10.4.2.1 Cables In Fire-Protection Wrapped Tray. Raceway systems in nuclear generating stations are often enclosed with protective wrap to meet regulatory or underwriter requirements. Fire- protective wrap systems are designed to reduce the heat transfer from a fire source to the raceway interior. The fire wrap may also impede the transfer of heat from the cables inside the raceway to the environment.

The energy balance equations (10.35) are also applicable in this case. Let θ_{FB} be the temperature at the bottom of the fire-protection material. Since the heat is transferred by conduction through the wrap, the following relationship holds at the bottom of the fire-protection barrier:

$$
\theta_{FB} = \theta_{sB} - \frac{W_{\mathrm{total}}}{2} \frac{\rho_W \cdot z_B}{w} \tag{10.38}
$$

At the top of the tray, the thermal resistances in equation (10.35) are modified to represent the combined thermal resistance of the tray cover and fire barrier material:

$$
T_4'' = \frac{\rho_C z_C + \rho_w z_T}{w} \tag{10.39}
$$

where
ρ_C = thermal resistivity of the cover material, K·m/W
ρ_W = thermal resistivity of the wrap material, K·m/W
z_B = thickness of the fire barrier material at the bottom of the tray, m
z_T = thickness of the fire barrier material at the top of the tray, m
z_C = thickness of the tray cover material, m
w = width of the tray, m.

The model is applicable to cables in ladders, troughs, or solid bottom trays. The cables may be installed randomly, without maintained spacing, and no maintained segregation of power and control cables.

The model is applicable to a fire-wrap system that is made of a single, relatively homogeneous wrap material. The wrap material support configuration (if any) need not be considered if the support is relatively thin or has relatively high thermal conductivity.

10.5 CABLES IN TUNNELS AND SHAFTS

10.5.1 Horizontal Tunnels

Cables are sometimes installed in tunnels provided for other purposes. In generating stations, short tunnels are often used to convey a large number of cable circuits. Long tunnels are built or existing tunnels adapted solely for the purpose of carrying major EHV transmission circuits which for various reasons cannot be carried overhead. River crossings are obvious cases where tunnels would be used either for technical or environmental reasons. The cost of such installations is very considerable, and it is desirable to optimize as far as possible the current-carrying capacity, groupings, and number of circuits to be installed to meet a given transmission capacity.

Detailed investigations into the rating of cables in tunnels have been given by Burrell (1951), Giaro (1960), Germay (1963), Kitagawa (1964), and Whitehead and Hutchings (1938). These are based on heat transfer evaluations from equations established for mechanical and chemical installations. Weedy and El Zayyat (1972, 1973) presented a method more suitable for cable installations. They performed measurements to determine suitable convection coefficients. Their results were later adopted by CIGRE and published in *Electra* (CIGRE 1992a, 1992b). Only the steady-state ratings with natural cooling are considered here. Transient analysis using numerical methods and forced cooling computations are described in the cited *Electra* papers.

10.5.1.1 Thermal Model. It may be desirable to rate cable circuits in horizontal tunnels on the basis of free convection only in view of the difficulty in assessing air velocities along the sides and flow of a tunnel, as distinct from the bulk velocity. In this case, since the equivalent diameter of the tunnel is assumed to be much greater than the diameter of the cable, the heat transfer to the outside wall of the tunnel can be described by the energy balance equations (10.1) and (10.7). The heat transfer for the tunnel outside surface is described by a conduction equation. Thus, the energy balance equations take the form

$$\begin{aligned} W_{\text{total}} &= h_s(\theta_s - \theta_{\text{gas}})A_s + A_{sr}F_{s,w}\sigma_B(\theta_s^{*4} - \theta_w^{*4}) \\ \frac{\theta_w - \theta_o}{T_4''} &= h_w(\theta_{\text{gas}} - \theta_w)A_w + A_{sr}F_{s,w}\sigma_B(\theta_s^{*4} - \theta_w^{*4}) \\ \frac{\theta_w - \theta_o}{T_4''} &= \frac{\theta_o - \theta_{\text{amb}}}{T_4'''} \end{aligned} \tag{10.40}$$

where T_4'' and T_4''' are the thermal resistances of the tunnel wall and the surrounding soil, respectively. For deep tunnels, adiabatic conditions are sometimes assumed, which means that $\theta_o = \theta_{\text{amb}}$ (CIGRE, 1992a).

The thermal resistance of the soil for circular tunnels is computed from equation (9.20) with the tunnel diameter replacing cable external diameter. For a square tunnel, the expression derived by Goldenberg for a buried square trough and reported by Symm (1969) can be used. This expression is

$$T_4''' = \frac{\rho_s}{2\pi}\ln\left(3.388\frac{L_T}{a}\right) \tag{10.41}$$

where ρ_s = thermal resistivity of the soil, K·m/W
L_T = depth of tunnel centerline, m
a = height and width of square cable tunnel, m.

The radiation shape factor is given by

$$F_{s,w} = (1 + \sigma_s/\varepsilon_s + \zeta A_s\sigma_w/A_w\varepsilon_w)^{-1}$$
$$\zeta = \frac{N}{1 - \dfrac{N(N-1)\pi A_s}{A_w}\dfrac{\varepsilon_s\sigma_w}{\varepsilon_w}} \tag{10.42}$$

where N is the number of cables. The mutual radiation area is computed from equation (10.16). For two cables, it can be shown to be

$$A_m = \sqrt{s^2 + D_e^2} - s \quad (s \neq D_e) \tag{10.43}$$

where s is the spacing between the axes of the cables. For three cables in trefoil, the mutual radiation area is $A_m = 0.618D_e$, and for three cables spaced horizontally, it is twice the value given by equation (10.43).

Since cables in tunnels are often installed in horizontal and vertical groups, the mutual heating effect reduces cable ratings. Taking this into account, the convection coefficient for groups of cables in horizontal tunnels is obtained from the correlations given in Table 10.3 with the modification shown in equation (10.44):

$$h_s = h_{s,n} = \eta_c \cdot \eta_{wh} \cdot \eta_{wv} \cdot \text{Nu} \cdot k_{\text{air}}/D_e$$
$$\text{Nu} = c(\text{Gr}\cdot\text{Pr})^n \tag{10.44}$$
$$\text{Gr} = g\beta(\theta_s - \theta_{\text{gas}})D_e^3/v^2$$

where η_c, η_{wh}, and η_{wv} are the correction factors for multicable, horizontal wall, and vertical wall effects, respectively. These correction factors are obtained from Table 9.11 and equation (9.93).

The values of coefficients c and n obtained from experimental results by Weedy and El Zayyat (1972a) are summarized in Table 10.6.

The heat convection coefficient for the tunnel wall h_w can be taken from the correlations describing a small heat source near a vertical wall:

$$h_w = \text{Nu} \cdot k_{\text{air}}/L$$
$$\text{Nu} = 0.59(\text{Gr}\cdot\text{Pr})^{0.25} \tag{10.45}$$
$$\text{Gr} = g\beta(\theta_{\text{gas}} - \theta_w)L^3/v^3$$

where L is the height of the tunnel. For circular tunnels, L can be replaced by the tunnel inside diameter. If the cables are located on one side of the tunnel only, then the surface area A_w in equations (10.40) should be equal to the area of the wall close to the cables. In the case of circular tunnels, half of the area of the tunnel should be used.

TABLE 10.6 Parameters for Convection Formulas for Cables in Tunnels (Weedy, 1988)

Grouping	Spacing	Cable of Group	c	n
∴ (three, triangular)	Not touching $s / D_e > 2$		0.725	0.24
•••	Not touching $s / D_e > 2$	Middle	0.785	0.21
		Outer	0.815	0.21
•••	Touching	Middle	0.55	0.20
		Outer	0.69	0.20
⋮	Touching	Middle	0.40	0.215
		Upper	0.54	0.215
		Lower	0.715	0.215
⋮	Not touching $s / D_e > 2$	Middle	0.60	0.225
		Upper	0.70	0.225
		Lower	0.78	0.225
∴	Touching	Upper	0.48	0.20
		Lower	0.68	0.20

EXAMPLE 10.2

We will compute the ratings of six circuits each in trefoil formation located in a horizontal tunnel as shown in Fig. 10.6. Cable model No. 1 will be used with the following parameters (see Table A.1): $D_e = 0.0358$ mm, $R = 0.0781 \cdot 10^{-3} \Omega/\text{m}$, $\lambda_1 = 0.09$, $T_1 = 0.214$ K·m/W, and $T_3 = 0.104$ K·m/W.

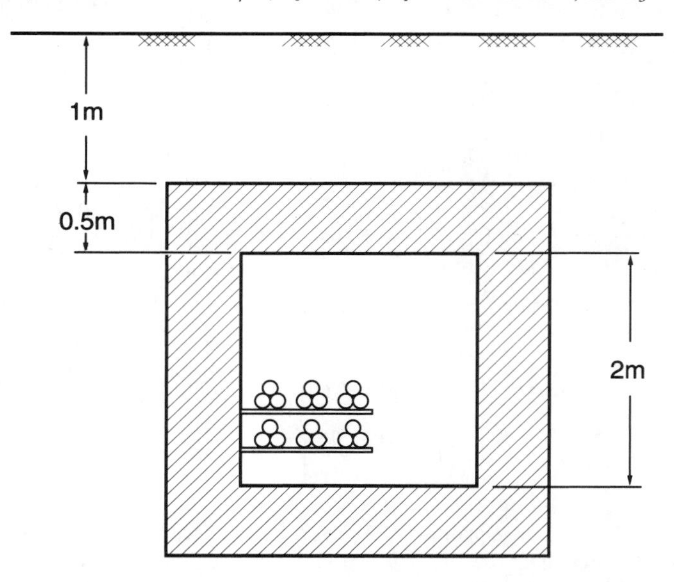

Figure 10.6 Installation of cables in tunnel in Example 10.6.

The tunnel has a square cross section with a height of 2.0 m, shallow installation with 0.5 m concrete walls, and 1 m of soil above the roof. The thermal resistivities of concrete and soil are 0.6 and 1.2 K·m/W, respectively. The soil ambient temperature is 15°C.

The external thermal resistances of the tunnel wall and the soil are obtained from equations (3.4) and (10.41), respectively, and are equal to

$$T_4'' = \frac{\rho_{th} \cdot l}{S} = \frac{0.6 \cdot 0.5}{2 \cdot 1} = 0.15 \text{ K·m/W}$$

$$T_4''' = \frac{\rho_s}{2\pi} \ln\left(3.388\frac{L_T}{a}\right) = \frac{1.2}{2\pi} \ln\left(3.338\frac{2.5}{2}\right) = 0.273 \text{ K·m/W}$$

The heat transfer coefficient for the hottest cable is obtained from equation (10.44). The air thermophysical properties are evaluated at the film temperature $\theta_f = (\theta_s + \theta_w)/2$. At this point, neither temperature is known; however, they can be approximated for the purpose of performing the calculations. Let us assume that $\theta_f = 30.5°\text{C}$. Since only one iteration of the process is illustrated in this example, if the film temperature computed at the end of the example is different from the one assumed here, computations should be repeated from this point on. At the assumed film temperature, the air thermophysical properties are (see Appendix D)

$$\begin{aligned} \beta &= 3.3 \cdot 10^{-3} \text{ 1/K} \\ \nu &= 16.1 \cdot 10^{-6} \text{ m}^2/\text{s} \\ \text{Pr} &= 0.71 \\ k_{\text{air}} &= 0.0265 \text{ W/K·m} \end{aligned}$$

The cable surface convection coefficient is obtained from equation (10.44) assuming that convection takes place on one wall only. The required constants are taken from Table 10.4.

$$\begin{aligned} \text{Gr} &= g\beta(\theta_s - \theta_{\text{gas}})D_e^3/\nu^2 = 9.81 \cdot 3.3 \cdot 10^{-3}\left(\theta_s - \frac{\theta_s + \theta_w}{2}\right) 0.0358^3/(16.1 \cdot 10^{-6})^2 \\ &= 5.7 \cdot 10^3 \left(\theta_s - \frac{\theta_s + \theta_w}{2}\right) \\ \text{Gr·Pr} &= 5.7 \cdot 10^3 \left(\theta_s - \frac{\theta_s + \theta_w}{2}\right) \cdot 0.71 = 4.0 \cdot 10^3 \left(\theta_s - \frac{\theta_s + \theta_w}{2}\right) \\ \text{Nu} &= c(\text{Gr·Pr})^n = 0.48\left[4.0 \cdot 10^3 \left(\theta_s - \frac{\theta_s + \theta_w}{2}\right)\right]^{0.20} = 2.52\left(\theta_s - \frac{\theta_s + \theta_w}{2}\right)^{0.2} \\ h_s &= \eta_c \cdot \eta_{wh} \cdot \eta_{wv} \cdot \text{Nu} \cdot k_{\text{air}}/D_e = \eta_c \cdot \eta_{wv} \cdot 2.52\left(\theta_s - \frac{\theta_s + \theta_w}{2}\right)^{0.2} \cdot 0.0265/0.0358 \\ &= \eta_c \cdot \eta_{wv} 1.87\left(\theta_s - \frac{\theta_s + \theta_w}{2}\right)^{0.2} \quad \text{W/m}^2\text{·K} \end{aligned}$$

From Table 9.11, and equation (9.93), the reduction factors for a group of cables are as follows: the mutual heating effect $\eta_c = 1/1.39 = 0.72$ and the vertical wall effect $\eta_{wv} = 1/1.23 = 0.81$. Thus,

$$h_s = 0.72 \cdot 0.81 \cdot 1.87\left(\theta_s - \frac{\theta_s + \theta_w}{2}\right)^{0.2} = 1.09\left(\theta_s - \frac{\theta_s + \theta_w}{2}\right)^{0.2} \quad \text{W/m}^2\text{·K}$$

The heat convection coefficient for the inside surface of the tunnel is obtained from the equations (10.45). Hence,

$$\mathrm{Gr} = g\beta(\theta_{\mathrm{gas}} - \theta_w)L^3/v^2 = 9.81 \cdot 3.3 \cdot 10^{-3}\left(\frac{\theta_s + \theta_w}{2} - \theta_w\right)2.0^3/(16.1 \cdot 10^{-6})^2$$

$$= 1.0 \cdot 10^9 \left(\frac{\theta_s + \theta_w}{2} - \theta_w\right)$$

$$\mathrm{Gr{\cdot}Pr} = 1.0 \cdot 10^9 \left(\frac{\theta_s + \theta_w}{2} - \theta_w\right) \cdot 0.71 = 7.1 \cdot 10^8 \left(\frac{\theta_s + \theta_w}{2} - \theta_w\right)$$

$$\mathrm{Nu} = 0.59(\mathrm{Gr{\cdot}Pr})^{0.25} = 0.59\left[7.1 \cdot 10^8 \left(\frac{\theta_s + \theta_w}{2} - \theta_w\right)\right]^{0.25} = 96.3\left(\frac{\theta_s + \theta_w}{2} - \theta_w\right)^{0.25}$$

$$h_w = \mathrm{Nu} \cdot k_{\mathrm{air}}/L = 96.3\left(\frac{\theta_s + \theta_w}{2} - \theta_w\right)^{0.25} \cdot 0.0265/2.0$$

$$= 1.28\left(\frac{\theta_s + \theta_w}{2} - \theta_w\right)^{0.25} \ \mathrm{W/m^2{\cdot}K}$$

The emissivity of the cable surface is 0.9 (see Table 9.10) and of the concrete wall is 0.63. Because the cable and the tunnel wall are opaque, we have

$$\sigma_s = 1 - \varepsilon_s = 1 - 0.9 = 0.1, \qquad \sigma_w = 1 - \varepsilon_w = 1 - 0.63 = 0.37$$

The thermal radiation shape factor for one circuit is obtained from equation (10.42) and is equal to

$$\zeta = \frac{N}{1 - \dfrac{N(N-1)\pi A_s}{A_w}\dfrac{\varepsilon_s\sigma_w}{\varepsilon_w}} = \frac{3}{1 - \dfrac{3(3-1)\pi^2 0.0358}{2.0 \cdot 1.0}\dfrac{0.9 \cdot 0.37}{0.63}} = 6.82$$

$$F_{s,w} = (1 + \sigma_s/\varepsilon_s + \zeta A_s\sigma_w/A_w\varepsilon_w)^{-1}$$

$$= [1 + 0.1/0.9 + 6.82 \cdot \pi \cdot 0.0358 \cdot 0.37/(2.0 \cdot 1 \cdot 0.63)]^{-1} = 0.75$$

The mutual radiation area for a single circuit is equal to $A_m = 3 \cdot 0.618 \cdot 0.0358 = 0.066\ \mathrm{m^2}$. Hence, $A_{sr} = 3 \cdot \pi \cdot 0.0358 - 0.066 = 0.271\ \mathrm{m^2}$ per unit length.

Assuming that the maximum allowable conductor temperature is 90°C, the total heat generated in the cable can be obtained from equation (4.1) and the constants taken from Table A.1. We have

$$W_t = I^2R(1 + \lambda_1 + \lambda_2) = I^2 0.0781 \cdot 10^{-3}(1 + 0.09) = 0.085 \cdot 10^{-3} I^2$$

The internal thermal resistance of the cable is obtained from equation (4.4) as

$$T = T_1 + (1 + \lambda_1)T_3 = 0.214 + 1.09 \cdot 0.104 = 0.327\ \mathrm{K{\cdot}m/W}$$

and the relationship between conductor and surface temperature is thus given by

$$R \cdot I^2 \cdot T = 0.085 \cdot 10^{-3} \cdot I^2 \cdot 0.327 = 2.8 \cdot 10^{-5} \cdot I^2 = 90 - \theta_s$$

The energy balance equations (10.40) together with the last equation can be written in the following form:

$$W_{\text{total}} = h_s(\theta_s - \theta_{\text{gas}})A_s + A_{sr}F_{s,w}\sigma_B(\theta_s^{*4} - \theta_w^{*4})$$

$$18 \cdot 2.8 \cdot 10^{-5} \cdot I^2 = 6 \cdot 1.09 \left(\theta_s - \frac{\theta_s + \theta_w}{2}\right)^{1.2} \pi \cdot 0.0358 + 6.0271 \cdot 0.75 \cdot 5.667 \cdot 10^{-8}(\theta_s^{*4} - \theta_w^{*4})$$

$$\frac{\theta_w - \theta_o}{T_4''} = h_w(\theta_{\text{gas}} - \theta_w)A_w + A_{sr}F_{s,w}\sigma_B(\theta_s^{*4} - \theta_w^{*4})$$

$$\frac{\theta_w - \theta_o}{0.15} = 6 \cdot 1.28 \left(\frac{\theta_s + \theta_w}{2} - \theta_w\right)^{1.25} 2.0 \cdot 1 + 6 \cdot 0.271 \cdot 0.75 \cdot 5.667 \cdot 10^{-8}(\theta_s^{*4} - \theta_w^{*4})$$

$$\frac{\theta_w - \theta_o}{T_4''} = \frac{\theta_o - \theta_{\text{amb}}}{T_4'''}$$

$$\frac{\theta_w - \theta_o}{0.15} = \frac{\theta_o - 15}{0.273}$$

$$2.8 \cdot 10^{-5} \cdot I^2 = 90 - \theta_s$$

The solution of these equation yields

$$I = 411 \text{ A}, \quad \theta_s = 85^\circ\text{C}, \quad \theta_w = 79^\circ\text{C}, \quad \theta_o = 56^\circ\text{C}$$

Since the film temperature is higher than assumed in the computations, the iterations should be repeated with new air properties until convergence is achieved.

10.5.2 Vertical Shafts

We will, again, consider installations with natural convection only. There is very little experimental evidence available to determine heat convection coefficients for this type of installation. As a first approximation, cables in vertical shafts can be studied as the cables on riser poles with the open top and closed bottom and adiabatic conditions at the shaft external surface. Hence, the energy balance equation will take the form

$$n[I^2R(1 + \lambda_1 + \lambda_2) + W_d] = h_s\left(\theta_s - \frac{\theta_s + \theta_{\text{amb}}}{2}\right)A_s + A_{sr}F_{s,w}\sigma_B(\theta_s^{*4} - \theta_{\text{amb}}^{*4}) \tag{10.46}$$

The heat convection coefficient is obtained from correlations given at the bottom of Table 10.1.

An alternative equation is given by Endacott *et al.* (1970). Assuming an effective emissivity of 0.9 and natural convection, the heat dissipated from the cable/pipe in a shaft is given by the following empirical formula (Weedy, 1988):

$$W_t = 3.23 \cdot 10^{-11} m \frac{D_e^*}{2}(\theta_s^{*4} - \theta_w^{*4}) + 1.21 \frac{D_e^*}{2}(\theta_s - \theta_{\text{gas}})^{1.25} \tag{10.47}$$

where m is the proportion of the cable/pipe surface available for radiation (affected by obstructions and the proximity of other cables).

10.6 CABLES IN BURIED TROUGHS

Surface trough types of installation has been widely used in North America and Europe for cable routes running alongside railways, canals, and in substations. The cables are normally placed with a small separation between phases at a depth of approximately 0.3 m in a concrete trough provided with a reinforced concrete lid to afford mechanical protection. The important thermal resistances external to the cable are those associated with the trough fill and between the trough surface and the air.

10.6.1 Buried Troughs Filled with Sand

When cables are installed in a sand-filled trough, either completely buried or with the cover flush with the ground surface, there is a danger that the sand will dry out and remain dry for long periods. The cable external thermal resistance may then be very high, and the cable may reach undesirably high temperatures.

For cables laid at depths of less than 0.6 m, cable temperatures may show significant daily or even hourly variations due to changes in solar radiation, wind velocity, and air temperature. The maximum cable surface temperature in a surface trough system derived from analog investigations can be written as (Endacott *et al.*, 1970)

$$\theta_s = W_t T_4 + \frac{0.33 W_t}{\nu^{0.74} L^{*0.2}} + \frac{0.29 H}{\nu^{0.89} L^{*0.07}} + \theta_{\text{amb}} \tag{10.48}$$

where θ_s = maximum cable surface temperature, °C

W_t = total cable heat dissipation, W/m

T_4 = thermal resistance from the hottest cable computed in accordance with equation (9.20), K·m/W

L^* = vertical depth below ground surface of the cable axis, m

ν = wind velocity at a height of 10 m, m/s

θ_{amb} = average ambient temperature, °C

H = steady net solar radiation to the ground surface, W/m^2.

From equation (10.48), the effective external thermal resistance is equal to

$$T_4^{\text{eff}} = \frac{\theta_s - \theta_{\text{amb}}}{W_t} = T_4 + \frac{0.33}{\nu^{0.74} L^{0.2}} + \frac{0.29 H}{\nu^{0.89} L^{0.07} W_t} \tag{10.49}$$

This value of the external thermal resistance is substituted in equation (4.3) for T_4. Since T_4^{eff} depends on W_t, equation (4.3) has to be solved iteratively in this case. Numerical studies performed by the author have shown that on some occasions, equation (10.49) may give very high external thermal resistance. Since the upper limit of the external thermal resistance is obtained when the sand is assumed completely dry, this value should be used when the iterative procedure results in T_4^{eff} reaching values which are too high.

A more approximate method, which is used in Great Britain, consists of calculating the temperature rise of the cable assuming the ground surface to be isothermal with an effective ground ambient temperature which takes account of wind velocity, solar radiation, and ambient air temperature effects. These temperatures, for different weather conditions considered for overhead lines specified in column 2, are given in column 4 of Table 10.7. Corresponding values used for directly buried systems at normal laying depths (i.e., $>$ 0.6 m) are given for comparison in column 3 (Endacott *et al.*, 1970).

An alternative simpler approach is suggested in IEC 287 (1982). The standard recommends that the cable rating be computed in the case of surface troughs filled with sand

TABLE 10.7 Equivalent Ground Temperatures for Shallow Filled Troughs

Overhead Line		Ground Temperature	
Rating Condition	Air Temperature (°C)	At Depth > 0.6 m	At Depth 0.3 m
Cold weather	< 4.5	10	10
Normal weather	4.5–18	15	30
Hot weather	> 18	15	40

using a value of 2.5 K·m/W for the thermal resistivity of the sand filling unless a specially selected filling has been used for which the dry resistivity is known.

The three approaches discussed above can lead to radically different cable ratings, as illustrated in Example 10.3. The most pessimistic rating is usually obtained by applying the IEC method.

EXAMPLE 10.3

Consider a three-phase cable system composed of model cable No. 1 laid in a surface trough filled with sand of thermal resistivity 0.6 K·m/W when moist and 2.5 K·m/W when dry. The laying information is shown in Fig. 10.7. We will assume wind velocity of 3 m/s and solar radiation intensity of 200 W/m^2. We will determine the rating of this cable system.

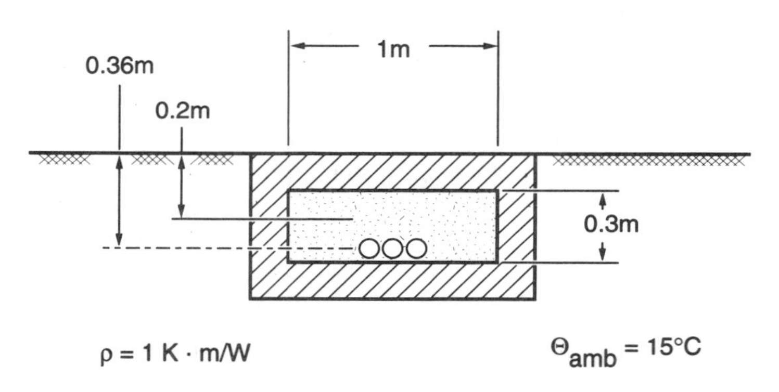

Figure 10.7 Three cables in a buried trough.

The thermal resistances of the internal parts of the cable are given in Table A1. The external thermal resistance of the middle cable is computed in two stages. From equation (9.32), we have

$$T_4(\text{wet}) = \rho_s[0.475\ln(2u) - 0.346] = 0.6\left(0.475\ln\frac{4\cdot 0.36}{0.0358} - 0.346\right) = 0.845 \text{ K·m/W}$$

$$T_4(\text{dry}) = \rho_s[0.475\ln(2u) - 0.346] = 2.5\left(0.475\ln\frac{4\cdot 0.36}{0.0358} - 0.346\right) = 3.52 \text{ K·m/W}$$

The equivalent radius of the trough (ignoring the concrete walls) is obtained from equation (9.61) as

$$\ln r_b = \frac{1}{2}\frac{x}{y}\left(\frac{4}{\pi} - \frac{x}{y}\right)\ln\left(1 + \frac{y^2}{x^2}\right) + \ln\frac{x}{2} = \frac{1}{2}\frac{0.3}{1.0}\left(\frac{4}{\pi} - \frac{0.3}{1.0}\right)\ln\left(1 + \frac{1.0^2}{0.3^2}\right) + \ln\frac{0.3}{2} = -1.53.$$

Hence, $r_b = 0.22$ m and $u = 0.2/0.22 = 0.91$.

Since $u < 1$, the correction factor in equation (9.60) is obtained from the table of extended values. For wet sand it is equal to 0.0464, and for dry sand is equal to −0.174. Thus, $T_4(\text{wet}) = 0.845 + 0.0464 = 0.891$ K·m/W and $T_4(\text{dry}) = 3.52 - 0.174 = 3.35$ K·m/W. In order to compute T_4^{eff}, we need the value of W_t. We will start the iterative process by neglecting first the effect of wind and solar radiation. The ac resistance of the conductor is in this case equal to 0.791 Ω/km, and the concentric neutral loss factor for the hottest cable is equal to 0.101. Therefore, from equation (4.3), we have

$$I = \left[\frac{90 - 15}{0.0791 \cdot 10^{-3} \cdot [0.214 + 1.101 \cdot (0.104 + 0.891)]}\right]^{0.5} = 851 \text{ A}$$

The total losses are equal to $1.101 \cdot 0.0000791 \cdot 851^2 = 63.1$ W/m. The effective external thermal resistance is now obtained from equation (10.49) including the effect of wind and solar radiation:

$$T_4^{\text{eff}} = T_4 + \frac{0.33}{\nu^{0.74} L^{0.2}} + \frac{0.29H}{\nu^{0.89} L^{0.07} W_t} = 0.883 + \frac{0.33}{3^{0.74} 0.36^{0.2}} + \frac{0.29 \cdot 200}{3^{0.89} 0.36^{0.07} \cdot 63.1} = 1.44 \text{ K·m/W}$$

The new ampacity is obtained again from equation (4.3), and is equal to 706 A. Corresponding losses are 43.4 W/m, and the new value of T_4^{eff} is equal to 1.6 K·m/W. Continuing the same way, the process ends after four iterations with the ampacity of 673 A and $T_4^{\text{eff}} = 1.7$ K·m/W. Since the value is smaller than the thermal resistance with completely dry sand, the calculations are completed.

For comparison, we observe that the ampacity of this system using the IEC method is 484 A, and using the equivalent soil ambient temperature of 30°C, taken from Table 10.6, the current rating is equal to 764 A.

10.6.2 Unfilled Troughs of Any Type, with the Top Flush with the Soil Surface and Exposed to Free Air

IEC 287 (1982) suggests using an empirical formula which gives the temperature rise of the air in the trough above the air ambient temperature as

$$\Delta\theta_{tr} = \frac{W_t}{3p} \tag{10.50}$$

where p (m) is that part of the trough perimeter which is effective for heat dissipation. Any portion of the perimeter which is exposed to sunlight is therefore not included in the value of p. The rating of a particular cable in the trough is then calculated as for a cable in free air, but the ambient temperature shall be increased by $\Delta\theta_{tr}$.

EXAMPLE 10.4

We will compute the rating of the cable system analyzed in Example 10.3 assuming now that the trough is unfilled. Since the trough is exposed to solar radiation, we will assume that the value of p is equal to 1.2 m.

The heat dissipation coefficient is obtained from equation (9.80) with the required constants given in Table 9.8.

$$h = \frac{Z}{(D_e^*)^g} + E = \frac{0.62}{0.0358^{0.25}} + 1.95 = 3.38 \text{ W/m}^2\text{·K}^{5/4}$$

To compute cable surface temperature rise, we use equations (9.83) and (9.86):

$$K_A = \frac{\pi D_e^* h T}{1 + \lambda_1 + \lambda_2} = \frac{\pi \cdot 0.0358 \cdot 3.38}{1.09}[0.214 + 1.09 \cdot 0.104] = 0.114$$

$$\frac{\sigma H K_A}{\pi h} = \frac{0.6 \cdot 200 \cdot 0.114}{\pi \cdot 3.38} = 1.29$$

We will start an iterative process by setting the value of $(\Delta\theta_s)^{1/4}$ in the denominator of equation (9.86) to 2. Thus, we have

$$(\Delta\theta_s)_1^{1/4} = \left[\frac{\Delta\theta + \Delta\theta_d + \sigma H K_A/\pi h}{1 + K_A(\Delta\theta_s)_n^{1/4}}\right] = \left[\frac{75 + 1.29}{1 + 0.114 \cdot 2}\right]^{1/4} = 2.8°\text{C}$$

In the second iteration, the value of $(\Delta\theta_s)_2^{1/4}$ is equal to 2.76°C. Convergence is achieved at this iteration. The external thermal resistance is obtained from equation (9.78) and is equal to

$$T_4 = \frac{1}{\pi D_e^* h(\Delta\theta_s)^{1/4}} = \frac{1}{\pi \cdot 0.0358 \cdot 3.38 \cdot 2.76} = 0.95\ \text{K·m/W}$$

The cable rating is given by equation (4.3) as

$$I = \left[\frac{75 - 0.6 \cdot 0.0358 \cdot 500 \cdot 0.95}{0.0781 \cdot 10^{-3}(0.214 + 1.09(0.104 + 0.95))}\right]^{0.5} = 780\ \text{A}$$

The corresponding total losses are equal to 51.8 W/m. Thus, the ambient temperature reduction obtained from equation (10.50) is equal to

$$\Delta\theta_{tr} = \frac{W_t}{3p} = \frac{3 \cdot 51.8}{3 \cdot 1.2} = 43.2°\text{C}$$

The new value of $(\Delta\theta_s)^{1/4}$ is 2.3°C, and the corresponding value of T_4 = 1.15 K·m/W. The permissible current rating is equal to 397 A, and the corresponding losses are 10.3 W/m per cable. The ambient temperature reduction is equal in this case to 10.3°C. Continuing the iterative process, the current rating is obtained at 24th iteration and is equal to 590 A.

REFERENCES

Al-Nimr, M. A. (1993), "Analytical solution for transient laminar fully developed free convection in vertical concentric annuli," *Int. J. Heat and Mass Transfer*, vol. 36, pp. 2385–2395.

Anders, G. J., Roiz, J., and Moshref, A. (July 1990), "Advanced computer programs for power cable ampacity calculations," *IEEE Comput. Appl. Power*, vol. 3, no. 3, pp. 42–45 (1996 revision).

Anders, G. J. (Jan. 1996), "A unified approach to rating of cables on riser poles, in trays, in tunnels and in shafts—A review," *IEEE Trans. Power Delivery*, vol. 11, no. 1, pp. 3–11.

Anders, G. J. (1995), "Rating of cables on riser poles," in *Proc. Jicable '95 Conf.*, Versailles, France, June 1995.

Burmeister, L. C. (1983), *Convective Heat Transfer*. New York: Wiley.

Burrell, R. W. *et al.* (1951), "Forced air cooling for station cables," *AIEE Trans.*, vol. III (70), pp. 1798–1803.

CIGRE (1992a), "Calculation of temperatures in ventilated cable tunnels—Part 1," *Electra*, no. 143, pp. 39–59.

CIGRE (1992b), "Calculation of temperatures in ventilated cable tunnels—Part 2," *Electra*, no. 144, pp. 97–105.

Cress, S. L., and Motlis, J. (Jan. 1991), "Temperature rise of submarine cable on riser poles," *IEEE Trans. Power Delivery*, vol. 6, no. 1, pp. 25–33.

Dryer, J. (Oct. 1978), "Natural convective flow through a vertical duct with restricted entry," *Int. J. Heat and Mass Transfer*, vol. 21, pp. 1344–1354.

El-Shaarawi, M. A., and Sarhan, A. A. (Nov. 1981), "Developing laminar free convection in a heated vertical open-ended concentric annulus," *Indust. & Eng. Chem., Fundamentals*, vol. 20, no. 4, pp. 388–394.

Endacott, J. D., Flack, H. W., Morgan, A. M., Holdup, H. W., Miranda, F. J., Skipper, D. J., and Thelwell, M. J. (1970), "Thermal design parameters used for high capacity EHV cable circuits," CIGRE, Report 21-03.

Engmann, G. (1984), "Ampacity of cable in covered tray," *IEEE Trans. Power App. Syst.*, vol. PAS-103, pp. 345–350.

Engmann, G. (1986), "Cable ampacity in tray with raised cover," *IEEE Trans. Energy Conversion*, vol. EC-1, pp. 113–119.

Germay, N. (1963), "Calculation of the temperature rise of cable in a gallery with forced ventilation," *Rev. Elec., Suppl. to Bull. Soc., Roy. Belge. Elec.*, vol. 4, no. 1, pp. 3–13.

Giaro, J. A. (1960), "Temperature rise of power cables in a gallery with forced ventilation," CIGRE, Report N1 213.

Harshe, B. L., and Black, W. Z. (1994), "Ampacity of cables in single open-top cable trays," *IEEE Trans. Power Delivery*, vol. PWRD-9, no. 4, pp. 1733–1740.

Harshe, B. L., and Black, W. Z. (1996), "Ampacity of cables in single covered trays," Paper 96 WM 209-7 presented at the Winter Meeting of PES in Baltimore, MD.

Hartlein, R. A., and Black, W. Z. (June 1983), "Ampacity of electric power cables in vertical protective risers," *IEEE Trans. Power App. Syst.*, vol. PAS-102, no. 6, pp. 1678–1686.

Holman, J. P. (1990), *Heat Transfer*. New York: McGraw-Hill.

ICEA/NEMA (1986), "Ampacities of cables in open-top cable trays," ICEA Publication No. P54-440, NEMA Publication No. WC51, Washington, DC.

IEC (1982), "Calculation of the continuous current rating of cables (100% load factor)," IEC Standard Publication 287, 2nd ed.

Incropera, F. P., and De Witt, D. P. (1990), *Introduction to Heat Transfer*. New York: Wiley.

Joshi, H. M. (Nov.–Dec. 1987), "Fully developed natural convection in an isothermal vertical annular duct," *Int. Commun. in Heat and Mass Transfer*, vol. 14, no. 6, pp. 657–664.

Joshi, H. M. (1988), "Numerical solutions for developing laminar free convection in vertical annular ducts open at both ends," *Numer. Heat Transfer*, vol. 13, no. 3, pp. 393–403.

Keyhani, M., and Kulacki, F.A. (1985), "Natural convection in enclosures containing tube bundles," in *Natural Convection*, S. KaKac, W. Aung, and R. Viskanta, Eds. New York: Hemisphere.

Keyhani, M., Kulacki, F. A., and Christensen, R. N. (Aug. 1985), "Experimental investigation of free convection in a vertical rod bundle—A general correlation for Nusselt numbers," *J. Heat Transfer, Trans. ASME*, vol. 107, no. 3, pp. 611–623.

Keyhani, M., Prasad, V., and Kulacki, F. A. (1986), "An approximate analysis for thermal convection with application to vertical annulus," *Chem. Eng. Commun.*, vol. 42, no. 4–6, pp. 281–289.

Kitagawa, K. (1964), "Forced cooling of power cables in Japan," CIGRE Paper 213.

Martin, B., and Cohen, M. (1954), "Heat transfer by free convection in an open thermosyphon tube," *Brit. J. Appl. Phys.*, vol. 5, pp. 91–95.

Martin, B. (1955), "Free convection in an open thermosyphon with special reference to turbulent flow," *Proc. Roy. Soc.*, vol. 230(ser. A), p. 502.

Morgan, V. T. (1982), "The thermal rating of overhead-line conductors, Part I. The steady-state thermal model," *Elec. Power Syst. Res.*, vol. 5, pp. 119–139.

Morgan, V. T. (Mar. 1993), "External thermal resistance of aerial bundled cables," *Proc. IEE*, vol. 140, part C, no. 2, pp. 65–62.

Morgan, V. T. (Mar. 1992), "Effect of mixed convection on the external resistance of single-core and multicore bundled cables in air," *Proc. IEE*, vol. 139, part C, no. 2, pp. 109–116.

Morgan, V. T. (1995), Discussion of Anders (1996); *IEEE Trans. Power Delivery*, vol. 11, no. 1, pp. 3–11.

Neher, J. H., and McGrath, M. H. (Oct. 1957), "The calculation of the temperature rise and load capability of cable systems," *AIEE Trans.*, vol. 76, part 3, pp. 752–772.

Raithby, G. D., and Hollands, K. G. T. (1985), "Natural convection," in *Handbook of Heat Transfer*, W. M. Rohsenow and J. P. Hartnett, Eds. New York: McGraw-Hill.

Save, P., and Engmann, G. (1989), "Fire protection wrapped cable tray ampacity," *IEEE Trans. Energy Conversion*, vol. 4, no. 4, pp. 575–584.

Seki, N., Fukusako, S., and Koguchi, K. (1980), "Single-phase heat transfer characteristics of concentric-tube thermosyphone," *Warme und Stoffubertragung*, vol. 4, pp. 189–199.

Stolpe, J. (1971), "Ampacities of cables in randomly filled trays," *IEEE Trans. Power App. Syst.*, vol. PAS-90, pp. 967–973.

Symm, G. T. (1969), "External thermal resistance of buried cables and troughs," *Proc. IEE*, vol. 166, no. 10, pp. 1696–1698.

Weedy, B. M., and El Zayyat, H. M. (1972), "Heat transfer from cables in tunnels and shafts," IEEE Conf. Paper C72506-4.

Weedy, B. M., and El Zayyat, H. M. (1973), "The current capacity of power cables in tunnels," *IEEE Trans. Power App. Syst.*, vol. PAS-92, pp. 298–307.

Weedy, B. M. (1988), *Thermal Design of Underground Systems.* Chichester, England: Wiley.

Whitehead, S., and Hutchings, E. E. (1938), "Current ratings of cables for transmission and distribution," *J. IEE*, vol. 38, pp. 517–557.

11

Ampacity Computations Using Numerical Methods

11.1 INTRODUCTION

We have shown in Chapter 2 that if the thermal resistance is constant, the heat conduction equation (2.12) can be written as

$$\frac{\partial^2 \theta}{\partial x^2} + \frac{\partial^2 \theta}{\partial y^2} + W_{\text{int}}\rho = \frac{1}{\delta} \frac{\partial \theta}{\partial t} \tag{11.1}$$

where $\delta = 1/\rho c$ is the thermal diffusivity of the medium (m^2/s) and the remaining symbols are defined in the list of symbols.

The boundary conditions associated with (11.1) can be expressed in two different forms. If the temperature is known along a portion of the boundary, then

$$\theta = \theta_B(s) \tag{11.2}$$

where θ_B is the known boundary temperature that may be a function of the surface length s. If heat is gained or lost at the boundary due to convection $h(\theta - \theta_{\text{amb}})$ or a heat flux q, then

$$\frac{1}{\rho} \frac{\partial \theta}{\partial n} + q + h\left(\theta - \theta_{\text{amb}}\right) = 0 \tag{11.3}$$

where n is the direction of the normal to the boundary surface, h is a convection coefficient, and θ is an unknown boundary temperature. The solution of these equations yields the temperatures at all points of the region, including the cable conductor.

As discussed in Chapter 2, if the medium surrounding the cables is not uniform, and the earth surface boundary is not isothermal, an analytical solution is generally not feasible and a numerical approach is indicated.

The usual ampacity problem is to compute the permissible conductor current so that the maximum conductor temperature does not exceed a specified value. When numerical methods are used to determine cable rating, an iterative approach has to be used for the purpose. This is accomplished by specifying a certain conductor current and computing the corresponding conductor temperature. Then, the current is adjusted until the specified temperature is found to converge within a specified tolerance.

The limitations of the classical methods will be apparent from a few examples. In the transient analysis discussed in Chapter 5, separate computations were performed for the internal and external parts of the cable. Coupling between internal and external circuits was achieved by assuming that the heat flow into the soil is proportional to the attainment factor of the transient between the conductor and the outer surface of the cable. The validity of the methods did not rest on an analytical proof, but on an empirical agreement of the responses given by the recommended circuits and the temperature transients calculated by more lengthy but more accurate computer-based methods.

In the analytical methods derived in Chapters 4 and 5, the case of a group of cables is dealt with on the basis of the restricted application of superposition. To apply this principle, it must be assumed that the presence of another cable, even if it is not loaded, does not disturb the heat flux path from the first cable, nor the generation of heat within it. This allows separate computations to be performed for each cable, with the final temperature rise being an algebraic sum of the temperature rises due to the cable itself and the rise caused by the other cables. Such a procedure is not theoretically correct, and for better precision, the temperature rise caused by simultaneous operation of all cables should be considered. A direct solution of the heat conduction equation employing numerical methods offers such a possibility.

Numerical methods allow not only better representation of the mutual heating effects, but also permit more accurate modeling of the region's boundaries (e.g., a convective boundary at the earth surface, constant heat flux circular boundaries for heat or water pipes in the vicinity of the cables, or an isothermal boundary at the water level at the bottom of a trench).

In what follows, we will develop a solution to equations (11.1)–(11.3) using finite-element and finite-difference methods. Even though both methods have been applied to solve the heat conduction problem around loaded power cables, our view is that the finite-element method is better suited for this application. Comparison of the two methods is given in Section 11.5.

11.2 GENERAL CHARACTERISTICS OF NUMERICAL METHODS

Both the finite-element and finite-difference methods discussed in this chapter require discretization of the partial differential equations in space and time domains. The finer this discretization is, the more accurate are the results, yet the heavier is the computational burden. The size of the region around power cables to be discretized is also of importance for both the accuracy and efficiency of the computations. The following sections give some guidance on these topics.

11.2.1 Selection of the Region to be Discretized

The location of boundaries is an important consideration in numerical studies. The objective is to select a large enough region so that the calculated values along the boundaries agree with those that exist in the physical problem. For the cable rating problem, the earth's surface is an obvious boundary and the side and bottom boundaries must be selected in such a way that the nodal temperatures at those boundaries all have the same value and the temperature gradient across the boundary is equal to zero.

Experience plus a study of how others modeled similar infinite regions is probably the best guide. In the experience of the author, a rectangular 10 m wide and 5 m deep field, with the cables located in the center, gives satisfactory results in the majority of practical cases.

For transient analysis, the radius of the soil, out to which heat disperses, will increase with time, and for practical purposes, it is sufficient to consider only that radius within which a sensible temperature rise occurs. This radius can be estimated from a simplified form of equation (5.14) assuming that each cable is a line source of heat:

$$\theta_{r,t} = \frac{W_I \rho_s}{4\pi}\left[-Ei\left(\frac{-r^2}{4\delta t}\right)\right] \tag{11.4}$$

where $\theta_{r,t}$ is the threshold temperature value at the distance r from the cable axis and the remaining quantities are defined in Chapter 5 and also in the list of symbols. This value can be taken as 0.1 K when the number of cables is not greater than 3 and suitably smaller for a large number of cables. Equation (11.4) is applied for each cable. The region to be discretized will be an envelope around all the circles representing the individual cables.

11.2.2 Representation of Cable Losses

Conductor, sheath, and dielectric losses are represented as heat sources, and provision should be made to vary these as required with time and/or temperature. Values of these losses are recalculated at each time step, using methods given in Chapters 4, 6, 7, and 8.

In finite-difference and finite-element methods, each node contains an initial temperature rise at $t = 0$. In the general case where the transient due to the dielectric loss may not have reached its steady state, these initial temperature rises (relative to the ambient temperature outside the mesh) must be obtained from a prior calculation. From the beginning of the transient, the computation must in all cases take account of both the joule and the dielectric losses. The two usual situations are

1. The transient due to dielectric losses has reached a steady state (the voltage has been applied for a very long time). The initial temperature rise at each node is put equal to the steady state temperature rise at that point caused by the dielectric losses only.
2. The voltage is applied to the cable at the same time as the load current. In this case, the initial temperature rises are zero, and the dielectric loss generators should be allocated their proper values from time zero.

11.2.3 Selection of a Time Step

Since, in general, the computations involve evaluation of temperatures in increments of time, care must be taken in the selection of the time step. In principle, one should select as large a time interval as possible to reduce the amount of computation. Unfortunately, too large a time step can compromise the accuracy of the computations. CIGRE (1983) and Libondi (1975) provide some guidance on the selection of suitable values. The duration of the time step $\Delta\tau$ will depend on: (1) the time constant $\Sigma T \cdot \Sigma Q$ of the network [defined as the product of its total thermal resistance (between conductor and outer surface) and its total thermal capacitance (whole cable)], (2) time elapsed from the beginning of the transient τ, and (3) the location of the time τ with relation to the shape of the load curve being applied. Requirement (3) can be illustrated as shown in Fig. 11.1 where the value of $\Delta\tau$ is selected to coincide with the change of the shape of the load curve.

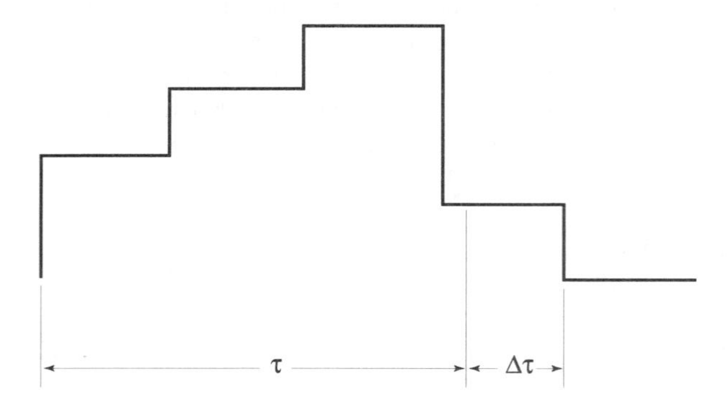

Figure 11.1 Relationship among the time step, the load curve, and the time elapsed from the beginning of the transient.

The following conditions are suggested for the selection of the time step $\Delta\tau$ (CIGRE, 1983):

$$\begin{aligned} \log_{10} \frac{\Delta\tau}{\Sigma T \cdot \Sigma Q} &= \frac{1}{3} \log_{10} \frac{\tau}{\Sigma T \cdot \Sigma Q} - 1.58 \qquad \text{for} \qquad \tau < \frac{1}{3} \Sigma T \cdot \Sigma Q \\ \log_{10} \frac{\Delta\tau}{\Sigma T \cdot \Sigma Q} &= \frac{1}{3} \log_{10} \frac{\tau}{\Sigma T \cdot \Sigma Q} - 1.25 \qquad \text{for} \qquad \tau > \frac{1}{3} \Sigma T \cdot \Sigma Q \end{aligned} \tag{11.5}$$

Adjusting the time step automatically during the computations is the preferable approach.

11.3 THE FINITE-ELEMENT METHOD

11.3.1 Overview

The finite-element method is a numerical technique for solving partial differential equations. Among the many physical phenomena described by such equations, the heat conduction problem and heat and mass transfer in the vicinity of power cables have been often addressed in the literature (Abdel-Hadi, 1978; Anders *et al.*, 1987; Anders and Radhakrishna, 1988a, 1988b; Anders *et al.*, 1988; El-Kady, 1985; El-Kady and Horrocks, 1985; Flatabo, 1973; Labridis and Dokopoulos, 1988; Konrad, 1982; Mitchell and Abdel-Hadi, 1979; Mushamalirwa *et al.*, 1988; Tarasiewicz *et al.*, 1987; Thomas *et al.*, 1980). The fundamental concept of the finite-element method is that temperature can be approximated by a discrete model composed of a set of continuous functions defined over a finite number of subdomains. The piecewise continuous functions are defined using the values of temperature at a finite number of points in the region of interest.

The discrete solution is constructed as follows:

1. A finite number of points in the solution region is identified. These points are called *nodal points* or *nodes*.
2. The value of the temperature at each node is denoted as a variable which is to be determined.
3. The region of interest is divided into a finite number of subregions called *elements*. These elements are connected at common nodal points, and collectively approximate the shape of the region.
4. Temperature is approximated over each element by a polynomial that is defined using nodal values of the temperature. A different polynomial is defined for each element, but the element polynomials are selected in such a way that continuity is maintained along the element boundaries. The nodal values are computed so that they provide the "best" approximation possible to the true temperature distribution. This selection is accomplished by minimizing some quantity associated with the physical problem (this is the so-called Rayleigh–Ritz method), or by using Galerkin's method (Zienkiewicz, 1971) which deals with the differential equations directly. Either approach results in a matrix equation whose solution vector contains coefficients of the approximating polynomials. The solution vector of the algebraic equations gives the required nodal temperatures. The answer is then known throughout the solution region.

In the following section, we will develop the algebraic equations from which nodal temperatures are obtained.

In cable rating applications, two-dimensional elements are most commonly used. The elements in the two-dimensional domain are functions of x and y and are generally either triangular or quadrilateral in shape. The element function becomes a plane (Fig. 11.2).

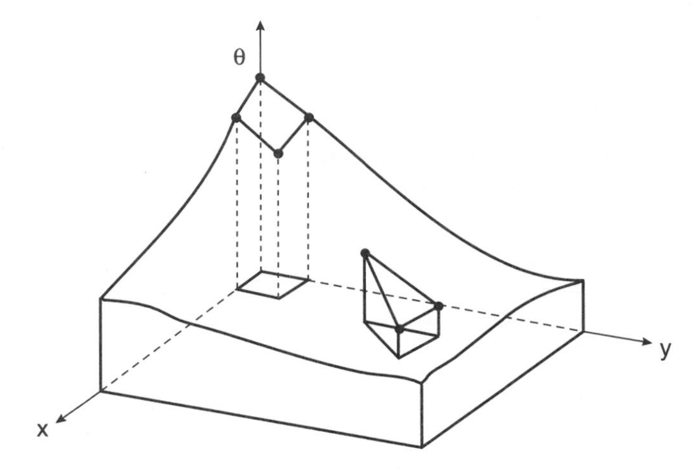

Figure 11.2 Modeling of a two-dimensional scalar function using triangular or quadrilateral elements (Segerlind, 1984).

The plane is associated with the minimum number of element nodes, which is three for the triangle and four for the quadrilateral.

The element function can be a curved surface when more than the minimum number of nodes are used. An excess number of nodes also allows the elements to have curved boundaries (Fig. 11.3). For the purpose of introducing the method and explaining how it is used in cable rating computations, we will use the simplest and the most common shape for two-dimensional elements, the triangle. In this chapter, the words "triangle," "element," and "finite element" will be used interchangeably.

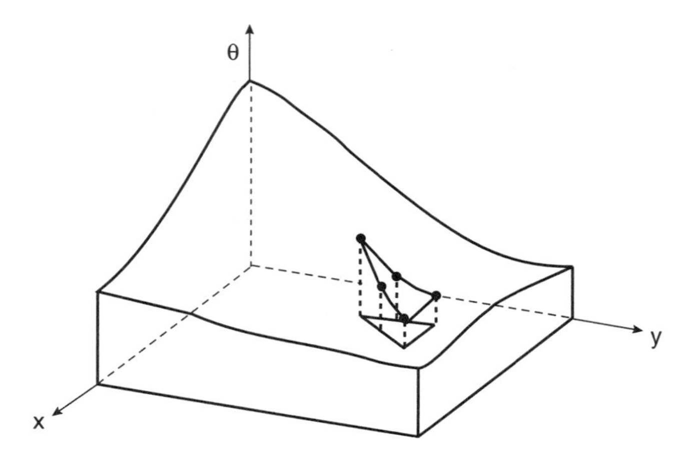

Figure 11.3 Modelling of a two-dimensional scalar function using a quadratic triangular element (Segerlind, 1984).

11.3.2 Approximating Polynomials

Consider a simple triangular element shown in Fig. 11.4. For this element, the temperature θ at any point inside can be uniquely specified as (Flatabo, 1972)

$$\theta = A\omega_i + B\omega_j + C\omega_m \tag{11.6}$$

where ω_i, ω_j, and ω_m are the area coordinates defined as in Fig. 11.4. These area coordinates uniquely define the position of any point P inside the triangle ijm. To determine the constant A, the temperature at node i is written as [equation (11.6)]:

$$\theta_i = 1 \times A + 0 \times B + 0 \times C$$

This gives $A = \theta_i$. Similarly, applying (11.6) for nodes j and m, we obtain $B = \theta_j$ and $C = \theta_m$. Therefore,

$$\theta = \omega_i\theta_i + \omega_j\theta_j + \omega_m\theta_m = \left[\omega_i, \omega_j, \omega_m\right] \cdot \begin{bmatrix} \theta_i \\ \theta_j \\ \theta_m \end{bmatrix} = \mathbf{N}^e \cdot \mathbf{\Theta}^e \tag{11.7}$$

where $\mathbf{N}^e = \left[\omega_i, \omega_j, \omega_m\right]$, $\mathbf{\Theta}^e = \left[\theta_i, \theta_j, \theta_m\right]^t$, and the superscript t denotes transposition.

Assuming that the time derivatives are prescribed functions of the space coordinates at any particular instant of time, we can write the time derivative for the temperature within

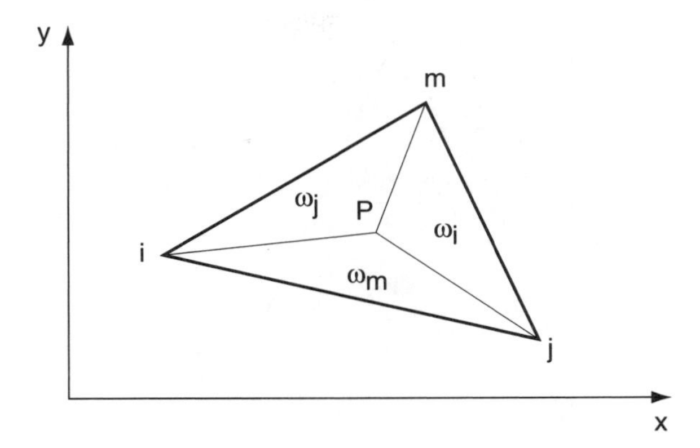

Figure 11.4 Area coordinates.

each element as

$$\frac{\partial\theta}{\partial t} = \omega_i \frac{\partial\theta_i}{\partial t} + \omega_j \frac{\partial\theta_j}{\partial t} + \omega_m \frac{\partial\theta_m}{\partial t} = \left[\omega_i, \omega_j, \omega_m\right] \cdot \begin{bmatrix} \dfrac{\partial\theta_i}{\partial t} \\ \dfrac{\partial\theta_j}{\partial t} \\ \dfrac{\partial\theta_m}{\partial t} \end{bmatrix} = \mathbf{N}^e \cdot \frac{\partial\mathbf{\Theta}^e}{\partial t} \tag{11.8}$$

because $\mathbf{N}^e$ is a function of the coordinate system, but not of the time.

The relationship between area coordinates and Cartesian coordinates is

$$\begin{bmatrix} x \\ y \\ 1 \end{bmatrix} = \begin{bmatrix} x_i & x_j & x_m \\ y_i & y_j & y_m \\ 1 & 1 & 1 \end{bmatrix} \begin{bmatrix} \omega_i \\ \omega_j \\ \omega_m \end{bmatrix}$$

The inverse relationship yields the coefficients of vector $\mathbf{N}^e$:

$$\begin{bmatrix} \omega_i \\ \omega_j \\ \omega_m \end{bmatrix} = \frac{1}{2A} \begin{bmatrix} (y_j - y_m) & (x_m - x_j) & (x_j y_m - x_m y_j) \\ (y_m - y_i) & (x_i - x_m) & (x_m y_i - x_i y_m) \\ (y_i - y_j) & (x_j - x_i) & (x_i y_j - x_j y_i) \end{bmatrix} \begin{bmatrix} x \\ y \\ 1 \end{bmatrix} \tag{11.9}$$

where A is the area of the triangle.

We can observe from equations (11.7) and (11.9) that the temperature is a linear function in x and y. This means that the gradients in either x or y directions are constant. A constant gradient within any element means that many small elements have to be used to approximate a rapid change in the value of θ.

EXAMPLE 11.1

Consider a triangular element shown in Fig. 11.5. We will evaluate the element equation and calculate the value of the temperature at point P for the following nodal temperature values: $\theta_i = 40°\text{C}$, $\theta_j = 34°\text{C}$, and $\theta_m = 46°\text{C}$. P is located at (2.0, 1.5).

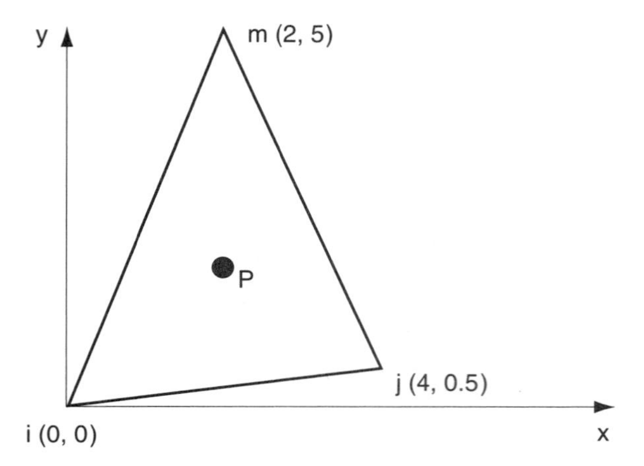

Figure 11.5 Illustration for Example 11.1.

The temperature θ is given by equation (11.7) with the shape function described by equation (11.9). First, we need to compute the area of the triangle. This is obtained from

$$2A = \begin{vmatrix} 1 & x_i & y_i \\ 1 & x_j & y_j \\ 1 & x_m & y_m \end{vmatrix} = \begin{vmatrix} 1 & 0 & 0 \\ 1 & 4 & 0.5 \\ 1 & 2 & 5 \end{vmatrix} = 19$$

From equation (11.9),

$$\begin{bmatrix} \omega_i \\ \omega_j \\ \omega_m \end{bmatrix} = \frac{1}{2A} \begin{bmatrix} (y_j - y_m) & (x_m - x_j) & (x_j y_m - x_m y_j) \\ (y_m - y_i) & (x_i - x_m) & (x_m y_i - x_i y_m) \\ (y_i - y_j) & (x_j - x_i) & (x_i y_j - x_j y_i) \end{bmatrix} \begin{bmatrix} x \\ y \\ 1 \end{bmatrix}$$

$$= \frac{1}{19} \begin{bmatrix} (0.5 - 5) & (2 - 4) & (4 \cdot 5 - 2 \cdot 0.5) \\ (5 - 0) & (0 - 2) & (2 \cdot 0 - 0 \cdot 5) \\ (0 - 0.5) & (4 - 0) & (0 \cdot 0.5 - 4 \cdot 0) \end{bmatrix} \begin{bmatrix} 2 \\ 1.5 \\ 1 \end{bmatrix} = \begin{bmatrix} 0.368 \\ 0.368 \\ 0.264 \end{bmatrix}$$

The temperature at point P is obtained from equation (11.7):

$$\theta = \left[\omega_i, \omega_j, \omega_m\right] \cdot \begin{bmatrix} \theta_i \\ \theta_j \\ \theta_m \end{bmatrix} = 0.368 \cdot 40 + 0.358 \cdot 34 + 0.264 \cdot 46 = 39.4°\text{C}$$

In triangular elements, the temperature varies linearly between any two nodes. Any line of constant temperature is a straight line and intersects two sides of the element. The only exceptions are when all nodes have the same value. These two properties make it easy to locate isothermal contour lines.

EXAMPLE 11.2

We will determine the 41°C contour line for the triangular element used in Example 11.1.

The temperature isotherm for 41°C intersects sides im and mj. The coordinates at which this isotherm intersects the sides of the triangle are obtained from the following simple ratios:

$$\frac{46 - 41}{46 - 34} = \frac{2 - x}{2 - 4} \quad \text{or} \quad x = 2.83$$

and

$$\frac{46-41}{46-34} = \frac{5-y}{5-0.5} \quad \text{or} \quad y = 3.12$$

The contour is shown in Fig. 11.6.

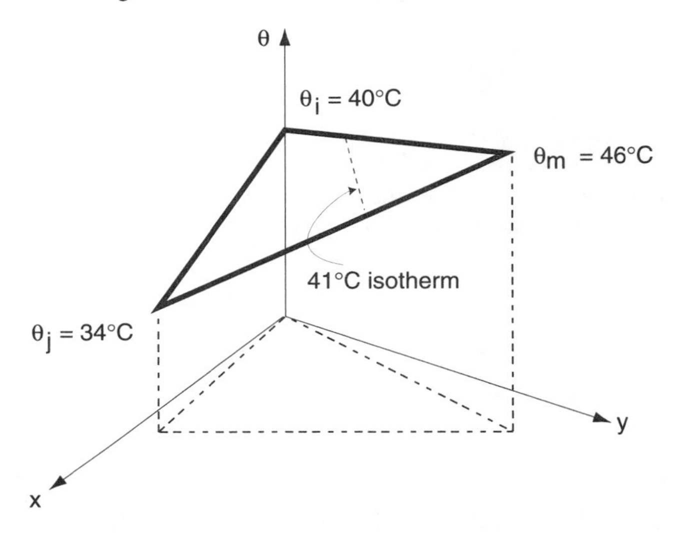

Figure 11.6 41°C isothermal contour.

11.3.3 Finite-element Equations

In the previous section, we have discussed how to compute the temperature at any point inside an element if the temperature values at the nodes are known. To obtain node temperatures, we use a property, known in the variational calculus, which states that the minimization of the functional

$$\chi = \int_S \frac{1}{2\rho}\left[(\nabla\theta)^t\nabla\theta + \left(W_{\text{int}} - c\frac{d\theta}{dt}\right)\theta\right]dS + \int_C \left[q\theta + \frac{1}{2}h\,(\theta - \theta_{\text{amb}})^2\right]dC \tag{11.10}$$

over the area S bounded by the closed curve C requires that the differential equation (11.1) with the boundary conditions (11.2) and (11.3) be satisfied. The temperature gradient is defined as

$$\nabla\theta = \begin{bmatrix} \dfrac{\partial\theta}{\partial x} \\ \dfrac{\partial\theta}{\partial y} \end{bmatrix}$$

Clearly, any temperature distribution that makes χ a minimum also satisfies the governing differential equations, and is therefore a solution to the problem being studied.

Equation (11.10) is a starting point for determining the temperature at each node. We minimize (11.10) by using our set of element functions, each defined over a single element and written in terms of the nodal values. The nodal values θ_n are the unknown values in our formulation. These values are obtained by taking derivatives of χ with respect to each θ_n and equating them to zero.

Recalling that the functions θ are defined over each individual element, the integrals in (11.10) must be separated into integrals over the individual elements and the derivatives computed for each element; that is,

$$\chi = \sum_{e=1}^{E} \chi^e \tag{11.11}$$

where χ^e is the functional defined for element e and E is the total number of elements.

Let us consider a single element first. As any element contributes to only three of the differentials associated with its nodes, these contributions can be listed as

$$\left(\frac{\partial \chi}{\partial \theta_n}\right)^e = \begin{bmatrix} \dfrac{\partial \chi^e}{\partial \theta_i} \\ \dfrac{\partial \chi^e}{\partial \theta_j} \\ \dfrac{\partial \chi^e}{\partial \theta_m} \end{bmatrix} \tag{11.12}$$

The derivatives in equation (11.12) cannot be evaluated until the integrals in (11.10) have been written in terms of the nodal values Θ^e. This is done by first computing the derivatives of θ with respect to x and y. Because the sum of the area coordinates equals one, only two of them are independent. Assuming that these are ω_i and ω_j, we have

$$\nabla\theta = \begin{bmatrix} \dfrac{\partial \theta}{\partial x} \\ \dfrac{\partial \theta}{\partial y} \end{bmatrix} = \mathbf{J} \begin{bmatrix} \dfrac{\partial \theta}{\partial \omega_i} \\ \dfrac{\partial \theta}{\partial \omega_j} \end{bmatrix}$$

$$= \frac{1}{2A} \begin{bmatrix} (y_j - y_m) & (y_m - y_i) \\ (x_m - x_j) & (x_i - x_m) \end{bmatrix} \begin{bmatrix} \dfrac{\partial \theta}{\partial \omega_i} \\ \dfrac{\partial \theta}{\partial \omega_j} \end{bmatrix} \tag{11.13}$$

$$= \frac{1}{2A} \begin{bmatrix} b_i & b_j \\ a_i & a_j \end{bmatrix} \begin{bmatrix} \dfrac{\partial \theta}{\partial \omega_i} \\ \dfrac{\partial \theta}{\partial \omega_j} \end{bmatrix}$$

where the Jacobian $\mathbf{J}$ is obtained by differentiating equation (11.9). Further, from equation (11.7) and the fact that $\omega_i + \omega_j + \omega_m = 1$, we obtain

$$\begin{bmatrix} \dfrac{\partial \theta}{\partial \omega_i} \\ \dfrac{\partial \theta}{\partial \omega_j} \end{bmatrix} = \begin{bmatrix} 1 & 0 & -1 \\ 0 & 1 & -1 \end{bmatrix} \begin{bmatrix} \theta_i \\ \theta_j \\ \theta_m \end{bmatrix} = \mathbf{V}\Theta^e \tag{11.14}$$

Thus, for a single element, we have

$$\nabla\theta = \mathbf{J} \cdot \mathbf{V} \cdot \Theta^e \tag{11.15}$$

Substituting (11.15) into (11.10), with S and C corresponding to a single element, and differentiating with respect to Θ^e, after some routine but tedious computations, equation (11.12) can be written as (Flatabo, 1972)

$$\left(\frac{\partial \chi}{\partial \theta_n}\right)^e = \mathbf{h}^e\Theta^e + \mathbf{q}^e\frac{\partial \Theta^e}{\partial t} - \mathbf{k}^e \tag{11.16}$$

Denoting by d_{ij}, d_{jm}, and d_{mi} the distance between nodes ij, jm, and mi, respectively, the element conductivity matrix $\mathbf{h}^e$ is equal to

$$\mathbf{h}^e = \frac{1}{4A\rho}\left\{\begin{bmatrix} a_i^2 & a_ia_j & a_ia_m \\ a_ia_j & a_j^2 & a_ja_m \\ a_ia_m & a_ja_m & a_m^2 \end{bmatrix} + \begin{bmatrix} b_i^2 & b_ib_j & b_ib_m \\ b_ib_j & b_j^2 & b_jb_m \\ b_ib_m & b_jb_m & b_m^2 \end{bmatrix}\right\}$$
$$+ \frac{hd_{ij}}{6}\begin{bmatrix} 2 & 1 & 0 \\ 1 & 2 & 0 \\ 0 & 0 & 0 \end{bmatrix} + \frac{hd_{jm}}{6}\begin{bmatrix} 0 & 0 & 0 \\ 0 & 2 & 1 \\ 0 & 1 & 2 \end{bmatrix} + \frac{hd_{mi}}{6}\begin{bmatrix} 2 & 0 & 1 \\ 0 & 0 & 0 \\ 1 & 0 & 2 \end{bmatrix} \tag{11.17}$$

$$a_i = x_m - x_j \quad a_j = x_i - x_m \quad a_m = x_j - x_i$$
$$b_i = y_j - y_m \quad b_j = y_m - y_i \quad b_m = y_i - y_j$$

If there is no convective boundary along any segment of the element, the relevant term in equation (11.17) is omitted (see Example 11.3).

The element capacity matrix is given by

$$\mathbf{q}^e = \frac{cA}{12}\begin{bmatrix} 2 & 1 & 1 \\ 1 & 2 & 1 \\ 1 & 1 & 2 \end{bmatrix} \tag{11.18}$$

and the element heat generation vector is equal to

$$\mathbf{k}^e = \frac{W_{\text{int}}A}{3}\begin{bmatrix} 1 \\ 1 \\ 1 \end{bmatrix} + \frac{(h\theta_{\text{amb}} + q)\,d_{ij}}{2}\begin{bmatrix} 1 \\ 1 \\ 0 \end{bmatrix} + \frac{(h\theta_{\text{amb}} + q)\,d_{jm}}{2}\begin{bmatrix} 0 \\ 1 \\ 1 \end{bmatrix}$$
$$+ \frac{(h\theta_{\text{amb}} + q)\,d_{mi}}{2}\begin{bmatrix} 1 \\ 0 \\ 1 \end{bmatrix} \tag{11.19}$$

Here, again, the last three terms apply only if the appropriate boundary exists along the element edge. Factor $W_{\text{int}}A$ represents the total heat in W/m generated in the element.

Performing computations given by equations (11.16)–(11.19) for each element, we finally obtain the following set of linear algebraic equations for the whole region:

$$\frac{\partial \chi}{\partial \Theta} = \sum_{e=1}^{E} \left(\frac{\partial \chi}{\partial \theta_n} \right)^e = \mathbf{H}\Theta + \mathbf{Q}\frac{\partial \Theta}{\partial t} - \mathbf{K} = \mathbf{0} \tag{11.20}$$

In this equation, **H** is the heat conductivity matrix, **Q** is the heat capacity matrix, Θ and $\partial\Theta/\partial t$ are vectors containing the nodal temperatures and their derivatives, and **K** is a vector which expresses the distribution of heat sources and heat sinks over the region under consideration.

In a steady-state analysis, equation (11.20) simplifies to

$$\mathbf{H}\Theta - \mathbf{K} = \mathbf{0} \tag{11.21}$$

EXAMPLE 11.3

Consider the element examined in Example 11.1. Assume that this element experiences convection on surface ij and a constant heat flux on surface mi. We will calculate element matrices given the numerical dimensions and properties shown in Fig. 11.7.

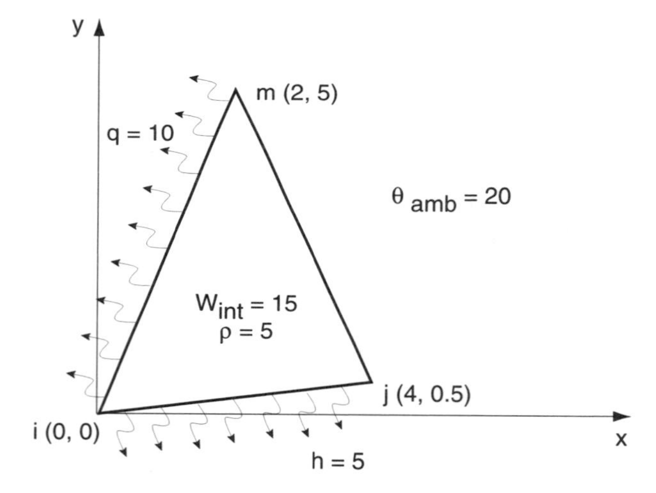

Figure 11.7 Illustration for Example 11.3.

The lengths of the boundary sides are

$$d_{ij} = \sqrt{(0-4)^2 + (0-0.5)^2} = 4.03, \quad d_{mi} = \sqrt{(2-0)^2 + (5-0)^2} = 5.39$$

Substituting the numerical values in equation (11.17), we obtain the following conductivity matrix:

$$a_i = x_m - x_j = 2 - 4 = -2 \quad a_j = x_i - x_m = 0 - 2 = -2 \quad a_m = x_j - x_i = 4 - 0 = 4$$

$$b_i = y_j - y_m = 0.5 - 5 = -4.5 \quad b_j = y_m - y_i = 5 - 0 = 5 \quad b_m = y_i - y_j = 0 - 0.5 = -0.5$$

$$\mathbf{h}^e = \frac{1}{4A\rho}\left\{\begin{bmatrix} a_i^2 & a_ia_j & a_ia_m \\ a_ia_j & a_j^2 & a_ja_m \\ a_ia_m & a_ja_m & a_m^2 \end{bmatrix} + \begin{bmatrix} b_i^2 & b_ib_j & b_ib_m \\ b_ib_j & b_j^2 & b_jb_m \\ b_ib_m & b_jb_m & b_m^2 \end{bmatrix}\right\}$$

$$+ \frac{hd_{ij}}{6}\begin{bmatrix} 2 & 1 & 0 \\ 1 & 2 & 0 \\ 0 & 0 & 0 \end{bmatrix} + \frac{hd_{jm}}{6}\begin{bmatrix} 0 & 0 & 0 \\ 0 & 2 & 1 \\ 0 & 1 & 2 \end{bmatrix} + \frac{hd_{mi}}{6}\begin{bmatrix} 2 & 0 & 1 \\ 0 & 0 & 0 \\ 1 & 0 & 2 \end{bmatrix}$$

$$= \frac{1}{4\cdot 9.5\cdot 5}\left\{\begin{bmatrix} (-2)^2 & (-2)(-2) & (-2)(4) \\ (-2)(-2) & (-2)^2 & (-2)4 \\ (-2)(4) & (-2)(4) & (4)^2 \end{bmatrix}\right.$$

$$\left. + \begin{bmatrix} (-4.5)^2 & (-4.5)(5) & (-4.5)(-0.5) \\ (-4.5)(5) & (5)^2 & (5)(-0.5) \\ (-4.5)(-0.5) & (5)(-0.5) & (-0.5)^2 \end{bmatrix}\right\}$$

$$+ \begin{bmatrix} \frac{5\cdot 4.03}{3} & \frac{5\cdot 4.03}{6} & 0 \\ \frac{5\cdot 4.03}{6} & \frac{5\cdot 4.03}{3} & 0 \\ 0 & 0 & 0 \end{bmatrix} = \begin{bmatrix} 6.85 & 3.26 & -0.03 \\ 3.26 & 6.87 & -0.06 \\ -0.03 & -0.06 & 0.09 \end{bmatrix}$$

The heat generation vector is obtained from equation (11.19):

$$\mathbf{k}^e = \frac{W_{\text{int}}A}{3}\begin{bmatrix} 1 \\ 1 \\ 1 \end{bmatrix} + \frac{(h\theta_{\text{amb}} + q)\,d_{ij}}{2}\begin{bmatrix} 1 \\ 1 \\ 0 \end{bmatrix} + \frac{(h\theta_{\text{amb}} + q)\,d_{jm}}{2}\begin{bmatrix} 0 \\ 1 \\ 1 \end{bmatrix} + \frac{(h\theta_{\text{amb}} + q)\,d_{mi}}{2}\begin{bmatrix} 1 \\ 0 \\ 1 \end{bmatrix}$$

$$= \frac{15\cdot 9.5}{3}\begin{bmatrix} 1 \\ 1 \\ 1 \end{bmatrix} + \frac{5\cdot 20\cdot 4.03}{2}\begin{bmatrix} 1 \\ 1 \\ 0 \end{bmatrix} + \frac{10\cdot 5.39}{2}\begin{bmatrix} 1 \\ 0 \\ 1 \end{bmatrix} = \begin{bmatrix} 276 \\ 249 \\ 74.5 \end{bmatrix}$$

EXAMPLE 11.4

We will consider now a domain composed of three elements, one being the same as examined in Example 11.3 and two adjacent elements as shown in Fig. 11.8. We will determine nodal temperatures in the steady state for this system, assuming that the other boundary surfaces have zero temperature gradient.

The matrix for element 1 was obtained in Example 11.3, and is equal to

$$\mathbf{h}_1^e = \begin{bmatrix} 6.85 & 3.26 & -0.03 \\ 3.26 & 6.87 & -0.06 \\ -0.03 & -0.06 & 0.09 \end{bmatrix}$$

The element matrices for elements 2 (nodes j, n, m) and 3 (nodes j, k, n) are obtained from equation (11.17):

$$\mathbf{h}_2^e = \begin{bmatrix} 0.33 & -0.22 & -0.11 \\ -0.22 & 0.90 & -0.68 \\ -0.11 & -0.68 & 0.79 \end{bmatrix} \quad \text{and} \quad \mathbf{h}_3^e = \begin{bmatrix} 0.5 & -0.5 & 0 \\ -0.5 & 0.63 & -0.13 \\ 0 & -0.13 & 0.13 \end{bmatrix}$$

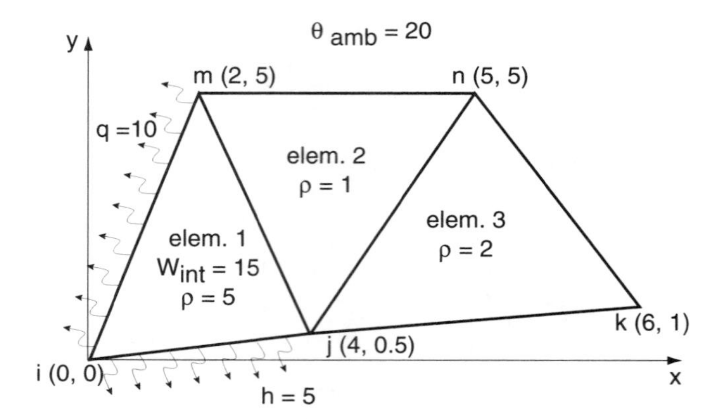

Figure 11.8 Illustration for Example 11.4.

Since there are five nodes in this system, the matrix **H** will have five rows and five columns:

$$\mathbf{H} = \begin{array}{c} \begin{matrix} i & j & m & n & k \end{matrix} \\ \begin{bmatrix} 6.85 & 3.26 & -0.03 & 0 & 0 \\ 3.26 & 6.87 & -0.06 & 0 & 0 \\ -0.03 & -0.06 & 0.09 & 0 & 0 \\ 0 & 0 & 0 & 0 & 0 \\ 0 & 0 & 0 & 0 & 0 \end{bmatrix} \end{array} + \begin{bmatrix} 0 & 0 & 0 & 0 & 0 \\ 0 & 0.33 & -0.11 & -0.22 & 0 \\ 0 & -0.11 & 0.79 & -0.68 & 0 \\ 0 & -0.22 & -0.68 & 0.90 & 0 \\ 0 & 0 & 0 & 0 & 0 \end{bmatrix}$$

$$+ \begin{bmatrix} 0 & 0 & 0 & 0 & 0 \\ 0 & 0.5 & 0 & 0 & -0.5 \\ 0 & 0 & 0 & 0 & 0 \\ 0 & 0 & 0 & 0.13 & -0.13 \\ 0 & -0.5 & 0 & -0.13 & 0.63 \end{bmatrix} = \begin{bmatrix} 6.85 & 3.26 & -0.03 & 0 & 0 \\ 3.26 & 7.7 & -0.17 & -0.22 & -0.5 \\ -0.03 & -0.17 & 0.88 & -0.68 & 0 \\ 0 & -0.22 & -0.68 & 1.03 & -0.13 \\ 0 & -0.5 & 0 & -0.13 & 0.63 \end{bmatrix}$$

Since elements 2 and 3 do not generate any heat and have zero temperature gradient, vector **K** is the same as obtained in Example 11.4 with the components corresponding to nodes n and k equal to zero; that is,

$$\mathbf{K} = \begin{bmatrix} 276 \\ 249 \\ 74.5 \\ 0 \\ 0 \end{bmatrix}$$

With the conductance matrix and heat generation vector given above, the following nodal temperatures are obtained by solving equations (11.21):

$$\theta_i = 24.6°\text{C}, \quad \theta_j = 34.9°\text{C}, \quad \theta_m = 211.9°\text{C}, \quad \theta_n = 154.9°\text{C}, \quad \theta_k = 59.7°\text{C}$$

The set of ordinary differential equations (11.20) which define the discretized problem can be solved using one of the many recursion schemes. There are two popular procedures for solving these equations to obtain the values of Θ at each point in time. The first is to approximate the time derivative using a finite-difference scheme. The alternate procedure is to use finite elements defined in the time domain. Flatabo (1972) used the midinterval Crank–Nicolson finite-difference algorithm for the solution of this equation. This method requires an iteration within each time step. Here, we propose to use Lees' (1966) three-level, time-stepping scheme in which the discretized equation is replaced by the recurrence relationship

$$\Theta^{n+1} = -\left[\frac{\mathbf{H}^n}{3} + \frac{\mathbf{Q}^n}{2\Delta\tau}\right]^{-1}\left[\frac{\mathbf{H}^n\Theta^n}{3} + \frac{\mathbf{H}^n\Theta^{n-1}}{3} - \frac{\mathbf{Q}^n\Theta^{n-1}}{2\Delta\tau} - \mathbf{K}^n\right] \tag{11.22}$$

where the superscript n refers to the time level and $\Delta\tau$ is the time step. The procedure is unconditionally stable, and has the advantage of producing the solution at time level $n+1$ without the need for any iteration as the coefficient matrices are evaluated at level n. The initial conditions have to be specified, and the first time step iteration is performed by a modified version of equation (11.22) requiring only one previous time step solution.

11.3.4 Some Comments on Computer Implementation

The implementation of the finite-element method proceeds in two stages. In the first stage, the region under consideration is discretized into a finite number of elements and relevant matrices are formed. In the second stage, the resulting set of linear equations is solved. The efficiency of the computations depends to a large extent on the discretization process. This process can be divided into two general parts: the division of the region into elements, and the labeling of the elements and the nodes. The latter sounds quite simple, but is complicated by the desire to increase the computational efficiency. We will briefly discuss this process using the example of triangular elements.

The division of any two-dimensional domain into elements should start with the division of the body into quadrilateral and triangular regions. These regions are then subdivided into triangles. The subdivision between regions should be located where there is a change in geometry, or material properties, or both.

In most cable rating computations, the region which is divided into elements is usually a rectangle. This is simply accomplished as illustrated in Example 11.5 below. When a rectangle is divided into two triangles, division using the shortest diagonal is preferable because elements close to an equilateral shape produce more accurate results than long narrow triangles. The spacing between boundary nodes can be varied to obtain desired element sizes (the elements should be smallest closer to the cables). There will be $2(n-1)(m-1)$ elements in a quadrilateral where n and m are the number of nodes on adjacent sides. The ability to vary the element size is an important advantage of the finite-element method.

The labeling of nodes (assigning numbers) influences computational efficiency. The conductance matrix in equation (11.20) is usually very sparse. Good general advice is to number the nodes in such a way that the largest difference between the node numbers in a single element is as small as possible. The numbering of elements does not affect the computational aspects of the problem.

There are commercially available computer programs which automatically generate finite-element meshes for thermal problems. However, we would recommend that a person wishing to use the finite-element method read some specialized books on the subject and do some experimentation on his/her own.

We will illustrate an application of the finite-element method to determine the geometric factor of a backfill or duct bank.

EXAMPLE 11.5

The following procedure to obtain the geometric factor for cables located in duct banks and backfills was proposed by El-Kady and Horrocks (1985).

Consider the thermal circuit configuration given in Fig. 11.9 where the cable bank is represented by a rectangular cross-sectional surface C of height h and width w. For this configuration, the total thermal resistance between the duct bank surface and the ground ambient is given by Goldenberg (1969):

$$T = -\frac{\rho_s(\theta_c - \theta_{\text{amb}})}{\displaystyle\int_C \frac{\partial\theta}{\partial n} ds} \tag{11.23}$$

where ρ_s is the thermal resistivity of the soil, C represents the duct bank surface, and $\partial\theta/\partial n$ denotes differentiation in the direction perpendicular to C.

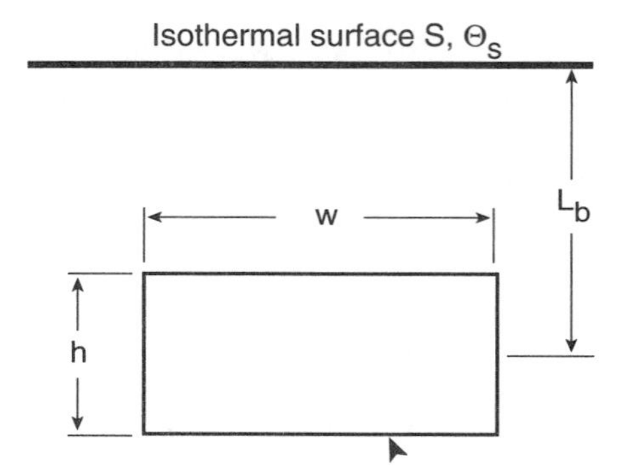

Figure 11.9 Thermal circuit configuration in Example 11.5.

In the finite-element solution, the medium surrounding the surface C is partitioned into small triangles constituting a finite-element grid such that the first grid layer, enclosing the bank surface C, is carefully structured, as shown in Fig. 11.10, to attain an efficient subsequent evaluation of equation (11.23).

The surface C is partitioned into K small segments, as shown in Fig. 11.10, where the temperatures $\theta_1, \theta_2, \cdots$ of the middle points of the first grid layer (which constitute nodes of the finite-element grid) are evaluated. The accuracy of the solution can be controlled by adjusting the size of the elements of the grid. Equation (11.23) can now be written in the discretized form

$$T = -\frac{\rho}{\displaystyle\sum_{i\in I_c} \frac{\Delta\theta_i}{\Delta n_i} \frac{\Delta S_i}{\theta_{ci} - \theta_{\text{amb}}}} \tag{11.24}$$

where, as shown in Fig. 11.10, θ_i is the temperature of segment i along the first finite-element grid layer surrounding the duct bank surface, θ_{ci} is the temperature at the duct bank surface C of segment

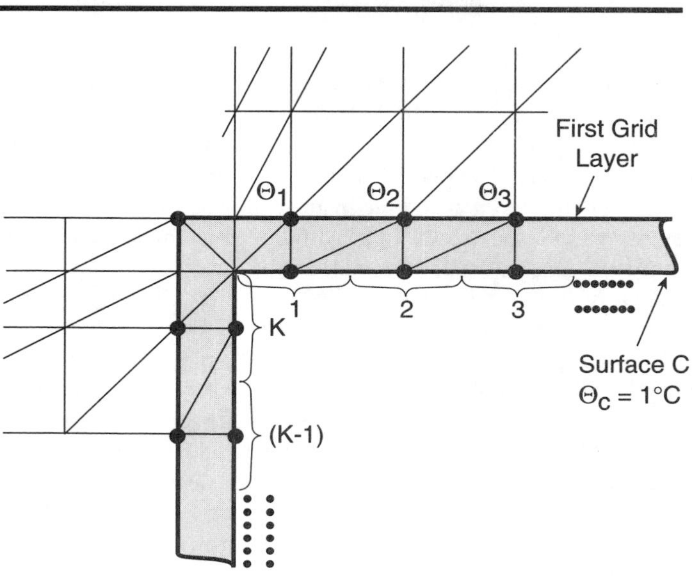

Figure 11.10 Finite-element grid structure for an outer layer of a duct bank.

i, and I_c is the index set of segments along C. By choosing $\Delta S_i / \Delta n_i = 1$ for all i, equation (11.24) reduces to (see Section 9.6.5.2 for the discussion of a geometric factor)

$$T = \frac{\rho}{2\pi} G = -\frac{\rho}{\sum_{i \in I_c} \frac{\theta_i - \theta_{ci}}{\theta_{ci} - \theta_{\text{amb}}}} \tag{11.25}$$

Hence,

$$G = \frac{2\pi}{\sum_{i \in I_c} \frac{\theta_{ci} - \theta_i}{\theta_{ci} - \theta_{\text{amb}}}} \tag{11.26}$$

Equation (11.26) provides the value of the geometric factor in terms of the temperature results from the finite-element analysis. We note that the surface C is not assumed to be isothermal. If, in fact, the duct bank surface is an isotherm, then $\theta_{ci} = \theta_c$ for all i in equation (11.26), leading to

$$G = \frac{2\pi \left(\theta_c - \theta_{\text{amb}}\right)}{\sum_{i \in I_c} \left(\theta_c - \theta_i\right)} \tag{11.27}$$

If we set $\theta_c = 1$ and $\theta_{\text{amb}} = 0$, equation (11.27) further simplifies to

$$G = \frac{2\pi}{K - \sum_{i \in I_c} \theta_i} \tag{11.28}$$

Equation (11.28) was used by El-Kady and Horrocks (1985) to obtain the extended values of the geometric factor for duct banks and backfills reported in Table 9.7.

11.4 THE FINITE-DIFFERENCE METHOD

11.4.1 Overview

Historically, numerical methods employing finite differences have been used more frequently than those based on finite elements to solve partial differential equations (Black and Park, 1983; Groenveld *et al.*, 1984; Hanna *et al.*, 1993; Hartley and Black, 1981; Hiramandani, 1991; Kellow, 1981; KEMA, 1981; Radhakrishna *et al.*, 1984). However, the reader should be aware of the manner in which the approximations used in the finite-difference methods are used. Finite-difference methods are approximate in the sense that derivatives at a point are approximated by difference quotients over a small interval. This is in contrast to the finite-element methods where the temperature as a function of space and/or time is approximated by polynomials.

In finite-difference methods (Fig. 11.11), the area of integration, that is, the area S bounded by the closed curve C, is overlaid by a system of rectangular meshes formed by two sets of equally spaced lines. An approximate solution to the differential equation (11.1) is found at the points of intersection of the parallel lines, the so-called nodes or mesh points. This solution is obtained by approximating the partial differential equation over the area S by n algebraic equations involving the values of θ at the n mesh points internal to C. The approximation consists of replacing each derivative of the partial differential equation at the point $P_{i,j}$ (say) by a finite-difference approximation in terms of the values of θ at $P_{i,j}$ and at neighboring mesh points and boundary points, and of writing down for each of the n internal mesh points the algebraic equation approximating the differential equation.

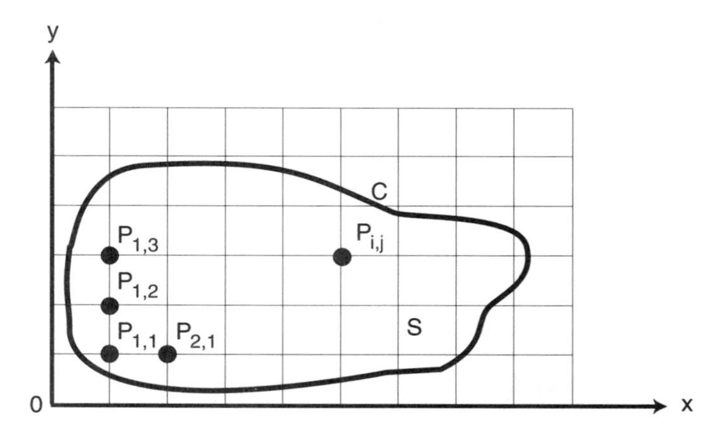

Figure 11.11 Discretization of region S by a rectangular mesh.

This process, illustrated in the next section, clearly gives n algebraic equations for the n unknowns $\theta_{1,1}, \theta_{1,2}, \cdots, \theta_{i,j}, \cdots$. Accuracy can usually be improved either by increasing the number of mesh points or by including "correction terms" in the approximations for the derivatives.

11.4.2 Finite-difference Approximations to Derivatives

The temperature θ is a function of space and time coordinates. The partial derivative with respect to time can be approximated as follows. Let $\Delta\tau$ be a small time interval.

Expanding the value of θ in a Taylor series, we have

$$\begin{aligned}\theta(t+\Delta\tau) &= \theta(t) + \Delta\tau\frac{\partial\theta}{\partial t} + \frac{1}{2}(\Delta\tau)^2\frac{\partial^2\theta}{\partial t^2} + \frac{1}{6}(\Delta\tau)^3\frac{\partial^3\theta}{\partial t^3} + \cdots \\ \theta(t-\Delta\tau) &= \theta(t) - \Delta\tau\frac{\partial\theta}{\partial t} + \frac{1}{2}(\Delta\tau)^2\frac{\partial^2\theta}{\partial t^2} - \frac{1}{6}(\Delta\tau)^3\frac{\partial^3\theta}{\partial t^3} + \cdots\end{aligned} \tag{11.29}$$

Adding up these expressions,

$$\theta(t+\Delta\tau) + \theta(t-\Delta\tau) = 2\theta(t) + (\Delta\tau)^2\frac{\partial^2\theta}{\partial t^2} + O\left[(\Delta\tau)^4\right] \tag{11.30}$$

where $O\left[(\Delta\tau)^4\right]$ denotes terms containing fourth and higher powers of $\Delta\tau$. Assuming that these are negligible in comparison with lower powers of $\Delta\tau$, we obtain from equation (11.30)

$$\frac{\partial^2\theta(t)}{\partial t^2} = \frac{\theta(t+\Delta\tau) - 2\theta(t) + \theta(t-\Delta\tau)}{(\Delta\tau)^2} \tag{11.31}$$

Substituting (11.31) into (11.29) and neglecting the terms of order $(\Delta\tau)^3$ and higher leads to

$$\frac{\partial\theta(t)}{\partial t} = \frac{\theta(t+\Delta\tau) - \theta(t-\Delta\tau)}{2\Delta\tau} \tag{11.32}$$

Equation (11.32) is called the *central-difference* approximation of the time derivative. Two other approximations often used in practice are

The *forward-difference* formula:

$$\frac{\partial\theta(t)}{\partial t} = \frac{\theta(t+\Delta\tau) - \theta(t)}{\Delta\tau} \tag{11.33}$$

and the backward-difference formula:

$$\frac{\partial\theta(t)}{\partial t} = \frac{\theta(t) - \theta(t-\Delta\tau)}{\Delta\tau} \tag{11.34}$$

Similar expansion into a Taylor's series can be performed for space coordinates. Let us assume, for the purpose of illustrating the concepts, that the region S is a rectangle, and we divide it as shown in Fig. 11.12.

Consider a representative mesh point P in Fig. 11.12. Denoting the value of θ at P by

$$\theta_P = \theta_{i,j}$$

the partial derivatives appearing in equations (11.1)–(11.3) can be approximated as follows:

$$\begin{aligned}\left(\frac{\partial^2\theta}{\partial x^2}\right)_P &= \left(\frac{\partial^2\theta}{\partial x^2}\right)_{i,j} = \frac{\theta_{i+1,j} - 2\theta_{i,j} + \theta_{i-1,j}}{\Delta x^2} \\ \left(\frac{\partial^2\theta}{\partial y^2}\right)_P &= \left(\frac{\partial^2\theta}{\partial y^2}\right)_{i,j} = \frac{\theta_{i,j+1} - 2\theta_{i,j} + \theta_{i,j-1}}{\Delta y^2}\end{aligned} \tag{11.35}$$

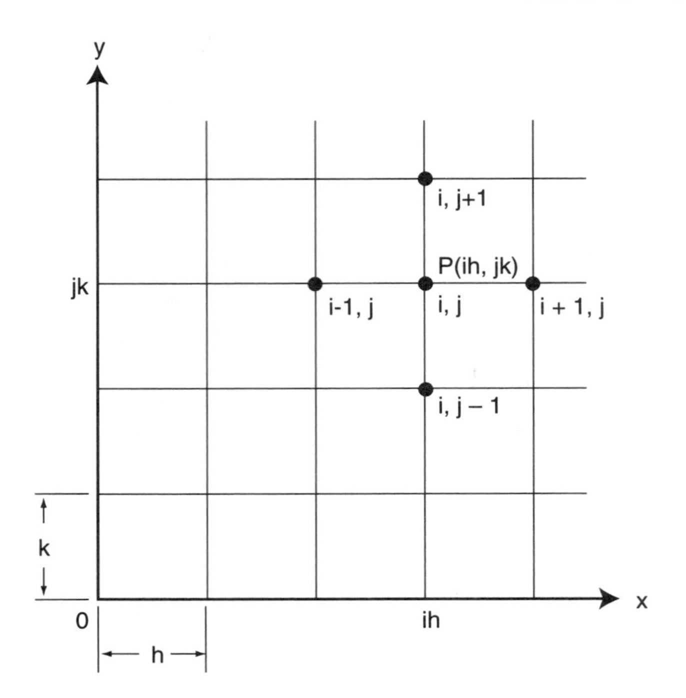

Figure 11.12 Rectangular finite-difference mesh.

We can combine equations (11.33) and (11.35) by introducing the notation

$$x = i\Delta x,\ y = j\Delta y,\ t = n\Delta\tau,\ \text{and}\ \theta(i\Delta x, j\Delta y, n\Delta\tau) = \theta_{i,j,n}$$

With this notation, equation (11.1) can be written as

$$\frac{1}{\delta_{i,j}}\frac{\theta_{i,j,n+1}-\theta_{i,j,n}}{\Delta\tau} = \frac{\theta_{i+1,j,n}-2\theta_{i,j,n}+\theta_{i-1,j,n}}{\Delta x^2} + \frac{\theta_{i,j+1,n}-2\theta_{i,j,n}+\theta_{i,j-1,n}}{\Delta y^2} + W_{\text{int},i,j,n}\rho_{i,j} \tag{11.36}$$

The step sizes Δx and Δy could be varied along the x and y axes. Also, in equation (11.36), the material properties can change in space and the heat generated can vary in space and time. Since the problem becomes computationally complex even for a small number of nodes, we will illustrate the above concepts for a steady-state heat transfer problem.

EXAMPLE 11.6

We will consider a single cable located 1 m below the ground and generating 12 W/m of heat. The earth surface is a convective boundary with convection coefficient 1 W/m^2 · K and the ambient air temperature is 0°C. We will also assume that the soil ambient temperature at the outer border of the mesh is also equal to 0°C.

Since the computations are performed without the aid of a computer, we will consider a small (2 m × 2 m) mesh with the cable located in the center. Placing the origin of the coordinates at the cable center, the region is bounded by straight lines $x = \pm 1$, $y = \pm 1$. We will use the division

$\Delta x = \Delta y = \frac{1}{2}$ and label the mesh points as indicated in Fig. 11.13, equal numbers denoting equal values of θ.

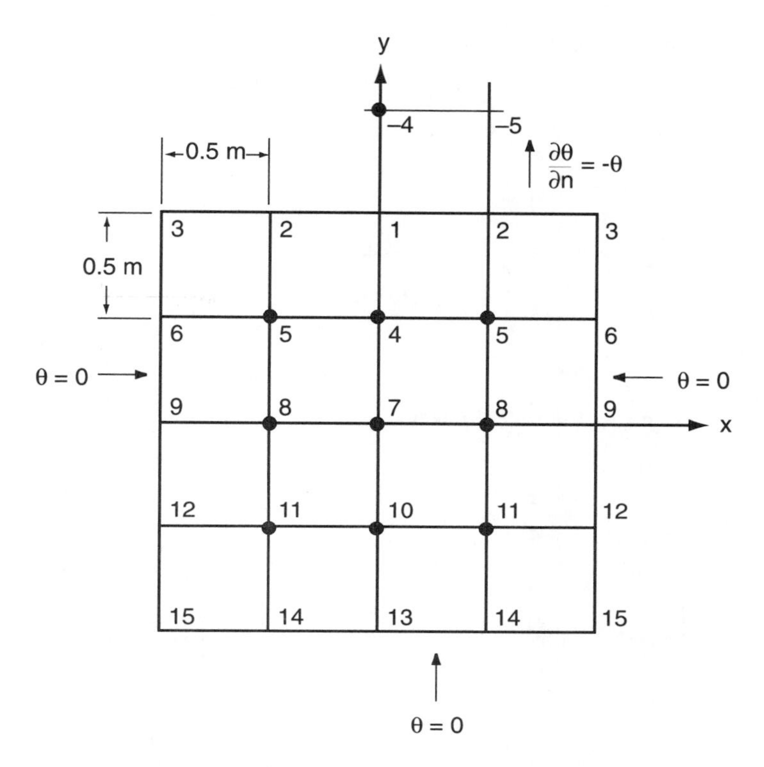

Figure 11.13 Finite-difference mesh for Example 11.6.

For simplicity, we will allocate the whole heat source at node (0, 0). Soil thermal resistance in each square is 1 K · m/W. The boundary conditions can be written as

$$\theta = 0 \text{ on } x = \pm 1 \text{ and } y = -1; \quad \frac{\partial \theta}{\partial n} + \theta = 0 \text{ on } y = 1$$

We will denote the mirror image of the point 2 (say) in $y = 1$ by -2, and the value of θ at point 2 by θ_2, and so on. One finite-difference representation of equation (11.1) is

$$\frac{1}{0.5^2}(\theta_{i+1,j} - 2\theta_{i,j} + \theta_{i-1,j}) + \frac{1}{0.5^2}(\theta_{i,j+1} - 2\theta_{i,j} + \theta_{i,j-1}) = g$$

or

$$\theta_{i+1,j} + \theta_{i,j+1} + \theta_{i-1,j} + \theta_{i,j-1} - 4\theta_{i,j} = g/4$$

where $g = -12$ for node 7 and zero otherwise. At the boundary point 1, we imagine the heat conducting area extended to the first row of external mesh points, and therefore we have

$$\theta_2 + \theta_{-4} + \theta_2 + \theta_4 - 4\theta_1 = 0$$

and from the boundary condition, using central differences,

$$(\theta_{-4} - \theta_4)/2\Delta y = (\theta_{-4} - \theta_4) = -\theta_1$$

Elimination of θ_{-4} gives the equation

$$-5\theta_1 + 2\theta_2 + 2\theta_4 = 0$$

In the same way, remembering that $\theta_3 = 0$, point 2 yields the equation

$$\theta_1 - 5\theta_2 + 2\theta_5 = 0$$

Similarly, points 4, 5, 7, 8, 10, and 11 yield the equations

$$\theta_1 - 4\theta_4 + 2\theta_5 + \theta_7 = 0$$

$$\theta_2 + \theta_4 - 4\theta_5 + \theta_8 = 0$$

$$\theta_4 - 4\theta_7 + 2\theta_8 + \theta_{10} = -3$$

$$\theta_5 + \theta_7 - 4\theta_8 + \theta_{11} = 0$$

$$\theta_7 - 4\theta_{10} + 2\theta_{11} = 0$$

$$\theta_8 + \theta_{10} - 4\theta_{11} = 0$$

Written in a matrix form, these equations become

$$\begin{bmatrix} -5 & 2 & 2 & & & & & \\ 1 & -5 & 0 & 2 & & & & \\ 1 & 0 & -4 & 2 & 1 & & & \\ & 1 & 1 & -4 & 0 & 1 & & \\ & & 1 & 0 & -4 & 2 & 1 & \\ & & & 1 & 1 & -4 & 0 & 1 \\ & & & & 1 & 0 & -4 & 2 \\ & & & & & 1 & 1 & -4 \end{bmatrix} \begin{bmatrix} \theta_1 \\ \theta_2 \\ \theta_4 \\ \theta_5 \\ \theta_7 \\ \theta_8 \\ \theta_{10} \\ \theta_{11} \end{bmatrix} = \begin{bmatrix} 0 \\ 0 \\ 0 \\ 0 \\ -3 \\ 0 \\ 0 \\ 0 \end{bmatrix} \tag{11.36a}$$

where the empty area in the matrix is filled with zeros. We can observe that the coefficient matrix is of the form

$$\begin{bmatrix} (\mathbf{B}-\mathbf{I}) & 2\mathbf{I} & & \\ \mathbf{I} & \mathbf{B} & \mathbf{I} & \\ & \mathbf{I} & \mathbf{B} & \mathbf{I} \\ & & \mathbf{I} & \mathbf{B} \end{bmatrix} \quad \text{where } \mathbf{B} = \begin{bmatrix} -4 & 2 \\ 1 & -4 \end{bmatrix}$$

The solution of (11.36a) is

$$\theta_1 = 0.26°\text{C}, \quad \theta_2 = 0.16°\text{C}, \theta_4 = 0.49°\text{C}, \quad \theta_5 = 0.27°\text{C},$$
$$\theta_7 = 1.18°\text{C}, \quad \theta_8 = 0.41°\text{C}, \theta_{10} = 0.40°\text{C}, \quad \theta_{11} = 0.20°\text{C}$$

In Example 11.6, uniform soil was assumed. When different soil layers are present, the grid lines should coincide with the boundaries between soils with different characteristics. In this case, it may be more convenient to write equation (11.36) in the form

$$Q_{i,j}\frac{\theta_{i,j,n+1} - \theta_{i,j,n}}{\Delta\tau} = \frac{k_{i+1,j}\theta_{i+1,j,n} - 2k_{i,j}\theta_{i,j,n} + k_{i-1,j}\theta_{i-1,j,n}}{\Delta x^2} + \frac{k_{i,j+1}\theta_{i,j+1,n} - 2k_{i,j}\theta_{i,j,n} + k_{i,j-1}\theta_{i,j-1,n}}{\Delta y^2} + W_{\text{int},i,j,n} \tag{11.37}$$

where $k_{i,j}$ and $Q_{i,j}$ denote the thermal conductance and thermal capacity of the material in which node ij is located. We will demonstrate how the node conductivities are computed by again considering Example 11.6.

EXAMPLE 11.7

Assume that cable in Example 11.6 is located in a thermal backfill with thermal resistance of 0.6 K·m/W, which encompasses six squares bounded by corner nodes 2, 2, 11, and 11 in Fig. 11.13. We will reformulate the finite-difference equations to reflect varying thermal resistivities.

Equation (3.37) will now take the form

$$k_{i+1,j}\theta_{i+1,j,n} + k_{i,j+1}\theta_{i,j+1,n} + k_{i-1,j}\theta_{i-1,j,n} + k_{i,j-1}\theta_{i,j-1,n} - 4k_{i,j}\theta_{i,j,n} = g/4$$

where the conductivity values are computed as follows.

The thermal conductivity associated with a node is an arithmetic average of the conductivities of the rectangles to which the node belongs. For the nodes inside the backfill (nodes 1, 4, and 7), $k_{i,j} = 1/0.6 = 1.67$. For the nodes located inside the native soil, $k_{i,j} = 1/1 = 1$. For nodes 2, 5, 8, and 10, on the perimeter of the backfill, $k_{i,j} = (1/0.6 + 1/1)/2 = 1.33$, and for node 11, $k_{i,j} = (1/0.6 + 3 \cdot 1/1)/4 = 1.17$.

At the boundary point 1, we have

$$1.33\theta_2 + 1.67\theta_{-4} + 1.33\theta_2 + 1.67\theta_4 - 4 \cdot 1.67\theta_1 = 0$$

and from the boundary condition, using central differences,

$$1.67(\theta_{-4} - \theta_4) = -\theta_1$$

Elimination of θ_{-4} gives the equation

$$-2.3\theta_1 + \theta_2 + \theta_4 = 0$$

Similarly, point 2 yields the equation

$$1.67\theta_1 - 6.32\theta_2 + 2.66\theta_5 = 0$$

Likewise, points 4, 5, 7, 8, 10, and 11 yield the equations

$$\begin{aligned}
\theta_1 - 4\theta_4 + 1.6\theta_5 + \theta_7 &= 0 \\
\theta_2 + 1.25\theta_4 - 4\theta_5 + \theta_8 &= 0 \\
1.25\theta_4 - 5\theta_7 + 2\theta_8 + \theta_{10} &= -3 \\
1.33\theta_5 + 1.67\theta_7 - 5.32\theta_8 + 1.17\theta_{11} &= 0 \\
1.67\theta_7 - 5.32\theta_{10} + 2.34\theta_{11} &= 0 \\
\theta_8 + \theta_{10} - 3.5\theta_{11} &= 0
\end{aligned}$$

Their solution is

$$\theta_1 = 0.27°\text{C}, \quad \theta_2 = 0.19°\text{C}, \quad \theta_4 = 0.43°\text{C}, \quad \theta_5 = 0.29°\text{C},$$
$$\theta_7 = 0.96°\text{C}, \quad \theta_8 = 0.42°\text{C}, \quad \theta_{10} = 0.40°\text{C}, \quad \theta_{11} = 0.24°\text{C}$$

Equation (11.36) is called the explicit finite-difference representation of equation (11.1). This is because the values of θ can be computed explicitly from this equation given the initial and boundary conditions. Although the explicit method is computationally simple, it has one serious drawback. The time step $\Delta\tau$ is necessarily very small because the process is valid only for

$$\rho\left\{\frac{1}{\Delta x^2}+\frac{1}{\Delta y^2}\right\}\Delta\tau \le \frac{1}{2} \tag{11.38}$$

and Δx and Δy must be kept small in order to attain reasonable accuracy. Crank and Nicolson (Smith, 1965) proposed a method that reduces the total volume of calculation and is stable. In this approach equation (11.36) is replaced by

$$\frac{1}{\delta}\frac{\theta_{i,j,n+1}-\theta_{i,j,n}}{\Delta\tau}=\frac{1}{2}\left[\left(\frac{\partial^2\theta}{\partial x^2}+\frac{\partial^2\theta}{\partial y^2}\right)_{i,j,n}+\left(\frac{\partial^2\theta}{\partial x^2}+\frac{\partial^2\theta}{\partial y^2}\right)_{i,j,n+1}\right]+W_{\text{int}}\rho \tag{11.39}$$

which yields somewhat more complex finite-difference equations.

When the cable is not a line heat source but has a specified diameter, the cable outer surface cuts the vertical and horizontal lines of the grid as shown in Fig. 11.14, producing an irregular boundary inside the trench. The finite-difference equation (11.37) has to be modified to reflect the new boundary points. For example, in the case shown in Fig. 11.14, we have

$$\begin{aligned}\left(\frac{\partial^2\theta}{\partial x^2}\right)_P&=\left(\frac{\partial^2\theta}{\partial x^2}\right)_{i,j}=\frac{1}{\Delta x^2}\left[\frac{2\theta_{i+1,j}}{\eta_{i,j}(1+\eta_{i,j})}-\frac{2\theta_{i,j}}{\eta_{i,j}}+\frac{2\theta_{i-1,j}}{(1+\eta_{i,j})}\right]\\ \left(\frac{\partial^2\theta}{\partial y^2}\right)_P&=\left(\frac{\partial^2\theta}{\partial y^2}\right)_{i,j}=\frac{\theta_{i,j+1}-2\theta_{i,j}+\theta_{i,j-1}}{\Delta y^2}\end{aligned} \tag{11.40}$$

The finite-difference equations become even more complex if the internal structure of the cable is to be represented properly or when normal derivatives at the cable surface are to be approximated by finite differences. When a circular symmetry with respect to the cable

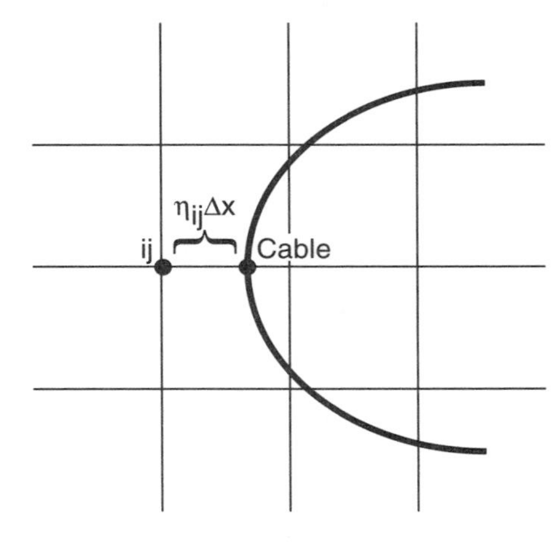

Figure 11.14 Cable outer diameter cuts the grid lines.

center can be assumed, a radial form of the finite-difference equations could be applied. In practice, only under certain restrictive assumptions can such a symmetry be assumed. One such case is a single-core cable located in a uniform soil with an isothermal boundary condition at the earth surface. This has been studied by CIGRE WG21.02 (CIGRE, 1983). We will discuss this case next.

11.4.3 Application of the Finite-Difference Method for the Calculation of the Response of Single-Core Cables to a Step Function Thermal Transient[1]

11.4.3.1 General Characteristics of the Method. The finite-difference method described in the previous section can be viewed as being equivalent to a representation of the cable and its surrounding by a ladder network of thermal resistances and capacitances, in the same general way as was described in Chapter 5, but with many more subdivisions.

To show this, we will assume that the heat flow is radial both inside and outside the cable so that it is possible to divide the cable and the soil into a number of concentric layers corresponding to elements in the ladder network. The outer radius of the soil where the heat disperses will increase with time, and for practical purposes, it is sufficient to consider only that radius within which a sensible temperature rise occurs. In this case, this radius can be estimated from equation (11.4).

For long-duration transients, this cylinder of soil may include adjacent cables. In the case of isothermal earth surface boundary, the radius will also include image sources to both the cable under consideration and adjacent cables (Fig. 11.15), dissipating (at each time step) losses of the same value but opposite sign. The time steps can be varied during the computation, and their durations should be governed by equations (11.5).

When calculating the thermal field of a cable by the method presented in CIGRE (1983), the space occupied by other cables is assumed to have the same thermal properties as the soil (restricted application of superposition).

11.4.3.2 Description of the Method. The insulation (including the semiconducting screen) should be divided into ten elementary layers of equal thickness.[2] The semiconducting screens are part of the thermal circuit, and are assumed to have the same thermal characteristics as the insulation. The dielectric loss is calculated disregarding the screens but, for simplicity, it is approportioned over the total thickness (i.e., insulation plus screen). A single layer is used for cable covering.

The number of elementary layers in the soil n_s must fulfill the following conditions:

$$\frac{2r - D_e}{2n_s D_e} < 0.16 \tag{11.41}$$

where r is given by equation (11.4). The division of the cable and the soil into layers is illustrated in Fig. 11.16.

[1] Some of the material in this section is based on an article published in *Electra* (CIGRE, 1983).

[2] An alternative method, which can offer advantages in computation, is to divide the thickness of each material so as to obtain equal ratios of the radii, that is, each layer in a particular material would then have the same thermal resistance.

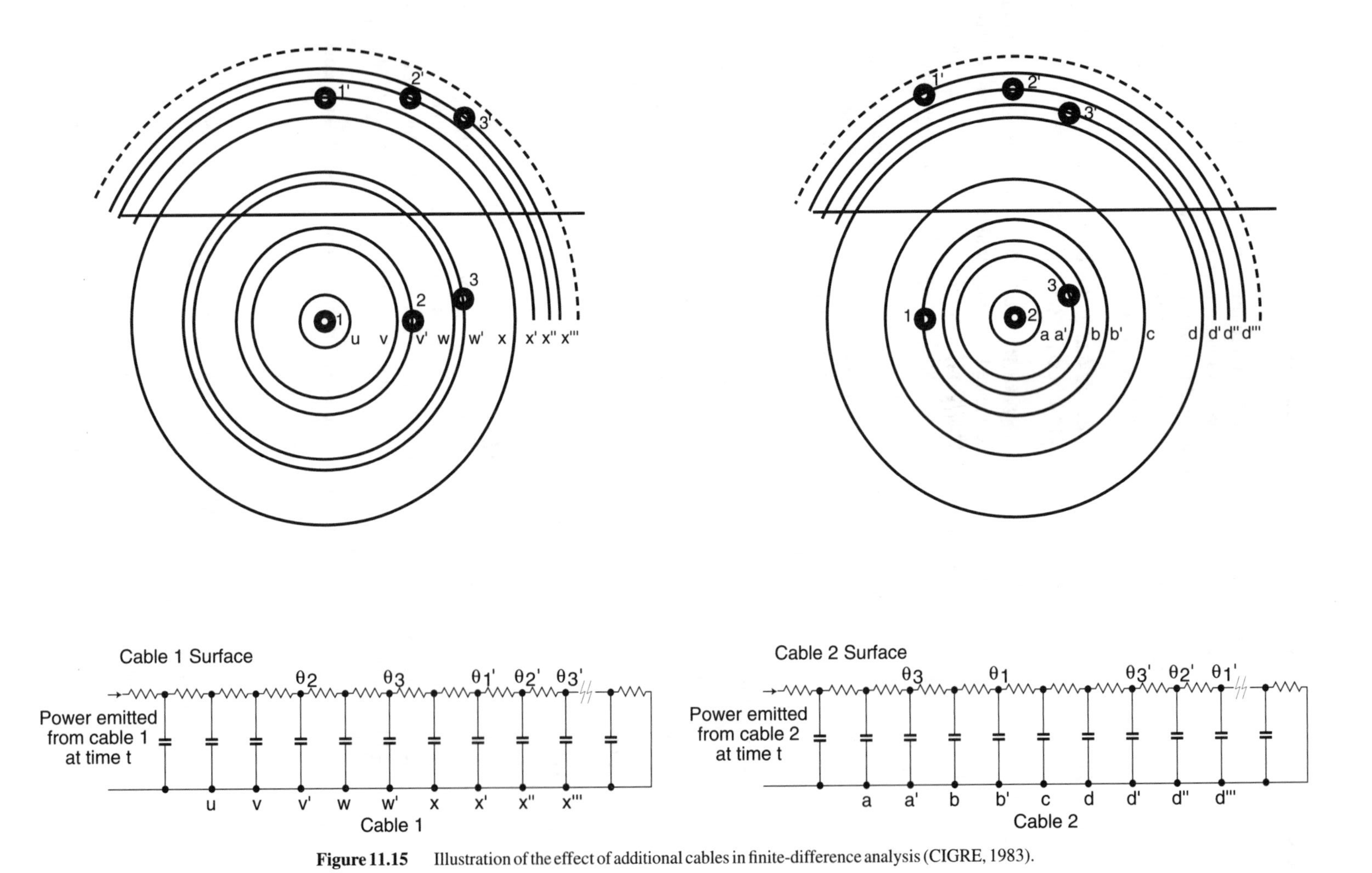

Figure 11.15 Illustration of the effect of additional cables in finite-difference analysis (CIGRE, 1983).

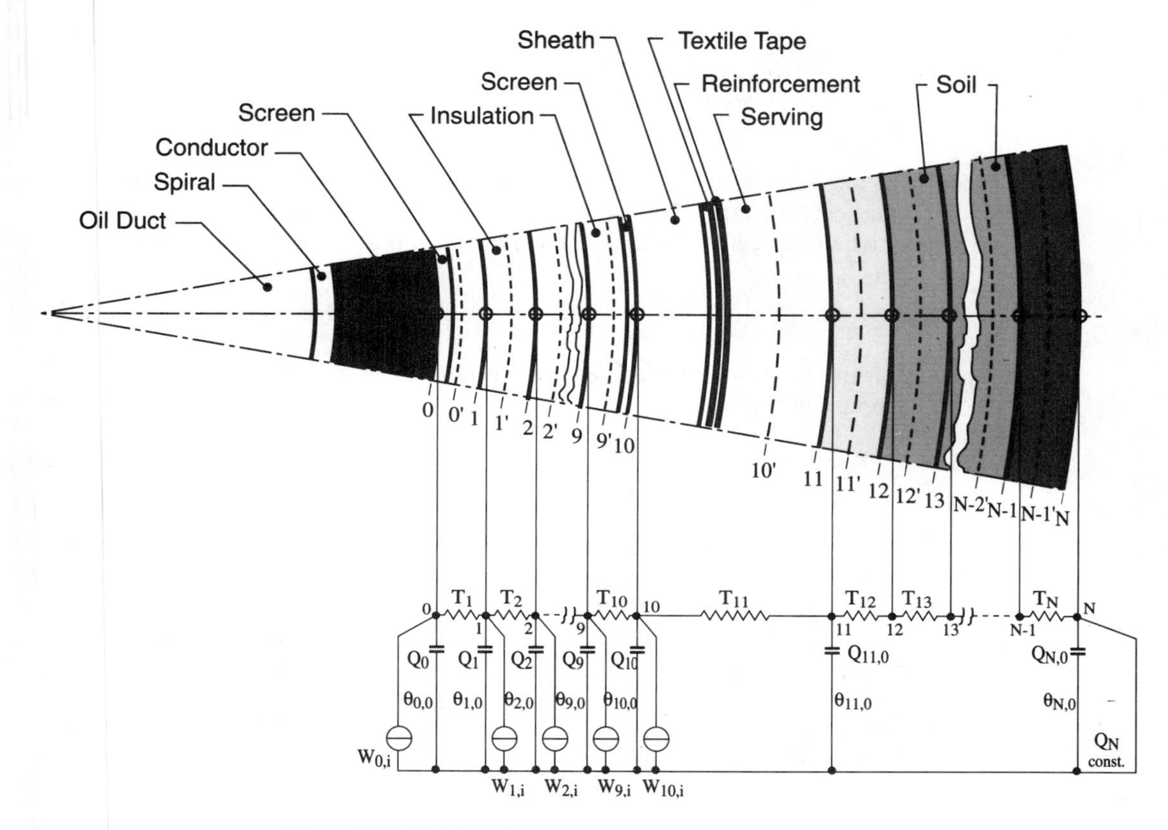

Figure 11.16 Division of the cable and the soil into layers (CIGRE, 1983).

Each layer has a thermal resistance obtained from applying equation (3.6):

$$T_j = \frac{\rho_T}{2\pi} \ln \frac{r_j}{r_{j-1}} \tag{11.42}$$

where r_j and r_{j-1} are the external and the internal radii of the layer, respectively. The thermal capacitance of each layer is divided at the midlayer radius, and allocated to the nodes at the inner and outer radii of the layer to give a π scheme for the equivalent circuit of each layer. The capacitance at each node is obtained from equation (3.14):

$$Q_j = \pi c \left[\left(\frac{r_j + r_{j+1}}{2} \right)^2 - \left(\frac{r_{j-1} + r_j}{2} \right)^2 \right] \tag{11.43}$$

In developing equations for temperature variations with time, we will consider first durations so short that no adjacent cables or image sources are involved. Next, we will discuss the case of longer duration transients.

11.4.3.3 Computation of Temperature Transient for a Single Cable. We will develop the finite-difference equations for the ladder network in Fig. 11.16 starting from the heat balance constraint at each node (Hanna *et al.*, 1993; Selsing, 1985; Weedy, 1988).

The heat balance equation, when differentiated, leads to the partial differential equation (11.1) (see Section 2.3.1). The heat input to the node j is equal to

$$k_j(\theta_{j-1,n} - \theta_{j,n}) + k_{j+1}(\theta_{j+1,n} - \theta_{j,n}) + q_j \tag{11.44}$$

where k_j is the thermal conductance from node $j-1$ to j, n is the index of elapsed time, and q_j is the heat generated in the volume represented by node j from the joule or dielectric loss. Over a finite time interval $\Delta\tau$, the average rate of flow to the region is given by

$$Q_j \frac{\theta_j(t+\Delta\tau) - \theta_j(t)}{\Delta\tau} = Q_j \frac{\theta_{j,n+1} - \theta_{j,n}}{\Delta\tau} \tag{11.45}$$

where Q_j is the thermal capacitance of an elementary layer between nodes $j-1$ and j.

Equating now (11.44) and (11.45), we obtain

$$\theta_{j,n+1} = \frac{k_j\Delta\tau}{Q_j}\theta_{j-1,n} + \frac{k_{j+1}\Delta\tau}{Q_j}\theta_{j+1,n} + \left[1 - \frac{\Delta\tau}{Q_j}(k_j + k_{j+1})\right]\theta_{j,n} + \frac{\Delta\tau q_j}{Q_j} \tag{11.46}$$

beginning at $j = N-1$ and ending at $j = 0$ where $\theta_{j,n}$ is the temperature rise at node j at time n and $\theta_{N,n}$ is always zero. N is the number of nodes, and we start with known or assumed initial temperatures.

Equation (11.41) can be applied to each core of a pipe-type cable. As Weedy (1988) pointed out, an accurate modeling of the oil in the pipe is much more difficult, and usually the oil is represented by a single lumped thermal capacity.

To assure numerical stability of equation (11.41), the coefficient of $\theta_{j,n}$ must be positive. This yields the following requirement for the time step at each node:

$$\Delta\tau < \frac{Q_j}{k_j + k_{j+1}} \tag{11.47}$$

The limitations of the time step in the explicit method may be overcome by the use of implicit equations such as given by Crank and Nicolson [equation (11.39)] by Lees' approach (1966), or by a Hopscotch method (Gourley, 1970).

CIGRE (1983) proposed the following algorithm equivalent to equation (11.46).

1. For each node, we compute

$$A_{j,n} = \left[(1 - A_{j-1,n})\cdot\left(\frac{T_{j+1}}{T_j}\right) + \left(\frac{Q_jT_{j+1}}{\Delta\tau_i}\right) + 1\right]^{-1} \tag{11.48}$$

beginning at $j = 0$ and ending at $j = N-1$, with $A_{-1,n} = 1$. Note that $A_{j,n}$ does not change during the computation unless the time step $\Delta\tau_i$ is changed after a given time τ during the computation.

2. Compute initial values for each node:

$$B_{j,n} = A_{j,n}T_{j+1}\left[W_{dj,n} + \theta_{j,n-1}\left(\frac{Q_j}{\Delta\tau_i}\right) + \frac{B_{j-1,n}}{T_j}\right] \tag{11.49}$$

beginning at $j = 0$ and ending at $j = N-1$, with $B_{-1,n} = 0$.

3. Compute the values of temperature rise from

$$\theta_{j,n} = (A_{j,n} + \theta_{j+1,n}) + B_{j,n} \tag{11.50}$$

beginning at $j = N - 1$ and ending at $j = 0$, where $\theta_{j,n}$ is the temperature rise at node j at time n and $\theta_{N,n}$ is always zero.

4. Repeat steps 2 and 3 for as many times as is necessary to complete the transient

$$\Delta\tau_i = \tau_i - \tau_{i-1}$$

The value of $\Delta\tau_i$ may be changed during the computations of the transient provided it remains with the limits specified by equation (11.5). Coefficient $A_{j,n}$ must then be recalculated.

11.4.3.4 Mutual Heating from Other Cables. When there is more than one cable, each will have its own thermal diagram and equivalent network from which the transient temperature rises caused by its own losses are calculated. If the duration of the transient is sufficiently long for mutual heating from other cables and image sources to be significant, this must be taken into account in the calculation of the true temperatures used to update cable losses at each time step. Since the presence of other cables violates cylindrical symmetry, a restricted application of superposition is applied.

The true temperature for each cable is obtained at each time step by adding to the cable's own temperature the temperature rise at the same location from every other cable and image source (see Fig. 11.15). These thermal fields are computed from the equations corresponding to the equivalent networks centered on each cable or image source. This process is carried out for all cables and image sources in the group. Note that the thermal diagram and network of an image source are identical to those of its corresponding cable except for the negative sign of its heat flux and temperature rise; separate networks for image sources are therefore not needed.

EXAMPLE 11.9 (CIGRE, 1983)

We will consider one cable in a circuit of three single-core cables with the duration of a transient lasting 4 h. This time is short enough for neglecting the effect of the two adjacent cables. The computations deal only with the self-heating of the center cable. The cables are 400 kV, 2000 mm^2 with paper/fluid insulation laid directly in the ground, 1 m deep, in flat configuration, spaced 70 cm apart, and with the sheaths cross bonded.

The additional information is as follows:

Conductor dc resistance at 20°C = 8.95 $\mu\Omega/m$

Diameter of the conductor = 57.5 mm

Diameter over conductor screen (semiconducting paper tapes) = 59.0 mm

Diameter over paper insulation = 105.0 mm

Relative permitivity of the insulation = 3.3

$\tan\delta$ = 0.0025

Diameter over semiconducting screen = 106.0 mm

Diameter over lead sheath = 114.0 mm

DC resistance of sheath at 20°C = 154.8 $\mu\Omega/m$

Diameter over polyethylene jacket = 122.0 mm.

Thermal resistivities of the metallic parts are assumed to be zero. All remaining thermal resistivities and heat capacities are as specified in Table 9.1.

The steady-state ampacity of this cable is equal to 1800 A. We will investigate variation of conductor temperature with time when an emergency current of 3600 A is applied to this cable for 4 h. Prior to the emergency condition, the cable was energized long enough for the temperature rise due to dielectric losses to reach its steady-state value.

Figure 11.17 shows the results of the computations.

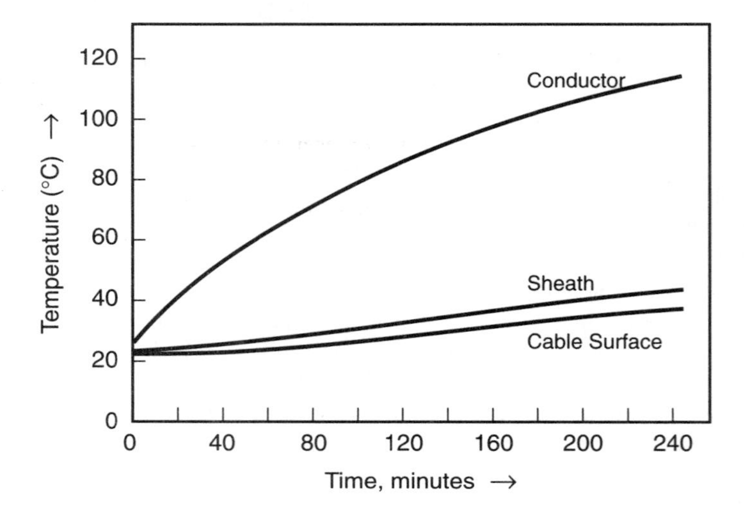

Figure 11.17 Calculated transient temperatures, 3600 A for 4 h.

We will close this chapter with a brief discussion of modeling and computational issues involved in the application of numerical methods.

11.5 MODELING AND COMPUTATIONAL ISSUES

Finite-difference and finite-element methods for solving the heat conduction differential equation often lead to a very large number of algebraic equations, and their solution is a problem in itself. Large sets would generally be solved iteratively and small sets by direct elimination methods. Iterative methods are more efficient then direct methods in that they take advantage of the large number of zero coefficients in the matrices. There is a vast literature on the subject dealing with sets of linear equations and the economics that can be achieved when the matrices are sparse, and the interested reader is referred to the many books on numerical methods.

The basic difference between finite-element and finite-difference methods lies in the manner in which approximations are performed. As stated earlier, in the finite-element method, the temperature is approximated by a discrete model composed of a set of continuous functions defined over a finite number of subdomains. The piecewise continuous functions are defined using the values of temperature at a finite number of points in the region of interest. In finite-difference methods, the derivatives at a point are approximated by difference quotients over a small interval. Both approximations can be quite accurate if the meshes are made suitably small.

Both methods allow a choice for the multinodal elementary subdomains. However, because of the nature of the approximations involved, a rectangular mesh with division lines parallel to the coordinate axes is usually chosen for the finite-difference method. This may create difficulty in representing circular boundaries which are quite common in cable representation. This difficulty does not exist in the finite-element method because curved boundaries can be well approximated by a suitable selection of elements. In addition, finite-difference equations may become very complex when nonuniform spacing is used along any of the axes, whereas the finite-element methods handle differing element sizes quite naturally.

Because of their flexibility in representing the region around the cables and the easy modeling of boundary conditions, finite-element methods appear to be more suitable for the numerical solution of the heat conduction equations considered in this chapter.

REFERENCES

Abdel-Hadi, O. N. (1978), "Flow of heat and water around underground power cables," Ph.D. Dissertation, University of California at Berkeley.

Anders, G. J., Bedard, N., Chaaban, M., and Ganton, R. W. (Oct. 1987), "New approach to ampacity evaluation of cables in ducts using finite element technique," *IEEE Trans. Power Delivery*, vol. PWRD-2, no. 4, pp. 969–975.

Anders, G. J., and Radhakrishna, H. S. (Oct. 1988a), "Power cable thermal analysis with consideration of heat and moisture transfer in the soil," *IEEE Trans. Power Delivery*, vol. 3, no. 4, pp. 1280–1288.

Anders, G. J., and Radhakrishna, H. S. (Jan. 1988b), "Computation on the temperature field and moisture content in the vicinity of current carrying underground power cables," *Proc. IEE*, Vol. 153, part C, no. 1, pp. 51–62.

Anders, G. J., El-Kady, M. A., Horrocks, D. J., and Motlis, J. (Oct. 1988), "Modified values for geometric factor of external thermal resistance of cables in ducts," *IEEE Trans. Power Delivery*, vol. 3, no. 4, pp. 1303–1309.

Black, W. Z., and Park, S. (1983), "Emergency ampacities of direct buried three-phase underground cable systems," *IEEE Trans. Power App. Syst.*, vol. PAS-102, no. 7, pp. 2124–2132.

CIGRE (1983), "Computer method for the calculation of the response of single-core cables to a step function thermal transient," *Electra*, no. 87, pp. 41–58.

El-Kady, M. A. (Aug. 1985), "Calculation of the sensitivity of power cable ampacity to variations of design and environmental parameters," *IEEE Trans. Power App. Syst.*, vol. PAS-103, no. 8, pp. 2043–2050.

El-Kady, M. A., and Horrocks, D. J. (1985), "Extended values of geometric factor of external thermal resistance of cables in duct banks," *IEEE Trans. Power App. Syst.*, vol. PAS-104, pp. 1958–1962.

EPRI (1982), "Thermal stability of soils adjacent to underground transmission power cables," Final Report of Research Project 7883.

Flatabo, N. (1973), "Transient heat conduction problem in power cables solved by the finite element method," *IEEE Trans. Power App. Syst.*, vol. PAS-92, pp. 161–168.

Gourlay, A. R. (1970), "Hopscotch: A fast second order partial differential equation solver," *J. Inst. Math. Appl.*, vol. 6, pp. 375–390.

Groeneveld, G. J., Snijders, A. L., Koopmans, G., and Vermeer, J. (1984), "Improved method to calculate the critical conditions for drying out sandy soils around power cables," *Proc. IEE*, vol. 131, part C, no. 2, pp. 42–53.

Hanna, M. A., Chikhani, A. Y., and Salama, M. M. A. (1993), "Thermal analysis of power cables in multi-layered soil. Parts 1 and 2," *IEEE Trans. Power Delivery*, vol. 8, no. 3, pp. 761–778.

Hartley, J. G., and Black, W. Z. (1981), "Transient simultaneous heat and mass transfer in moist unsaturated soils," *ASME Trans.*, vol. 103, pp. 376–382.

Hiramandani, A. (1991), "Calculation of conductor temperatures and ampacities of cable systems using a generalized finite difference model," *IEEE Trans. Power Delivery*, vol. 6, no. 1, pp. 15–24.

Kellow, M. A. (1981), "A numerical procedure for the calculation of temperature rise and ampacity of underground cables," *IEEE Trans. Power App. Syst.*, vol. PAS-100, no. 7, pp. 3322–3330.

KEMA/HEIDEMIJ Report (1981), "Moisture migration and drying-out in sand around heat dissipating cables and ducts. A theoretical and experimental study."

Konrad, A. (1982), "Integrodifferential finite element formulation of two-dimensional steady-state skin effect problem," *IEEE Trans. Magn.*, vol. MAG-18, pp. 284–292.

Labridis, D., and Dokopoulos, P. (1988), "Finite element computation of field, losses and forces in a three-phase gas cable with non-symmetrical conductor arrangement," *IEEE Trans. Power Delivery*, vol. 3, no. 4, pp. 1326–1333.

Lees, M. (1966), "A linear three-level difference scheme for quasilinear equations," *Math. Comp.*, vol. 20, pp. 516–622.

Libondi, L. (Feb. 1975), "Calcolo numerico di transitori termici in cavi unipolari," *Elettrotecnica*, vol. LXII, no. 2.

Mitchell, J. K., and Abdel-Hadi, O. N. (1979), "Temperature distributions around buried cables," *IEEE Trans. Power App. Syst.*, vol. PAS-98, no. 4, pp. 1158–1166.

Mushamalirwa, D., Germay, N., and Steffens, J. C. (1988), "A 2-D finite element mesh generator for thermal analysis of underground power cables," *IEEE Trans. Power Delivery*, vol. 3, no. 1, pp. 62–68.

Radhakrishna, H. S., Lau, K. C., and Crawford, A. M. (1984), "Coupled heat and moisture flow through soils," *J. Geotech. Eng.*, vol. 110, no. 12, pp. 1766–1784.

Segerlind, L. J. (1984), *Applied Finite Element Analysis*, 2nd ed. New York: Wiley.

Selsing, J. (1985), "A versatile computer method for computation of conductor temperatures in cable terminations and pipe cable systems," *IEEE Trans. Power App. Syst.*, vol. PAS-104, no. 4, pp. 768–774.

Smith, G. D. (1965), *Numerical Solution of Partial Differential Equations.* London: Oxford University Press.

Tarasiewicz, E., El-Kady, M. A., and Anders, G. J. (Jan. 1987), "Generalized coefficients of external thermal resistance for ampacity evaluation of underground multiple cable systems," *IEEE Trans. Power Delivery*, vol. PWRD-2, no. 1, pp. 15–20.

Thomas, H. R., Morgan, K., and Lewis, R. W. (1980), "A fully nonlinear analysis of heat and mass transfer problems in porous bodies," *Int. J. Numer. Methods in Eng.*, vol. 15, pp. 1381–1393.

Weedy, B. M. (1988), *Thermal Design of Underground Systems.* Chichester, England: Wiley.

Zienkiewicz, O. C. (1971), *The Finite Element Method in Engineering Science.* New York: McGraw-Hill.

12

Economic Selection of Conductor Cross Section

12.1 INTRODUCTION

Selection of cable sizes is currently based on ampacity considerations; that is, a cable with minimum admissible cross-sectional area is usually selected without consideration of the cost of the losses that will occur during the life of the cable. Choosing a minimal cross section results in a minimal initial cost investment. However, the cost of losses over the lifetime of the cable may be quite substantial. Selection of a larger cable size than required for ampacity consideration will often result in a smaller value of losses, and hence may lead to a lower overall cost.

The importance of the cost of losses in selecting the most economical cable size has been discussed by Parr (1989), and an international standard sponsored by the IEC has been issued (IEC 1059, 1991).

In this chapter, we will introduce a mathematical model for the selection of the most economical cable size for a particular application. The model includes the estimation of the energy losses over the life of a cable, making allowance for growth in load and usage, and calculates the present value of these losses using projected interest rates. This present value of the cost of losses is added to the primary investment to give the total cost. To obtain the most economic conductor size, the total cost is minimized.

We will consider the representation of changes in the load pattern during the life span of the cable, as well as the effect of charging current and dielectric losses. A full coupling with the ampacity computation techniques described in earlier chapters is included in the optimization process.

To calculate the present value of the costs of the losses, the following parameters are required:

- Basic economic data: cost of installation, unit cost of losses, forecast of discount rate, and unit demand charges
- Basic construction of the cable: number of conductors, insulation type, outer protection type
- Basic installation information: depth of burial and spacing
- Basic cable operational data: voltage level, estimated economic life, current in the first year of operation, annual load growth, load curve characteristics, operating and ambient temperatures, soil thermal resistivity.

The mathematical models described below are partially based on the IEC Standard 1059 (1991). The optimum cross-sectional area of the conductor for the required load is first computed, and then the closest standard conductor size is selected.

12.2 COST OF JOULE LOSSES

The optimization model proposed below considers the cost of purchase and installation of the cable and the cost of losses over its economic life. Both costs are expressed in "*present values.*" The *future* costs of the energy losses are converted to their *present values* by utilization of a discount rate which is linked to the cost of borrowing money.

In the procedure given below, inflation has been omitted on the grounds that it will affect both the cost of borrowing money and the cost of energy. If these items are considered over the same period of time, and the effect of inflation is approximately the same for both, the choice of an economic size can be made satisfactorily without introducing the added complication of inflation.

The total cost of installing and operating a cable during its economic life, expressed in present values, is calculated as follows.

The total cost

$$CT = CI + CL \tag{12.1}$$

where CI = cost of the installed length of cable, \$

CL = equivalent cost, at the date the installation was purchased, of the losses during economic life of N years, \$.

The overall cost of the first year's joule losses is

$$I_0^2 \cdot R \cdot L \cdot N_p \cdot N_c \cdot (T \cdot P + D)$$

If these costs are paid at the end of the year, then at the date of the purchase of the installation, their present value is

$$\frac{I_0^2 \cdot R \cdot L \cdot N_p \cdot N_c \cdot (T \cdot P + D)}{1 + i}$$

where D = demand charge per year, \$/W·year

I_0 = maximum load on the cable during the first year, A

i = discount rate, not including the effect of inflation

L = length of cable, m

N_p = number of phase conductors per circuit

N_c = number of circuits carrying the same value and type of load

P = cost of 1 watt hour of energy at the relevant voltage level, \$/Wh

R = apparent ac resistance of a conductor per unit length, taking into account both skin and proximity effects and losses in metal screens and armor; since the resistance is dependent on the cable operating temperature, its value should be computed iteratively or an approximate mean value should be used as discussed in Section 12.2.2

T = operating time at maximum joule loss, hours/year

= number of hours per year that the maximum current I_0 would need to flow in order to produce the same total yearly energy losses as the actual, variable, load current:

$$T = \int_0^{8760} \frac{I^2(t)\,dt}{I_0^2}$$

where $I(t)$ is a load current as a function of time (A).

12.2.1 Constant Load Factor

If the load loss factor (μ) is known and can be assumed to be constant during the economic life, then

$$T = \mu \cdot 8760 \tag{12.2}$$

The load loss factor μ can be approximated using the Neher/McGrath (1957) approach as

$$\mu = p \cdot LF + (1 - p) \cdot LF^2 \tag{12.3}$$

where LF is the load factor and the value of constant p can be taken as equal to 0.3 for transmission circuits and equal to 0.2 for distribution networks (IEEE, 1990).

Similarly, the present value of the cost of losses during N years of operation discounted to the date of purchase is

$$\frac{I_0^2 \cdot R \cdot L \cdot N_p \cdot N_c \cdot [T \cdot P \cdot Q_P(N) + D \cdot Q_D(N)]}{1 + i}$$

where $Q_P(N)$ and $Q_D(N)$ are coefficients taking into account the increase in load, the increase in cost of energy over the N years, and the discount rate.

$$Q_P(N) = \sum_{n=1}^{N} r_P^{(n-1)} = \frac{1 - r_P^N}{1 - r_P}, \qquad r_P = \frac{(1+a)^2(1+b)}{1+i} \tag{12.4}$$

$$Q_D(N) = \sum_{n=1}^{N} r_D^{(n-1)} = \frac{1 - r_D^N}{1 - r_D}, \qquad r_D = \frac{(1+a)^2(1+c)}{1+i} \tag{12.5}$$

a = increase in load per year

b = increase in cost of energy per year, not including the effect of inflation

c = demand charge escalation factor.

Where a number of calculations involving different sizes of conductor is required, it is advantageous to express all the parameters except conductor current, resistance, and length in two coefficients F_1 and F_2, where

$$F_1(N) = \frac{T \cdot P \cdot N_p \cdot N_c \cdot Q_P(N)}{1+i}, \qquad F_2(N) = \frac{D \cdot N_p \cdot N_c \cdot Q_D(N)}{1+i} \tag{12.6}$$

The present value CL of the cost of joule losses is then obtained from

$$CL = I_0^2 \cdot R \cdot L \cdot [F_1(N) + F_2(N)] \tag{12.7}$$

Denoting $F(N) = F_1(N) + F_2(N)$, the total cost of investment and joule losses is obtained from equation (12.7):

$$CT = CI + I_0^2 \cdot R \cdot L \cdot F(N) \tag{12.8}$$

In this model, the installed cost CI is assumed to be paid fully at the time when the cable is energized for the first time. In reality, utilities have to borrow money to pay for cable purchase and installation, and this money is recovered through the rate structure. Since the cable installed costs are usually small in comparison with other capital costs (notably the costs of new generating facilities), utilities often adopt the assumption used in this model.

12.2.2 Mean Conductor Temperature and Resistance

It is convenient, and usually sufficiently accurate, to assume that conductor resistance is constant during the life of the cable. A simple formula for making an estimate of conductor operating temperature, and hence its resistance, is given in IEC 1059 (1991). This is based on observations of typical calculations that the average operating temperature rise of an economic size of conductor, taken over economic life, is in the region of one third of the rise occurring with its maximum permissible thermal rating. Thus, the average conductor temperature can be taken as

$$\theta_m = \theta_{amb} + \frac{\theta - \theta_{amb}}{3} \tag{12.9}$$

Here, θ is the maximum rated conductor temperature for the type of cable concerned and θ_{amb} is the ambient average temperature.

In general, a more precise value of conductor resistance will affect the selection of an economic size only in very marginal cases. If greater accuracy is desired for particular cases, refined values for conductor temperature and resistance can be made. The following derivation gives the formulas for one such estimate.

Conductor temperature is computed as a mean of the values θ_0 and θ_f during the first and the last years of an economic period. Assuming that the temperature rise is proportional to the conductor losses, we can write

$$\theta_0 = (\theta - \theta_{amb}) \left(\frac{I_0}{I_z} \right)^2 \frac{R_0}{R_z} + \theta_{amb} \tag{12.10}$$

where I_z is the current-carrying capacity, for a maximum permitted temperature rise of $\theta - \theta_{\text{amb}}$. If the average load factor during the economic life of the cable is different from unity, the value of I_z is computed by either using the Neher/McGrath modification of the external thermal resistance as discussed in Section (9.6.7) or by multiplying the current corresponding to 100% load factor by the cyclic rating factor M as discussed in Section 5.6.

Also, we have

$$R_0 = R_{20}\frac{\beta + \theta_0}{\beta + 20}, \qquad R_z = R_{20}\frac{\beta + \theta}{\beta + 20} \tag{12.11}$$

where β is the reciprocal of the temperature coefficient of resistance of the conductor material in degrees Kelvin. For aluminum, $\beta = 228$; for copper, $\beta = 234.4$.

Substituting (12.11) into (12.10), we obtain

$$\theta_0 = (\theta - \theta_{\text{amb}})\left(\frac{I_0}{I_z}\right)^2\left(\frac{\beta + \theta_0}{\beta + \theta}\right) + \theta_{\text{amb}} \tag{12.12}$$

Denoting

$$\gamma = \left(\frac{I_0}{I_z}\right)^2\left(\frac{\theta - \theta_{\text{amb}}}{\beta + \theta}\right), \tag{12.13}$$

equation (12.12) can be written as

$$\theta_0 = \frac{\gamma\beta + \theta_{\text{amb}}}{1 - \gamma} \tag{12.14}$$

The equations are similar for the final year temperature:

$$\theta_f = (\theta - \theta_{\text{amb}})\left(\frac{I_f}{I_z}\right)^2\frac{R_f}{R_z} + \theta_{\text{amb}}$$

$$I_f^2 = I_0^2 \cdot g$$

where

$$g = (1 + a)^{2(N-1)} \tag{12.15}$$

Thus,

$$\theta_f = \frac{\gamma\beta g + \theta_{\text{amb}}}{1 - \gamma g} \tag{12.16}$$

From equations (12.14) and (12.16), we obtain the mean conductor temperature as

$$\theta_m = \frac{\theta_0 + \theta_f}{2} = \frac{1}{2}\left(\frac{2\theta_{\text{amb}} - 2\gamma^2 g\beta - \gamma g\theta_{\text{amb}} + \gamma g\beta - \gamma\theta_{\text{amb}} + \gamma\beta}{(1 - \gamma)(1 - g\gamma)}\right) \tag{12.17}$$

Adding β inside the brackets and subtracting it outside the brackets, equation (12.17), after simplification, can be written as

$$\theta_m = \frac{\beta + \theta_{\text{amb}}}{2}\left(\frac{1}{1 - \gamma} + \frac{1}{1 - g\gamma}\right) - \beta \tag{12.18}$$

The mean conductor resistance is obtained as a mean of the first and last year resistances; that is,

$$R_m = \frac{R_0 + R_f}{2} \tag{12.19}$$

Combining equations (12.11) and (12.14), we obtain

$$R_0 = \frac{R_{20}\left(\beta + \dfrac{\gamma\beta + \theta_{\text{amb}}}{1-\gamma}\right)}{\beta + 20} = \frac{R_{20}(\beta + \theta_{\text{amb}})}{\beta + 20}\frac{1}{1-\gamma} \tag{12.20}$$

Similarly,

$$R_f = \frac{R_{20}(\beta + \theta_{\text{amb}})}{\beta + 20}\frac{1}{1-\gamma g} \tag{12.21}$$

Substituting (12.20) and (12.21) into (12.19), we obtain

$$R_m = \frac{R_{20}}{2}\left(\frac{\beta + \theta_{\text{amb}}}{\beta + 20}\right)\left(\frac{1}{1-\gamma} + \frac{1}{1-g\gamma}\right) \tag{12.22}$$

In order to take into account the skin and proximity effects, as well as the losses in sheath and armor, we multiply the value of R_m obtained in equation (12.22) by a factor B given by

$$B = (1 + y_s + y_p)(1 + \lambda_1 + \lambda_2) \tag{12.23}$$

Then, the value of R_m can be substituted directly in equations (12.7) and (12.8).

Similarly, the following equation can be used to obtain a value of ρ_m which can be substituted for $\rho_{20}[1 + \alpha_{20}(\theta_m - 20)]$:

$$\rho_m = \frac{\rho_{20}}{2}\left(\frac{\beta + \theta_{\text{amb}}}{\beta + 20}\right)\left(\frac{1}{1-\gamma} + \frac{1}{1-g\gamma}\right) \tag{12.24}$$

where ρ_{20} = resistivity of the conductor at 20°C, $\Omega \cdot$ m.

The economic conductor size is unlikely to be identical to a standard size, and so it is necessary to provide a continuous relationship between resistance and size. This is done by assuming a value of resistivity for each conductor material. The values recommended in IEC 1059 (1991) for ρ_{20} are $18.35 \cdot 10^{-9}$ for copper and $30.3 \cdot 10^{-9}$ for aluminum. These values are not the actual values for the materials, but are compromise values chosen so that conductor resistances can be calculated directly from nominal conductor sizes, rather than from the actual effective cross-sectional area.

EXAMPLE 12.1

Consider cable model No. 1 with installation conditions as specified in Appendix A. The ampacity of this cable with unity load factor is equal to 629 A. The following are the relevant parameters of this cable taken from Table A1: $R_{20} = 0.0601 \cdot 10^{-3}$ Ω/m, $y_s = 0.014$, $y_p = 0.0047$, and $\lambda_1 = 0.09$. We will determine the present value of the cost of joule losses for this cable with the following operational and economic parameters:

$N_c = 1$
$N_p = 3$
$I_0 = 160$ A
$N = 30$ years
$LF = 0.75$
$p = 0.2$
$P = 0.05$ \$/kWh
$D = 0.003$ \$/W·year
$a = 2\%$
$b = 3\%$
$c = 3\%$
$i = 5\%$
$L = 500$ m

First, we will compute the average conductor temperature and resistance during the economic life. This resistance will depend on the cable rating, including the effect of a nonunity load factor. With a load factor of 75%, the cable ampacity is increased to 740 A (see computation of the external thermal resistance in Section 9.6.7). Also, from equations (12.3) and (12.2), we have

$$\mu = p \cdot LF + (1 - p) \cdot LF^2 = 0.2 \cdot 0.75 + (1 - 0.2) \cdot 0.75^2 = 0.6$$
$$T = \mu \cdot 8760 = 0.6 \cdot 8760 = 5256 \text{ h}$$

The mean value of the temperature is obtained from equations (12.13), (12.15), and (12.18):

$$\gamma = \left(\frac{I_0}{I_z}\right)^2 \left(\frac{\theta - \theta_{\text{amb}}}{\beta + \theta}\right) = \left(\frac{160}{740}\right)^2 \left(\frac{90 - 15}{234.4 + 90}\right) = 0.0108,$$
$$g = (1 + a)^{2(N-1)} = (1 + 0.02)^{2(30-1)} = 3.15$$
$$\theta_m = \frac{\beta + \theta_{\text{amb}}}{2}\left(\frac{1}{1 - \gamma} + \frac{1}{1 - g\gamma}\right) - \beta$$
$$= \frac{234.4 + 15}{2}\left(\frac{1}{1 - 0.0108} + \frac{1}{1 - 3.15 \cdot 0.0108}\right) - 234.4 = 20.75°\text{C}$$

The mean conductor resistance is computed from equation (12.22), which yields

$$R_m = \frac{R_{20}}{2}\left(\frac{\beta + \theta_{\text{amb}}}{\beta + 20}\right)\left(\frac{1}{1 - \gamma} + \frac{1}{1 - g\gamma}\right)$$
$$= \frac{0.0601 \cdot 10^{-3}}{2}\left(\frac{234.4 + 15}{234.4 + 20}\right)\left(\frac{1}{1 - 0.0108} + \frac{1}{1 - 3.15 \cdot 0.0108}\right) = 0.0603\ \Omega/\text{km}$$

To include the effect of joule losses in the screen wires, we multiply this resistance by the factor B given by equation (12.23). For this example, we will use the value of the screen loss factor given in Table A1. In a computer program, the screen and armor loss factors should be computed for the average temperature obtained above. We have

$$R_m = 0.0603 \cdot (1 + y_p + y_s)(1 + \lambda_1 + \lambda_2) = 0.0603 \cdot (1 + 0.014 + 0.0047)(1 + 0.09) = 0.0669\ \Omega/\text{km}$$

Next, we will compute the auxiliary quantities required in equation (12.6). Since, in this example, $b = c$, we have from equations (12.4) and (12.5)

$$r_P = r_D = \frac{(1+a)^2(1+b)}{1+i} = \frac{(1+0.02)^2(1+0.03)}{1+0.05} = 1.021$$

$$Q_P(30) = Q_D(30) = \frac{1-r_P^N}{1-r_P} = \frac{1-1.021^{30}}{1-1.021} = 41.21$$

Factors F are obtained from equation (12.6):

$$F_1(30) = \frac{T \cdot P \cdot N_p \cdot N_c \cdot Q_P(N)}{1+i} = \frac{5256 \cdot 0.05 \cdot 10^{-3} \cdot 3 \cdot 1 \cdot 41.21}{1.05} = 30.94$$

$$F_2(30) = \frac{D \cdot N_p \cdot N_c \cdot Q_D(N)}{1+i} = \frac{0.003 \cdot 3 \cdot 1 \cdot 41.21}{1.05} = 0.35$$

Finally, the cost of joule losses is obtained from equation (12.7) as

$$CL = I_0^2 \cdot R \cdot L \cdot [F_1(N) + F_2(N)] = 160^2 \cdot 0.0669 \cdot 10^{-3} \cdot 500 \cdot (30.94 + 0.35) = \$26\,794$$

12.2.3 Growing Load Factor. In many systems, the load factor will grow with time due to many reasons such as increases in load diversity with load growth, increases in energy consumption per kW of connected load with time, measures taken by the utilities to flatten the load curves for improving system efficiency and to curb the growth of peak demand, and so on. The load growth on a particular cable circuit is limited by the circuit ampacity, and when this value is reached, new transmission facilities have to be installed. On some occasions, the load growth may stop even before circuit ampacity is reached; this may be the case, for example, when the cables are installed to transport power from a generating station whose output will grow in the future until the full station is completed. These considerations may be taken into account in the above model (Anders *et al.*, 1991).

Adopting an assumption made by Sheer (1966), that the system load factor grows, reducing the difference between an ultimate load factor LF_u and the present load factor LF_p by one half over a period of 16 years, the load factor at any year n is given by

$$\begin{aligned} LF_n &= LF_u - (0.5)^{n/16} \cdot (LF_u - LF_p) \\ &= LF_u - (0.9576)^n \cdot (LF_u - LF_p) \end{aligned} \tag{12.25}$$

Assuming that the annual growth in the load factor is achieved during the last day of the year, the value of T becomes

$$\begin{aligned} T_n &= 8760 \cdot \{p[LF_u - (0.9576)^{n-1} \cdot \Delta LF] \\ &\qquad + (1-p)[LF_u - (0.9576)^{n-1} \cdot \Delta LF]^2\} \\ &= c - 0.9576^{n-1} \cdot d + 0.9170^{n-1} \cdot e \end{aligned} \tag{12.26}$$

where

$$\begin{aligned} c &= 8760 \cdot [pLF_u + (1-p)LF_u^2] \\ d &= 8760 \cdot [p\Delta LF + (1-p)2LF_u \cdot \Delta LF] \\ e &= 8760 \cdot (1-p) \cdot \Delta LF^2 \end{aligned} \tag{12.27}$$

with

$$\Delta LF = LF_u - LF_p.$$

The factor $F_1(N)$ then becomes

$$F_1(N) = \frac{P \cdot N_p \cdot N_c \cdot Q'_P(N)}{1+i} \tag{12.28}$$

where

$$Q'_P(N) = cQ_P(N) - dQ_{P1}(N) + eQ_{P2}(N) \tag{12.29}$$

and

$$Q_{Pk}(N) = \frac{1 - r_{Pk}^N}{1 - r_{Pk}}, \qquad k = 1, 2 \tag{12.30}$$

Factors r_{Pk} are given by

$$r_{P1} = \frac{0.9576(1+a)^2(1+b)}{1+i}$$
$$r_{P2} = \frac{0.9170(1+a)^2(1+b)}{1+i} \tag{12.31}$$

EXAMPLE 12.2

Let us assume that the cable system studied in Example 12.1 experiences a growing load factor with an initial value of 75% and a final value of 95%. We will determine the cost of joule losses in this case.

The increase in the load factor is equal to

$$\Delta LF = 0.95 - 0.75 = 0.2$$

The constants c, d, and e are computed from equation (12.27):

$$c = 8760 \cdot [pLF_u + (1-p)LF_u^2] = 8760[0.2 \cdot 0.95 + (1-0.2)0.95^2] = 7989.12$$
$$d = 8760 \cdot [p\Delta LF + (1-p)2LF_u \cdot \Delta LF] = 8760[0.2 \cdot 0.2 + (1-0.2)2 \cdot 0.95 \cdot 0.2] = 3013.44$$
$$e = 8760 \cdot (1-p) \cdot \Delta LF^2 = 8760 \cdot (1-0.2)0.2^2 = 280.32$$

From equations (12.31), (12.30), and (12.29), we have

$$r_{P1} = \frac{0.9576(1+a)^2(1+b)}{1+i} = \frac{0.9576(1+0.02)^2(1+0.03)}{1+0.05} = 0.9773$$
$$r_{P2} = \frac{0.9170(1+a)^2(1+b)}{1+i} = \frac{0.9170(1+0.02)^2(1+0.03)}{1+0.05} = 0.9359$$
$$Q_{P1}(30) = \frac{1 - r_{P1}^N}{1 - r_{P1}} = \frac{1 - 0.9773^{30}}{1 - 0.9773} = 21.93$$
$$Q_{P2}(30) = \frac{1 - r_{P2}^N}{1 - r_{P2}} = \frac{1 - 0.9359^{30}}{1 - 0.9359} = 13.46$$
$$Q'_P(30) = cQ_P(N) - dQ_{P1}(N) + eQ_{P2}(N)$$
$$= 7989.12 \cdot 41.21 - 3013.44 \cdot 21.93 + 280.32 \cdot 13.46 = 266\,920.$$

The new factor F_1 is computed from equation (12.28):

$$F_1(30) = \frac{P \cdot N_p \cdot N_c \cdot Q'_P(N)}{1+i} = \frac{0.05 \cdot 10^{-3} \cdot 3 \cdot 1 \cdot 266\,920}{1.05} = 38.13$$

The cost of joule losses in this case is obtained from equation (12.7):

$$CL = I_0^2 \cdot R \cdot L \cdot [F_1(N) + F_2(N)] = 160^2 \cdot 0.0699 \cdot 10^{-3} \cdot 500(38.13 + 0.35) = \$34\,429$$

This value is 28% higher than the cost computed in Example 12.1.

If we assume that the load growth on the circuit of interest stops after N_0 years ($N_0 \leq N$), then the growth of the LF on this circuit is limited to N_0 years; the increase in LF due to other factors beyond N_0 years is ignored. In this case,

$$F_1(N) = \frac{P \cdot N_p \cdot N_c \cdot Q^*_P(N)}{1+i}, \qquad F_2(N) = \frac{D \cdot N_p \cdot N_c \cdot Q^*_D(N)}{1+i} \tag{12.32}$$

where

$$\begin{aligned} Q^*_P(N) &= Q'_P(N_0) + f[Q''_P(N) - Q''_P(N_0)] \\ Q^*_D(N) &= Q_D(N_0) + (1+a)^{2(N_0-1)}[Q''_D(N) - Q''_D(N_0)] \end{aligned} \tag{12.33}$$

with

$$f = 8760(1+a)^{2(N_0-1)}\mu_{N_0} \tag{12.34}$$

$$\begin{aligned} Q''_P(N_0) &= \frac{1 - r'^{N_0}_P}{1 - r'_P}, \qquad r'_P = \frac{1+b}{1+i} \\ Q''_D(N_0) &= \frac{1 - r'^{N_0}_D}{1 - r'_D}, \qquad r'_D = \frac{1+c}{1+i} \end{aligned} \tag{12.35}$$

where μ_{N_0} is the ultimate load loss factor.

EXAMPLE 12.3

Suppose that in Example 12.2 the load factor growth is achieved in a period of ten years. We will compute the cost of joule losses in this case.

If the load stops growing after ten years, the final load factor is obtained from equation (12.25):

$$LF_{10} = LF_u - (0.9576)^n \cdot (LF_u - LF_p) = 0.95 - 0.9576^{10} \cdot 0.2 = 0.82$$

and the ultimate load loss factor is computed from equation (12.3):

$$\mu_{10} = p \cdot LF + (1-p) \cdot LF^2 = 0.2 \cdot 0.82 + (1 - 0.2) \cdot 0.82^2 = 0.70$$

The auxiliary quantities are obtained from equations (12.35) and (12.34):

$$r'_P = r'_D = \frac{1+b}{1+i} = \frac{1+0.03}{1+0.05} = 0.98,$$

$$f = 8760(1+a)^{2(N_0-1)}\mu_{N_0} = 8760(1+0.02)^{2(10-1)} \cdot 0.70 = 8758.0$$

$$Q''_P = Q''_D(10) = \frac{1 - r'^{N_0}_P}{1 - r'_P} = \frac{1 - 0.98^{10}}{1 - 0.98} = 9.15,$$

$$Q''_P(30) = Q''_D(30) = \frac{1 - r'^{N_0}_D}{1 - r'_D} = \frac{1 - 0.98^{30}}{1 - 0.98} = 22.73$$

Also, from equations (12.4), (12.30), and (12.29), we have

$$Q_P(10) = Q_D(10) = \frac{1 - r_P^N}{1 - r_P} = \frac{1 - 1.021^{10}}{1 - 1.021} = 11.0$$

$$Q_{P1}(10) = \frac{1 - r_{P1}^N}{1 - r_{P1}} = \frac{1 - 0.9773^{10}}{1 - 0.9773} = 9.04$$

$$Q_{P2}(10) = \frac{1 - r_{P2}^N}{1 - r_{P2}} = \frac{1 - 0.9359^{10}}{1 - 0.9359} = 7.56$$

$$Q'_P(10) = cQ_P(N) - dQ_{P1}(N) + eQ_{P2}(N) = 7989.12 \cdot 11.0 - 3013.44 \cdot 9.04 + 280.32 \cdot 7.56$$
$$= 62\,758.$$

The new values of constants F_1 and F_2 are obtained from equations (12.33) and (12.32):

$$Q_P^*(N) = Q'_P(N_0) + f[Q''_P(N) - Q''_P(N_0)] = 62\,758 + 8758 \cdot (22.73 - 9.15) = 181\,692$$

$$Q_D^*(N) = Q_D(N_0) + (1+a)^{2(N_0-1)}[Q''_D(N) - Q''_D(N_0)] = 11 + (1+0.02)^{2(10-1)}(22.73 - 9.15)$$
$$= 30.4$$

$$F_1(30) = \frac{P \cdot N_p \cdot N_c \cdot Q_P^*(N)}{1+i} = \frac{0.05 \cdot 10^{-3} \cdot 3 \cdot 1 \cdot 181\,692}{1.05} = 26.0$$

$$F_2(30) = \frac{D \cdot N_p \cdot N_c \cdot Q_D^*(N)}{1+i} = \frac{0.003 \cdot 3 \cdot 1 \cdot 30.4}{1.05} = 0.26.$$

Assuming that the average resistance is the same as computed in Example 12.1, the present value of the cost of joule losses is equal to

$$CL = I_0^2 \cdot R \cdot L \cdot [F_1(N) + F_2(N)] = 160^2 \cdot 0.0669 \cdot 10^{-3} \cdot 500(26.0 + 0.26) = \$22\,487.$$

This cost is substantially lower than the value obtained in Example 12.2.

EXAMPLE 12.4

We will revisit Example 12.3, but we will change the assumption on the load growth. Let us assume that the load factor is constant and equal to 0.75 during the entire 30 year period, but that the load stops growing after the first ten years and then remains constant for the remaining 20 years. We will compute the present value of the cost of losses.

From Example 12.3, we have

$$Q''_P(10) = Q''_D(10) = 9.15, \qquad Q''_P(30) = Q''_D(30) = 22.73$$

Factor f will now be equal to

$$f = 8760(1+a)^{2(N_0-1)}\mu_{N_0} = 8760(1+0.02)^{2(10-1)} \cdot 0.6 = 7506.9$$

Since, in this example, $\Delta LF = 0$, we have $d = 0$ and $e = 0$. From Example 12.3, $Q_P(10) = 11$, and from equation (12.29), we have

$$Q'_P(10) = CQ_P(N) - dQ_{P1}(N) + eQ_{P2}(N) = 8760 \cdot 0.6 \cdot 11 = 57\,816$$

From equation (12.33), we obtain

$$Q_P^*(30) = Q'_P(N_0) + f[Q''_P(N) - Q''_P(N_0)] = 57\,816 + 7506.9 \cdot (22.73 - 9.15) = 159\,760$$

$$Q_D^*(30) = Q_D(N_0) + (1+a)^{2(N_0-1)}[Q''_D(N) - Q''_D(N_0)] = 11 + (1+0.02)^{2(10-1)}(22.73 - 9.15)$$
$$= 30.4$$

From equation (12.32), we have

$$F_1(30) = \frac{P \cdot N_p \cdot N_c \cdot Q_P^*(N)}{1+i} = \frac{0.05 \cdot 10^{-3} \cdot 3 \cdot 1 \cdot 159\,760}{1.05} = 22.82$$

$$F_2(30) = \frac{D \cdot N_p \cdot N_c \cdot Q_D^*(N)}{1+i} = \frac{0.003 \cdot 3 \cdot 1 \cdot 30.4}{1.05} = 0.26$$

Assuming again that the average resistance is the same as computed in Example 12.1, the present value of the cost of joule losses is equal to

$$CL = I_0^2 \cdot R \cdot L \cdot [F_1(N) + F_2(N)] = 160^2 \cdot 0.0669 \cdot 10^{-3} \cdot 500(22.82 + 0.26) = \$19\,764$$

12.3 EFFECT OF CHARGING CURRENT AND DIELECTRIC LOSSES

Dielectric losses and the losses due to charging current are always present when the cable is energized, and therefore operate at 100% load factor. Both types of losses are significant only at high-voltage levels and are dependent on cable capacitance. Evaluation of transmission cable systems often assumes the placement of shunt reactors at the ends of the cable system to supply the reactive vars required by the cable. The reactors have losses equal to about 0.8% of power rating. Those losses should be considered in the evaluation of cable system losses, and the cost of the reactors should be added to the cable purchase cost.

The dielectric and charging current losses are sometimes referred to as voltage-dependent losses, in contrast to the joule losses which are referred to as current-dependent losses.

Cable capacitance C is given by equation (6.2):

$$C = \frac{\varepsilon}{18 \ln\left(\dfrac{D_i}{d_c}\right)} \cdot 10^{-9} \tag{12.36}$$

where ε is the relative permitivity of insulation, d_c (mm) is the diameter of the conductor including the screen, and D_i (mm) is the diameter over insulation.

Charging current is calculated from

$$I_c = 2\pi f C U_0 \tag{12.37}$$

where f is the system frequency and U_0 (V) is phase-to-ground voltage.

Charging current is not uniform along the cable. In a cable with all charging current flowing from one end, the charging current losses are computed from (IEEE, 1990)

$$W_{ch} = \tfrac{1}{3} I_c^2 \cdot L^3 \cdot R \tag{12.38}$$

If the system has equal charging current flowing from each end, either due to natural system conditions or to the addition of reactors to force the equal flow, the losses per phase will be calculated from

$$W_{ch} = 2\left[\frac{1}{3} I_c^2 \left(\frac{L}{2}\right)^3 \cdot R\right] \tag{12.39}$$

Thus, in general, the charging current will be obtained from

$$W_{ch} = g \cdot I_c^2 \cdot L^3 \cdot R \tag{12.40}$$

where $g = 1/3$ or $1/12$, depending on whether equation (12.38) or (12.39) applies.

The dielectric losses are proportional to the square of the voltage [equation (6.4)]:

$$W_d = 2\pi f \cdot L \cdot C \cdot U_o^2 \cdot \tan\delta \tag{12.41}$$

where $\tan\delta$ is the loss factor of the insulation.

The total cost, including the effect of charging current and dielectric losses, can be represented by extending equation (12.8) to

$$CT = CI + I_0^2 \cdot R \cdot L \cdot F(N) + (g \cdot I_c^2 \cdot R \cdot L^3 + W_d) \cdot F_3(N) \tag{12.42}$$

with

$$F_3(N) = \frac{N_p \cdot N_c \cdot [8760 \cdot P \cdot Q_P''(N) + D \cdot Q_D''(N)]}{1+i} \tag{12.43}$$

where $Q_P''(N)$ and $Q_D''(N)$ are defined in equation (12.26).

EXAMPLE 12.5

We will consider a 5.8 km long cable (model No. 3) with laying conditions as described in Appendix A. The parameters of this cable are: $R_{20} = 0.1817 \cdot 10^{-4}$ Ω/m, $y_s = 0.05$, $y_p = 0.054$, $\lambda_1 = 0.01$, and $\lambda_2 = 0.311$. The load is assumed to grow from 250 to 795 A during the 40 year economic life. The charging current is assumed to flow from one end only (the worst case scenario). We will compute the cost of current- and voltage-dependent losses for the following nominal conditions:

$N_c = 1$

$N_p = 3$

$I_0 = 250$ A

$N = 40$ years

$LF = 0.85$

$p = 0.3$

$P = 0.05$ \$/kWh

$D = 1.0$ \$/W·year

$a = 3\%$

$b = 7.8\%$

$c = 7.8\%$

$i = 11\%$

$L = 5800$ m.

Since we assume that the load grows steadily over the 40 year economic life period, the load growth factor is obtained by solving the following equation:

$$795 = 250 \cdot (1 + a)^{39}$$

which yields $a = 3.0\%$.

First, we will compute the average conductor temperature and resistance during the economic life. This resistance will depend on the cable rating, including the effect of a nonunity load factor. With a load factor of 85%, the cable ampacity is equal to 902 A. Also, from equations (12.3) and (12.2), we have

$$\mu = p \cdot LF + (1-p) \cdot LF^2 = 0.3 \cdot 0.85 + (1-0.3) \cdot 0.85^2 = 0.761$$
$$T = 8760 \cdot \mu = 8760 \cdot 0.761 = 6664 \text{ h}$$

The mean value of the temperature is obtained from equations (12.13), (12.15), and (12.18):

$$\gamma = \left(\frac{I_0}{I_z}\right)^2 \left(\frac{\theta - \theta_{\text{amb}}}{\beta + \theta}\right) = \left(\frac{250}{902}\right)^2 \left(\frac{70-20}{234.4+70}\right) = 0.0114,$$
$$g = (1+a)^{2(N-1)} = (1+0.03)^{2(40-1)} = 10.03$$
$$\theta_m = \frac{\beta + \theta_{\text{amb}}}{2}\left(\frac{1}{1-\gamma} + \frac{1}{1-g\gamma}\right) - \beta$$
$$= \frac{234.4+25}{2}\left(\frac{1}{1-0.0114} + \frac{1}{1-10.03 \cdot 0.0114}\right) - 234.4 = 43.2^\circ\text{C}$$

The mean conductor resistance is computed from equation (12.22), which yields

$$R_m = \frac{R_{20}}{2}\left(\frac{\beta + \theta_{\text{amb}}}{\beta + 20}\right)\left(\frac{1}{1-\gamma} + \frac{1}{1-g\gamma}\right)$$
$$= \frac{18.35 \cdot 10^{-9}}{2 \cdot 0.00101}\left(\frac{234.4+25}{234.4+20}\right)\left(\frac{1}{1-0.0114} + \frac{1}{1-10.03 \cdot 0.0114}\right) = 0.0198\ \Omega/\text{km}$$

To include the effect of joule losses in the screen wires, we multiply this resistance by the factor B given by equation (12.23). The values of y_s, y_p, λ_1, and λ_2 are given in Table A1. Thus, we have

$$R_m = 0.0198 \cdot (1+y_s+y_p)(1+\lambda_1+\lambda_2) = 0.0198 \cdot (1+0.05+0.054)(1+0.01+0.311) = 0.0289\ \Omega/\text{km}$$

Next, we will compute the auxiliary quantities required in equation (12.6). Since, in this example, $b = c$, we have from equations (12.4) and (12.5)

$$r_P = r_D = \frac{(1+a)^2(1+b)}{1+i} = \frac{(1+0.03)^2(1+0.078)}{1+0.11} = 1.03$$
$$Q_P(40) = Q_D(40) = \frac{1-r_P^N}{1-r_P} = \frac{1-1.03^{40}}{1-1.03} = 75.4$$

Factors F are obtained from equation (12.6):

$$F_1(40) = \frac{T \cdot P \cdot N_p \cdot N_c \cdot Q_P(N)}{1+i} = \frac{6664 \cdot 0.05 \cdot 10^{-3} \cdot 3 \cdot 1 \cdot 75.4}{1.11} = 67.9$$
$$F_2(40) = \frac{D \cdot N_p \cdot N_c \cdot Q_D(N)}{1+i} = \frac{1.0 \cdot 3 \cdot 1 \cdot 75.4}{1.11} = 203.8$$

The cost of joule losses is obtained from equation (12.7) as

$$CL = I_0^2 \cdot R \cdot L \cdot [F_1(N) + F_2(N)] = 250^2 \cdot 0.0289 \cdot 10^{-3} \cdot 5800 \cdot (67.9 + 203.8) = \$2\,846\,397.$$

The dielectric capacitance of the insulation is computed from equation (12.36) as

$$C = \frac{\varepsilon}{18 \ln\left(\frac{D_i}{d_c}\right)} \cdot 10^{-9} = \frac{3.5}{18 \ln \frac{67.11}{41.45}} \cdot 10^{-9} = 0.404 \cdot 10^{-9} \text{ F/m}$$

The charging current and charging current losses are obtained from equations (12.37) and (12.38), respectively:

$$I_c = 2\pi f C U_0 = 2\pi \cdot 60 \cdot 0.404 \cdot 10^{-9} \cdot 138\,000/\sqrt{3} = 0.012 \text{ A}$$

$$W_{ch} = \tfrac{1}{3} I_c^2 \cdot L^3 \cdot R = \tfrac{1}{3} \cdot 0.012^2 \cdot 5800^3 \cdot 0.0298 \cdot 10^{-3} = 271 \text{ W}$$

The dielectric losses are given by equation (12.41):

$$W_d = 2\pi f \cdot L \cdot C \cdot U_0^2 \cdot \tan\delta = 2\pi \cdot 60 \cdot 5800 \cdot 0.404 \cdot 10^{-9} \cdot (138\,000/\sqrt{3})^2 \cdot 0.005 = 28\,038 \text{ W}.$$

From equation (12.35), we have

$$r'_P = r'_D = \frac{1+b}{1+i} = \frac{1+0.078}{1+0.11} = 0.97$$

$$Q''_P(40) = Q''_D(40) = \frac{1 - r_P'^{N_0}}{1 - r'_P} = \frac{1 - 0.97^{40}}{1 - 0.97} = 23.48$$

Factor $F_3(N)$ is obtained from equation (12.43):

$$F_3(40) = \frac{N_p \cdot N_c \cdot [8760 \cdot P \cdot Q''_P(N) + D \cdot Q''_D(N)]}{1+i}$$

$$= \frac{3 \cdot 1 \cdot (8760 \cdot 0.05 \cdot 10^{-3} \cdot 23.48 + 1 \cdot 23.48)}{1.11} = 91.3$$

Finally, the cost of charging current and dielectric losses is computed from equation (12.42) and is equal to

$$CL_{ch+d} = (g \cdot I_c^2 \cdot R \cdot L^3 + W_d) \cdot F_3(N) = (271 + 28\,038) \cdot 91.3 = \$2\,584\,612$$

In this example, the cost of voltage-dependent losses is comparable to the cost of joule losses, mainly because of the very high dielectric losses.

12.4 SELECTION OF THE ECONOMIC CONDUCTOR SIZE

The economic conductor size S_{ec} is determined in two stages. First, we neglect voltage-dependent losses and find the cross section that minimizes the cost function:

$$\boxed{CT = CI(S) + I_0^2 \cdot R(S) \cdot F(N) \cdot L} \qquad (12.44)$$

where $CI(S)$ and $R(S)$ are expressed as functions of the conductor cross section S.

The equation for the relationship between $CI(S)$ and conductor size can be derived from the known costs of standard cable sizes. In general, if a reasonably linear relationship can be fitted to the costs, possibly over a restricted range of conductor sizes, it should be used. This will cause little error in the results in view of the possible uncertainties in the assumed financial parameters for the economic life period chosen. If a linear model can be fitted to the values of initial cost for the type of cable and installation under consideration, then

$$CI(S) = L(A \cdot S + G) \tag{12.45}$$

where A = variable component of cost related to conductor size, \$/mm²/m
G = constant component of cost unaffected by size of cable, \$/m.

Then, the optimum size S_{ec} (mm²) can be obtained by equating to zero the derivative of CT with respect to S, giving for the case when voltage-dependent losses are neglected

$$S_{ec} = 1000\sqrt{\frac{I_0^2 \cdot F(N) \cdot \rho_{20} \cdot B[1 + \alpha_{20}(\theta_{\text{av}} - 20)]}{A}} \tag{12.46}$$

B is an auxiliary value which takes into account skin and proximity effects as well as sheath and armor losses. As the economic size is unknown, it is necessary to make assumptions as to the probable cable size in order that reasonable values of y_p, y_s, λ_1, and λ_2 can be calculated. Recalculation may be necessary if the economic size is too different.

We can observe that S_{ec} does not depend on the value of the constant component G of the cost which is unaffected by the size of the cable. Therefore, in performing a comparative analysis for the selection of an optimal conductor size for a specific installation, we can compare only the size-dependent component $CT - CI$ of the total cost. This approach is used in the numerical example discussed below. However, in the computer program, the user should be able to enter the value of G so that the true total cost can be computed.

S_{ec} is unlikely to be exactly equal to a standard size, and so the cost for the adjacent larger and smaller standard sizes must be calculated and the most economical one chosen.

EXAMPLE 12.6

We will compute the optimal conductor size for the cable system studied in Examples 12.1–12.3. We will assume that the coefficient of that part of the installation cost which depends on conductor size is equal to 0.1133 \$/m/mm². We will also assume that the installation cost is independent of the conductor size. This is normally the case when we consider the narrow range of cable sizes.

First, we have to assume an initial value of the conductor cross section. For all the cases studied, we will start with S = 300 mm². Table 12.1 summarizes cable information for three standard conductor sizes which will be considered in the example.

TABLE 12.1 Cable Data

Conductor Size (mm²)	Ampacity (A)	Resistance at 90°C (Ω/km)	y_s	y_p	λ_1
300	629	0.0781	0.014	0.005	0.090
400	862	0.0599	0.023	0.006	0.098
500	971	0.467	0.037	0.011	0.119

(1) Constant load factor

From equation (12.24), the mean electrical resistivity is equal to

$$\rho_m = \frac{\rho_{20}}{2}\left(\frac{\beta+\theta_{\text{amb}}}{\beta+20}\right)\left(\frac{1}{1-\gamma}+\frac{1}{1-g\gamma}\right)$$
$$= \frac{18.35\cdot 10^{-9}}{2}\left(\frac{234.4+15}{234.4+20}\right)\left(\frac{1}{1-0.0108}+\frac{1}{1-3.15\cdot 0.0108}\right) = 18.4\cdot 10^{-9}\ \Omega\cdot\text{m}$$

We will take factor B as given in Table 12.1 with the loss factor values corresponding to 90°C. A small error introduced by this assumption will have no effect on the selection of the economic conductor size. To be consistent, in the remainder of this example, the values of B will be computed for the conductor temperature of 90°C.

$$B = (1+y_s+y_p)(1+\lambda_1+\lambda_2) = (1+0.014+0.005)(1+0.090) = 1.11.$$

The optimal conductor cross section is obtained from equation (12.46), with the factor F obtained in Example 12.1, and is equal to

$$S_{ec} = 1000\sqrt{\frac{I_0^2\cdot F(N)\cdot\rho_{20}\cdot B[1+\alpha_{20}(\theta_{\text{av}}-20)]}{A}}$$
$$= 1000\sqrt{\frac{160^2\cdot(30.94+0.35)\cdot 18.4\cdot 10^{-9}\cdot 1.11}{0.1133}} = 380.0\ \text{mm}^2$$

The closest standard conductor sizes are 300 and 400 mm^2. The factor B for the 400 mm^2 conductor is recomputed using equation (8.62) for the middle cable and applying equation (12.23):

$$B = (1+y_s+y_p)(1+\lambda_1+\lambda_2) = (1+0.023+0.006)(1+0.098) = 1.13$$

The auxiliary variables in equations (12.13) and (12.15) are equal to

$$\gamma = \left(\frac{I_0}{I_z}\right)^2\left(\frac{\theta-\theta_{\text{amb}}}{\beta+\theta}\right) = \left(\frac{160}{862}\right)^2\left(\frac{90-15}{234.4+90}\right) = 0.008,$$
$$g = (1+a)^{2(N-1)} = (1+0.02)^{2(30-1)} = 3.15$$

From equation (12.24), the mean electrical resistivity is equal to

$$\rho_m = \frac{\rho_{20}}{2}\left(\frac{\beta+\theta_{\text{amb}}}{\beta+20}\right)\left(\frac{1}{1-\gamma}+\frac{1}{1-g\gamma}\right)$$
$$= \frac{18.35\cdot 10^{-9}}{2}\left(\frac{234.4+15}{234.4+20}\right)\left(\frac{1}{1-0.008}+\frac{1}{1-3.15\cdot 0.008}\right) = 18.3\cdot 10^{-9}\ \Omega\cdot\text{m}$$

The revised optimal conductor cross section is obtained again from equation (12.46):

$$S_{ec} = 1000\sqrt{\frac{I_0^2\cdot F(N)\cdot\rho_{20}\cdot B[1+\alpha_{20}(\theta_{\text{av}}-20)]}{A}}$$
$$= 1000\cdot\sqrt{\frac{160^2\cdot(30.94+0.35)\cdot 18.3\cdot 10^{-9}\cdot 1.13}{0.1133}} = 382.4\ \text{mm}^2$$

This is also within the 300–400 mm^2 range.

The total cost of cable and joule losses for each of the possible conductor sizes is now calculated with the aid of equations (12.44) and (12.45). For the 300 mm^2 cable, the joule losses were calculated in Example 12.1. Thus,

$$CT_{300} = A \cdot L \cdot S + CT_{300\text{ joule}} = 0.1133 \cdot 500 \cdot 300 + 26\,794 = \$43\,789$$

$$CT_{400} = CI(S) + I_0^2 \cdot R(S) \cdot F(N) \cdot L$$
$$= 0.1133 \cdot 500 \cdot 400 + 160^2 \cdot 18.3 \cdot 10^{-9} \cdot (30.94 + 0.35) \cdot 500/0.0004 = \$40\,983$$

Therefore, the 400 mm^2 conductor is the more economical size.

(2) Growing load factor

The factor F computed in Example 12.2 is equal in this case to 38.48. Hence, the optimal conductor size is

$$S_{ec} = 1000\sqrt{\frac{I_0^2 \cdot F(N) \cdot \rho_{20} \cdot B[1 + \alpha_{20}(\theta_{av} - 20)]}{A}}$$

$$= 1000 \cdot \sqrt{\frac{160^2 \cdot 38.48 \cdot 18.4 \cdot 10^{-9} \cdot 1.11}{0.1133}} = 421.4 \text{ mm}^2$$

The two neigboring standard conductor sizes are 400 and 500 mm^2. For a conductor size of 500 mm^2, the revised value of factor B is 1.17 and the mean resistivity is equal to $18.2 \cdot 10^{-9}\Omega\cdot$m. This increases S_{ec} to 430.6 mm^2. The present values of the cost of cable and joule losses of the two cables are equal to

$$CT_{400} = CI(S) + I_0^2 \cdot R(S) \cdot F(N) \cdot L$$
$$= 0.1133 \cdot 500 \cdot 400 + 160^2 \cdot 18.3 \cdot 10^{-9} \cdot 500 \cdot 38.48/0.0004 = \$45\,194$$

$$CT_{500} = CI(S) + I_0^2 \cdot R(S) \cdot F(N) \cdot L$$
$$= 0.1133 \cdot 500 \cdot 500 + 160^2 \cdot 18.2 \cdot 10^{-9} \cdot 500 \cdot 38.48/0.0005 = \$46\,254$$

Since the installation costs were assumed to be the same for all standard conductor sizes, we select the optimal conductor size equal to 400 mm^2.

(3) Load growth stops after ten years

In this case, $F = 26.26$ and

$$S_{ec} = 1000\sqrt{\frac{I_0^2 \cdot F(N) \cdot \rho_{20} \cdot B[1 + \alpha_{20}(\theta_{av} - 20)]}{A}}$$

$$= 1000 \cdot \sqrt{\frac{160^2 \cdot 26.26 \cdot 18.4 \cdot 10^{-9} \cdot 1.11}{0.1133}} = 348.1 \text{ mm}^2$$

The closest standard conductor sizes are 300 and 400 mm^2 with the total costs equal to

$$CT_{300} = CI(S) + I_0^2 \cdot R(S) \cdot F(N) \cdot L$$
$$= 0.1133 \cdot 500 \cdot 300 + 160^2 \cdot 18.4 \cdot 10^{-9} \cdot 500 \cdot 26.26/0.0003 = \$37\,611$$

$$CT_{400} = CI(S) + I_0^2 \cdot R(S) \cdot F(N) \cdot L$$
$$= 0.1133 \cdot 500 \cdot 400 + 160^2 \cdot 18.3 \cdot 10^{-9} \cdot 500 \cdot 26.26/0.0004 = \$38\,038$$

Thus, in this case, the optimal conductor size is 300 mm^2.

EXAMPLE 12.7[1]

A 10 kV cable circuit has to be sized to supply ten 10 kV/0.4 kV substations equally spaced along a route from a 150 kV/10 kV station (see Fig. 12.1). Our aim is to select a conductor size for each section based on the following considerations: 1) different economic conductor size for each section, 2) conductor size based on thermal current-carrying capability, and 3) the same most economical conductor size for all sections.

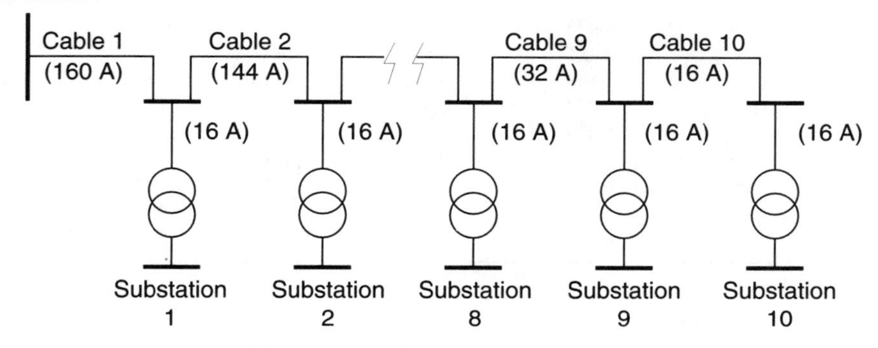

Figure 12.1 Distribution feeder supplying ten stations. (IEC Standard 1059, 1991)

There is only one three-phase circuit, so $N_c = 1$ and $N_p = 3$. The cable length between substations is 500 m.

The highest hourly mean value of current I_0 in the first year in the first section of the route is 160 A. In subsequent sections, the current is reduced by 16 A at each substation; thus, in the second section, the current is equal to 144 A and in the final section 16 A. The cyclic rating factor for all loads is 1.11. It is assumed that this factor remains constant during the economic life of the cable.

The financial data are as follows:

$N = 30$

$P = 0.0609$ \$/kWh

$D = 0.003$ \$/W·year

$A = 0.1133$ \$/m·mm^2

$a = 0.5\%$

$b = 2.0\%$

$c = 2.0\%$

$i = 5\%$

For the purpose of this example, a fictional three-core 6/10 kV type of cable has been assumed. The ac resistances of the conductors at 40 and 80°C are given in columns (2) and (3) of Table 12.2, and the financial details are given in columns (4) and (5). The cable has a permissible maximum conductor temperature of 80°C, and when laid in the ground, the steady-state ratings at this temperature, for a 20°C ground ambient temperature, are given in Table 12.3.

[1] This example is extracted from IEC Standard 1059 (1991).

TABLE 12.2 Cable Details

Cable Size (mm^2)	Resistance by Phase 40°C (Ω/km)	Resistance by Phase 80°C (Ω/km)	Primary Cost Cable ($/m)	Primary Cost Laying ($/m)	Primary Cost Sum ($/m)
(1)	(2)	(3)	(4)	(5)	(6)
25	1.298	1.491	10.62	17.23	27.85
35	0.939	1.079	11.65	17.33	28.98
50	0.694	0.798	13.19	17.49	30.68
70	0.481	0.553	15.24	17.71	32.95
95	0.348	0.400	17.81	17.97	35.78
120	0.277	0.318	20.37	18.24	38.61
150	0.226	0.259	23.45	18.55	42.00
185	0.181	0.208	27.04	18.92	45.96
240	0.140	0.161	32.69	19.51	52.20
300	0.114	0.131	38.85	20.14	58.99
400	0.091	0.104	49.11	21.20	70.31

TABLE 12.3 Cable Ratings

Nominal Size (mm^2)	25	35	50	70	95	120	150	185	240	300	400
Current-Carrying Capacity(A)	103	125	147	181	221	255	281	328	382	429	482

(1) Economic conductor size for each section
Since, in this example, $b = c$, we have from equations (12.4) and (12.5)

$$r_P = r_D = \frac{(1+a)^2(1+b)}{1+i} = \frac{(1+0.05)^2(1+0.02)}{1+0.05} = 0.98117$$

$$Q_P(30) = Q_D(30) = \frac{1-r_P^N}{1-r_P} = \frac{1-0.9812^{30}}{1-0.9812} = 23.081$$

Factors F are obtained from equation (12.6):

$$F_1(30) = \frac{T \cdot P \cdot N_p \cdot N_c \cdot Q_P(N)}{1+i} = \frac{2250 \cdot 0.0609 \cdot 10^{-3} \cdot 3 \cdot 1 \cdot 23.081}{1.05} = 9.0363$$

$$F_2(30) = \frac{D \cdot N_p \cdot N_c \cdot Q_D(N)}{1+i} = \frac{0.003 \cdot 3 \cdot 1 \cdot 23.081}{1.05} = 0.1978$$

Assuming initially that a conductor size of 185 mm^2 could be the economic optimum, the factor B is equal to 1.023. The average conductor temperature is computed from equation (12.9) as

$$\theta_m = \theta_{\text{amb}} + \frac{\theta - \theta_{\text{amb}}}{3} = 20 + \frac{80-20}{3} = 40^\circ\text{C}$$

From equation (12.46), we have

$$S_{ec} = 1000\sqrt{\frac{I_0^2 \cdot F(N) \cdot \rho_{20} \cdot B[1 + \alpha_{20}(\theta_{av} - 20)]}{A}}$$

$$= 1000 \cdot \sqrt{\frac{160^2 \cdot 9.2341 \cdot 30.3 \cdot 10^{-9} \cdot 1.023[1 + 0.0039(40 - 20)]}{0.1133}} = 264 \text{ mm}^2$$

Thus, either a 240 mm² or a 300 mm² conductor size could be chosen.

The initial choice of a 185 mm² conductor for the estimation of B can now be improved. Recalculating with the value of $B = 1.057$ for a 300 mm² conductor gives a value of S_{ec} of 269 mm², which is also within the 240–300 mm² range.

The total cost for each of the possible conductor sizes is now calculated with the aid of equation (12.8):

$$CT_{240} = CI + I_0^2 \cdot R \cdot L \cdot F(N) = [52.2 \cdot 500] + [160^2 \cdot (0.140/1000) \cdot 500 \cdot 9.2341 = \$42\,648$$

$$CT_{300} = CI + I_0^2 \cdot R \cdot L \cdot F(N) = [58.99 \cdot 500] + [160^2 \cdot (0.114/1000) \cdot 500 \cdot 9.2341 = \$42\,969$$

The 240 mm² conductor is therefore the more economical size.

Sizes and costs for the other sections have been calculated in a similar manner. The values of optimal sizes and corresponding costs are given in Table 12.4.

TABLE 12.4 Economic Loading

Section Number	1	2	3	4	5	6	7	8	9	10	Sum
Load											
I_0 (A)	160	144	128	112	96	80	64	48	32	16	—
Cable											
Size (mm²)	240	240	185	185	150	120	95	70	50	25	—
Capacity (A)	382	382	328	328	281	255	221	181	147	103	—
Cost per Section and total											
Cable ($)	16 345	16 345	13 520	13 520	11 725	10 185	8905	7620	6595	5310	110 070
Laying ($)	9755	9755	9460	9460	9275	9120	8985	8855	8745	8615	92 025
CI ($)	26 100	26 100	22 980	22 980	21 000	19 305	17 890	16 475	15 340	13 925	202 095
CL ($)	16 548	13 403	13 692	10 483	9616	8185	6581	5117	3281	1534	88 440
CT ($)	42 648	39 503	36 672	33 463	30 616	27 490	24 471	21 592	18 621	15 459	290 535

(2) Conductor size based on maximum load—choice based on thermal ratings

We will now select a cable size for each section so it can carry the anticipated maximum load for the last year of the economic life and not exceed the maximum permissible conductor temperature.

For section 1:

$$I_0 \text{ (first year)} = 160 \text{ A}$$

$$\text{Maximum current in last year} = 160 \cdot (1 + 0.005)^{30-1} = 185 \text{ A}$$

The required current-carrying capacity (100% load factor) I for the final year shall be not less than

$$185/1.11 = 167 \text{ A}$$

where the number 1.11 is the cyclic rating factor assumed above. From Table 12.3, the required conductor size is 70 mm^2. In order to make a fair comparison with the losses and financial figures calculated for the economic choice of conductor size, we have to assume an appropriate conductor temperature at which to calculate the losses. For the economic choice, we assumed that the temperature of the conductor would be about 40°C. We propose that a comparable assumption for the temperature of conductors chosen on the basis of thermal ratings would be the maximum permissible value of 80°C. The conductor resistances at a temperature of 80°C are given in Table 12.2.

The total cost of section 1 during the 30 year period is obtained from equation (12.44):

$$CT = CI(S) + I_0^2 \cdot R(S) \cdot F(N) \cdot L = (32.95 \cdot 500) + (160^2 \cdot 0.000553 \cdot 500 \cdot 9.2336) = \$81\,834.$$

Comparison with the cost for this section when using the economic size of conductor shows that the saving in cost for this section is $(81\,834 - 42\,648) \cdot 100/82\,834 = 48\%$. Similar calculations using sizes based on maximum thermal current-carrying capacity have been made for all the sections and are given in Table 12.5. The total saving for the ten sections is $(547\,864 - 290\,535) \cdot 100/547\,864 = 47\%$.

TABLE 12.5 Current-Carrying Capacity Criteria

Section Number	1	2	3	4	5	6	7	8	9	10	Sum
Load											
I_0 (A)	160	144	128	112	96	80	64	48	32	16	
I_{end} (A)	185	166	148	129	111	92	74	55	37	18	
$I_{end}/1.11$ (A)	167	150	133	117	100	83	67	50	33	17	
Cable											
Size (mm^2)	70	70	50	35	25	25	25	25	25	25	
Capacity (A)	181	181	147	125	103	103	103	103	103	103	
Cost per Section and total											
Cable ($)	7620	7620	6595	5852	5310	5310	5310	5310	5310	5310	59 520
Laying ($)	8855	8855	8745	8665	8615	8615	8615	8615	8615	8615	86 810
CI ($)	16 475	16 475	15 340	14 490	13 925	13 925	13 925	13 925	13 925	13 925	146 330
CL ($)	65 363	52 944	60 365	62 492	63 443	44 058	28 197	15 861	7049	1762	401 534
CT ($)	81 834	69 419	75 705	76 982	77 368	57 983	42 122	29 786	20 974	15 687	547 864

(3) Calculations based on the use of one standard conductor size for all sections

It is first necessary to assume a probable conductor size, and the total cost is calculated with equation (12.44) using this size for all sections. Then, costs assuming the use of the next smaller and larger size of conductor are calculated in order to confirm that the assumed size is indeed the most economical. For the purpose of this example, we assume that a 185 mm^2 conductor would be the best choice.

Although only one conductor size is used, the current is different for each cable section so that the average losses must be computed (all sections are assumed to operate at the same temperature and hence the same conductor resistance).

$$\frac{\text{average losses}}{\text{maximum losses}} = \frac{500 \cdot 160^2 + 500 \cdot 144^2 + \cdots + 500 \cdot 16^2}{10 \cdot 500 \cdot 160^2} = 0.385$$

From equation (12.46), using B for a 185 mm^2 conductor, we obtain

$$S_{ec} = 1000\sqrt{\frac{I_0^2 \cdot F(N) \cdot \rho_{20} \cdot B[1 + \alpha_{20}(\theta_{av} - 20)]}{A}}$$

$$= 1000 \cdot \sqrt{\frac{160^2 \cdot 1.023 \cdot 30.3 \cdot 10^{-9} \cdot 9.2341 \cdot 0.385 \cdot [1 + 0.00403(40 - 20)]}{0.1133}} = 164 \text{ mm}^2$$

so that either 150 or 185 mm^2 conductors could prove to be the most economic.

The total costs for each of these conductors are

$$CT_{150} = 42.00 \cdot 500 \cdot 10 + 160^2 \cdot (0.226/1000) \cdot 500 \cdot 10 \cdot 9.2341 \cdot 0.385 = \$312\,843$$

$$CT_{185} = 45.96 \cdot 500 \cdot 10 + 160^2 \cdot (0.181/1000) \cdot 500 \cdot 10 \cdot 9.2341 \cdot 0.385 = \$312\,165$$

Thus, the 185 mm^2 conductor is confirmed as the most economic size to use if only one conductor size is to be used throughout the route.

It is clear, by comparison with the sizes chosen in Table 12.4, that a 185 mm^2 conductor is thermally adequate to carry the maximum load at the end of the 30 year period.

Summary of results

The summary of the results of the calculations for the cable and conditions studied in this example is given in Table 12.6.

TABLE 12.6 Summary of Costs

	CI	*CL*	Total	
Basis of Costing	\$	\$	\$	%
Thermal current-carrying capacity for each section	146 330	401 534	547 864	100
Economic size for each section	202 095	88 440	290 535	53
Economic size using one standard size of 185 mm^2 throughout	229 800	82 365	312 165	57

Equation (12.44) does not reflect the effect of voltage-dependent losses. The capacitance of the cable depends on the ratio of insulation thickness t_i and conductor diameter d_c. For a given voltage level, an increase of conductor diameter results in an increase of cable capacitance (the insulation thickness usually remains unchanged or may even decrease), and hence an increase in voltage-dependent losses. Because of this, the optimization procedure will tend to decrease conductor diameter as opposed to the effect of current-dependent losses. Since the dielectric losses in high-voltage cables can sometimes be higher than the conductor losses, this effect may be quite significant. The losses caused by charging current I_c are proportional to $I_c^2 R$, where R is the same as that used to calculate the losses due to the load current, and this will tend to increase conductor cross section. Since, on the other hand, I_c depends on cable capacitance, the increase in the conductor size will tend to increase the value of the charging current, and this will increase the charging current losses, forcing a lower optimal conductor cross section. The influence of dielectric and charging current losses is assessed through an iterative procedure in the second stage of the optimization process.

EXAMPLE 12.8

We will reconsider Example 12.5, and we will include the effect of dielectric and charging current losses. The cost coefficient A is equal to 0.22 \$/m/mm^2.

The smallest standard conductor size which satisfies ampacity requirements for this system is 887 mm^2 (1750 kcmil). The economic conductor size is first computed from equation (12.46) with the required parameters computed in Example 12.5:

$$\rho_m = \frac{\rho_{20}}{2}\left(\frac{\beta+\theta_{\text{amb}}}{\beta+20}\right)\left(\frac{1}{1-\gamma}+\frac{1}{1-g\gamma}\right)$$

$$= \frac{18.35\cdot 10^{-9}}{2}\left(\frac{234.4+25}{234.4+20}\right)\left(\frac{1}{1-0.0114}+\frac{1}{1-10.03\cdot 0.0114}\right) = 20.0\cdot 10^{-9}\ \Omega\text{·m}$$

$$S_{ec} = 1000\sqrt{\frac{I_0^2\cdot F(N)\cdot \rho_{20}\cdot B[1+\alpha_{20}(\theta_{\text{av}}-20)]}{A}}$$

$$= 1000\cdot\sqrt{\frac{250^2\cdot(67.9+203.8)\cdot 20.0\cdot 10^{-9}\cdot 1.46}{0.22}} = 1501\ \text{mm}^2$$

This is very close to the standard conductor size of 1520 mm^2 (3000 kcmil). The ampacity of a 1520 mm^2 cable with 85% load factor is equal to 943 A. We will compute the mean resistance and the factor B for this cable. From equations (12.13), (12.22), and (12.24), we have

$$\gamma = \left(\frac{I_0}{I_z}\right)^2\left(\frac{\theta-\theta_{\text{amb}}}{\beta+\theta}\right) = \left(\frac{250}{943}\right)^2\left(\frac{70-25}{234.4+70}\right) = 0.0104$$

$$R_m = \frac{R_{20}}{2}\left(\frac{\beta+\theta_{\text{amb}}}{\beta+20}\right)\left(\frac{1}{1-\gamma}+\frac{1}{1-g\gamma}\right)$$

$$= \frac{18.35\cdot 10^{-9}}{2\cdot 0.00152}\left(\frac{234.4+25}{234.4+20}\right)\left(\frac{1}{1-0.0104}-\frac{1}{1-10.03\cdot 0.0104}\right) = 0.0129\ \Omega/\text{km}$$

$$\rho_m = \frac{\rho_{20}}{2}\left(\frac{\beta+\theta_{\text{amb}}}{\beta+20}\right)\left(\frac{1}{1-\gamma}+\frac{1}{1-g\gamma}\right)$$

$$= \frac{18.35\cdot 10^{-9}}{2}\left(\frac{234.4+25}{234.4+20}\right)\left(\frac{1}{1-0.0104}+\frac{1}{1-10.03\cdot 0.0104}\right) = 19.6\cdot 10^{-9}\ \Omega\text{·m}$$

Factor B is equal to 1.78. This gives $S_{ec} = 1641$ mm^2. This is very close to a standard conductor size of 1647 mm^2 (3250 kcmil). Neglecting the effect of charging current and dielectric losses, we would select the 3250 kcmil conductor as the most economical one.

To account for the effect of charging current and dielectric losses, we should compute the values of these losses for several conductor sizes in the neighborhood of the selected optimal cross section. We will start with the 1520 mm^2 conductor cable.

The dielectric capacitance of the insulation is equal to

$$C = \frac{\varepsilon}{18\ln\left(\frac{D_i}{d_c}\right)}\cdot 10^{-9} = \frac{3.5}{18\ln\frac{77.12}{52.22}}\cdot 10^{-9} = 0.499\cdot 10^{-9}\ \text{F/m}$$

Charging current and dielectric losses are obtained from equations (12.38) and (12.41), respectively:

$$I_c = 2\pi f C U_0 = 2\pi 60\cdot 0.499\cdot 10^{-9}\cdot 138\,000/\sqrt{3} = 0.015\ \text{A}$$

$$W_{ch} = \tfrac{1}{3}I_c^2\cdot L^3\cdot R = \tfrac{1}{3}\cdot 0.015^2\cdot 5800^3\cdot(0.0129\cdot 10^{-3}\cdot 1.78) = 336\ \text{W}$$

$$W_d = 2\pi f\cdot L\cdot C\cdot U_o^2\cdot \tan\delta = 0.015\cdot 5800\cdot 0.005\cdot 138\,000/\sqrt{3} = 34\,658\ \text{W}$$

The costs of cable and losses are computed from equations (12.45) and (12.42):

$$CI = A \cdot L \cdot S = 0.22 \cdot 5800 \cdot 1520 = \$1\,939\,520$$

$$CL_{\text{joule}} = I_0^2 \cdot R(S) \cdot L \cdot F(N) = 250^2 \cdot 0.023 \cdot 10^{-3} \cdot 5800 \cdot 271.7 = \$2\,265\,299$$

$$CL_{ch+d} = (336 + 34\,658) \cdot 91.3 = \$3\,194\,952$$

Thus, the total cost is equal to

$$CT = 1\,939\,520 + 2\,265\,299 + 3\,194\,952 = \$7\,399\,771$$

We observe that this cost is greater than the total cost of the cable with a 1010 mm^2 conductor. Indeed, the cost of losses for this cable was computed in Example 12.5; thus, the total cost for the 1010 mm^2 cable is equal to

$$CT_{1010} = 0.22 \cdot 5800 \cdot 1010 + 2\,846\,397 + 2\,584\,612 = \$6\,719\,769$$

Hence, in spite of the fact that the cost of joule losses is higher for the smaller conductor, the cost of dielectric losses forces the selection of a smaller conductor size. Figure 12.2 shows the total cost for standard conductor sizes. In this case, the computations start at the 887 mm^2 (1750 kcmil) conductor size since this is the smallest cross section to satisfy ampacity requirements.

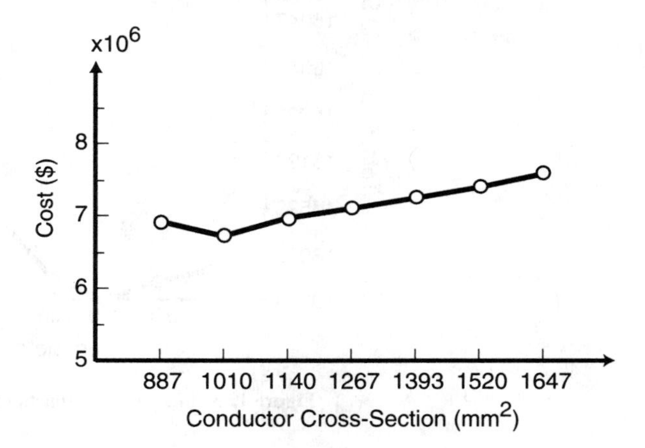

Figure 12.2 Total cost as a function of conductor cross section.

Thus, in this case, the optimal conductor size is 1010 mm^2.

The voltage-dependent losses can be neglected for distribution voltages, and the limiting voltage levels for each insulation type can be those given in IEC Publication 287 (1982) and listed in Table 6.2.

12.5 PARAMETERS AFFECTING ECONOMIC SELECTION OF CABLE SIZES

Sensitivity analysis forms an important part of engineering studies. In the actual cable system design process, several parameters are either unknown or can be predicted with limited accuracy. Some of these parameters can have a profound influence on the selection of the optimal cable size and the associated cost of losses, whereas for others, approximate values can be safely entered since they will have little influence on the final results. The following study examines the effect of two of the parameters discussed in the previous section. A more complete study is given in Anders *et al.* (1991).

EXAMPLE 12.9[2]

We will consider three self-contained, insulating liquid-filled cables with paper insulation and lead sheath with copper reinforcing tape. The system is operated at 230 kV. Cables are located 1 m below the ground with soil thermal resistivity of 1 K·m/W. Ambient soil temperature is 20°C. All the economic and laying parameters are the same as studied in Examples 12.5 and 12.8. The minimal standard conductor size to meet ampacity requirements is equal to 460 mm^2 (900 kcmil).

For the nominal parameters specified in Examples 12.5 and 12.8, the most economic conductor size is 630 mm^2 as shown in Fig. 12.3, and the optimal cable dimensions are shown in Fig. 12.4. The optimal cable has an ampacity of 982 A with losses split as follows: conductor joule losses = 34.41 W/m, sheath joule losses = 3.0 W/m, dielectric losses = 7.52 W/m.

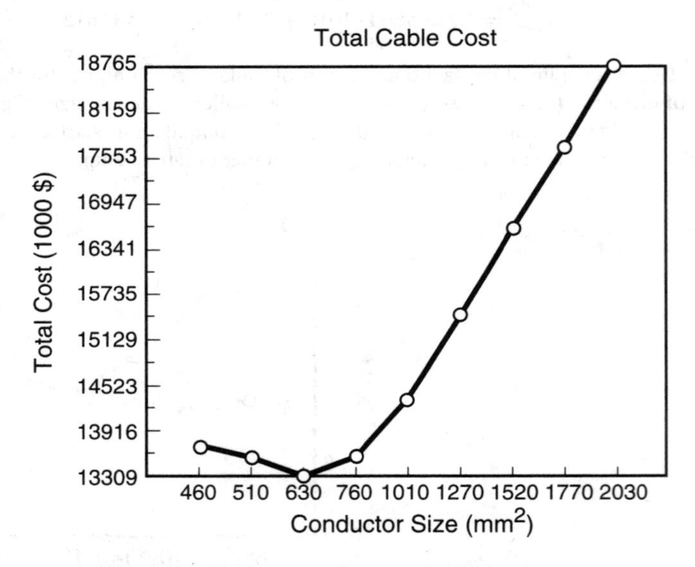

Figure 12.3 Total cost as a function of cable cross section.

In the studies presented below, we will vary one parameter at a time, and the influence of this parameter on the cost of losses and the selection of the optimal conductor size will be investigated. Even though the computer program[3] used for this study computes the theoretically optimum economic cross section of the conductor, results giving standard sizes only are presented below.

Case 1: Variation of the Load Growth Factor

The load growth factor a was varied between 1 and 5%. Factors F defined in equations (12.6) and the optimal cross section in equation (12.42) were recomputed for each value of a, and the results plotted in Fig. 12.5. Intuitively, the rate that the load grows over the economic life of the cable should have a significant effect on the selection of the economic size of the conductor cross section. The

[2] This example is extracted from Anders *et al.* (1993).

[3] The ECOPT computer program used in this study is a part of the Cable Ampacity Program (CAP) developed by Ontario Hydro for Canadian Electrical Association (Anders *et al.*, 1990, 1991).

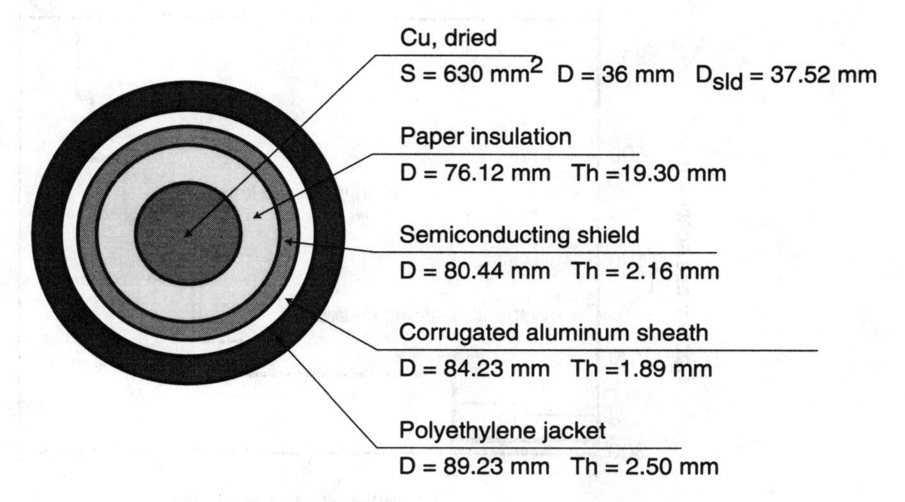

Figure 12.4 Optimal cable dimensions for nominal conditions.

larger the load growth factor, the higher are the conductor losses. This, in turn, will favor larger conductor sizes for the most economic cable selection. The effect is very dramatic, as illustrated in Fig. 12.5. The step function is a result of the fact that only standard cross sections are shown.

Since at a low load growth rate (1–2%) the current carried through the cable at the end of the economic life will be smaller than 550 A, the optimal conductor size to meet ampacity requirements is smaller than 460 mm^2. The analysis indicated that for the voltage level of 230 kV, the dielectric losses can become significant as shown in the example, and their influence on the total cost of losses is considerable. In this case, the voltage-dependent losses result in the selection of a smaller conductor size for some values of the factor a in comparison with the case when the dielectric losses are neglected.

Case 2: The Effect of Changing Financial Factors

The rate at which future energy costs will increase may have a significant effect on the future cost of losses. If the discount rate grows at the same time, the effect of the growth in factor b may be somewhat mitigated. In this study, the value of factor b is varied between 1 and 10% and the discount rate is kept constant at 11%. The economic life is equal to 40 years, and so is the period over which the load growth occurs. The results are shown in Fig. 12.6.

As expected, the increase in the cost of energy has quite a significant effect on the selection of the conductor optimal cross section and on the cost of losses over the economic life of the cable. The costs increase from about \$8 500 000 when b is equal to 1% to over \$17 000 000 with $b = 10\%$. Since the cost of losses is affected significantly when the value of b is large, the standard optimal cable cross section changes by several sizes from 460 to 630 mm^2 in the present case. This effect is particularly evident when b exceeds 5%. Note the effect of dielectric losses on the optimal conductor size shown in Fig. 12.6. For values of b between 3.5 and 10%, the cost of dielectric losses becomes so significant that it forces the selection of the conductor cross section one size lower in comparison with the case when the dielectric losses are neglected. The presence of dielectric losses has the effect of shifting down the selection of a larger conductor cross section for larger values of b. This effect of dielectric losses on the selection of the optimal conductor size was explained in Section 12.4.

An increase in the discount rate has the opposite effect to an increase in energy cost. Since discounting has the effect of reducing future costs, the higher the discount rate, the lower is the cost of losses. Therefore, increasing the discount rate will result in the selection of a smaller conductor size. This effect is illustrated in Fig. 12.7.

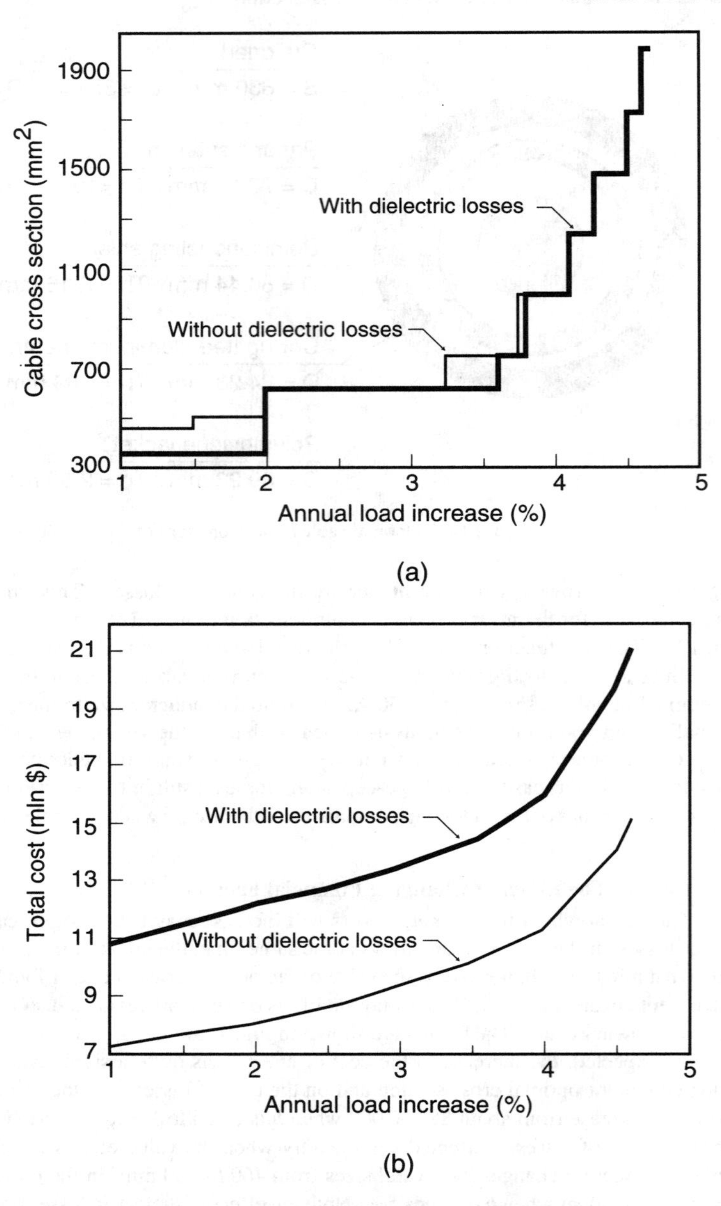

Figure 12.5 The effect of varying the load growth on the optimal cable cross section and costs.

We can observe from Fig. 12.7 that the effect is quite dramatic. Here, again, inclusion of the dielectric losses has a significant effect on the selection of the most economic cable cross section.

Some of the important conclusions drawn from the studies reported in Anders *et al.* (1991) can be summarized as follows. The optimal conductor size for a high-voltage cable is, under almost all conditions, greater than the minimal conductor cross section needed to

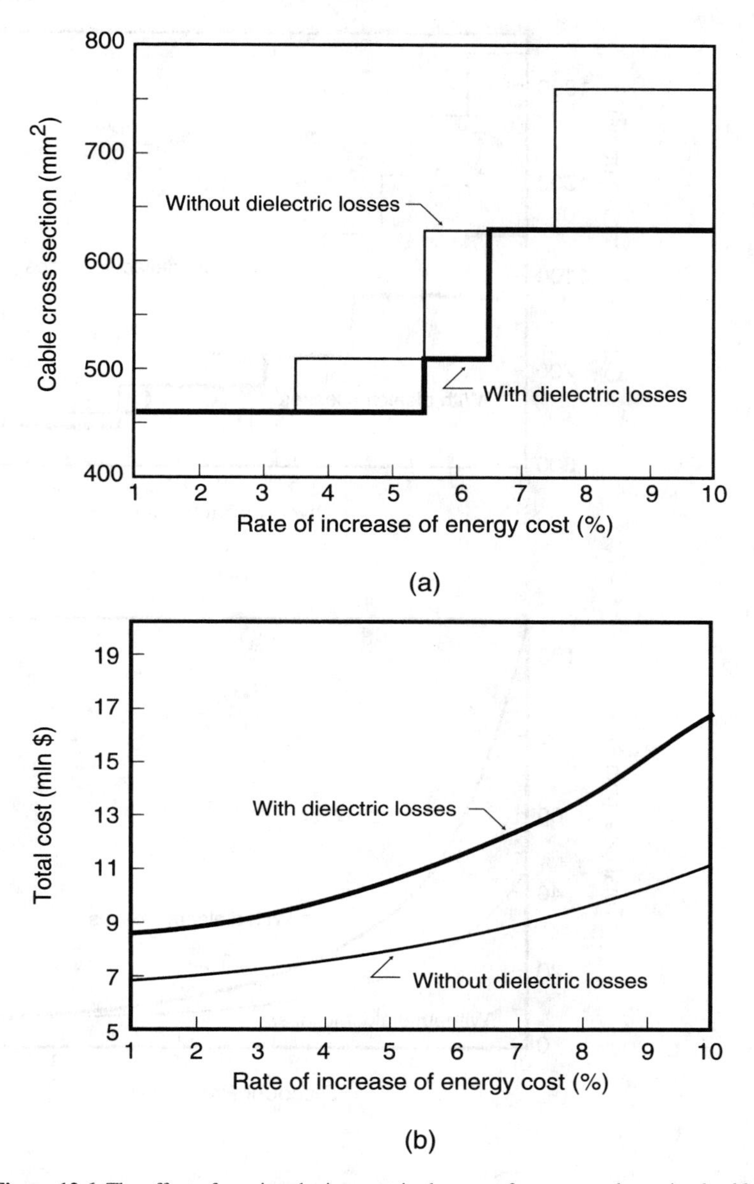

Figure 12.6 The effect of varying the increase in the cost of energy on the optimal cable cross section and cost.

meet ampacity requirements only. This confirms the conclusion reached in studies reported by Parr (1989) for a low-voltage cable. In most of the cases studied, an increase in the economic conductor size caused by including the influence of joule losses was tempered by the associated cost of dielectric losses which may force the selection of a smaller conductor size in comparison with the case when these losses are neglected. The effects of various parameters on the cost of losses and the selection of the optimal conductor size are as follows.

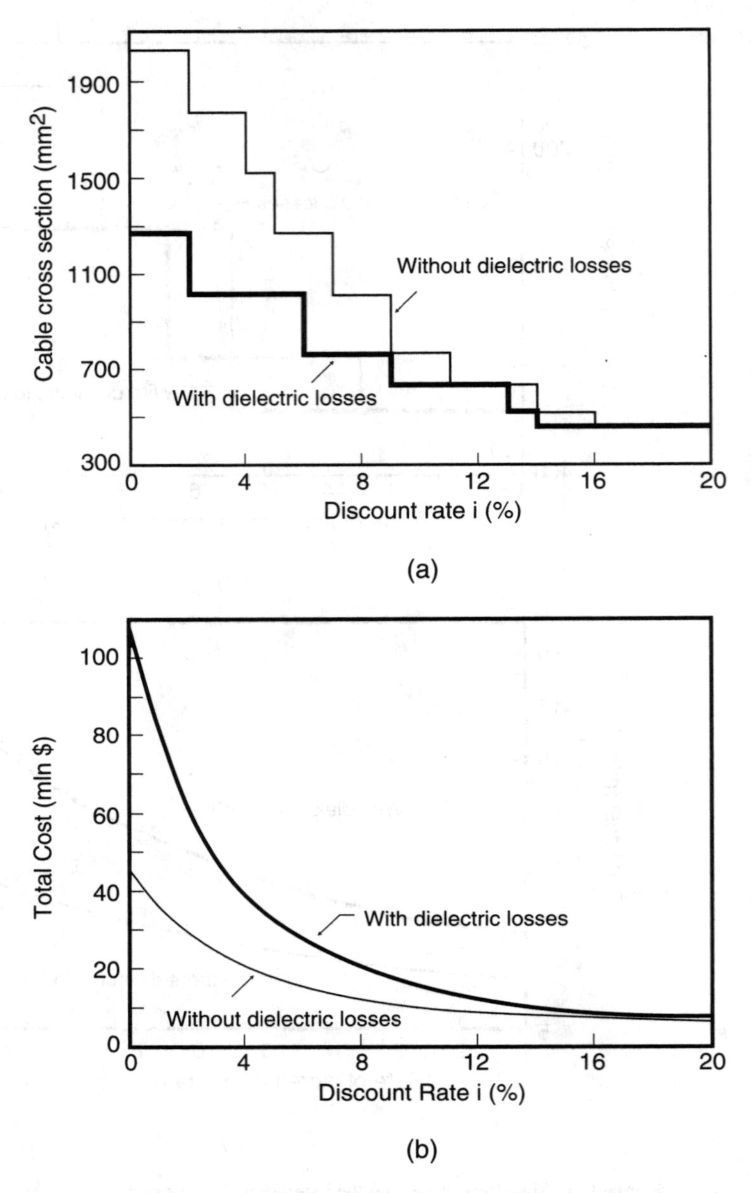

Figure 12.7 The effect of varying discount rate on the optimal cable cross section and cost.

1. The rate at which the load grows has a significant effect on the selection of the most economic conductor cross section. In the example studied, when the load growth changes in a fairly small range between 1 and 5% over a period of 40 years, the optimal conductor cross section increases by 11 standard sizes from 350 to 2030 mm^2.
2. The length of the economic life of the cable is another factor which can have a significant effect on the selection of the conductor size. In some countries, this value is set to as low as ten years, and this favors selection of smaller conductors.

In most cases, the longer the economic life, the larger the size of the most economic conductor. We should perhaps point out that the length of economic life may be selected quite independently of the actual service life of the cable, and this selection may be based on financial or regulatory considerations and not on operational expectations.

3. Financial factors can have a significant effect on the final conductor selection. An increase in the discount rate favors smaller conductor sizes as the initial cost becomes more significant. A faster increase in the energy cost, on the other hand, will favor larger conductor sizes. These two effects may almost cancel each other if the growth in the discount rate is approximately equal to the growth in the energy costs.
4. When the full economic impact of cable losses on the selection of the optimal cable cross section is studied, the importance of proper representation of the load growth increases. If the load grows initially very fast and then levels off, a much larger conductor size may be required as compared to the case when the load growth rate is moderate but extends over a longer period of time. The change in the ultimate load factor, on the other hand, has little influence on the selection of optimal conductor size, but it should be modeled if the cost of losses is to be evaluated correctly. The same conclusion can be drawn with regard to the representation of charging current losses. Only for long distances and high voltages can an uncompensated cable system have significant losses caused by charging current, and in that case, the cost of these losses should be evaluated.

REFERENCES

Anders, G. J., Moshref, A., and Roiz, J. (July 1990), "Advanced computer programs for power cable ampacity calculations," *IEEE Comput. Appl. in Power*, vol. 3, no. 3, pp. 42–45.

Anders, G. J., Vainberg, M., Horrocks, D. J., Foty, M., and Motlis, J. (1991), "A user-friendly computer program for economic selection of cable sizes," in *Proc. 3rd Int. Conf. on Extruded Dielectric Power Cables—Jicable-91*, Versaille, France, June 1991.

Anders, G. J., Vainberg, M., Horrocks, D. J., Motlis, J., Foty, M., and Jarnicki, J. (1993), "Parameters affecting economic selection of cable sizes," *IEEE Trans. Power Delivery*, vol. 8, no. 4, pp. 1661–1667.

IEC 287 (1982), "Calculation of the continuous current rating of cables (100% load factor)," IEC Publication 287, 2nd ed.

IEC 1059 (1991), "Economic optimization of power cable size," IEC Publication 1059.

IEEE (Oct. 1990), "Loss evaluation for underground transmission and distribution cable systems," Insulated Conductors Committee—Task Group 7-39; *IEEE Trans. Power Delivery*, vol. 5, no. 4, pp. 1652–1659.

Neher, J. H., and McGrath, M. H. (Oct. 1957), "The calculation of the temperature rise and load capability of cable systems," *AIEE Trans.*, vol. 76, part 3, pp. 752–772.

Parr, R. G. (1989), "The economic choice of conductor size," *Revue General de l'Electricite*, no. 10, Paris, pp. 45–50.

Scheer, C. B. (Feb. 1966), "Future power prediction," *Energy Int.*, pp. 14–16.

PART IV APPENDIXES

A

Model Cables

Five model cables are described in this Appendix. The cables are used throughout the book to illustrate various concepts as they are developed. Design and computed parameters for the model cables are summarized in Table A.1.

The installation conditions of the model cables are also described below. However, the installation information may vary in the examples to show the sensitivity of cable rating to variations in various laying parameters.

A.1 MODEL CABLE NO. 1

This is a 10 kV single conductor XLPE cable. Conductor resistance at 20°C is taken from IEC 228 (1978). The cable has copper screen wires with a given initial electrical resistance (at 20°C) and a PVC jacket. All thermal and electrical parameters are as specified in IEC 287 (1982), and these values are given in the various tables in this book. The cable cross section is shown in Fig. A.1.

The laying conditions are assumed as follows: cables are located 1 m below the ground in a flat configuration. Uniform soil properties are assumed throughout. Spacing between cables is equal to one cable diameter (spacing between centers equals to two cable diameters). Ambient soil temperature is 15°C. The thermal resistivity of the soil is equal to 1.0 K · m/W. The cables are solidly bonded and not transposed.

A.2 MODEL CABLE NO. 2

This is a 10 kV three-core XLPE cable. The conductor resistance at 20°C is taken from IEC 228 (1978). The cable has copper tape screen around each core with a given initial

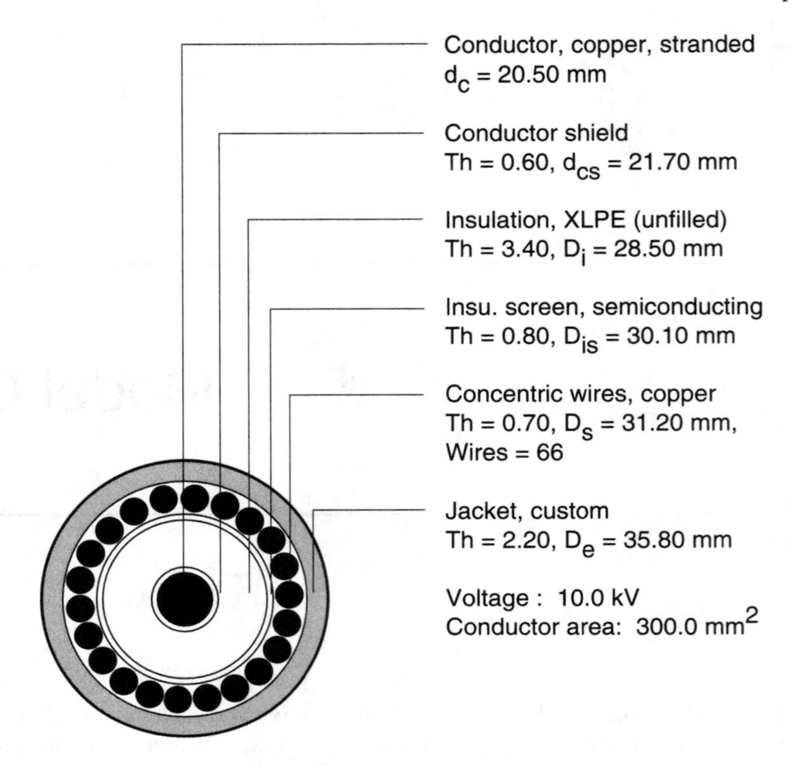

Figure A.1 Cross section of the model cable No. 1.

electrical resistance (at 20°C) and a PVC jacket. All thermal and electrical parameters are as specified in IEC 287 (1982). The cable cross section is shown in Fig. A.2.

The installation conditions are assumed as follows: the cable is located in free air (not clipped to a vertical wall). The ambient air temperature is 25°C, and the cable is shaded from solar radiation.

A.3 MODEL CABLE NO. 3

This is a 138 kV high-pressure liquid-filled (HPLF) cable. All parameters are the same as in the Neher/McGrath (1957) paper (see Table A.1). The cable shield consists of an intercalated 7/8(0.003) inches bronze tape—1 in lay, and a single 0.1 (0.2) in D-shaped bronze skid wire—1.5 in lay. The cables lie in a cradle configuration and operate at 85% load-loss factor. Several parameters are different from those used in IEC 287 (1982). The ones which are different are given as follows:

$$\text{thermal resistivity of the insulation} = 5.5 \text{ K} \cdot \text{m/W}$$
$$\text{thermal resistivity of pipe coating} = 1.0 \text{ K} \cdot \text{m/W}$$
$$\text{dielectric constant} = 3.5$$
$$\tan \delta = 0.005$$

The cable cross section is shown in Fig. A.3.

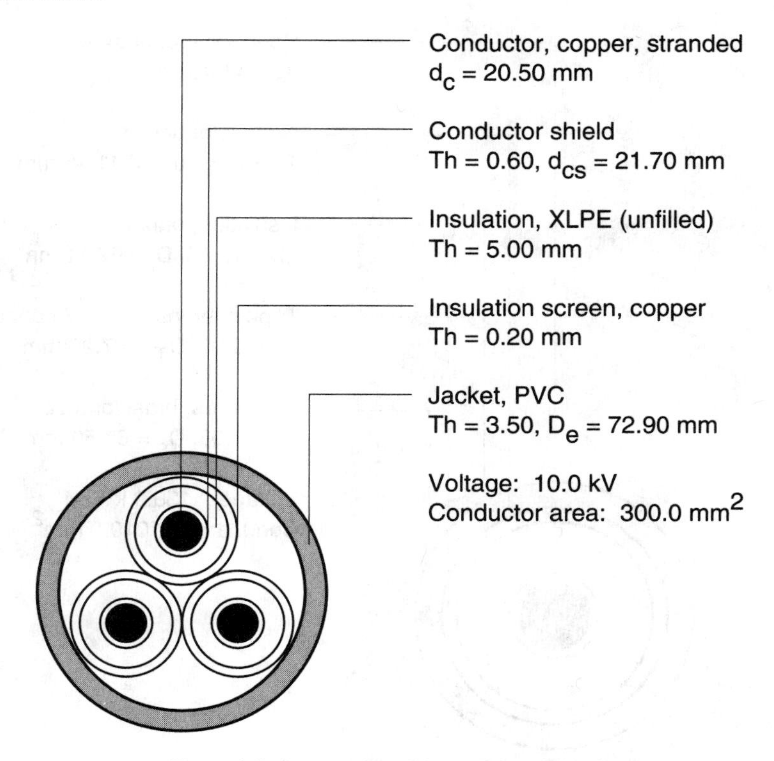

Figure A.2 Cross section of the model cable No. 2.

The laying conditions are assumed as follows. The cables are located in a steel pipe of 8.625 in outside diameter. The pipe is covered with an asphalt mastic covering 0.5 in thick. The center of the pipe is located 3 ft below the ground. The soil is uniform throughout. The ambient soil temperature is 25°C. The thermal resistivity of the soil is equal to 0.8 K · m/W.

A.4 MODEL CABLE NO. 4

This is a 230 kV low-pressure, oil-filled cable with 1250 kcmil (633 mm^2) copper hollow core conductor. The cable has paper insulation with an insulation screen composed of four layers of carbon tapes. On top of insulation screen is a lead sheath reinforced with three layers of copper tapes, 50% overlap. The tapes are 25 mm wide and 1.3 mm thick. The lay of tapes is equal to 115.2 mm. Armor is composed of 51, #4 AWG copper wires with a lay of length of 121.8 mm. The armor bedding is a saturated jute and a layer of polyethylene with the equivalent thermal resistivity 4.27 K · m/W. The armor serving is a saturated jute. All remaining thermal and electrical parameters are as specified in IEC 287 (1982).

The cable cross section is shown in Fig. A.4.

Three cables are laid in a thermal backfill as shown in Fig. A.5.

Soil ambient temperature is 20°C. Thermal resistivities of the soil and backfill are shown in Fig. A.5. Cable sheaths are two-point bonded.

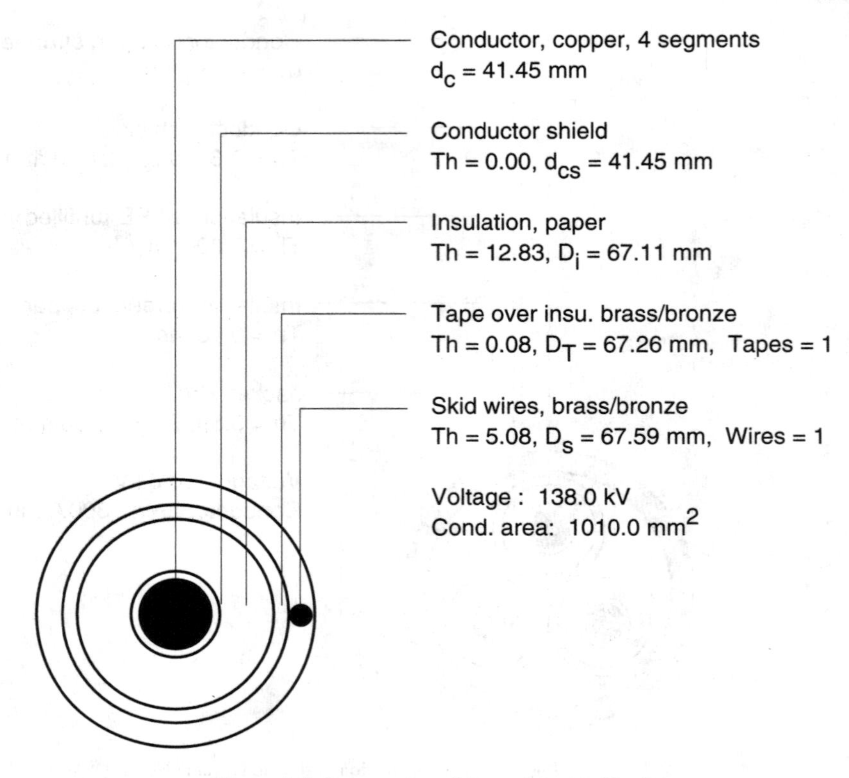

Figure A.3 Cross section of the model cable No. 3.

A.5 MODEL CABLE NO. 5

This is a 400 kV paper–polypropylene–paper cable with 2000 mm^2 copper segmental conductor and aluminum corrugated sheath. The outer covering is a PVC jacket.

The cable cross section is shown in Fig. A.6.

The cables are laid in a flat formation without transposition, directly in the soil with thermal resistivity of 1 K · m/W and ambient temperature equal to 25°C. The sheaths are cross bonded with unknown minor section lengths. The centers of the cables are 1.8 m below the ground and phases are 0.5 m apart.

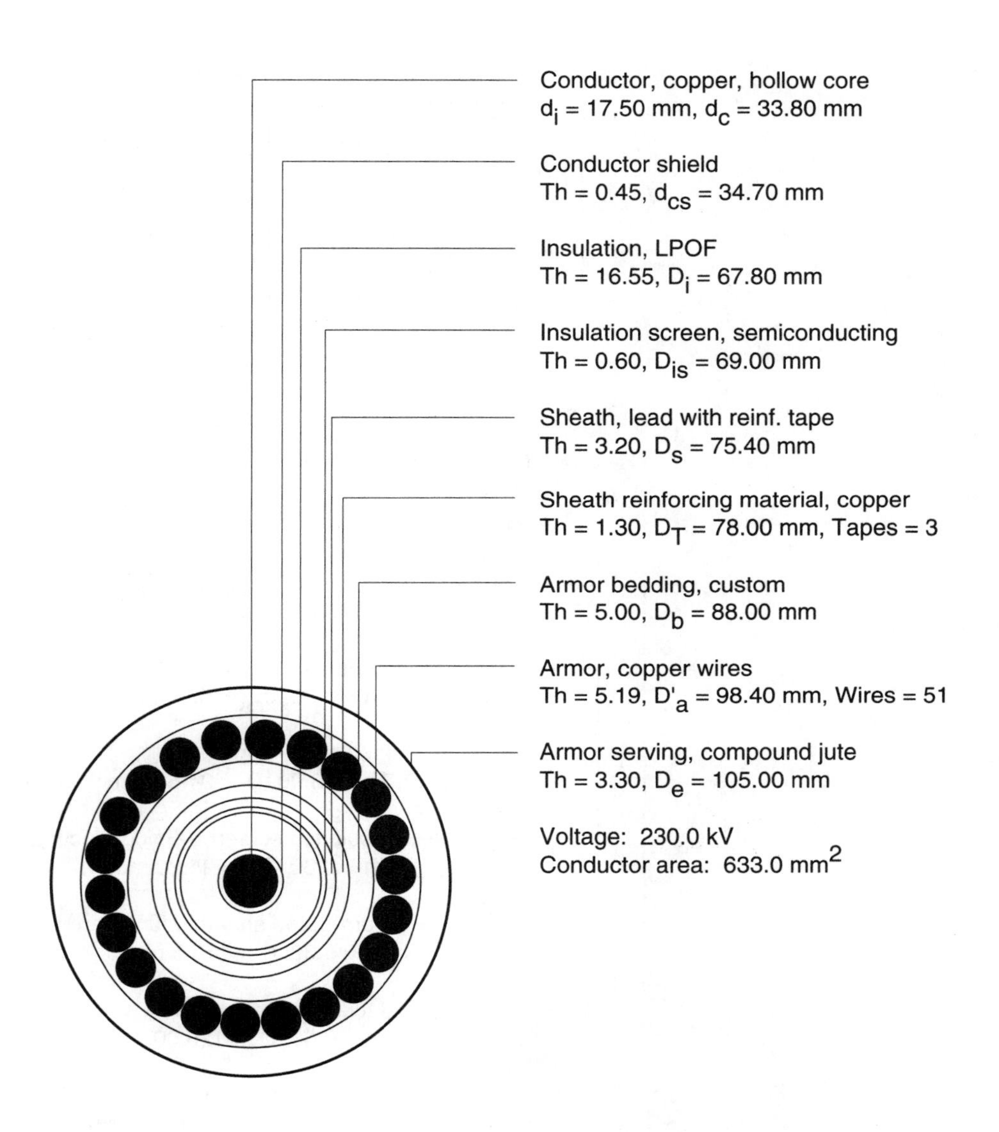

Figure A.4 Cross section of the model cable No. 4.

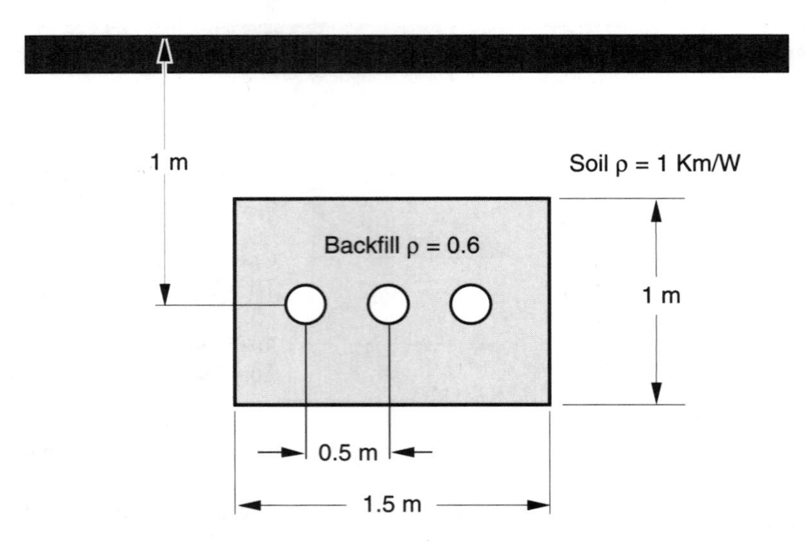

Figure A.5 Laying conditions for model cable No. 4.

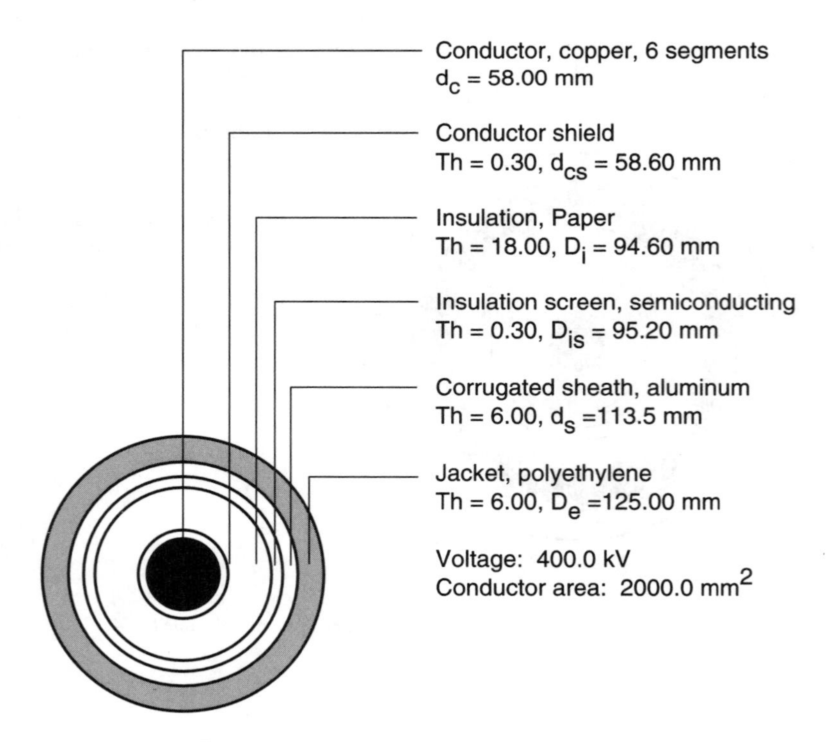

Figure A.6 Cross section of the model cable No. 5.

TABLE A.1 Parameters of Model Cables

	Cable No. 1	Cable No. 2	Cable No. 3	Cable No. 4	Cable No. 5
			Construction		
Conductor	Cu stranded	Cu stranded	Cu segmental	Cu hollow core	Cu segmental
S (mm^2)	300	300	1010	633	2000
d_c (mm)	20.5	20.5	41.45	33.8	58.0
d_{cs} (mm)	21.7	21.7	41.45	34.7	58.6
d_i (mm)				17.5	
c (mm)		17.6			
s (mm)		30.5	67.59		
Insulation	XLPE	XLPE	Paper	Paper	PPLP
D_i (mm)	28.5	28.5	67.11	67.8	94.6
D_{is} (mm)	30.1	30.1	67.26	69.0	95.2
t_i (mm)		5.3			
t (mm)		9.6			
t_1 (mm)		4.8[1]			
Screen/sheath	*Cu wires*	*Cu tape*	Brass/bronze Tape and 1 skid wire	Lead with 3 Cu tapes	Corrug'd. aluminum
D_s (mm)	31.4		67.59	75.4	102
D_{oc} (mm)					113
D_{it} (mm)					
d (mm)				73.8	
d_2 (mm)		30.5		76.0/ (78 over tape)	
δ_1 (mm)		0.2			
t_s (mm)				3.2	6.0
Armor				Copper	
D_a (mm)				98.4	
δ (mm)				5.189	
Jacket/bedding	PVC	PVC	Somastic pipe coating	Jute/poly-ethylene jute	PVC
t_2 (mm)				5.0	
t_3 (mm)				3.3	
t_j (mm)	2.2	3.5			6.0
D_e (mm)	35.8	72.9	244.48	105	125.2
Duct or pipe			Steel		
D_d (mm)			206.38		
D_0 (mm)			219.08		

[1] Insulation thickness includes the thickness of semiconducting screens at the conductor and above the insulation.

TABLE A.1 *Continued*

			Given Cable Parameters		
θ (°C)	90	90	70	85	85
f (Hz)	50	50	60	50	60
ε			3.5		2.8
$\tan\delta$			0.005		0.001
R_0 (Ω/km)	0.0601	0.0601			
R_{s0} (Ω/km)	0.759	0.906			
			Computed Parameters (Hottest Cable)		
I (A)	629	713	902	771	1365
R (Ω/km)	0.0781	0.0798	0.0245	0.0356	0.0126
y_s	0.014	0.014	0.050	0.021	0.132
y_p	0.0047	0.027	0.054	0	0.005
X (Ω/km)	0.096	0.044	0.052	0.154	0.168
X_1 (Ω/km)	0.102			0.170	0.186
X_m (Ω/km)	0.044			0.044	0.052
λ_1	0.090	0.0218	0.010	0.325	0.140
λ_2			0.311	0.955	
T_1 (K·m/W)	0.214	0.307[2]	0.422	0.568	0.579
T_2 (K·m/W)			0.082	0.082	
T_3 (K·m/W)	0.104	0.078	0.017	0.066	0.056
T_4 (K·m/W)	1.933	0.345	0.343 (0.289)[3]	0.814	1.276
Q_c (J/K·m)	1035	1035[4]	3484.5	2183.9	6900
Q_i (J/K·m)	915.6	915.6	1458.4	5684.0	8952
Q_o (J/K·m)			38570	1995.1[5]	
Q_s (J/K·m)	4.0	4.0	1.2	2133.2[6]	3916.3
Q_b (J/K·m)				613.3	
Q_j (J/K·m)	394.8	432.4	31847[7]	2108.7	5392.5
$\Sigma T \cdot \Sigma Q$ (h)	1.47	1.04	18.1	6.26	13.4
W_c (W/m)	30.85	40.58	19.93	21.16	23.40
W_d (W/m)			4.83	6.62	6.53
W_I (W/m)	33.59	41.45	26.32	48.24	26.92
W_t (W/m)	33.59	41.45	31.15	54.86	33.45

[2] For cable No. 2, all computed values except T_1 are given on a per-core basis. T_1 for a single core is one third of the value in Table A.1.

[3] The second T_4 value corresponds to the Neher/McGrath method with load-loss factor $\mu = 0.85$.

[4] The conductor and insulation capacitances are given on a per-core basis. See Example 3.8 for representation of this cable by an equivalent single-core cable.

[5] This is oil in the conductor.

[6] Capacitance of sheath and reinforcement.

[7] This value includes capacitance of the pipe and pipe covering.

B

An Algorithm to Calculate the Coefficients of the Transfer Function Equation

The Laplace transform of a ladder network transfer function is given by a ratio [see equation (5.1)]:

$$H(s) = \frac{P(s)}{Q(s)}$$

$P(s)$ and $Q(s)$ are polynomials, their forms depending on the number of loops in the network. An algorithm for the computation of the coefficients of these polynomials is described below.

B.1 DENOMINATOR EQUATION

The degree of the denominator equation will be the same as the number of loops of the equivalent network. This equation is common for all the nodes of the circuit and has the form

$$Q(s) = s\left(b_n s^n + b_{n-1} s^{n-1} + b_1 s + b_0\right) \tag{B.1}$$

where n is the number of loops of equivalent network.

We define

$I = \{1, 2, \cdots, n\}$ set of indices

$I_k = \{i_1, i_2, \cdots, i_k\}$ any combination of k elements of I

$I'_k = I - I_k = \{i_{k+1}, \cdots, i_n\}$ complement of I_k in I

$B_k = \{\beta_1, \cdots, \beta_{n-k}\}$ set of sequences in which β are equal to zero or one

$I_k'' = \{i_{k+1} - \beta_1, i_{k+2} - \beta_2, \cdots, i_n - \beta_{n-k}\}$ sequences of B_k over I_k' where all elements of I_k'' are different.

The coefficients b in equation (B.1) are obtained from

$$b_k = \sum_{I_k} \prod_{j=1}^{k} Q_{ij} \left(\prod_{m \in I_k'} G_m + \sum_{B_k} \prod_{l \in I_k''} G_l \right) \tag{B.2}$$

where Q_t = equivalent capacitances, J/K · m

$G_x = 1/T_x$ = equivalent conductances, W/K · m.

The first summation in (B.2) includes all the combinations of k elements out of n. There are $\binom{n}{k}$ terms. The second summation includes all the sequences B_k, except $B_k = \{0, 0, \cdots, 0\}$. There are $2^{(n-k)} - 1$ terms.

B.2 NUMERATOR EQUATION

For each node i, the numerator equation $P(s)$ is different. The degree of $P(s)$ is equal to the number of loops minus the node index

$$m = n - i \tag{B.3}$$

and

$$P(s) = a_m a_{mi} s^m + a_{(m-1)i} s^{m-1} + \cdots + a_{1i} s + a_{0i} \tag{B.4}$$

We define

$I = \{1, 2, \cdots, n\}$ set of indices

$I_k = \{i_1, i_2, \cdots, i_k\}$ any combination of k elements of I with $i_p > i$ for $p = 1, \cdots, k$.

$I_k' = I - I_k = \{i_{k+1}, \cdots, i_n\}$ complement of I_k in I but index i excluded, that is, $i_p \neq i$ for $p = k+1, \cdots, n$.

$B_k = \{\beta_1, \cdots, \beta_{n-k}\}$ set of sequences in which β are equal to zero or one excluding $B_0 = \{0, 0, \cdots, 0\}$

$I_k'' = \{i_{k+1} - \beta_1, i_{k+2} - \beta_2, \cdots, i_n - \beta_{n-k}\}$ sequences of B_k over I_k' where all elements of I_k'' are different.

Similarly to the denominator, the coefficients a for the numerator equation (B.4) are given by

$$a_{ki} = \sum_{I_k} \prod_{j=1}^{k} Q_{ij} \left(\prod_{m \in I_k'} G_m + \sum_{B_k} \prod_{l \in I_k''} G_l \right) \tag{B.5}$$

C

Digital Calculation of Quantities Given Graphically in Figs. 9.1–9.7

This Appendix gives formulas and methods suitable for digital calculations for several quantities given graphically in Chapter 9 (Figs. 9.1–9.7). The method used is the approximation by algebraic equations, followed by quadratic or linear interpolation where necessary. The maximum percentage error prior to interpolation is given in each case. The formulas were first published in IEC Publication 287.

C.1 GEOMETRIC FACTOR G FOR TWO-CORE BELTED CABLES WITH CIRCULAR CONDUCTORS (FIG. 9.1)

Denote

$$X = t_1/t, \qquad Y = (2t_1/t) - 1 \tag{C.1}$$

Then

$$G = MG_s \tag{C.2}$$

where

$$M = \text{ Mie formula } = \ln\left[\frac{1 - \alpha\beta + \left[\left(1-\alpha^2\right)\left(1-\beta^2\right)\right]^{1/2}}{\alpha - \beta}\right] \tag{C.3}$$

$$\alpha = \frac{1}{\left[1 + \dfrac{X}{1 + X/(1+Y)}\right]^2}; \qquad \frac{\beta}{\alpha} = \frac{\dfrac{X}{1+Y} - \dfrac{1}{2}}{\dfrac{X}{1+Y} + \dfrac{3}{2}} \tag{C.4}$$

$G_s = G_s(X, Y)$; that is, G_s is a function of both X and Y. To obtain this function, we first compute

$$\begin{aligned} G_s(X, 0) &= 1.06719 - 0.0671778X + 0.0179521X^2 \\ G_s(X, 0.5) &= 1.06798 - 0.0661648X + 0.0158125X^2 \\ G_s(X, 1) &= 1.06700 - 0.0557156X + 0.0123212X^2 \end{aligned} \tag{C.5}$$

Then $G_s(X, Y)$ is obtained by quadratic interpolation using the following formula:

$$\begin{aligned} G_s(X, Y) = G_s(X, 0) &+ Y[-3G_s(X, 0) + 4G_s(X, 0.5) - G_s(X, 1)] \\ &+ Y^2[2G_s(X, 0) - 4G_s(X, 0.5) + 2G_s(X, 1)] \end{aligned} \tag{C.6}$$

The maximum percentage error in the calculation of $G_s(X, 0)$, $G_s(X, 0.5)$, and $G_s(X, 1)$ is less than 0.5% compared with corresponding graphical values.

C.2 GEOMETRIC FACTOR G FOR THREE-CORE BELTED CABLES WITH CIRCULAR CONDUCTORS (FIG. 9.2)

Equations (C.1)–(C.3) are also applicable in this case with the following parameters:

$$\alpha = \frac{1}{\left[1 + \dfrac{2X}{1 + \dfrac{2}{\sqrt{3}}\left(1 + \dfrac{2X}{1+Y}\right)}\right]^3}; \qquad \frac{\beta}{\alpha} = \frac{\dfrac{2}{\sqrt{3}}\left(1 + \dfrac{2X}{1+Y}\right) - 3}{\dfrac{2}{\sqrt{3}}\left(1 + \dfrac{2X}{1+Y}\right) + 3} \tag{C.7}$$

As before, G_s is a function of both X and Y. To compute this function, we first compute

$$\begin{aligned} G_s(X, 0) &= 1.09414 - 0.0944045X + 0.0234464X^2 \\ G_s(X, 0.5) &= 1.09605 - 0.0801857X + 0.0176917X^2 \\ G_s(X, 1) &= 1.09831 - 0.0720631X + 0.0145909X^2 \end{aligned} \tag{C.8}$$

Then $G_s(X, Y)$ is obtained by quadratic interpolation using equation (C.6). The maximum percentage error in the calculation of $G_s(X, 0)$, $G_s(X, 0.5)$, and $G_s(X, 1)$ is less than 0.5% compared with corresponding graphical values.

C.3 THERMAL RESISTANCE OF THREE-CORE SCREENED CABLES WITH CIRCULAR CONDUCTORS COMPARED TO THAT OF A CORRESPONDING UNSCREENED CABLE (FIG. 9.4)

Denote

$$X = (\delta_1 \rho_T / d_c \rho_m), \qquad Y = t_1 / d_c \tag{C.9}$$

The screening factor K is a function of both X and Y. To calculate it, we first compute $K(X, 0.2)$, $K(X, 0.6)$, and $K(X, 1)$ from the following formulas according to whether $0 < X \leq 6$ or $6 < X \leq 25$.

$$
\begin{aligned}
0 < X \le 6 \qquad & K(X, 0.2) = 0.998095 - 0.123369X \\
& \qquad + 0.0202620X^2 - 0.00141667X^3 \\
& K(X, 0.6) = 0.999452 - 0.0896589X \\
& \qquad + 0.0120239X^2 - 0.000722228X^3 \\
& K(X, 1) = 0.997976 - 0.0528571X \\
& \qquad + 0.00345238X^2 \\
6 < X \le 25 \qquad & K(X, 0.2) = 0.824160 - 0.0288721X \\
& \qquad + 0.000928511X^2 - 0.0000137121X^3 \\
& K(X, 0.6) = 0.853348 - 0.0246874X \\
& \qquad + 0.000966967X^2 - 0.0000159967X^3 \\
& K(X, 1) = 0.883287 - 0.0153782X \\
& \qquad + 0.000260292X^2
\end{aligned}
\tag{C.10}
$$

Then, $K(X, Y)$ is obtained by quadratic interpolation using the following formula:

$$
\begin{aligned}
K(X, Y) = K(X, 0.2) &+ Z[-3K(X, 0.2) + 4K(X, 0.6) - K(X, 1)] \\
&+ Z^2[2K(X, 0.2) - 4K(X, 0.6) + 2K(X, 1)]
\end{aligned}
\tag{C.11}
$$

where $Z = 1.25Y - 0.25$.

The maximum percentage error in the calculation of the reduction factor is less than 0.5% compared with the corresponding graphical values.

C.4 THERMAL RESISTANCE OF THREE-CORE SCREENED CABLES WITH SECTOR-SHAPED CONDUCTORS COMPARED TO THAT OF A CORRESPONDING UNSCREENED CABLE (FIG. 9.5)

Denote

$$
X = (\delta_1 \rho_T / d_x \rho_m), \qquad Y = t_1/d_x \tag{C.12}
$$

The screening factor K is a function of both X and Y. To calculate it, we first compute $K(X, 0.2)$, $K(X, 0.6)$, and $K(X, 1)$ from the following formulas according to whether $0 < X \le 3$, $3 < X \le 6$, or $6 < X \le 25$.

$$
\begin{aligned}
0 < X \le 3 \qquad & K(X, 0.2) = 1.00169 - 0.0945X + 0.00752381X^2 \\
& K(X, 0.6) = 1.00171 - 0.0769286X + 0.00535714X^2 \\
& K(X, 1) = K(X, 0.6)
\end{aligned}
\tag{C.13}
$$

$3 < X \le 6$ $\quad$ $K(X, 0.2)$ and $K(X, 0.6)$ are given by the same formula as for $0 < X \le 3$.

$$K(X, 1) = 1.00117 - 0.0752143X + 0.00533334X^2$$

$$6 < X \leq 25 \qquad K(X, 0.2) = 0.811646 - 0.0238413X + 0.000994933X^2 - 0.0000155152X^3$$

$$K(X, 0.6) = 0.833598 - 0.0223155X + 0.000978956X^2 - 0.0000158311X^3$$

$$K(X, 1) = 0.842875 - 0.0227255X + 0.00105825X^2 - 0.0000177427X^3$$

For $0 < X \leq 3$ and $0.2 < Y \leq 0.6$, $K(X, Y)$ is obtained by linear interpolation between $K(X, 0.2)$ and $K(X, 0.6)$ as

$$K(X, Y) = K(X, 0.2) + 2.5(Y - 0.2)[K(X, 0.6) - K(X, 0.2)] \qquad \text{(C.14)}$$

For $3 < X \leq 25$, $K(X, Y)$ is obtained by quadratic interpolation between the three calculated values from equation (C11). The maximum percentage error in the calculation of the sector correction factor is less than 1% compared with graphical values.

C.5 CURVE $\bar{G}$ FOR OBTAINING THE THERMAL RESISTANCE OF THE FILLING MATERIAL BETWEEN THE SHEATHS AND ARMOR OF SL TYPE CABLES (FIG. 9.7)

We will denote by X the thickness of material between sheaths and armor expressed as a function of outer diameter of the sheath. The lower curve is given by

$$0 < X \leq 0.03 \qquad \bar{G} = 2\pi(0.000202380 + 2.03214X - 21.6667X^2)$$

$$0.03 < X \leq 0.15 \qquad \bar{G} = 2\pi(0.0126529 + 1.101X - 4.56104X^2 + 11.5093X^3) \qquad \text{(C.15)}$$

The maximum percentage error in the calculation of $\bar{G}$ is less than 1%.

The upper curve is given by

$$0 < X \leq 0.03 \qquad \bar{G} = 2\pi(0.00022619 + 2.11429X - 20.4762X^2)$$

$$0.03 < X \leq 0.15 \qquad \bar{G} = 2\pi(0.0142108 + 1.17533X - 4.49737X^2 + 10.6352X^3) \qquad \text{(C.16)}$$

The maximum percentage error in the calculation of $\bar{G}$ is again less than 1%.

D

Properties of Air at Atmospheric Pressure

The air thermodynamical and transport properties used in this book are from the U.S. National Bureau of Standards, cited in Holman (1990). All of the properties are temperature and pressure dependent. For the convenience of computer programming, formulas are generated by curve fitting to relate the air density (ρ), thermal conductivity (k), viscosity (ν), and Prandtl Number (Pr) for temperature (θ^*) in the range of 250–450 K:

$$\begin{aligned} \rho &= 352.64/\theta, && \text{kg/m}^3 \\ k_{\text{air}} &= 10^{-8}(-27997.7 + 989.998\theta^* - 3.54283\theta^{*2}), && \text{W/K}\cdot\text{m} \\ \nu &= 10^{-11}(-376936 + 3780.05\theta^* + 9.11422\theta^{*2}), && \text{m}^2/\text{s} \\ \text{Pr} &= 0.833209 - 0.582349\theta^* \cdot 10^{-3} + 0.552336\theta^{*2} \cdot 10^{-6} \end{aligned} \tag{D.1}$$

where the temperature is in Kelvin. Similar formulas are widely used in thermal engineering. Morgan (1982) gives the following simpler relationships:

$$\begin{aligned} k &= 2.42 \cdot 10^{-2} + 7.2 \cdot 10^{-5}(\theta^* - 273) \\ \nu &= 1.32 \cdot 10^{-5} + 9.5 \cdot 10^{-8}(\theta^* - 273) \\ \text{Pr} &= 0.715 - 0.00025(\theta^* - 273) \end{aligned} \tag{D.2}$$

The accuracy of Morgan's formulas is compared with the equations (D.1) in Anders and Gu (1995).

REFERENCES

Anders, G. J., and Gu, N. (1995), "Energy conservation equations for cables in air," Canadian Electrical Association Report 138 D 375E.

Holman, J. P. (1990), *Heat Transfer*. New York: McGraw-Hill.

Morgan, V. T. (1982), "The thermal rating of overhead-line conductors, Part I. The steady-state thermal model," *Elec. Power Sys. Res.*, vol. 5, pp. 119–139.

E

Calculation Sheets for Steady-State Cable Ratings

The calculation procedures set up in this Appendix are applicable to single- and three-conductor cables. They are based on the material presented in this book, and include the basic computations covered in IEC Standard 287 and the Neher–McGrath paper.

E.1 GENERAL DATA

$f\,(\text{Hz}) =$ system frequency

$U\,(\text{V}) =$ cable operating voltage (phase-to-phase)

$LF =$ daily load factor

$\theta =$ conductor temperature[1]

$\theta_{\text{amb}} =$ ambient temperature

E.2 CABLE PARAMETERS

E.2.1 Conductor

$S\,(\text{mm}^2) =$ cross-sectional area of conductor

$D_e\,(\text{mm}) =$ external diameter of cable

$D_e^*\,(\text{m}) =$ external diameter of cable, or equivalent diameter of cable (cables in air)

[1] Subscripts t, s, and a will be used to denote tape, sheath, and armor, respectively.

d_c (mm) = external diameter of conductor

d'_c (mm) = conductor diameter of equivalent solid conductor having the same central oil duct

d_i (mm) = conductor inside diameter

n = number of conductors in a cable

Three-conductor Cables.

d_x (mm) = diameter of an equivalent circular conductor having the same cross-sectional area and degree of compactness as the shaped one

c (mm) = distance between the axes of conductors and the axis of the cable for three-core cables ($= 0.55r_1 + 0.29t$ for sector-shaped conductors)

r_1 (mm) = circumscribing radius of three-sector shaped conductors in three-conductor cable

E.2.2 Insulation

D_i (mm) = diameter over insulation

t_1 (mm) = insulation thickness between conductors and sheath

ρ (K · m/W) = thermal resistivity of the material[2]

Three-conductor Cables.

t (mm) = insulation thickness between conductors

t_i (mm) = thickness of core insulation, including screening tapes plus half the thickness of any nonmetallic tapes over the laid up cores

E.2.3 Sheath

D_s (mm) = sheath diameter

d (mm) = sheath mean diameter

t_s (mm) = sheath thickness

$\left.\begin{matrix} p_2 \\ q_2 \end{matrix}\right\}$ = ratios of minor section lengths, where minor section lengths are a, p_2a, q_2a and a is the shortest section

ζ ($\Omega \cdot$ m) = electrical resistivity of sheath material at operating temperature

[2] The same symbol is used for thermal resistivity of various materials. The appropriate numerical value taken from the table on page 400 will correspond to the material considered.

E.2.4 Armor or Reinforcement

A (mm^2) = cross-sectional area of the armor
D_a (mm) = external diameter of armor
d_a (mm) = mean diameter of armor
d_f (mm) = diameter of armor wires
d_2 (mm) = mean diameter of reinforcement
n_a = number of armor wires
n_t = number of tapes
ℓ_a (mm) = length of lay of a steel wire along a cable
ℓ_T (mm) = length of lay of a tape
t_t (mm) = thickness of tape
w_t (mm) = width of tape
β = angle between axis of armor wire and axis of cable

E.2.5 Jacket/serving

t_J (mm) = thickness of the jacket
t_3 (mm) = thickness of the serving

E.3 CABLE PARAMETERS—PIPE-TYPE CABLES

D_o (mm) = external diameter of the pipe
D_e (mm) = external diameter of the pipe coating
D_{sm} (mm) = moisture barrier mean diameter
D_{sw} (mm) = diameter of skid wire
D_t (mm) = diameter of moisture barrier assembly
D_d (mm) = internal diameter of pipe
n_{sw} = number of skid wires
n_t = number of moisture barrier metallic tapes
ℓ_{sw} (mm) = lay of length of skid wires
ℓ_T (mm) = lay of moisture barrier metallic tapes
t_3 (mm) = thickness of pipe coating
t_t = thickness of moisture barrier metallic tape in pipe-type cable
w_t (mm) = width of moisture barrier metallic tape

E.4 CABLE PARAMETERS—INSTALLATION CONDITIONS

E.4.1 Cables in Air

H (W/m^2) = solar radiation

E.4.2 Buried Cables

L (mm) = depth of burial of cables

D_x (mm) = fictitious diameter at which effect of loss factor commences

ρ_s (K · m/W) = thermal resistivity of soil

s (mm) = spacing between conductors of the same circuit

s_2 (mm) = axial separation of cables; for cables in flat formation, s_2 is the geometric mean of the three spacings

E.4.3 Duct Bank/Thermal Backfill

L_G (mm) = distance from the soil surface to the center of a ductbank

x, y (mm) = sides of duct bank/backfill ($y > x$)

N = number of loaded cables in a duct bank/backfill

ρ_c(K · m/W) = thermal resistivity of concrete used for a duct bank or of the backfill

ρ_e (K · m/W) = thermal resistivity of earth surrounding a duct bank/backfill

E.4.4 Cables in Ducts

D_d (mm) = internal diameter of the duct

D_o (mm) = external diameter of the duct

θ_m(°C) = mean temperature of duct filling medium

E.5 CONDUCTOR AC RESISTANCE

Material	Resistivity (ρ_{20}) · 10^{-8} Ω · m at 20°C	Temperature Coefficient (α_{20}) · 10^{-3} per K at 20°C
Copper	1.7241	3.93
Aluminum	2.8264	4.03

$$R' = \frac{1.02 \cdot 10^6 \rho_{20}}{S}[1 + \alpha_{20}(\theta - 20)]$$

$R' =$ ______ Ω/m

Type of Conductor	Whether Dried and Impregnated or Not	k_s	k_p
Copper			
Round, stranded	Yes	1	0.8
Round, stranded	No	1	1
Round, compact	Yes	1	0.8
Round, compact	No	1	1
Round, segmental		0.435	0.37
Hollow, helical stranded	Yes	eq. (7.12)	0.8
Sector-shaped	Yes	1	0.8
Sector-shaped	No	1	1
Aluminum			
Round, stranded	Either	1	*
Round, 4 segment	Either	0.28	
Round, 5 segment	Either	0.19	
Round, 6 segment	Either	0.12	
Segmental with peripheral strands	Either	eq. (7.17)	

*Since there are no accepted experimental results dealing specifically with aluminum stranded conductors, IEC 287 recommends that the values of k_p given in the above table for copper conductors also be applied to aluminum stranded conductor of similar design as the copper ones.

$k_s =$ ____________

$k_p =$ ____________

$$F_k = \frac{8\omega f \cdot 10^{-7}}{R'}$$

$F_k =$ ____________

$$x_s^2 = F_k \cdot k_s$$

$x_s^2 =$ ____________

If $x_s \leq 2.8$ (a majority of the cases), then the following equations apply. Otherwise, use equations (7.10) or (7.11).

$$y_s = \frac{x_s^4}{192 + 0.8x_s^4}$$

$y_s =$ ____________

$$\boxed{x_p^2 = F_k \cdot k_p}$$

$x_p^2 =$ ____________

If $x_p \leq 2.8$ (a majority of the cases), then the following equations apply. Otherwise, use equations (7.26) or (7.27).

$$\boxed{F_p = \frac{x_p^4}{192 + 0.8x_p^4}}$$

$F_p =$ ____________

$$\boxed{y_p = F_p \left(\frac{d_c}{s}\right)^2 \left[0.312 \left(\frac{d_c}{s}\right)^2 + \frac{1.18}{F_p + 0.27}\right]}$$

$y_p =$ ____________

For sector-shaped conductors:

$$s = d_x + t$$
$$d_c = d_x$$
$$y_p = 2y_p/3$$

For oval conductors:

$$d_c = \sqrt{d_{\text{c minor}} \cdot d_{\text{c major}}}$$

$$\boxed{R = R'(1 + y_s + y_p)}$$

$R =$ ____________ Ω/m

For cables in magnetic pipes and conduits:

$$\boxed{R = R'[1 + 1.5(y_s + y_p)]}$$

$R =$ ____________ Ω/m

E.6 DIELECTRIC LOSSES

Type of Cable	ε	tan δ
Cables insulated with impregnated paper		
Solid type, fully impregnated, preimpregnated or mass-impregnated nondraining	4	0.01
Oil-filled, low-pressure		
up to $U_0 = 36$ kV	3.6	0.0035
up to $U_0 = 87$ kV	3.6	0.0033
up to $U_0 = 160$ kV	3.5	0.0030
up to $U_0 = 220$ kV	3.5	0.0028
Oil-pressure, pipe-type	3.7	0.0045
Internal gas-pressure	3.4	0.0045
External gas-pressure	3.6	0.0040
Cables with other kinds of insulation		
Butyl rubber	4	0.050
EPR—up to 18/30 kV	3	0.020
EPR—above 18/30 kV	3	0.005
PVC	8	0.1
PE (HD and LD)	2.3	0.001
XLPE up to and including 18/30 (36) kV—unfilled	2.5	0.004
XLPE above 18/30 (36) kV—unfilled	2.5	0.001
XLPE above 18/30 (36) kV—filled	3	0.005
Paper–polypropylene–paper (PPL)*	2.8	0.001

*The dielectric constant and the loss factor of PPL insulation has not been standardized yet.

$\varepsilon =$ ____________

$\tan\delta =$ ____________

$$C = \frac{\varepsilon}{18 \ln\left(\frac{D_i}{d_c}\right)} \cdot 10^{-9}$$

$C =$ ____________ F/m

$$U_0 = \frac{U}{\sqrt{3}}$$

$U_0 =$ ____________ V

$$W_d = 2\pi f \cdot C \cdot U_o^2 \cdot \tan\delta$$

$W_d =$ ____________ W/m

E.7 SHEATH LOSS FACTOR

E.7.1 Sheath Resistance

Material	Resistivity (ρ_{20}) $\cdot 10^{-8}$ $\Omega \cdot$m at 20°C	Temperature Coefficient (α_{20}) $\cdot 10^{-3}$ per K at 20°C
Lead or lead alloy	21.4	4.0
Steel	13.8	4.5
Bronze	3.5	3.0
Stainless steel	70	Negligible
Aluminum	2.84	4.03

$$R_s = \frac{\rho_{20} \cdot 10^6}{\pi \cdot d \cdot t_s} [1 + \alpha_{20} (\theta_s - 20)]$$

$R_s =$ ____________ Ω/m

For lead sheath reinforced with nonmagnetic tapes:

$$R_t = \frac{\rho_{20} \cdot 10^6}{w_t \cdot n_t \cdot t_t} \left[1 + \left(\frac{\pi d}{\ell_T} \right)^2 \right]^{1/2} [1 + \alpha_{20} (\theta_t - 20)]$$

$R_t =$ ____________ Ω/m

If $\dfrac{d_s}{\ell_T} \geq 0.44$, $R_t = 2R_t$ computed above.

To calculate sheath losses, use the combined resistance of sheath and reinforcement.

$$R_{st} = \frac{R_s R_t}{R_s + R_t}$$

$R_{st} =$ ____________ Ω/m

Substitute R_{st} for R_s in what follows.

E.7.2 Sheath Reactances

For single-conductor and pipe-type cables:

$$X = 4\pi f \cdot 10^{-7} \cdot \ln \frac{2s}{d}$$

$X =$ ____________ Ω/m

For single-conductor cables in flat formation, regularly transposed, sheaths bonded at both ends:

$$X_1 = 4\pi f \cdot 10^{-7} \cdot \ln\left[2 \cdot \sqrt[3]{2}\left(\frac{s}{d}\right)\right]$$

$X_1 =$ ______________ Ω/m

For single-conductor cables in flat configuration with sheaths solidly bonded at both ends, the sheath loss factor depends on the spacing. If it is not possible to maintain the same spacing in the electrical section (i.e., between points at which the sheaths of all cables are bonded), the following allowances should be made:

(1) If the spacings are known, the value of X is computed from

$$X = \frac{l_a X_a + l_b X_b + \cdots + l_n X_n}{l_a + l_b + \cdots + l_n}$$

$X =$ ______________ Ω/m

where $l_a, l_b, \cdots, l_n$ are lengths with different spacing along an electrical section

$X_a, X_b, \cdots, X_n$ are the reactances per unit length of cable, given by equations for X or X_1 above

where appropriate values of spacing $s_a, s_b, \cdots, s_n$ are used.

(2) If the spacings are not known, the value of λ_1' calculated below should be increased by 25%.

$$X_m = 8.71 \cdot 10^{-7} \cdot f$$

$X_m =$ ______________ Ω/m

E.7.3 Single-conductor Cables

(1) Sheath bonded both ends—triangular configuration:

$$\lambda_1' = \frac{R_s}{R} \cdot \frac{1}{1 + \left(\frac{R_s}{X_1}\right)^2}$$

$\lambda_1' =$

$\lambda_1'' = 0$

(2) Sheath bonded both ends—flat configuration, regular transposition:

$$\lambda_1' = \frac{R_s}{R} \cdot \frac{1}{1 + \left(\frac{R_s}{X}\right)^2}$$

$\lambda_1' =$

$\lambda_1'' = 0$

(3) Sheath bonded both ends—flat configuration, no transposition. Center cable equidistant from other cables:

$$P = X_m + X$$

$P =$ ________ Ω/m

$$Q = X - X_m/3$$

$Q =$ ________ Ω/m

$$RP = R_s^2 + P^2$$

$RP =$

$$RQ = R_s^2 + Q^2$$

$RQ =$

$$A = \frac{0.75P^2}{RP}$$

$A =$

$$B = \frac{0.25Q^2}{RQ}$$

$B =$

$$C = \frac{2R_s \cdot P \cdot Q \cdot X_m}{\sqrt{3} \cdot RP \cdot RQ}$$

$C =$ ____________

$$R_{cs} = \frac{R_s}{R}$$

$R_{cs} =$ ____________

$$\lambda'_{11} = R_{cs}(A + B + C)$$

Outer cable carrying lagging phase.

$\lambda'_{11} =$ ____________

$$\lambda'_{1m} = R_{cs}\frac{Q^2}{RQ}$$

Center cable.

$\lambda'_{1m} =$ ____________

$$\lambda'_{12} = R_{cs}(A + B - C)$$

Outer cable carrying leading phase.

$\lambda'_{12} =$ ____________

$\lambda'_1 =$

$\lambda''_1 = 0$ ____________

Ratings for cables in air should be calculated using λ'_{11}; λ'_1 is equal to λ'_{11} or λ'_{12} or λ'_{1m} depending on which cable is the hottest.

E.7.4 Large Segmental Conductors

When conductor proximity effect is reduced, for example, by large conductors having insulated segments, λ''_1 cannot be ignored and is calculated as follows:

$$M = N = \frac{R_s}{X}$$

Cables in trefoil formation.

$M = N =$ ________

and

$$M = \frac{R_s}{X + X_m}$$

$$N = \frac{R_s}{X - \frac{X_m}{3}}$$

Cables in flat formation with equidistant spacing.

$M =$

$N =$ ________

$$F = \frac{4M^2N^2 + (M + N)^2}{4(M^2 + 1)(N^2 + 1)}$$

$F =$ ________

λ_1'' is calculated by multiplying the value of the eddy current sheath loss factor calculated below by F.

E.7.5 Sheaths Single-point Bonded or Cross Bonded

Lead-sheathed cables.

$$\beta_1 = 0$$
$$g_s = 1$$

For corrugated sheaths, the mean outside diameter shall be used.

$$\beta_1 = \sqrt{\frac{4\pi\omega}{10^7\zeta}}$$

$\beta_1 =$ ________

$$\omega = 2\pi f$$

$\omega =$ ________

$$g_s = 1 + \left(\frac{t_s}{D_s}\right)^{1.74} (\beta_1 D_s \cdot 10^{-3} - 1.6)$$

$g_s =$ ____________

$$m = \frac{2\pi f}{R_s} \cdot 10^{-7}$$

$m =$ ____________

If $m \leq 0.1$, $\Delta_1 = 0$, $\Delta_2 = 0$.

E.7.6 Three Single-conductor Cables in Triangular Configuration

$$\lambda_0 = 3\left(\frac{d}{2s}\right)^2 \frac{m^2}{1+m^2}$$

$\lambda_0 =$ ____________

$$\Delta_1 = (1.14m^{2.45} + 0.33)\left(\frac{d}{2s}\right)^{0.92m+1.66}$$

$\Delta_1 =$ ____________

$$\Delta_2 = 0$$

$\Delta_2 = 0$ ____________

E.7.7 Three Single-conductor Cables in Flat Configuration

(1) Center cable:

$$\lambda_0 = 6\left(\frac{d}{2s}\right)^2 \frac{m^2}{1+m^2}$$

$\lambda_0 =$ ____________

$$\boxed{\Delta_1 = 0.86m^{3.08}\left(\frac{d}{2s}\right)^{1.4m+0.7}}$$

$\Delta_1 =$ ________

$$\boxed{\Delta_2 = 0}$$

$\Delta_2 = 0$ ________

(2) Outer cable leading phase:

$$\boxed{\lambda_0 = 1.5\left(\frac{d}{2s}\right)^2 \frac{m^2}{1+m^2}}$$

$\lambda_0 =$ ________

$$\boxed{\Delta_1 = 4.7m^{0.7}\left(\frac{d}{2s}\right)^{0.16m+2}}$$

$\Delta_1 =$ ________

$$\boxed{\Delta_2 = 21m^{3.3}\left(\frac{d}{2s}\right)^{1.47m+5.06}}$$

$\Delta_2 =$ ________

(3) Outer cable lagging phase:

$$\boxed{\lambda_0 = 1.5\left(\frac{d}{2s}\right)^2 \frac{m^2}{1+m^2}}$$

$\lambda_0 =$ ________

$$\boxed{\Delta_1 = -\frac{0.74(m+2)m^{0.5}}{2+(m-0.3)^2}\left(\frac{d}{2s}\right)^{m+1}}$$

$\Delta_1 =$ ________

$$\Delta_2 = 0.92m^{3.7}\left(\frac{d}{2s}\right)^{m+2}$$

$\Delta_2 =$ ____________

$$\lambda_1'' = \frac{R_s}{R}\left[g_s\lambda_0(1 + \Delta_1 + \Delta_2) + \frac{(\beta_1 t_s)^4}{12} \cdot 10^{-12}\right]$$

$\lambda_1'' =$ ____________

E.7.8 Sheaths Cross Bonded

The ideal cross-bonded system will have equal lengths and spacing in each of the three sections. If the section lengths are different, the induced voltages will not sum to zero and circulating currents will be present. These circulating currents are taken account of by calculating the circulating current loss factor λ_1' assuming the cables were not cross bonded, and multiplying this value by a factor to take account of the length variations. This factor F_c is given by

$$F_c = \left[\frac{p_2 + q_2 - 2}{p_2 + q_2 + 1}\right]^2$$

where $p_2 a$ = length of the longest section

$q_2 a$ = length of the second longest section

a = length of the shortest section.

This formula deals only with differences in the length of minor sections. Any deviations in spacing must also be taken into account.

Where lengths of the minor sections are not known, IEC 287-2-1 (1994) recommends that the value for λ_1' based on experience with carefully installed circuits be

$\lambda_1' = 0.03$ for cables laid directly in the ground

$\lambda_1' = 0.05$ for cables installed in ducts.

E.7.9 Three-conductor Cables

$$\lambda_1' = 0$$

$\lambda_1' = 0$ ____________

1) Round or oval conductors in common sheath, no armor:
$R_s \leq 100\ \mu\Omega/\text{m}$

$$\lambda_1'' = \frac{3R_s}{R}\left[\left(\frac{2c}{d}\right)^2 \frac{1}{1+\left(\frac{R_s 10^7}{\omega}\right)^2} + \left(\frac{2c}{d}\right)^4 \frac{1}{1+4\left(\frac{R_s 10^7}{\omega}\right)^2}\right]$$

$\lambda_1'' =$ __________

$R_s > 100\ \mu\Omega/\text{m}$

$$\lambda_1'' = \frac{3.2\omega^2}{RR_s}\left(\frac{2c}{d}\right)^2 10^{-14}$$

$\lambda_1'' =$ __________

Sector-shaped conductors

$$\lambda_1'' = 0.94\frac{R_s}{R}\left(\frac{2r_1 + t}{d}\right)^2 \frac{1}{1+\left(\frac{R_s 10^7}{\omega}\right)^2}$$

$\lambda_1'' =$ __________

E.7.10 Three-conductor Cables with Steel Tape Armor

The value for λ_1'' calculated above should be multiplied by the factor F_t.

$$\delta_0 = \frac{A}{\pi d_a}$$

$\delta_0 =$ __________ mm

$$F_t = \left[1 + \left(\frac{d}{d_a}\right)^2 \frac{1}{1+\frac{d_a}{\mu\delta_0}}\right]$$

$F_t =$ __________

μ is usually taken as 300.

$$\lambda_1'' = \lambda_1'' F_t$$

$\lambda_1'' =$ __________

E.7.11 Cables with Separate Lead Sheath (SL Type) with Armor

$$\lambda_1' = \frac{R_S}{R} \frac{1.5}{1 + \left(\frac{R_S}{X}\right)^2}$$

$\lambda_1' =$

$\lambda_1'' = 0$

E.8 SHIELDING/SKID WIRE LOSS FACTOR

E.8.1 Pipe-type Cables

$$R_t = \frac{\rho_{20} \cdot 10^6}{w_t \cdot n_t \cdot t_t} \left[1 + \left(\frac{\pi D_t}{\ell_T}\right)^2\right]^{1/2} [1 + \alpha_{20} (\theta_t - 20)]$$

$R_t =$ ______ Ω/m

$$R_{sw} = \frac{4 \cdot \rho_{20} \cdot 10^6}{\pi \cdot n_{sw} \cdot D_{sw}^2} \left[1 + \left(\frac{\pi D_{sm}}{\ell_{sw}}\right)^2\right]^{1/2} [1 + \alpha_{20} (\theta_{sw} - 20)]$$

$R_{sw} =$ ______ Ω/m

$$R_s = \frac{R_{sw} \cdot R_t}{R_{sw} + R_t}$$

$R_s =$ ______ Ω/m

$$\lambda_1' = \frac{R_S}{R} \frac{1.5}{1 + \left(\frac{R_S}{X}\right)^2}$$

$\lambda_1' =$

$\lambda_1'' = 0$

E.8.2 Total Sheath Loss Factor

$$\lambda_1 = \lambda_1' + \lambda_1''$$

$\lambda_1 =$

E.9 ARMOR LOSS FACTOR

E.9.1 Single-conductor Cables

(1) With nonmagnetic wire armor:

$$R_A = \frac{4\rho_{20} \cdot 10^6}{\pi \cdot n_a \cdot d_f^2} \left[1 + \left(\frac{\pi D_a}{\ell_a}\right)^2\right]^{1/2} [1 + \alpha_{20}(\theta_a - 20)]$$

$R_A =$ ________ Ω/m

$$\chi = \frac{R_s R_A}{(R_s + R_A)^2}$$

$\chi =$ ________

$$\lambda_1 = \frac{R_A}{R} \cdot \chi$$

$\lambda_1 =$ ________

$$\lambda_2 = \frac{R_s}{R} \cdot \chi$$

$\lambda_2 =$ ________

(2) With magnetic wire armor:

The following applies to cables spaced at least 10 m apart. The ac resistance of the armor wires will vary between about 1.2 and 1.4 times its dc resistance, depending on the wire diameter, but this variation is not critical because the sheath resistance is generally considerably lower than that of the armor wires.

$$R_e = \frac{R_s R_A}{R_s + R_A}$$

$R_e =$ ________ Ω/m

Average values for magnetic properties of armor wires with diameters in the range of 4–6 mm and tensile strengths on the order of 40 MPa:

$\mu_e = 400$

$\mu_t = 10$ for armor wires in contact

$\mu_t = 1$ for armor wires which are spaced

$\gamma = \pi/4$.

If a more precise calculation is required and the wire properties are known, then it is initially necessary to know an approximate value for the magnetizing force H in order to find the appropriate magnetic properties.

$$H = \frac{\bar{I} + \bar{I}_s}{\pi d_A} \text{ A/m}$$

where $\bar{I}$ and $\bar{I}_s$ are the vectorial values of the conductor and sheath currents, respectively. For the initial choice of magnetic properties, it is usually satisfactory to assume that $\bar{I} + \bar{I}_s = 0.8I$, and to repeat the calculations if it is subsequently established that the calculated value is significantly different.

$$H_s = 2 \times 10^{-7} \ln\left(\frac{2s_2}{d}\right)$$

$H_s =$ ______________ H/m

$$H_1 = \pi \mu_e \left(\frac{n_a d_f^2}{\ell_a d_a}\right) 10^{-7} \sin\beta \cos\gamma$$

$H_1 =$ ______________ H/m

$$H_2 = \pi \mu_e \left(\frac{n_a d_f^2}{\ell_a d_a}\right) 10^{-7} \sin\beta \sin\gamma$$

$H_2 =$ ______________ H/m

$$H_3 = 0.4\left(\mu_t \cos^2\beta - 1\right)\left(\frac{d_f}{d_a}\right) 10^{-6}$$

$H_3 =$ ______________ H

$$B_1 = \omega(H_s + H_1 + H_3)$$

$B_1 =$ ______________ Ω/m

$$B_2 = \omega H_2$$

$B_2 =$ ______________ Ω/m

$$\lambda_1' = \lambda_2 = \frac{R_e}{2R}\left(\frac{B_2^2 + B_1^2 + R_e B_2}{(R_e + B_2)^2 + B_1^2}\right)$$

$\lambda_1' =$ ____________

$\lambda_2 =$ ____________

E.9.2 Three Conductor Cables—Steel Wire Armor

$$R_A = \frac{4\rho_{20} \cdot 10^6}{\pi \cdot n_a \cdot d_f^2}\left[1 + \left(\frac{\pi D_a}{\ell_a}\right)^2\right]^{1/2} [1 + \alpha_{20}(\theta_a - 20)]$$

$R_A =$ ____________ Ω/m

(1) Round conductor cable:

$$\lambda_2 = 1.23\frac{R_A}{R}\left(\frac{2c}{d_a}\right)^2 \frac{1}{\left(\frac{2.77 R_A 10^6}{\omega}\right)^2 + 1}$$

$\lambda_2 =$ ____________

(2) SL type cables:

λ_2 calculated above should be multiplied by $(1 - \lambda_1')$, where λ_1' is calculated in the section on the sheath loss factor for SL type cables.

(3) Sector-shaped conductors:

$$\lambda_2 = 0.358\frac{R_A}{R}\left(\frac{2r_1}{d_a}\right)^2 \frac{1}{\left(\frac{2.77 R_A 10^6}{\omega}\right)^2 + 1}$$

$\lambda_2 =$ ____________

E.9.3 Three Conductor Cables—Steel Tape Armor or Reinforcement

$$\delta_0 = \frac{A}{\pi d_a}$$

$\delta_0 =$ ____________ mm

$$k = \frac{1}{1 + \dfrac{d_a}{\mu \delta_0}} \times \left(\frac{f}{50}\right)$$

μ is usually taken as 300.

$k =$ ______

$$\lambda_2' = \frac{s^2 k^2 10^{-7}}{R d_a \delta_0}$$

$\lambda_2' =$ ______

$$\lambda_2'' = \frac{2.25 s^2 k^2 \delta_0 10^{-8}}{R d_a}$$

$\lambda_2'' =$ ______

$$\lambda_2 = \lambda_2' + \lambda_2''$$

$\lambda_2 =$ ______

E.9.4 Pipe-type Cables in Steel Pipe—Pipe Loss Factor

Configuration	*A*	*B*
Cradled	0.00438	0.00226
Triangular bottom of pipe	0.0115	–0.001485
Mean between trefoil and cradled	0.00794	0.00039
Three-core cable	0.0199	–0.001485

$A =$ ______

$B =$ ______

$$\lambda_2 = \left(\frac{f}{60}\right)^{1.5} \left(\frac{A \cdot s + B \cdot D_d}{R}\right) 10^{-5}$$

$\lambda_2 =$ ______

E.10 THERMAL RESISTANCES

Material	Thermal Resistivity (ρ) ($tK \cdot$ m/W)	Thermal Capacity ($c \cdot 10^{-6}$) [$J/(m^3 \cdot K)$]
*Insulating materials**		
Paper insulation in solid type cables	6.0	2.0
Paper insulation in oil-filled cables	5.0	2.0
Paper insulation in cables with external gas pressure	5.5	2.0
Paper insulation in cables with internal gas pressure		
preimpregnated	6.5	2.0
mass-impregnated	6.0	2.0
PE	3.5	2.4
XLPE	3.5	2.4
Polyvinyl chloride		
up to and including 3 kV cables	5.0	1.7
greater than 3 kV cables	6.0	1.7
EPR		
up to and including 3 kV cables	3.5	2.0
greater than 3 kV cables	5.0	2.0
Butyl rubber	5.0	2.0
Rubber	5.0	2.0
Paper–polypropylene–paper (PPL)	6.5	2.0
Protective coverings		
Compounded jute and fibrous materials	6.0	2.0
Rubber sandwich protection	6.0	2.0
Polychloroprene	5.5	2.0
PVC		
up to and including 35 kV cables	5.0	1.7
greater than 35 kV cables	6.0	1.7
PVC /bitumen on corrugated aluminum sheaths	6.0	1.7
PE	3.5	2.4
Materials for duct installations		
Concrete	1.0	2.3
Fiber	4.8	2.0
Asbestos	2.0	2.0
Earthenware	1.2	1.8
PVC	6.0	1.7
PE	3.5	2.4

*For the purpose of current rating computations, the semiconducting screening materials are assumed to have the same thermal properties as the adjacent dielectric materials

E.11 INSULATION THERMAL RESISTANCE

For oval-shaped conductors, the diameter over the insulation is the geometric mean of the minor and major diameters over the insulation.

E.11.1 Single-conductor Cables

$$T_1 = \frac{\rho}{2\pi} \ln\left(1 + \frac{2t_1}{d_c}\right)$$

$T_1 =$ ____________ $\text{K} \cdot \text{m/W}$

E.11.2 Three-conductor Belted Cables

(1) With round or oval conductors:
G is obtained from the figure below.

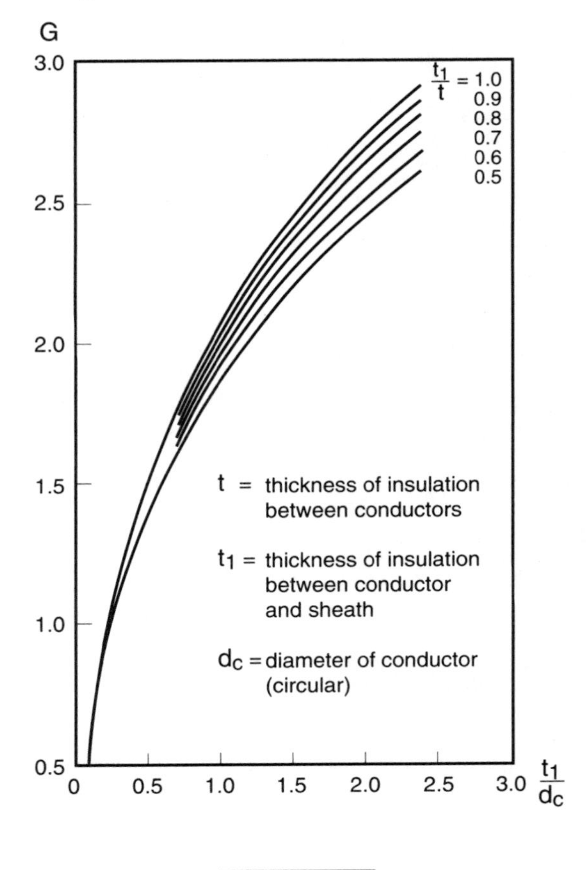

$$T_1 = \frac{\rho}{2\pi} G$$

$G =$ ____________

$T_1 =$ ____________ $\text{K} \cdot \text{m/W}$

(2) With sector-shaped conductors:

$$F_1 = 3 + \frac{9t}{2\pi(d_x + t) - t}$$

$F_1 =$ ______

$d_a =$ external diameter of the belt insulation, mm

$$G = F_1 \ln\left(\frac{d_a}{2r_1}\right)$$

$G =$ ______

$$T_1 = \frac{\rho}{2\pi} G$$

$T_1 =$ ______ $\text{K} \cdot \text{m/W}$

E.11.3 Three-conductor Shielded Cables

(1) With round or oval conductors:

$$t_1 = t/2$$

$t_1 =$ ______ mm

The geometric factor is read from the preceding figure. The screening factor is read from the figure below.

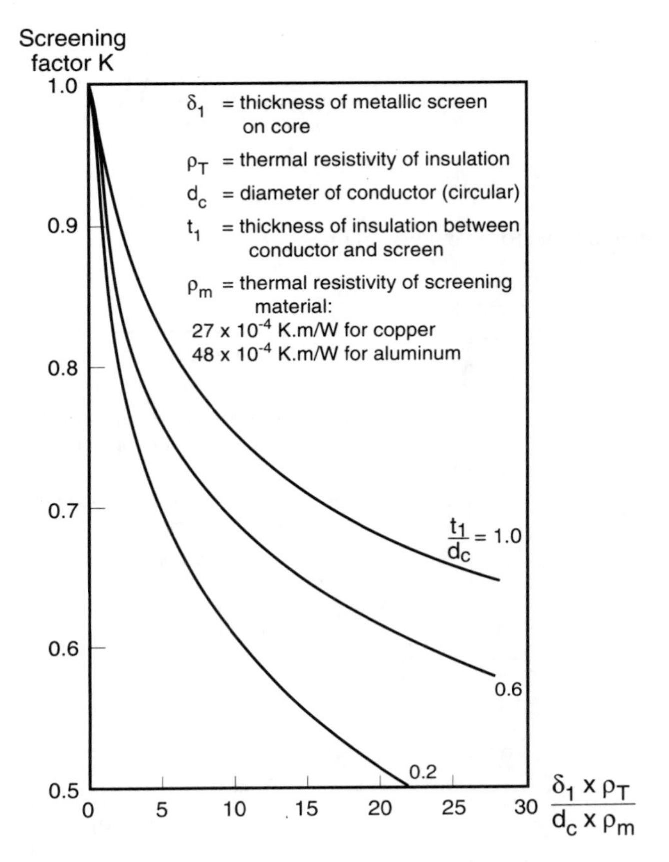

$G =$ ____________

$K =$ ____________

$$T_1 = K \frac{\rho}{2\pi} G$$

$T_1 =$ ____________ $\text{K} \cdot \text{m/W}$

(2) With sector-shaped conductors:

$$F_1 = 3 + \frac{9t}{2\pi(d_x + t) - t}$$

$F_1 =$ ____________

$$G = F_1 \ln\left(\frac{d_a}{2r_1}\right)$$

$G =$ ____________

The screening factor is read from the figure below.

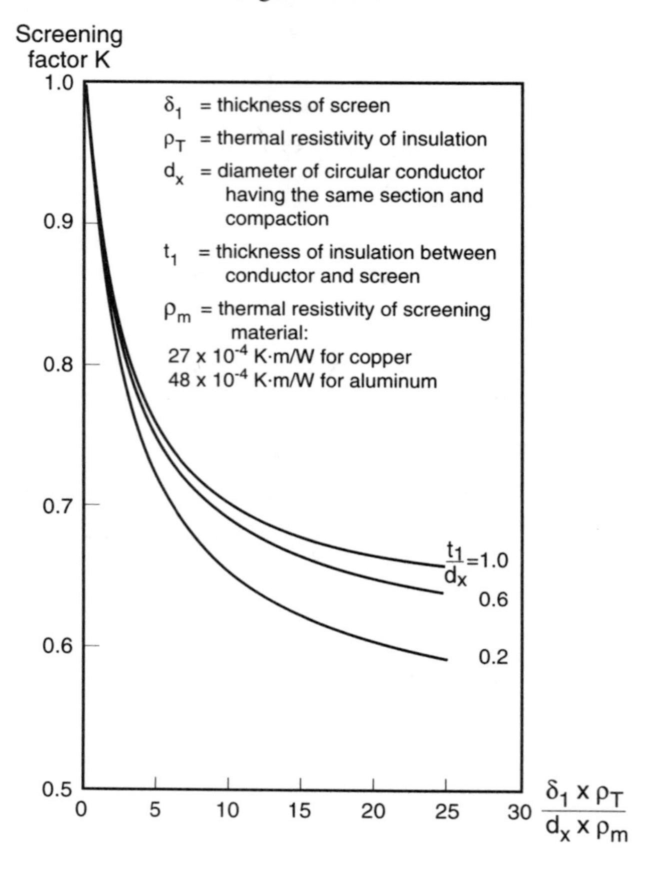

$G =$ ____________

$K =$ ____________

$$T_1 = K \frac{\rho}{2\pi} G$$

$T_1 =$ ____________ K · m/W

E.11.4 Oil-filled Cables with Round Conductors and Round Oil Ducts Between Cores

(1) *Metallized paper insulation shield:*

$$T_1 = 0.358\rho\left(\frac{2t_i}{d_c + 2t_i}\right)$$

$T_1 =$ ________ $\text{K} \cdot \text{m/W}$

(2) *Metal tape insulation shield:*

$$T_1 = 0.35\rho\left(0.923 - \frac{2t_i}{d_c + 2t_i}\right)$$

$T_1 =$ ________ $\text{K} \cdot \text{m/W}$

E.11.5 SL Type Cables

In SL type cables, the lead sheath around each core may be assumed isothermal. The thermal resistance T_1 is calculated in the same way as for single-core cables.

E.12 THERMAL RESISTANCE OF PIPE-FILLING MEDIUM

Installation Condition	U	V	Y
Gas pressure cable in pipe	0.95	0.46	0.0021
Oil pressure pipe-type cable	0.26	0.0	0.0026

For three cables in a pipe, $D_e = 2.15\times$ single-cable external diameter.

$$T_2 = \frac{U}{1 + 0.1(V + Y\theta_m)D_e}$$

$T_2 =$ ________ $\text{K} \cdot \text{m/W}$

E.13 JACKET THERMAL RESISTANCE

$$T_2 = \frac{\rho}{2\pi}\ln\left(1 + \frac{2t_J}{D_s}\right)$$

$T_2 =$ ________ $\text{K} \cdot \text{m/W}$

E.13.1 SL Type Cables

$\bar{G}$ is the geometric factor and is obtained from the figure below.

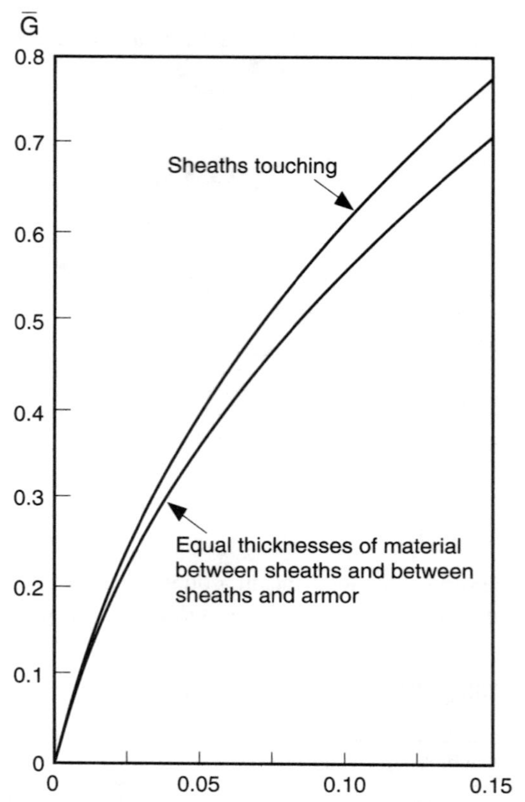

$\bar{G} =$ ____________

$$T_2 = \frac{\rho}{6\pi}\bar{G}$$

$T_2 =$ ____________ $\mathrm{K \cdot m/W}$

E.14 SERVING THERMAL RESISTANCE

$$T_3 = \frac{\rho}{2\pi}\ln\left(1 + \frac{2t_3}{D_a}\right)$$

$T_3 =$ ____________ $\mathrm{K \cdot m/W}$

For pipe-type cables:

$$T_3 = \frac{\rho}{2\pi} \ln\left(\frac{D_e}{D_o}\right)$$

$T_3 =$ ____________ $\text{K} \cdot \text{m/W}$

E.15 EXTERNAL THERMAL RESISTANCE OF BURIED CABLES

For buried cables, two values of the external thermal resistance are calculated: T_4 corresponding to dielectric losses (100% load factor), and $T_{4\mu}$—the thermal resistance corresponding to the joule losses, where allowance is made for the daily load factor (LF) and the corresponding loss factor μ.

$$\mu = 0.3 \cdot (LF) + 0.7 \cdot (LF)^2$$

$\mu =$ ____________

The effect of the loss factor is considered to start outside a diameter D_x defined as $D_x = 61\ 200\sqrt{\delta \text{ (length of cycle in hours)}}$ where δ is soil diffusivity ($\text{m}^2\text{/h}$). For a daily load cycle and typical value of soil diffusivity of $0.5 \cdot 10^{-6}$ $\text{m}^2\text{/s}$, D_x is equal to 211 mm (or 8.3 in). The value of D_x is valid even when the diameter of the cable or pipe is greater than D_x.

E.15.1 Mutual Heating Effect

A factor F accounts for the mutual heating effect of the other cables or cable pipes in a system of *equally loaded, identical cables or cable pipes*. The distances needed to compute factor F are defined in the diagram below. These are center-to-center distances.

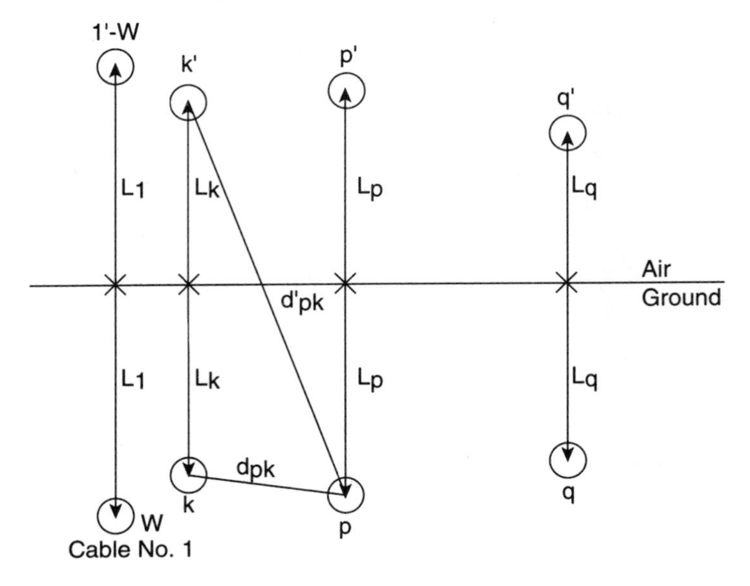

For cable p:

$$F = \left(\frac{d'_{p1}}{d_{p1}}\right)\left(\frac{d'_{p2}}{d_{p2}}\right)\cdots\left(\frac{d'_{pk}}{d_{pk}}\right)\cdots\left(\frac{d'_{pq}}{d_{pq}}\right)$$

$F =$ ____________

There are $(q-1)$ terms, with the term d'_{pp}/d_{pp} excluded. The rating of the cable system is determined by the rating of the hottest cable or cable pipe, usually the cable with the largest ratio L/D_o. For a single isolated cable or cable pipe, $F = 1$.

E.15.2 Single-conductor Cables

When the losses in the sheaths of single-core cables laid in a horizontal plane are appreciable, and the sheaths are laid without transposition and/or the sheaths are bonded at all joints, the inequality of losses affects the external thermal resistance of the cables. In such cases, the value of the F factor used to calculate $T_{4\mu}$ is modified by first computing the sheath factor (SHF):

$$(SHF) = \frac{1 + 0.5(\lambda'_{11} + \lambda'_{12})}{1 + \lambda'_m}$$

$(SHF) =$ ____________

and then calculating

$$F' = F^{(SHF)}$$

$F' =$ ____________

(1) Equally loaded similar cables:
Directly buried cables or cable pipes.
For cables in ducts, use D_o in place of D_e in the following formulas.

$$T_4 = \frac{\rho_s}{2\pi} \ln \frac{4L \cdot F}{D_e}$$

$T_4 =$ ____________ K · m/W

$$T_{4\mu} = \frac{\rho_s}{2\pi}\left\langle \ln \frac{D_x}{D_e} + \mu \cdot \ln \frac{4L \cdot F}{D_x} \right\rangle$$

$T_{4\mu} =$ ____________ K · m/W

(2) Cables or cable pipes buried in thermal backfill:[3]

$$\ln r_b = \frac{1}{2}\frac{x}{y}\left(\frac{4}{\pi} - \frac{x}{y}\right)\ln\left(1 + \frac{y^2}{x^2}\right) + \ln\frac{x}{2}$$

$r_b =$ __________ mm

$$u = \frac{L_G}{r_b}$$

$u =$ __________

$$G_b = \ln\left[u + \sqrt{(u^2 - 1)}\right]$$

$G_b =$ __________

$$T_4 = \frac{\rho_c}{2\pi}\ln\frac{4L \cdot F}{D_e} + \frac{N}{2\pi}(\rho_e - \rho_c)G_b$$

$T_4 =$ __________ $\text{K} \cdot \text{m/W}$

$$T_{4\mu} = \frac{\rho_c}{2\pi}\left(\ln\frac{D_x}{D_e} + \mu\ln\frac{4L \cdot F}{D_x}\right) + \mu\frac{N}{2\pi}(\rho_e - \rho_c)G_b$$

$T_{4\mu} =$ __________ $\text{K} \cdot \text{m/W}$

(3) Cables in ducts:

Installation Condition	U	V	Y
In metallic conduit	5.2	1.4	0.011
In fiber duct in air	5.2	0.83	0.006
In fiber duct in concrete	5.2	0.91	0.010
In asbestos cement			
duct in air	5.2	1.2	0.006
duct in concrete	5.2	1.1	0.011
Earthenware ducts	1.87	0.28	0.0036

[3] Valid for $y/x < 3$.

$$T_4' = \frac{U}{1 + 0.1(V + Y\theta_m)D_e}$$

$T_4' =$ ______ K · m/W

$$T_4'' = \frac{\rho}{2\pi} \ln \frac{D_o}{D_d}$$

ρ is the thermal resistivity of duct material.
For metal ducts, $T_4'' = 0$.

$T_4'' =$ ______ K · m/W

$$\ln r_b = \frac{1}{2} \frac{x}{y} \left(\frac{4}{\pi} - \frac{x}{y} \right) \ln \left(1 + \frac{y^2}{x^2} \right) + \ln \frac{x}{2}$$

$r_b =$ ______ mm

$$u = \frac{L_G}{r_b}$$

$u =$ ______

$$G_b = \ln \left[u + \sqrt{(u^2 - 1)} \right]$$

$G_b =$ ______

$$T_4''' = \frac{\rho_c}{2\pi} \ln \frac{4L \cdot F}{D_o} + \frac{N}{2\pi} (\rho_e - \rho_c) G_b$$

$T_4''' =$ ______ K · m/W

$$T_{4\mu}''' = \frac{\rho_c}{2\pi} \left(\ln \frac{D_x}{D_o} + \mu \ln \frac{4L \cdot F}{D_x} \right) + \mu \frac{N}{2\pi} (\rho_e - \rho_c) G_b$$

$T_{4\mu}''' =$ ______ K · m/W

$$T_4 = T_4' + T_4'' + T_4'''$$

$T_4 =$ ______________ K · m/W

$$T_{4\mu} = T_4' + T_4'' + T_{4\mu}'''$$

$T_{4\mu} =$ ______________ K · m/W

E.15.3 Unequally Loaded or Dissimilar Cables

In this case, heating between cables is accounted for by calculating the temperature rise at the surface of the cable being studied caused by other cables in the same group. To do this, the losses in each cable must first be estimated; these estimates are amended as necessary by the results of the calculation.

For cable j, the losses are

$$W_j = n\left[I_j^2 R_j(1 + \lambda_1 + \lambda_2)\mu_j + W_{dj}\right]$$

$W_j =$ ______________ W/m

The thermal resistance between cable j and cable i, the cable being studied, is for directly buried cables

$$T_{ij} = \frac{\rho_s}{2\pi} \ln \frac{d'_{ij}}{d_{ij}}$$

$T_{ij} =$ ______________ K · m/W

for cables in backfill or duct bank.

$$T_{ij} = \frac{\rho_c}{2\pi} \ln \frac{d'_{ij}}{d_{ij}} + \frac{N}{2\pi}(\rho_e - \rho_c)G_b$$

$T_{ij} =$ ______________ K · m/W

The temperature rise at the surface of cable i due to the losses in cable j:

$$\Delta\theta_{ij} = W_j \cdot T_{ij}$$

$\Delta\theta_{ij} =$ ______________ °C

The temperature rise at the surface of cable i due to all other cables in the group:

$$\Delta\theta_{\text{int}} = \sum_{\substack{j=1 \\ j \neq i}}^{N} \Delta\theta_{ij}$$

$\Delta\theta_{\text{int}} =$ ____________ °C

E.16 EXTERNAL THERMAL RESISTANCE OF CABLES IN AIR

Cables with jackets or other nonmetallic surfaces should be considered to have a black surface. Plain lead or unarmored cables should be assigned a value of h equal to 80% of the value for a cable with a black surface.

Material	σ
Bitumen/jute serving	0.8
Polychloroprene	0.8
PVC	0.6
PE	0.4
Lead	0.6

E.16.1 Heat Transfer Coefficient

$$h = \frac{Z}{(D_e^*)^g} + E$$

$h =$ ____________ $\text{W/m}^2 \cdot {}^\circ\text{C}^{5/4}$

	Installation	Z	E	g	Mode
1	Single cable*	0.21	3.94	0.60	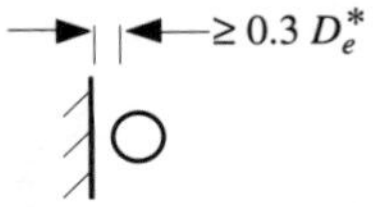 $\geq 0.3\,D_e^*$
2	Two cables touching, horizontal	0.21	2.35	0.50	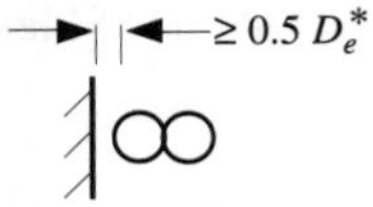 $\geq 0.5\,D_e^*$
3	Three cables in trefoil	0.96	1.25	0.20	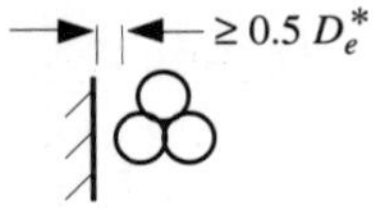 $\geq 0.5\,D_e^*$
4	Three cables touching, horizontal	0.62	1.95	0.25	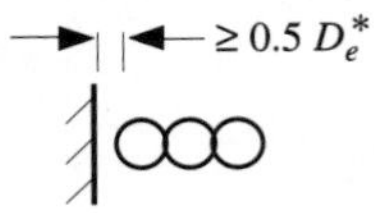 $\geq 0.5\,D_e^*$
5	Two cables touching, vertical	1.42	0.86	0.25	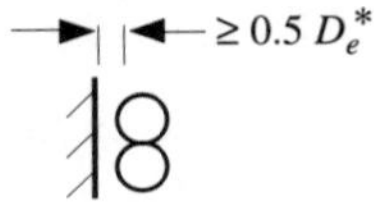 $\geq 0.5\,D_e^*$
6	Two cables spaced D_e^*, vertical	0.75	2.80	0.30	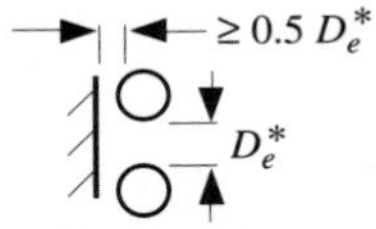 $\geq 0.5\,D_e^*$; D_e^*
7	Three cables touching, vertical	1.61	0.43	0.20	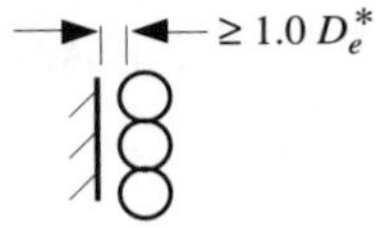 $\geq 1.0\,D_e^*$
8	Three cables spaced D_e^*, vertical	1.31	2.00	0.20	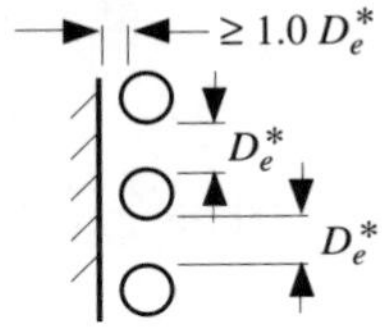 $\geq 1.0\,D_e^*$; D_e^*; D_e^*
9	Single cable	1.69	0.63	0.25	
10	Three cables in trefoil	0.94	0.79	0.20	

*Values for a "single cable" also apply to each cable of a group when they are spaced horizontally with a clearance between cables of at least 0.75 times the cable overall diameter.

$$K_A = \frac{\pi D_e^* h}{1 + \lambda_1 + \lambda_2} \left[\frac{T_1}{n} + (1 + \lambda_1) T_2 + (1 + \lambda_1 + \lambda_2) T_3 \right]$$

$K_A =$ ________

(1) Shielded from solar radiation:

$$\Delta\theta_d = W_d \left[\left(\frac{1}{1 + \lambda_1 + \lambda_2} - \frac{1}{2} \right) T_1 - \frac{n \lambda_2 T_2}{1 + \lambda_1 + \lambda_2} \right]$$

$\Delta\theta_d =$ ________ °C

$$(\Delta\theta_s)_{n+1}^{1/4} = \left[\frac{\Delta\theta + \Delta\theta_d}{1 + K_A (\Delta\theta_s)_n^{1/4}} \right]^{1/4}$$

$(\Delta\theta_s)_{n+1}^{1/4} =$ ________ $°C^{1/4}$

Set the initial value of $(\Delta\theta_s)^{1/4} = 2$. Reiterate until $(\Delta\theta_s)_{n+1}^{1/4} - (\Delta\theta_s)_n^{1/4} \leq 0.001$

$$T_4 = \frac{1}{\pi D_e^* h (\Delta\theta_s)^{1/4}}$$

$T_4 =$ ________ K · m/W

(2) Exposed to solar radiation:

$$(\Delta\theta_s)_{n+1}^{1/4} = \left[\frac{\Delta\theta + \Delta\theta_d + \sigma H K_A / \pi h}{1 + K_A (\Delta\theta_s)_n^{1/4}} \right]^{1/4}$$

$(\Delta\theta_s)_{n+1}^{1/4} =$ ________ $°C^{1/4}$

Set the initial value of $(\Delta\theta_s)^{1/4} = 2$. Reiterate until $(\Delta\theta_s)_{n+1}^{1/4} - (\Delta\theta_s)_n^{1/4} \leq 0.001$

$$T_4 = \frac{1}{\pi D_e^* h (\Delta\theta_s)^{1/4}}$$

$T_4 =$ ________ K · m/W

Ambient temperature should be increased by $\Delta\theta_{sr}$:

$$\Delta\theta_{sr} = \sigma D_e^* H T_4$$

$\Delta\theta_{sr} =$ ________ °C

E.17 AMPACITY

E.17.1 Buried Cables

$$I = \left[\frac{\Delta\theta - W_d\,[0.5T_1 + n(T_2 + T_3 + T_4)] - \Delta\theta_{\text{int}}}{RT_1 + nR(1+\lambda_1)T_2 + nR(1+\lambda_1+\lambda_2)(T_3 + T_{4\mu})} \right]^{0.5}$$

$I =$ ________ A

E.17.2 Cables in Air

$$I = \left[\frac{\Delta\theta - W_d\,[0.5T_1 + n(T_2 + T_3 + T_4)] - \Delta\theta_{sr}}{RT_1 + nR(1+\lambda_1)T_2 + nR(1+\lambda_1+\lambda_2)(T_3 + T_4)} \right]^{0.5}$$

$I =$ ________ A

E.17.3 Temperature Rise of Cable Components

buried cable

$$\Delta\theta_a = \Delta\theta_{\text{int}} + n\left\{\left[W_c(1+\lambda_1+\lambda_2)(T_3 + T_{4\mu})\right] + W_d(T_3 + T_4)\right\}$$

$\Delta\theta_a =$ ________ °C

cable in air

$$\Delta\theta_a = \Delta\theta_{sr} + n\left\{[W_c(1+\lambda_1+\lambda_2) + W_d]\,(T_3 + T_4)\right\}$$

$\Delta\theta_a =$ ________ °C

$$\Delta\theta_s = \Delta\theta_a + n\,[W_c(1+\lambda_1) + W_d]\,T_2$$

$\Delta\theta_s =$ ________ °C

$$\Delta\theta_c = \Delta\theta_s + (W_c + 0.5W_d)T_1$$

$\Delta\theta_c =$ ________ °C

F

Differences between the Neher/McGrath and IEC 287 Methods

The two methods are, in principle, the same, with the IEC method incorporating several new developments which took place after the publication of the Neher/McGrath (NM) paper. Similarities in the approaches are not surprising since, during the preparation of the standard, Mr. McGrath was in touch with the Chairman of Working Group 10 of IEC Subcommittee 20A (responsible for the preparation of ampacity calculation standards). The major difference between the two approaches is the use of metric units in IEC 287 and imperial units in the NM paper (the same equations look completely different because of this). Even though the methods are similar in principle, the IEC document is more comprehensive than the NM paper. IEC 287 not only contains all the formulas (with minor exceptions listed below) of the NM paper, but in several cases, it makes a distinction between different cable types and installation conditions where the NM paper would not make such a distinction. Also, the constants used in the IEC document are more up to date.

The following is a list of the most important differences between the two approaches:

Load Factor

1. The Neher/McGrath paper considers a nonunity load factor (see Section 9.6.7), whereas the IEC 287 document assumes a unity load factor. Another IEC document (IEC 853, 1985, 1989) deals with cyclic and emergency ratings (see Chapter 5).

Circulating and Eddy Current Losses

2. Equation 30 in the Neher/McGrath paper for the eddy current effect of single-conductor cables with single-point bonding applies only when the cables are arranged in an equilateral configuration. The IEC document, in addition to the

equilateral configuration, provides formulas for calculating the eddy current effect in the more usual flat configuration. In addition, the IEC document considers separately two- and three-core cables with steel tape armor which are not discussed in the NM paper.

3. Equation 27 in the Neher/McGrath paper for the circulating current effect of single-conductor cables with two-point bonding applies only when the cables are arranged in an equilateral configuration. The IEC document, in addition to this configuration, provides formulas for calculating the circulating current effect in the more usual flat configuration with and without transposition. In addition, the IEC document accounts for the effect of variation of spacing of single-core cables between sheath bonding points. The NM paper refers the reader to the Simmons (1932)[1] paper for computation of circulating current losses for cables in ducts.
4. For cables with large segmental-type conductors and sheaths bonded at both ends, IEC 287 provides an expression for eddy current computations. In the NM method, this contribution is ignored.
5. Calculation of losses in magnetic armor is treated only qualitatively in the NM paper with references to the literature for complex computational methods. Relevant approximations are proposed in IEC 287.

Calculation of Thermal Resistances

6. IEC 287 gives analytical expressions for the computation of the geometric factor of three-core cable insulation, whereas the NM paper makes a reference to the paper by Simmons (1932). In addition, the IEC document differentiates between various cable constructions, e.g., belted versus screened three-core cables, oil-filled, SL type, and so on. The NM paper does not provide this information.
7. The values of thermal resistivities specified in the NM paper are outdated (in view of the research which was carried out after the publication of the paper). Also, several new insulating materials are not listed.
8. The external thermal resistance of cables in air is somewhat more accurately computed in the IEC method (both methods are similar, the major difference being the formula for computation of the cable surface temperature). In the NM paper, the approximation is used that the heat transfer coefficient due to convection is independent of the cable/duct surface temperature (see Section 9.6.8.5), whereas the IEC document provides an iterative method for evaluating this coefficient as a function of the cable surface temperature rise (a more accurate assumption). Also, the IEC method distinguishes between various arrangements for cables in air, whereas the NM paper does not.
9. The NM paper considers the effect of wind on cable ampacity (see Section 9.6.8.5), whereas the IEC document assumes the worst case scenario with no wind. In this book, a general method of dealing with the effect of the wind in discussed in Chapter 10.
10. The IEC document distinguishes between trefoil and flat configurations for the computation of the external thermal resistances. The NM paper uses one formula only, which in reality is the same as the flat configuration formula in the IEC

[1] See references at the end of Chapter 9.

document. Formulas for the calculation of the external thermal resistance of cables in touching configurations are provided in the IEC document, and not in the NM paper.

11. The treatment of different cables types or unequally loaded cables in one installation is discussed in detail in the IEC document, and only qualitatively in the NM paper.
12. Consideration of the drying up of soil in the vicinity of loaded power cables is included in the IEC document, but not considered in the NM paper.

Emergency Ratings

13. The NM paper provides a formula for this rating. A corresponding formula is given in IEC 853-2 (1989).

In this book, the IEC 287 and 853 approaches are used as a basis for the presentation. The material which is included in the NM paper and not covered by the IEC documents is also discussed (see, for example, several sections in Chapter 9).

Index

A

B

C

D

E

T

V

W

About the Author

Employed by Ontario Hydro since 1975, George J. Anders is presently a Principal Engineer/Scientist in the Electrical Systems Technology Unit of Ontario Hydro Technologies. For many years, Dr. Anders has been responsible for Ontario Hydro's development of power cable calculation methods and tools.

Throughout his 22 years with Ontario Hydro, Dr. Anders has been involved in several aspects of power system analysis and design. His principal activities have been concentrated in three areas: (1) ampacity computations of electric power cables, (2) the application of probability methods in power system analysis and design, and (3) the application of novel techniques in electric power utility practice. He is the author of a book and has written over 70 papers published in several international journals. He has been conducting seminars on power cable ampacity issues in Canada and the United States as well as in Rio de Janeiro, Brazil; Warsaw, Poland; Bogota and Cali, Colombia; Porto, Portugal; Sydney and Melbourne, Australia; Santiago de Chile; and Hong Kong.

In the field of thermal analysis of electric power cables, Dr. Anders has made major contributions in three areas: (1) the development of computational techniques using the finite-element method to evaluate heat and moisture transfer in the vicinity of loaded power cables, (2) the development of optimization techniques for selection of the most economic conductor sizes, and (3) the development of new algorithms for transient ratings of buried power cables. He has published over 20 papers on the subjects dealing with thermal analysis of underground systems. As recognition of his work in this field, he received a New Technology Award from Ontario Hydro in 1990.

Dr. Anders is a Canadian representative in Working Group 10 (ampacity computations of power cables) and WG15 (short circuit temperatures of power cables) of the International Electrotechnical Commission. These working groups develop new computational techniques and new standards for power cable ampacity computations. He has also been a project leader on a number of projects dealing with ampacity computations sponsored by the Canadian Electrical Association. In the course of these projects, a series of highly successful computer programs were developed for CEA. These programs are in use by over 200 institutions in 33 countries on 5 continents.

In the field of application of probability methods in power system engineering, Dr. Anders has been involved in developing new methods and applications of probabilistic techniques to power system problems since 1975. He has published over 50 papers dealing with various topics on probability and optimization applications. Some of the problems he has been working on involved the development of mathematical techniques to model operator's action in probabilistic load flow analysis; a new technique to assess the importance of measures in power system reliability studies; a novel approach to frequency and duration analysis and uncertainty considerations for radial and two interconnected systems; a new method to evaluate the frequency of severe power system faults; and models to represent human failure in reliability analysis. He recently has been a project leader of several large projects on the probabilistic estimation of the remaining life of electrical equipment.

For several years, Dr. Anders has been teaching a course in the Faculty of Applied Science and Engineering at the University of Toronto on the application of probability methods in engineering. His book, *Probability Concepts in Electric Power Systems* (Wiley, New York, 1990), is well recognized around the world as a unique reference on the application of probability methods in power system planning, design, and operation. Dr. Anders is a Canadian representative in CIGRE WG 37.06.11 whose task is to develop methods for the reliability assessment of interconnected power systems. He is also a member of the IEEE Task Force on the Impact of Maintenance Strategies on Power System Reliability.

Dr. Anders also has been involved in developing new applications of optimization methods and novel techniques in power system problems. He started by developing an interval programming technique for application in reliability studies. Later, he was involved in developing models for optimal economic power transfer. He was also involved in publishing new techniques for the selection of the most economic cable sizes. Recently, he developed a procedure for the optimal construction of rigid-bus stations. He is also working on the application of multiobjective decision models for selecting an optimal maintenance strategy for power equipment. He has been a project leader of a large project undertaken by Ontario Hydro Technologies to develop procedures, methods, and computer tools for decision support and risk assessment.

Dr. Anders received the Master's degree in electrical engineering from the Technical University of Lodz, Poland in 1973, and the M.Sc. degree in mathematics and Ph.D. degree in power system reliability from the University of Toronto in 1977 and 1980, respectively. He is a Registered Professional Engineer in the Province of Ontario and a Senior Member of IEEE.